Low Back Pain (LBP)

Low Back Pain (LBP)

Editors

Vincenzo Denaro
Sergio Iavicoli
Fabrizio Russo
Gianluca Vadalà

MDPI • Basel • Beijing • Wuhan • Barcelona • Belgrade • Manchester • Tokyo • Cluj • Tianjin

Editors
Vincenzo Denaro
Fondazione Policlinico
Universitario Campus
Bio-Medico
Italy

Sergio Iavicoli
INAIL—Italian Workers'
Compensation Authority
Italy

Fabrizio Russo
Fondazione Policlinico
Universitario Campus
Bio-Medico
Italy

Gianluca Vadalà
Fondazione Policlinico
Universitario Campus
Bio-Medico
Italy

Editorial Office
MDPI
St. Alban-Anlage 66
4052 Basel, Switzerland

This is a reprint of articles from the Special Issue published online in the open access journal *International Journal of Environmental Research and Public Health* (ISSN 1660-4601) (available at: https://www.mdpi.com/journal/ijerph/special_issues/Low_Back_Pain_LBP).

For citation purposes, cite each article independently as indicated on the article page online and as indicated below:

LastName, A.A.; LastName, B.B.; LastName, C.C. Article Title. *Journal Name* **Year**, *Volume Number*, Page Range.

ISBN 978-3-0365-4455-7 (Hbk)
ISBN 978-3-0365-4456-4 (PDF)

Contents

Preface to "Low Back Pain (LBP)"

Low back pain (LBP) is a major public health problem, being the most commonly reported musculoskeletal disorder (MSD) and the leading cause of compromised quality of life and work absenteeism. Indeed, LBP is the leading worldwide cause of years lost to disability, and its burden is growing alongside the increasing and aging population.

The etiology, pathogenesis, and occupational risk factors of LBP are still not fully understood. It is crucial to give a stronger focus to reducing the consequences of LBP, as well as preventing its onset. Primary prevention at the occupational level remains important for highly exposed groups. Therefore, it is essential to identify which treatment options and workplace-based intervention strategies are effective in increasing participation at work and encouraging early return-to-work to reduce consequences of LBP.

The present Special Issue updates many of the recent advances and perspectives of this health problem. A number of topics are covered here, including the following major areas: prevalence and epidemiological data, etiology, prevention, assessment and treatment approaches, and health promotion strategies for LBP.

Vincenzo Denaro, Sergio Iavicoli, Fabrizio Russo, and Gianluca Vadalà
Editors

International Journal of
Environmental Research and Public Health

MDPI

Review

Does Workers' Compensation Status Affect Outcomes after Lumbar Spine Surgery? A Systematic Review and Meta-Analysis

Fabrizio Russo [1,*], Sergio De Salvatore [1], Luca Ambrosio [1], Gianluca Vadalà [1], Luca Fontana [2], Rocco Papalia [1], Jorma Rantanen [3], Sergio Iavicoli [2] and Vincenzo Denaro [1]

1 Department of Orthopaedic and Trauma Surgery, Campus Bio-Medico University of Rome, 00128 Rome, Italy; s.desalvatore@unicampus.it (S.D.S.); l.ambrosio@unicampus.it (L.A.); g.vadala@unicampus.it (G.V.); r.papalia@unicampus.it (R.P.); denaro@unicampus.it (V.D.)
2 Department of Occupational and Environmental Medicine, Epidemiology and Hygiene, Italian Workers' Compensation Authority (INAIL), 00078 Rome, Italy; lfontana73@yahoo.it (L.F.); s.iavicoli@inail.it (S.I.)
3 Department of Public Health/Occupational Health, University of Helsinki, 00250 Helsinki, Finland; jorma.h.rantanen@gmail.com
* Correspondence: fabrizio.russo@unicampus.it

Citation: Russo, F.; De Salvatore, S.; Ambrosio, L.; Vadalà, G.; Fontana, L.; Papalia, R.; Rantanen, J.; Iavicoli, S.; Denaro, V. Does Workers' Compensation Status Affect Outcomes after Lumbar Spine Surgery? A Systematic Review and Meta-Analysis. *IJERPH* **2021**, *18*, 6165. https://doi.org/10.3390/ijerph18116165

Academic Editor: Paul B. Tchounwou

Received: 28 April 2021
Accepted: 5 June 2021
Published: 7 June 2021

Abstract: Low back pain (LBP) is currently the leading cause of disability worldwide and the most common reason for workers' compensation (WC) claims. Studies have demonstrated that receiving WC is associated with a negative prognosis following treatment for a vast range of health conditions. However, the impact of WC on outcomes after spine surgery is still controversial. The aim of this meta-analysis was to systematically review the literature and analyze the impact of compensation status on outcomes after lumbar spine surgery. A systematic search was performed on Medline, Scopus, CINAHL, EMBASE and CENTRAL databases. The review included studies of patients undergoing lumbar spine surgery in which compensation status was reported. Methodological quality was assessed through ROBINS-I and quality of evidence was estimated using the GRADE rating. A total of 26 studies with a total of 2668 patients were included in the analysis. WC patients had higher post-operative pain and disability, as well as lower satisfaction after surgery when compared to those without WC. Furthermore, WC patients demonstrated to have a delayed return to work. According to our results, compensation status is associated with poor outcomes after lumbar spine surgery. Contextualizing post-operative outcomes in clinical and work-related domains helps understand the multifactorial nature of the phenomenon.

Keywords: disability; insurance; low back pain; lumbar decompression; lumbar fusion; musculoskeletal disorders; occupational health; pain; return to work; satisfaction

1. Introduction

Musculoskeletal disorders (MSDs) are the highest contributor to global disability and represent a substantial portion of occupational injury claims with a steadily rising incidence [1]. Low back pain (LBP) is the single worldwide leading cause of disability, has a strong relationship with years lived with disability (YLDs) and, since it was first measured in 1990 [1], it is the most common reason for workers' compensation (WC) claims [2]. It causes limitations of daily activity and work capacity, with high rates of work absenteeism and considerable economic and health consequences, therefore representing a major critical issue in the context of occupational medicine and public health [3].

Surgical procedures are quite commonly used as a treatment for LBP unresponsive to conservative treatments or associated with worsening neurological deficits [4]. The success of a surgical intervention in orthopaedic medicine is influenced by several key factors, the most important of which are the appropriateness of the surgical indication and surgeon's experience with the specific procedure. However, in this regard, the patient's compensation status has also been suggested as a potential factor influencing surgical outcomes. Indeed, additional elements including demographic and socioeconomic variables, such as lower degree of education, higher body mass index, smoking and lower annual wages, have been described to negatively impact outcomes following surgery [5].

In countries with modern social safety and welfare systems, an integrated compensation policy is guaranteed for disabled people or workers who experience accidents at work or occupational diseases. Compensation strategies and methodologies are extremely variable among nations, but commonly all of them provide workers with healthcare services, wage-replacement support, and other social benefits [6]. Usually, a government authority or a private sector organization acting on its behalf, carry out the administrative decision-making process which, after verifying the possession of eligibility criteria for claims, certifies the release of the different compensation services. Compensation approaches can be basically divided into two broad categories: cause-based systems typically require a correlation between occupational risk factors or work environment/activities and the resulting adverse health effects, whereas disability-based approaches provide benefits and services regardless of cause [7]. Therefore, WC benefits support the injured/sick workers by providing temporary aid, although in the most serious cases involving a high disability degree the type of compensation can also be permanent, until they can meet their respective clinical goals and return to work (RTW) as soon as possible with the least amount of disability. In this regard, it is important to note that the ability to RTW is one of the most clinically important outcomes in workers, in association with scores for disability, satisfaction and pain.

Nevertheless, it should be considered that available literature data provided evidence that the nature of compensation services and related methods of administration might adversely impact on health and work outcomes [8]. Indeed, several studies have demonstrated that receiving WC is associated with a negative prognosis following treatment for a vast range of health conditions [9–14]. Moreover, interactions of claimants with compensation authorities are often referred to by workers as stressful experiences that might induce poor mental health [8]. On the other hand, several procedural and bureaucratic features (e.g., delays in the claim processing times, strict and rigid procedures, lack of communication between workers and authorities) of the WC administrative process can increase the disability duration, thus delaying the reintegration of people into the workforce [15].

However, the influence of WC on the treatment of LBP is still controversial. Indeed, only a few studies have analyzed the impact of WC on outcomes after spine surgery, highlighting the importance of considering WC as a determining factor when evaluating outcomes of different spinal procedures [5,6,16]. Indeed, the reported strength of this association has widely varied from odds ratios of 1.31 to 7.22 among published studies [8,17,18]. The purpose of this meta-analysis is to systematically review the literature and analyze the impact of compensation status on lumbar spine surgery outcomes.

2. Materials and Methods

We focused our research on studies concerning the effect of WC on outcomes after lumbar spine surgery, comparing them to non-workers' compensation (NWC) patients. The Preferred Reporting Items for Systematic reviews and Meta-Analyses (PRISMA) guidelines were used to improve the reporting of the review.

2.1. Eligibility Criteria

The research question was formulated using a PICOS-approach: Patient (P); Intervention (I); Comparison (C); Outcome (O) and Study design (S). The aim of this systematic review was to select those articles which described "if patients undergoing lumbar spine surgery (P) with a known WC status (I) have worse results in terms of LBP, disability, satisfaction and time to RTW (O) compared to the NWC population (C)". For this purpose, only randomized control trials (RCT) and non-randomized controlled studies (NRCT) such as prospective (PS), retrospective (RS) observational studies (OS), case-series (CS) and case-control (CC) studies were included.

Inclusion and Exclusion Criteria

Inclusion criteria were:

- Peer-reviewed studies of every level of evidence according to the Oxford Classification. We included in our research RCT and NRCT.
- Studies including working patients with no limitations of age and type of work.
- Studies that reported outcomes for patients undergoing any type of surgical procedure involving the lumbar spine.
- Studies that included at least one assessment for each type of outcome (LBP, disability, satisfaction after surgery and RTW). The pain outcome had to be evaluated by one or more of the following scales: numerical pain rating scale (NRS) and visual analogue scale (VAS). The disability outcome needed to be evaluated by one or more of the following scales: Oswestry Disability Index (ODI); 36-Item Short Form Health Survey (SF-36); 12-Item Short Form Health Survey (SF-12); Roland Morris Disability Questionnaire (RMDQ); functional status (FS) and Lumbar Back Outcome Scale (LBOS). RTW was evaluated as the number of patients that went back to their previous working activities at the time of the last follow-up. The satisfaction rate after surgery was assessed in patients as follows: "Excellent", "Good", "Almost complete relief", "Good deal of relief", and "Satisfied" were considered as satisfactory outcomes, whereas "Fair", "Poor", "Only a little relief", "No relief or worse" and "Unsatisfied" were considered unsatisfactory. Moreover, in studies where the satisfaction rate was expressed in a numeric scale, values between 0 and 4 were considered unsatisfactory, whilst values between 5 and 10 were considered satisfactory.
- Only articles written in English and Italian were included.

We excluded case reports, technical notes, letters to editors, instructional courses, in vitro and cadaver studies, as well as studies including cervical or thoracic spine procedures.

2.2. Search

The articles included in the study were screened from inception to May 2020 through a systematic search of Medline, Scopus, CINAHL, EMBASE and CENTRAL databases. For the search strategy we decided to use the following string: (workers compensation [MeSH Terms]) AND ((spine) OR (lumbar) OR (spine surgery)). We used the keywords isolated or combined. We searched for more studies among the reference lists of the selected papers and systematic reviews.

2.3. Study Selection

We accepted only English and Italian publications. The initial search of the article was conducted by two reviewers (S.D.S. and L.A.). In case of disagreements, the consensus of a third reviewer (F.R.) was asked. The research was conducted using the CADIMA software [19]. The researchers used the following research order: titles were screened first, then abstracts and full papers. A paper was considered potentially relevant, and its full text reviewed, if following a discussion between the two independent reviewers, it could not be unequivocally excluded based on its title and abstract. The full text of all papers not excluded on the basis of abstract or title, was evaluated. The number of articles excluded or included were registered and reported in a PRISMA flowchart (Figure 1). For designing the PRISMA we followed the rules by Moher et al. [20].

Figure 1. PRISMA flowchart.

2.4. Data Extraction

General study characteristics extracted were author, year of publication, country of origin, type of study, level of evidence [21] (LOE), sample size (divided in WC and NWC), mean age (divided in WC, NWC and mean of both groups), last or average follow-up (in case of multiple time points, only the last follow-up was considered), type of surgery, type of comparison group (NWC), outcome measures (LBP, disability, satisfaction and RTW) and differences between groups.

2.5. Individual Study Quality

Given the observational design of included studies, we used the Risk of Bias in Non-randomized Studies of Interventions (ROBINS-I) tool to assess the quality of each study [22] (Figure 2). In order to avoid imprecisions, selected papers were rated independently by two reviewers (S.D.S. and L.A.) and verified by a third one (F.R.).

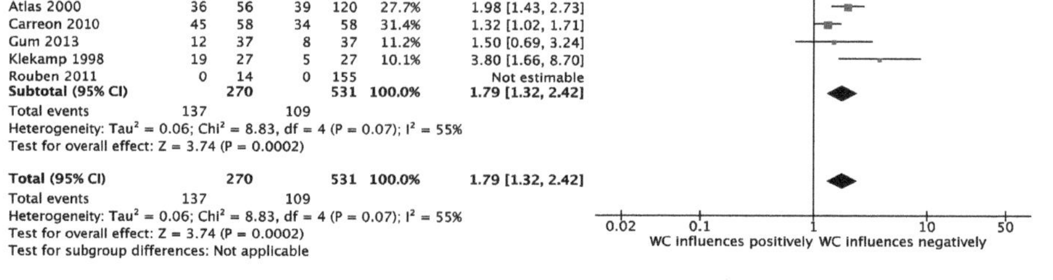

(A)

Figure 2. *Cont.*

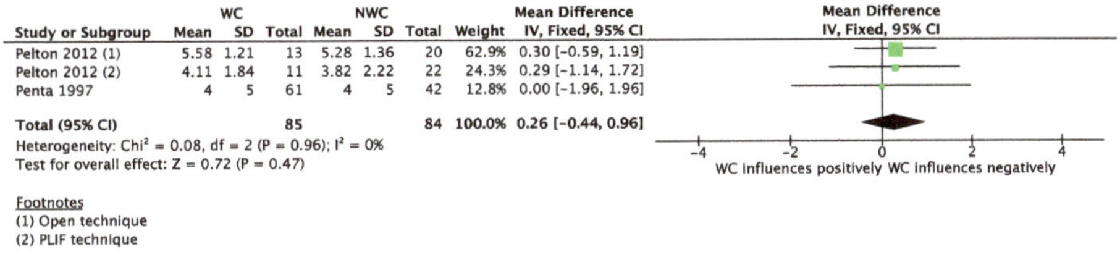

(B)

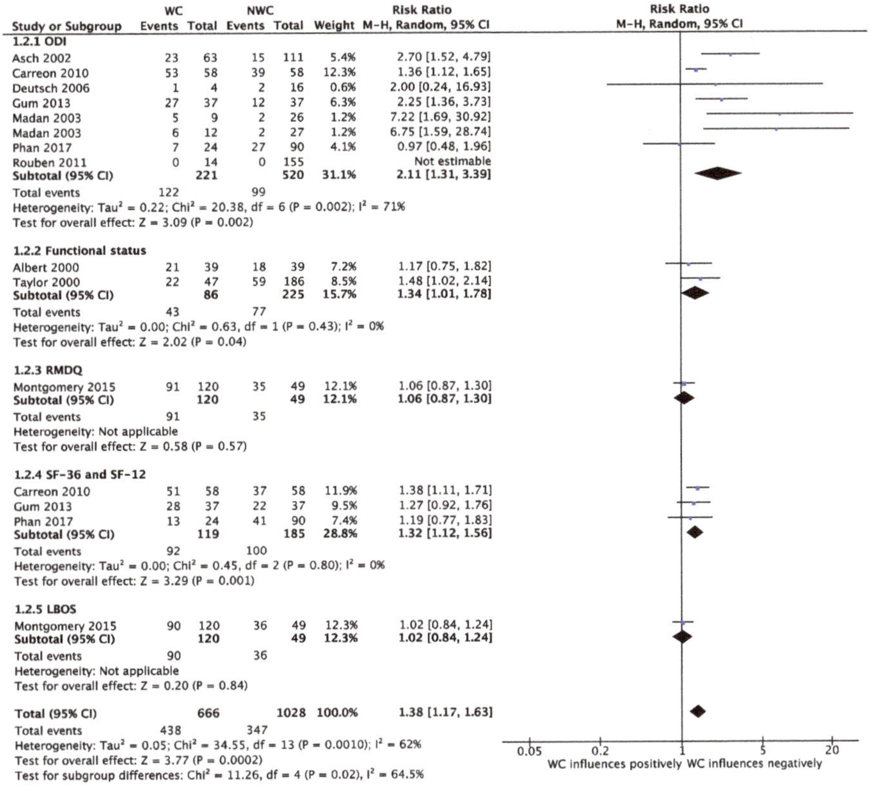

(C)

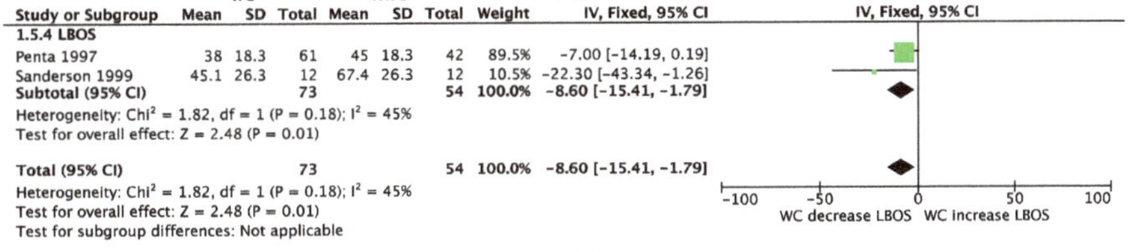

(D)

Figure 2. Forest plots depicting the effect of WC on post-operative pain measured by dichotomous (**A**) and continuous data (**B**) and the effect of WC on post-operative disability measured by dichotomous (**C**) and continuous data (**D**).

2.6. Quality of Evidence

We used the GRADE approach (Tables S1 and S2) to rate the overall quality of evidence. The GRADE approach classifies the quality of evidence for each outcome by grading the following domains: study design, risk of bias, inconsistency, indirectness, imprecision, publication bias, magnitude of the effect. The quality of evidence was then classified as follows:

- High Quality of Evidence: among 75% of articles included are considered with a low risk bias. Further research is useful to change either the estimate or confidence in results.
- Moderate Quality of Evidence: one of the GRADE domains is not met. Further studies are required to improve the quality of the study and the evidence.
- Low Quality of Evidence: two of the GRADE domains are not met. Further research is critical.
- Very Low Quality of Evidence: three of the GRADE domains are not met. The results of the study are very uncertain. In the case of studies with a sample size inferior to 300 subjects, the quality of the study is considered very low if there was also a high risk of bias (assessed with the ROBINS-I in the present study).

The outcomes assessed were LBP, disability, satisfaction after surgery and RTW evaluated at the end of the treatment. Furthermore, the outcomes were subgrouped per scales. To avoid imprecisions and considering the limited number of studies with continuous data, we considered for GRADE analysis only studies with dichotomous data.

2.7. Summary Measures

The summary measures of effect size considered in the study were the risk ratio (RR) for dichotomous data and the mean difference (MD) for continuous variables of data on outcome after surgery in terms of LBP, disability, satisfaction and RTW in WC and NWC populations.

2.8. Synthesis of Results

The Mantel–Haenszel method of meta-analysis was performed using Review Manager Software 5.0 (RevMan 5.0, Cochrane Collaborations, London, UK). For dichotomous data, risk ratio was applied using a 5% level of significance. Heterogeneity was assessed by a funnel plot and chi-square test, and inconsistency across studies was quantified using the I^2 statistic. An $I^2 > 50\%$ or a p value of chi-squared test > 0.05 were suggestive of a substantial heterogeneity. Random effects model was used in all analyses.

3. Results

3.1. Study Selection

We found a total of 592 studies (no additional studies were found in gray literature and no unpublished studies were retrieved). We obtained 335 studies following duplicate removal, 282 of which were excluded through title and abstract screening. Then, 53 full-text articles were screened. Out of these studies, 27 were excluded (no lumbar surgery, $n = 3$; no surgical intervention, $n = 3$; not defined WC group, $n = 5$; sample population including non-operative treatment, $n = 1$; unclear outcomes, $n = 5$ and results not estimable, $n = 10$). After this process, 26 articles were included in our study [16,18,23–46].

3.2. Study Characteristics

A summary of the characteristics of the included studies is reported in Tables 1 and 2 (see abbreviations explained below each Table). We did not find any RCTs eligible for this study. The articles selected included 26 NRCT (9 RS, LOE 3; 14 PS, LOE 2; 2 CC, LOE 3 and 1 CS, LOE 4). Studies were published between 1994 [29] and 2017 [40]. A total of 2668 patients (1045 WC and 1623 NWC) were assessed for outcomes after lumbar spine surgery. Of these studies, 3 were performed in Australia, 1 in New Zealand, 1 in Switzerland, 3 in the United Kingdom and 18 in the United States. Pain evaluation in these studies was performed using

NRS (3 studies [26,27,32]) and VAS (5 studies [16,25,38,39,41]) scores. The disability outcome was evaluated by one or more of the following scales: ODI (7 studies [16,18,25,27,28,40,41]); SF-36 (2 studies [16,27]); SF-12 (1 study [40]); RMDQ (1 study [36]); FS (2 studies [24,45]) and LBOS (3 studies [36,39,42]). RTW was evaluated in 9 studies [25,26,28,29,33,41,43,44,46] and satisfaction rate in 15 studies [23,25,26,29–31,33–35,37,39,43–46]. The studies cited in this review show a moderate heterogeneity between groups (50% < I^2 < 70%, except for the ODI subgroup with a I^2 = 71%) and differences in terms of study design, interventions, and outcome variables. Follow-ups were different and ranged from 6 months [38] to 16 years [36].

3.3. Methodological Quality

The ROBINS-I tool for NRCT was used to assess the methodological quality of each study. We found 7 studies with an overall risk of bias identified as "low" [18,25,29,30,34,35,40], 14 studies with a "moderate" risk [16,24,26–28,31–33,38,41–43,45,46] and 5 studies with a "serious" risk [23,36,37,39,44]. The quality of evidence of the studies included in the GRADE was classified as "low". Methodological quality assessments of each study are summarized in Supplementary Figure S1. The quality of evidence of full data was performed using the GRADE approach (Supplementary Tables S1 and S2). The analysis of the data of the study was reported using the RR for studies included dichotomous data and using the MD for studies with continues data. RevMan5 (version 5.3) was used to calculate the RR the MD of the included studies and the heterogeneity between studies using I^2 and Chi-squared test. The results of the meta-analysis are summarized using forest plots.

3.4. Results of Individual Studies

The intervention methods were usually well described in all the included studies. Moderate heterogeneity in the length of follow-up and the surgical procedure were reported in all the studies. We included all types of lumbar spine surgery: discectomy [25,26,31–35,45], laminectomy [31,45], hemilaminotomy [34], lumbar spine fusion [36], minimally invasive surgery or open approach for transforaminal lumbar interbody fusion (TLIF) [16,28,38,41], posterior lumbar interbody fusion (PLIF) [18,23], posterior lumbar fusion (PLF) [18,23,27,30,37,43], anterior lumbar interbody fusion (ALIF) [29,39,40,44], anteroposterior fusion [24] and uninstrumented posterolateral fusion [46]. The authors divided the description of intervention per outcome (LBP, disability, satisfaction rate and RTW). Disability outcomes were subgrouped per measure scale: ODI, SF-12 and SF-36, LBOS and FS. The results of each outcome are reported in Table 2.

3.5. Outcome: Pain

Eight observational studies were included (4 PS [25,26,38,39], 2 RS [32,41] and 2 CC [16,27]). They examined the influence of WC on pain modifications in patients undergoing lumbar surgery. Three studies used the NRS scale [26,27,32] and five studies used the VAS scale [16,25,38,39,41] to assess pain. Single studies were assessed for risk of bias using ROBINS-I tool. One study was classified as "serious risk" [39], six as "moderate" [16,26,27,32,38,41] and one as "low risk" [25].

The overall quality of evidence in these studies was assessed as "low" according to GRADE. The quantitative effect estimate was reported as RR in studies with dichotomous data (Figure 2A) and as the MD between and within studies (when possible) in case of continuous data (Figure 2B). The overall RR was 1.79, 95% CI 1.32 to 2.42; I^2 = 55%. 2 studies [38,39] reported the pain outcome as continuous data with a MD between WC and NWC of 0.26, 95% (CI −0.44 to 0.96; I^2 = 0%), showing a moderate negative influence of WC on pain improvement.

Table 1. Main characteristics of the studies included in the meta-analysis.

Author	Year	Country	Type of Study	LOE	Sample Size WC	Sample Size NWC	Mean Age WC (y)	Mean Age NWC (y)	Mean Age (y)
Agazzi et al. [23]	1999	Switzerland	Retrospective	3	34	37	-	-	-
Albert et al. [24]	2000	USA	Retrospective	3	28	9	-	-	40.3 ± 10.3
Asch et al. [25]	2002	USA	Prospective	2	80	132	-	-	41 ± 11.3
Atlas et al. [26]	2000	USA	Prospective	2	56	120	38.7	41.2	-
Carreon et al. [27]	2010	USA	Case-control	3	58	58	47.9 ± 9.4	47.8 ± 9.4	-
Deutsch et al. [28]	2006	USA	Prospective	2	4	16	-	-	49
Greenough et al. [29]	1994	UK	Prospective	2	106	45	-	-	-
Greenough et al. [30]	1998	UK	Prospective	2	66	62	-	-	-
Gum et al. [16]	2013	USA	Case-control	3	38	38	42	42.2	-
Herron et al. [31]	1996	USA	Prospective	2	89	186	-	-	43 (15–83)
Klekamp et al. [32]	1998	USA	Retrospective	3	23	27	35.0 ± 7.1	39.5 ± 12.0	-
Lew et al. [33]	2001	USA	Retrospective	3	10	37	49.7 ± 9.8	50.7 ± 10.8	-
MacKay et al. [34]	1995	USA	Prospective	2	46	108	-	-	40 (20–79)
Madan et al. [18]	2003	UK	Prospective	2	12	27	-	-	43 (25–67)
Marks et al. [35]	2000	USA	Retrospective	3	51	51	-	-	-
Montgomery et al. [36]	2015	New Zealand	Prospective	2	120	49	53 (24–81)	61 (31–82)	-
Parker et al. [37]	1996	USA	Prospective	2	10	23	-	-	41 (22–56)
Pelton et al. [38] §	2012	USA	Prospective	2	11	22	-	-	51.7 ± 12.2
Pelton et al. [38] §	2012	USA	Prospective	2	13	20	-	-	49.9 ± 10.7
Penta et al. [39]	1997	Australia	Prospective	2	61	42	-	-	48 (28–73)
Phan et al. [40]	2017	Australia	Prospective	2	24	90	46.3 ± 10.4	60.2 ± 12.9	-
Rouben et al. [41]	2011	USA	Retrospective	3	14	155	-	-	44.5 ± 10.9
Sanderson et al. [42]	1999	Australia	Retrospective	3	12	12	-	-	33.1 ± 14.2
Schnee et al. [43]	1997	USA	Retrospective	3	20	32	-	-	53.4 (24–77)
Slosar et al. [44]	2000	USA	Retrospective	3	73	60	-	-	38.8 (21–58)
Taylor et al. [45]	2000	USA	Prospective	2	47	189	-	-	46
Vaccaro et al. [46]	1997	USA	Case series	4	13	11	37	39	38 (24–50)

§ The study from Pelton et al. is composed of two different cohorts as illustrated in the table. LOE = level of evidence; WC = workers' compensation; NWC = non-worker' compensation.

Table 2. Types of lumbar spine surgery, outcomes measured in WC and NWC and main findings obtained by the studies included in the meta-analysis.

Author	Last Follow-Up	Type of Surgery	Comparison	Outcomes Measures	Conclusions
Agazzi et al. [23]	2 y	PLIF	Yes (internal)	Prolo economic and functional scale WC: 16/34 (47%); NWC: 31/37 (84%) RTW WC 2/33; NWC 24/30	Socioeconomic factors and WC issues seem to be significant prognostic indicators of outcome.
Albert et al. [24]	2 y	Anteroposterior fusion	Yes (internal)	Functional status Success: WC 18/27; NWC 9/27 Failure: WC 10/12; NWC 2/12	WC increased the chance of functional failure, though this correlation was not statistically significant.
Asch et al. [25]	3 y	Microdiscectomy	Yes (internal)	Pain relief success rate WC 67.9%; NWC 82.8% ($p < 0.05$) ODI success rate (< 40%) WC 63.5%; NWC 86.5% ($p < 0.001$) Satisfaction after surgery WC 63.2%; NWC 83.5% ($p < 0.001$) RTW WC 42.1%; NWC 71.5% ($p < 0.001$)	Progressively poorer outcomes occur with increasing patient age up to the late-50 s and confirms the disparity in outcomes between cases in which WC is being sought and those in which it is not.
Atlas et al. [26]	4 y	Open discectomy, percutaneous discectomy	Yes	Relief from pain WC 20 (36%); NWC 81 (68%) RMDQ WC −9.3; NWC −12.5 Satisfaction WC 24 (43%); NWC 85 (71%) RTW WC 17/133 (13%); NWC 7/190 (4%)	Patients who have been receiving WC at baseline were more likely to be receiving disability benefits and were less likely to report relief from symptoms and improvement in quality of life at the time of the four-year follow-up than patients who had not been receiving WC at baseline. Nonetheless, most patients returned to work regardless of their initial disability status, and those who had been receiving WC at baseline were only slightly less likely to be working after four years.
Carreon et al. [27]	2 y	PLF	Yes	NRS back WC 1.7 ± 3.1; NWC 2.5 ± 2.7 ($p = 0.073$) ODI WC 4.9 ± 14.1; NWC 13.3 ± 17.1 ($p = 0.009$) SF-36 WC −1.3 ± 9.7; NWC 3.9 ± 8.9 ($p = 0.007$)	Patients on WC have significantly less improvement of clinical outcomes in both mean change in ODI and SF-36, as well as the number of patients achieving substantial clinical benefit.

Table 2. *Cont.*

Author	Last Follow-Up	Type of Surgery	Comparison	Outcomes Measures	Conclusions
Deutsch et al. [28]	1 y	Unilateral TLIF with PLF	Yes, but not clearly defined	ODI WC: 3/4 patients improved at 6 months RTW WC: 2/4	No differences between WC and NWC were reported concerning to disability and RTW.
Greenough et al. [29]	2 y	ALIF	Yes	Satisfaction after surgery ($p < 0.05$) 8–10: WC (35; 37%); NWC (28; 67%) 5–7: WC (37; 40%); NWC (8; 19%) 2–4: WC (18; 19%); NWC (3; 7%) 0–1: WC (4; 4%); NWC (3; 7%) LBOS ($p < 0.01$) Excellent: WC (8; 10%); NWC (13; 43%) Good: WC (19; 22%); NWC (10; 25%) Fair: WC (34; 40%); NWC (10; 25%) Poor: WC (24; 28%); NWC (7; 17%)	The rate of fusion was influenced by the presence of a WC claim. WC status and psychological disturbance at presentation were significant prognostic factors. Psychological disturbance at review had a profound effect on the outcome and patient satisfaction ratings.
Greenough et al. [30]	2 y	PLF	Yes	LBOS WC 25 (7–72, $n = 57$); NWC 35 (7–75, $n = 63$) $p < 0.001$ Satisfaction after surgery WC 2 (0–3, $n = 56$); NWC 2 (0–3, $n = 59$) $p < 0.02$ VAS WC 6 (1–10, $n = 57$); NWC 5 (0–10, $n = 62$) $p < 0.02$	Results of instrumented PLF are poor and indications for the procedure need careful consideration. The results are significantly influenced by WC but not by technical success.
Gum et al. [16]	2 y	TLIF or PLF	Yes	VAS WC 0.94; NWC 2.51 ($p = 0.011$) ODI mean change WC 5.54; NWC 15.17 ($p = 0.009$) SF-36 mean change WC 1.69; NWC 4.09 ($p = 0.235$)	Patients receiving WC have the perception of poor clinical outcomes after lumbar fusion.
Herron et al. [31]	4 y	Laminectomy and discectomy	Yes (internal)	Surgical outcome ($p = 0.00$) Good: WC 52 (58%); NWC 174 (94%) Fair: WC 16 (18%); NWC 10 (5%) Poor: WC 21 (24%); NWC 2 (1%)	Patients with WC or litigation issues were significantly more likely to have poor outcomes.

Table 2. *Cont.*

Author	Last Follow-Up	Type of Surgery	Comparison	Outcomes Measures	Conclusions
Klekamp et al. [32]	11 m	Discectomy	Yes	WC: 29% of patients achieved good results NWC: 81% of patients achieved good results	WC group achieved worse results compared to NWC group.
Lew et al. [33]	18 m (4–51 m)	Discectomy	Yes	Satisfaction after surgery (%) Excellent or good WC: 7 (70); NWC: 33 (89) $p = 0.12$ Excellent WC: 5 (50); NWC: 22 (60) $p = 0.24$ Good WC: 2 (20); NWC: 11 (30) $p = 0.27$ Fair WC: 0 (0); NWC: 2 (5.4) $p = 0.62$ Poor WC: 3 (30); NWC: 2 (5.4) $p = 0.05$ RTW WC: 90%; NWC: 93% $p = 0.45$	WC recipients experienced significantly worse outcomes than the other patients in this study. Nevertheless, a high RTW rate was maintained (90%) in both groups.
MacKay et al. [34]	1 y	Hemilaminotomy, discectomy	Yes	Prolo scale Satisfactory: WC: 63%; NWC: 92% ($p < 0.0001$) Unsatisfactory: WC: 37%; NWC: 8%	WC group had a lower success rate compared to NWC group.
Madan et al. [18]	2.4 y (2–3.1 y)	PLF and PLIF	Yes (internal)	ODI ALIF group ($p = 0.0056$) Satisfied WC: 6 (50); NWC: 25 (92.6) Unsatisfied WC: 6 (50); NWC: 2 (7.4) PLIF ($p = 0.0064$) Satisfied WC: 4 (45); NWC: 24 (92.3) Unsatisfied WC: 5 (55.5); NWC 2 (7.7)	There were no differences between WC and NWC groups concerning to disability.
Marks et al. [35]	30.7 ± 17.9 m	Percutaneous discectomy	Yes, but not clearly defined	Pain, Job function, Physical restrictions, medications WC and NWC no differences (data not available), $p > 0.05$	WC status does not influence the outcomes.

Table 2. *Cont.*

Author	Last Follow-Up	Type of Surgery	Comparison	Outcomes Measures	Conclusions
Montgomery et al. [36]	8 y (4–14 y)	Lumbar spinal fusion	Yes	RMDQ 1-year postoperative WC: 8.0, 6.8–9.2/NWC: 4.6, 2.8–6.5 ($p <$ 0.05) At long-term follow-up WC: 5.9, 4.7–7.1/NWC: 3.8, 1.9–5.8 ($p >$ 0.05) LBOS 1 year post-operative WC: 43.9, 39.9–48.0/NWC: 54.1, 48.4–59.9 ($p < 0.05$) Long term follow-up WC: 47.0, 43.5–50.4/NWC: 55.4, 49.3–61.6 ($p > 0.05$) SF12 Long term follow up WC 41.6 ± 11.5/NWC 44.0 ± 13.0 ($p > 0.05$)	ACC patients achieved equivalent improvements compared to non-ACC patients and NWC patients as per in the published literature. They also achieve function that is considerably better than that achieved in WC patients in adversarial compensation jurisdictions.
Parker et al. [37]	47 m (27–84 m)	PLF	Yes	Clinical outcome pain, medications, and resume of previous activities WC: 1/10 good, 9/10 poor results NWC: 9/23 good or excellent, 3/23 fair, 11/23 poor results	Patients in WC group showed worse clinical outcomes compared to NWC group.
Pelton et al. [38]	6 m	MIS-TLIF and open TLIF	Yes	VAS (MIS-TLIF cohort) Differences between WC and NWC ($p = 0.712$) VAS (open TLIF cohort) Differences between WC and NWC ($p = 0.241$)	Immediate outcomes and hospitalizations between NWC and WC populations did not differ regardless of surgical technique (MIS/open). Differences occurred in improved outcomes with an MIS-TLIF versus an open TLIF even in a WC environment.
Penta et al. [39]	10 y	ALIF	Yes	LBOS WC: 38 (4–74); NWC: 45 (11–75) ($p = 0.06$) VAS WC: 4 (0–9), NWC: 4 (0–10) ($p = 0.29$)	WC had a negative effect on outcomes only in the first period (two years). After 10 years of follow up this effect disappeared.
Phan et al. [40]	2 y	ALIF	Yes	SF-12 WC: 11.3; NWC: 9.1 ($p = 0.691$) ODI WC: 26.3; NWC: 33.4 ($p = 0.232$)	No significant differences found between WC and NWC patients in terms of fusion rates, complications, clinical outcomes.

Table 2. *Cont.*

Author	Last Follow-Up	Type of Surgery	Comparison	Outcomes Measures	Conclusions
Rouben et al. [41]	50 m	MIS-TLIF	Yes (internal)	RTW 57% of WC patients (mean time: 17 weeks) ODI Mean change of 34% ($p < 0.001$) Post-operative VAS Significant improvement WC patients ($p < 0.001$)	WC patients responded well to surgical treatment.
Sanderson et al. [42]	3.1 y	Short segment fixation	Yes	LBOS WC: 45.1; NWC: 67.4 ($p < 0.05$)	The presence of a WC claim positively influenced the outcomes after surgery.
Schnee et al. [43]	18.6 m (6–36.7 m)	PLF	Not clearly defined	RTW WC: not defined; NWC: 84% of cases; Prolo scale Significant adverse effects of WC (p = 0.0001). Good pain results were seen in 81% of NWC	WC claims and smoking had very significant adverse impacts on both employment and pain results despite high fusion rates, particularly in patients under the age of 55.
Slosar et al. [44]	37.2 m	ALIF + PLF	Yes, but not clearly defined	Satisfaction after surgery ($p > 0.05$) 1 (best): WC 7 (9.6%); NWC 7 (11.7%) 2: WC 36 (49.3); NWC 32 (53.3%) 3: WC 14 (19.1%); NWC 12 (20%) 4: WC 16 (22%); NWC 9 (15%)	There was not a statistically significant difference in terms of satisfaction following surgery between WC and NWC patients.
Taylor et al. [45]	18 m	Discectomy, laminectomy, or fusion	Not clearly defined	Much better functioning WC: 52%; NWC: 68% $p < 0.05$ Very positive about the treatment WC: 57%; NWC: 71% $p < 0.05$	The study results indicate that WC payments and litigation are two important predictors of poor outcomes after low back surgery in community practice.
Vaccaro et al. [46]	37 m (18–64 m)	Uninstrumented PLF	No	Satisfaction after surgery Fair/poor results: WC: 13; NWC: 2 RTW None of the WC patients returned to work	WC is strongly associated with poor results of operative management of LBP in adult patients with low-grade spondylolisthesis.

ACC = Accident Compensation Corporation; ALIF = anterior lumbar interbody fusion; LBOS = low back outcome score; MIS = minimally invasive surgery; NRS = numeric rating scale; NWC = non-worker' compensation; ODI = Oswestry Disability Index; PLF = posterolateral fusion; PLIF = posterolateral interbody fusion; RMDQ = Roland and Morris Disability Questionnaire; RTW = return to work; SF-12 = 12-item Short Form Health Survey; SF-36 = 36-item Short Form Health Survey; TLIF = transforaminal lumbar interbody fusion; VAS = visual analogue scale; WC = workers' compensation.

3.6. Outcome: Disability

Twelve observational studies were included (7 PS [18,25,28,36,39,40,45]; three RS [24,41,42] and two CC [16,27]). They examined the influence of WC on disability modifications in patients undergoing lumbar surgery. Seven studies [16,18,25,27,28,40,41] used the ODI scale, two [16,27] used the FS, one [36] used the RMDQ, one [40] used the SF-12, two [16,27] used the SF-36 and three [36,39,42] used the LBOS to assess disability. Single studies were assessed for risk of bias using ROBINS-I tool. Two studies were classified as "serious risk" [36,39], seven as "moderate" [16,24,27,28,41,42,45] and three as "low risk" [18,25,40].

The overall quality of evidence in these studies was assessed as "low" according to GRADE. The quantitative effect estimate was reported as RR in studies with dichotomous data (Figure 2C) and as MD between and within studies (when possible) in case of continuous data (Figure 2D).

The overall RR was 1.38 (95% CI 1.17 to 1.63; I^2 = 62%), suggesting an overall negative influence of WC on disability improvement. The ODI subgroup had a RR of 2.11 (95% CI 1.31 to 3.39; I^2 = 71%); the FS subgroup reported a RR of 1.34 (95% CI 1.01 to 1.78; I^2 = 0%); the RMDQ subgroup reported a RR of 1.06 (95% CI 0.87 to 1.30; I^2 = 0%); the SF-12 and SF-36 subgroup showed a RR of 1.32 (95% CI 1.12 to 1.56; I^2 = 0%); the LBOS subgroup reported a RR of 1.02 (95% CI 0.84 to 1.24; I^2 = 0%). 2 studies [39,42] reported the disability outcome as continuous data with a MD between WC and NWC of −8.60 (95% CI −15.41 to −1.79; I^2 = 45%); showing that WC decreased LBOS postoperative values (the lower the value of LBOS, the higher disability of the patient).

3.7. Outcome: Return to Work

Nine observational studies were included (four PS [25,26,28,29]; four RS [33,41,43,44] and one CS [46]). They examined how WC influence RTW in patients after lumbar surgery (Figure 3A). RTW was considered at the time of the last follow-up. Single studies were assessed for risk of bias using ROBINS-I tool. One study was classified as "serious risk" [44], 6 as "moderate" [26,28,33,41,43,46] and 2 as "low risk" [25,29].

Study or Subgroup	WC Events	WC Total	NWC Events	NWC Total	Weight	Risk Ratio M-H, Fixed, 95% CI
Asch 2002	44	76	37	130	24.4%	2.03 [1.46, 2.84]
Atlas 2000	17	80	99	459	26.2%	0.99 [0.62, 1.56]
Deutsch 2006	2	4	0	0		Not estimable
Greenough 1994	42	85	10	40	12.1%	1.98 [1.11, 3.52]
Lew 2001	1	10	2	28	0.9%	1.40 [0.14, 13.82]
Rouben 2011	6	14	5	155	0.7%	13.29 [4.63, 38.09]
Schnee 1997	0	0	27	32		Not estimable
Slosar 2000	57	73	26	60	25.5%	1.80 [1.32, 2.47]
Vaccaro 1997	13	13	10	11	10.1%	1.10 [0.87, 1.40]
Total (95% CI)		**355**		**915**	**100.0%**	**1.68 [1.41, 1.99]**
Total events	182		216			

Heterogeneity: Chi^2 = 33.96, df = 6 (P < 0.00001); I^2 = 82%
Test for overall effect: Z = 5.81 (P < 0.00001)

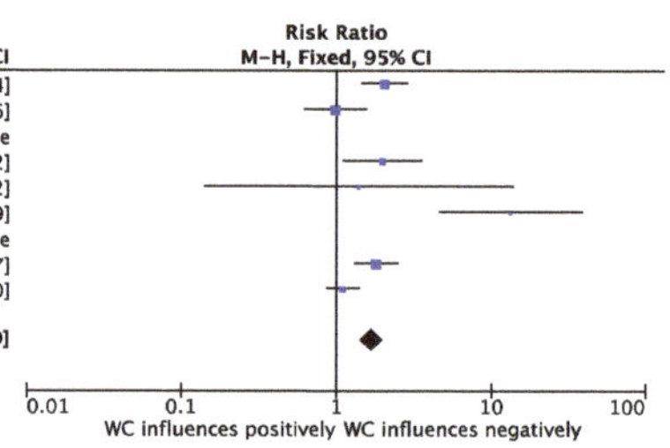

(A)

Study or Subgroup	WC Events	WC Total	NWC Events	NWC Total	Weight	Risk Ratio M-H, Fixed, 95% CI
Agazzi 1999	18	34	6	37	3.5%	3.26 [1.47, 7.25]
Asch 2002	28	76	22	133	9.7%	2.23 [1.38, 3.61]
Atlas 2000	32	56	35	120	13.6%	1.96 [1.37, 2.81]
Greenough 1994	27	85	8	40	6.6%	1.59 [0.79, 3.18]
Greenough 1998	24	56	18	59	10.7%	1.40 [0.86, 2.29]
Herron 1996 (1)	37	89	12	186	4.7%	6.44 [3.54, 11.74]
Lew 2001	3	10	4	37	1.0%	2.77 [0.74, 10.43]
Mackay 1995	17	46	9	108	3.3%	4.43 [2.14, 9.21]
Marks 2000	23	51	14	51	8.5%	1.64 [0.96, 2.82]
Parker 1996	9	10	9	13	4.8%	1.30 [0.86, 1.97]
Penta 1997	10	61	5	42	3.6%	1.38 [0.51, 3.74]
Schnee 1997	10	20	2	32	0.9%	8.00 [1.95, 32.82]
Slosar 2000	32	73	21	60	14.0%	1.25 [0.81, 1.93]
Taylor 2000	20	47	54	185	13.3%	1.46 [0.98, 2.18]
Vaccaro 1997	13	13	2	11	1.6%	4.63 [1.53, 14.01]
Total (95% CI)		**727**		**1114**	**100.0%**	**2.10 [1.82, 2.44]**
Total events	303		221			

Heterogeneity: Chi^2 = 42.91, df = 14 (P < 0.0001); I^2 = 67%
Test for overall effect: Z = 9.96 (P < 0.00001)

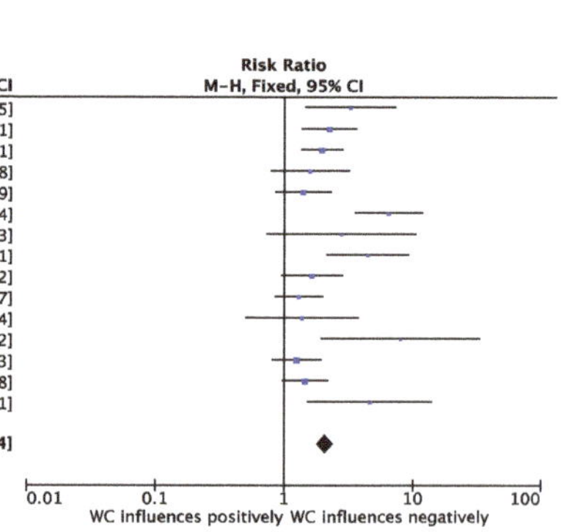

(B)

Figure 3. Forest plots depicting the effect of WC on return to work (**A**) and satisfaction (**B**) following lumbar spine surgery.

The overall quality of evidence in these studies was assessed as "low" according to GRADE. The quantitative effect estimate was reported as RR. The overall RR was 1.68 (95% CI 1.41 to 1.99; I^2 = 82%). The studies reported an overall negative influence of WC on RTW in patients after lumbar surgery.

3.8. Outcome: Satisfaction

Fifteen observational studies were included (nine PS [25,26,29–31,34,37,39,45]; five RS [23,33,35,43,44] and one CS [46]). They examined the influence of WC on satisfaction modifications in patients undergoing lumbar surgery (Figure 3B). Single studies were assessed for risk of bias using ROBINS-I tool. Four studies were classified as "serious risk" [23,37,39,44], six as "moderate" [26,31,33,43,45,46], and five as "low risk" [25,29,30,34,35].

The overall quality of evidence in these studies was assessed as "low" according to GRADE. The quantitative effect estimate was reported as RR. The overall RR was 2.10 (95% CI 1.82 to 2.44; I^2 = 67%). The studies reported an overall negative influence of WC on satisfaction of patients after lumbar surgery.

4. Discussion

The association between compensation status and poor clinical outcomes after orthopaedic surgery has already been described in the literature. In a meta-analysis from Harris et al. [5], WC patients presented with an approximately four times higher odds of worse outcomes after common orthopaedic procedures including shoulder acromioplasty, carpal tunnel release, lumbar fusion and lumbar discectomy compared to NWC patients. Similarly, in a recent meta-analysis from Cheriyan and colleagues [47], outcomes related to patient satisfaction and RTW were investigated in WC and NWC subjects after spine surgery. In this study, authors concluded that WC patients showed a 2.10 RR of unsatisfactory outcomes and a 1.68 RR of delayed RTW after surgical procedures involving the cervical, thoracic, and lumbar spine. These data are congruous with the meta-analysis of de Moraes et al. [48], who reported that compensated patients undergoing lumbar discectomy with or without fusion presented a 1.90 RR of unsatisfactory outcomes after surgery.

In the present study, we analyzed the effect of WC on clinical (pain, disability, and patient satisfaction) and work-related outcomes (RTW) following lumbar spine surgery. Consistently with previous studies, we reported that WC patients tended to exhibit higher post-operative pain (RR = 1.79) and disability (RR = 1.38) as well as lower satisfaction after surgery (RR = 2.10) compared to NWC patients. WC patients demonstrated also a delayed RTW (RR = 1.68) with a significant socioeconomic burden on both work insurances and employers [49]. This latter data is particularly important when considering that the annual expenditure for treating LBP in the United States is greater than $100 billion, with lost wages and reduced productivity accounting for approximately two thirds of the amount [50]. Furthermore, lumbar injuries resulting in spine surgery are among the most expensive WC claims [51]. However, the total cost may not be strictly related to the type of surgery alone but seems also affected by the time between the injury and the surgical treatment. Indeed, Lavin et al. have found that more prolonged and costly WC claims were associated with an interval of more than a year between injury and surgery, hence concluding that timeline of surgical indication is equally important in this subset of patients [52].

It is also important to note that several studies have demonstrated that lumbar spine surgery and particularly fusion procedures are characterized by a variable rate of success [53–56]. Therefore, inadequate patient selection and/or surgical indication may negatively affect patients' outcomes independent of their compensation status.

Differences between clinical and work-related outcomes among WC and NWC patients may have multiple explanations and depend on both clinical and nonclinical factors. First, work accidents and/or occupational diseases usually have particularly serious adverse health consequences, and they are associated with high and severe degrees of temporary or permanent disability [57,58]. For example, WC patients are more likely to depend on opioids for pain relief [59] and present with worse symptoms, probably due to the increased injury severity in work environments [60]. The use of narcotics after occupational acute low back injury has been associated with an increased risk of chronic disability [2]. In a retrospective study by Anderson et al. [61], only 11% of WC subjects assuming chronic opioids (>1 year after surgery) sustainedly returned to work compared to individuals using opioids in the short post-operative term. Moreover, these patients showed an increased risk of psychiatric comorbidities, failed-back syndrome, and additional surgery, with substantially higher medical costs. In a recent study conducted by Kukreja and colleagues, 41.3% patients within a WC cohort underwent reoperation after lumbar discectomy and/or laminectomy following an on-the-job injury [62]. Thence, increased reoperation rate may additionally contribute to worsen surgical outcome and satisfaction in this population.

Moreover, the relevance of the psychological status in patients undergoing lumbar spine surgery has been outlined by recent studies and may thus have a significant role in this specific subset of patients [63]. Indeed, WC subjects undergoing lumbar fusion and diagnosed with depression demonstrated higher rates of other psychiatric disorders, narcotic utilization and additional lumbar surgery compared to patients without depression. These individuals required significantly higher medical expenses due to their condition, with a very low RTW rate [64]. However, the aforementioned clinical factors are not sufficient on their own to explain why in WC subjects are observed worse results both in clinical and work-related terms.

Indeed, in this regard, the available literature data call into question also numerous nonclinical factors that mainly include demographic and socioeconomic variables such as male gender [65], lower degree of education [66], higher body mass index [67], smoking history [68], longer working hours [65], higher physical demands [69], civil litigation, legal representation [50,61,64], lower annual income and need for financial assistance [70,71]. Furthermore, longer compensation periods and higher compensation costs in WC patients may also depend on the fact that these subjects are more likely to conduct risky activities with higher chances of injury. A recent study by Khor et al. [72] proposed a prediction model for pain and functional outcomes following lumbar spine fusion surgery. Interestingly, they found that patients with worse improvements in pain and disability were more likely covered by WC and presenting with better preoperative ODI and NRS scores. In this regard, identifying presurgical risk factors and optimizing subject selection criteria for lumbar spine surgery in WC patients may help provide the most appropriate care for these individuals as well as to reduce the economic burden on national institutions providing WC.

At the same time, disputed and complex claims also represent an impeding condition for a prompt RTW. Indeed, they induce a sort of conflict of interest in workers since it is not in the claimant's interest to resume his working activity until the claim is resolved [70]. Several studies showed that a WC claim delays RTW [73,74]. In detail, data provided by our meta-analysis are in good agreement with previous published findings supporting the evidence that NWC returned fully to work at a faster rate than workers with recognized claims, especially after the request is denied [73]. However, studies on this topic commonly refer to NWC patients simply as individuals with no form of compensation, without specifying they did not possess the eligibility criteria or if, despite having made a claim, it was denied by the compensation authority. This is a substantial element to adequately understand the complex interaction between compensation status and health or work-related outcomes. Therefore, rather than comparing workers solely based on their compensation status, it would be useful to consider also claim processing time or any possible appeals made by workers in case of claim rejection. Indeed, some studies suggested that the observed negative association with the recognition of a compensation state could depend on an inefficient, long, and overly bureaucratic claim management [75]. Furthermore, claim processing times (and consequently RTW) might be also influenced by other factors related to the worker, workplace or the nature/severity of the work accident or occupational disease. For example, in the case of cause-based system compensations, it is not always easy or obvious to define a link between adverse effects suffered by workers and their working activities or exposure to certain occupational risk factors, especially when workers are elderly and have often important comorbidities [76,77].

On the other hand, it can be postulated that these patients, thanks to the financial support provided by WC and prolonged abstention from work, may be more likely to experience a full recovery without undertaking harmful activities.

This study has some limitations. Firstly, the overall level of evidence of the studies included is low due to the absence of RCTs comparing WC and NWC populations. Moreover, the NRCTs included were classified as "low quality" according to GRADE and single studies ranged from "low" to "high" risk of bias according to ROBINS-I. The small sample size of some included articles and the high heterogeneity among studies (I^2 = 55%, 62%, 82% and 67% for pain, disability, RTW and satisfaction outcomes, respectively), downgraded the overall quality of our results and may have led to an overestimation of their effects. As observational studies constituted the main source of our analysis, selection bias and confounding due to diverse expectations in WC patients should be taken in consideration. In addition, the different definition of RTW and heterogeneous lengths of follow-up in the examined studies may generate further inconsistencies. Moreover, as regulations of WC in terms of expense coverage, compensation amount, claim duration profoundly differ among countries, it is difficult to generalize our results to all compensation systems [78]. This is particularly true when considering the extreme fragmentation of the American compensation systems, especially in terms of coverage, benefit adequacy, disability determination and complexity of claims [79]. Furthermore, having excluded studies in languages other than English and Italian could have limited our understanding of the relationship between WC and surgical outcomes in different nations.

5. Conclusions

To our knowledge, this is the first systematic review and meta-analysis totally focused on the effect of WC on patients after lumbar spine surgery and the most updated report on the topic. Differently from previous studies, we have stratified post-operative outcomes in clinical (pain, disability, and satisfaction) and work-related (RTW) domains. This reflects the multifactorial nature of the phenomenon and may contribute to clarify which factors (and to what extent) are likely involved in reducing the clinical efficacy of surgery in such a specific population. Indeed, our findings are in good agreement with those already published in the literature, further confirming that the compensation status negatively affects both clinical and work-related outcomes. In this regard, the confounding bias induced by subjects receiving a compensation is a quite common drawback in lumbar spine surgery research investigating the effectiveness and the results of the therapeutic interventions adopted to deal with work-related diseases, conditions, and injuries. However, it is important to underline that it is not yet clear whether the negative effects on the different outcomes are a direct consequence of the compensation status itself or rather are more related to some specific aspects that are necessary to obtain the compensation status (i.e., time, claims, administrative and bureaucratic process). Therefore, it would be necessary to obtain a better understanding of the different aspects and intrinsic characteristics that govern the compensation recognition. In this regard, future studies on this topic should in our opinion focus not so much on the comparison between WC and NWC but rather on the analysis (within the WC group) of the different variables that can influence the timing and modalities with which the compensation status is recognized or not.

Supplementary Materials: The following are available online at https://www.mdpi.com/article/10.3390/ijerph18116165/s1, Figure S1: Methodological quality assessments of each study assessed using ROBINS, Table S1: GRADE profile of evidence, Table S2: GRADE Summary of findings.

Author Contributions: Conceptualization, F.R., G.V. and S.I.; methodology, F.R., G.V., S.D.S., L.A.; formal analysis, F.R., S.D.S., L.A.; data curation, F.R., S.D.S., L.A.; writing—original draft preparation, F.R., S.D.S., L.A., G.V., L.F.; writing—review and editing, F.R., S.D.S., L.A., G.V., L.F., R.P., J.R., S.I.,

V.D.; supervision, F.R., G.V., R.P., J.R., S.I., V.D. All authors have read and agreed to the published version of the manuscript.

Funding: This research was funded by the Research Grant (BRiC-2018 ID3) of the Italian Workers' Compensation Authority (INAIL).

Institutional Review Board Statement: Not applicable.

Informed Consent Statement: Not applicable.

Data Availability Statement: The datasets used and/or analyzed during the current study are available from the corresponding author on reasonable request.

Conflicts of Interest: The authors declare no conflict of interest.

References

1. Disease, G.B.D.; Injury, I.; Prevalence, C. Global, regional, and national incidence, prevalence, and years lived with disability for 354 diseases and injuries for 195 countries and territories, 1990–2017: A systematic analysis for the Global Burden of Disease Study 2017. *Lancet* **2018**, *392*, 1789–1858. [CrossRef]
2. Franklin, G.M.; Wickizer, T.M.; Coe, N.B.; Fulton-Kehoe, D. Workers' compensation: Poor quality health care and the growing disability problem in the United States. *Am. J. Ind. Med.* **2015**, *58*, 245–251. [CrossRef] [PubMed]
3. Hartvigsen, J.; Hancock, M.J.; Kongsted, A.; Louw, Q.; Ferreira, M.L.; Genevay, S.; Hoy, D.; Karppinen, J.; Pransky, G.; Sieper, J.; et al. What low back pain is and why we need to pay attention. *Lancet* **2018**, *391*, 2356–2367. [CrossRef]
4. Kwon, B.K.; Vaccaro, A.R.; Grauer, J.N.; Beiner, J. Indications, techniques, and outcomes of posterior surgery for chronic low back pain. *Orthop. Clin. N. Am.* **2003**, *34*, 297–308. [CrossRef]
5. Harris, I.; Mulford, J.; Solomon, M.; van Gelder, J.M.; Young, J. Association between compensation status and outcome after surgery: A meta-analysis. *JAMA* **2005**, *293*, 1644–1652. [CrossRef]
6. Lippel, K.; Lotters, F. Public insurance systems: A comparison of cause-based and disability-based income support systems. In *Handbook of Work Disability*; Loisel, P., Anema, J., Eds.; Springer: New York, NY, USA, 2013; pp. 183–203.
7. Gray, S.; Lane, T.; Sheehan, L.; Collie, A. Association between workers' compensation claim processing times and work disability duration: Analysis of population level claims data. *Health Policy* **2019**, *123*, 982–991. [CrossRef] [PubMed]
8. Kilgour, E.; Kosny, A.; McKenzie, D.; Collie, A. Interactions between injured workers and insurers in workers' compensation systems: A systematic review of qualitative research literature. *J. Occup. Rehabil.* **2015**, *25*, 160–181. [CrossRef]
9. Anderson, P.A.; Subach, B.R.; Riew, K.D. Predictors of outcome after anterior cervical discectomy and fusion: A multivariate analysis. *Spine* **2009**, *34*, 161–166. [CrossRef]
10. Hou, W.H.; Tsauo, J.Y.; Lin, C.H.; Liang, H.W.; Du, C.L. Worker's compensation and return-to-work following orthopaedic injury to extremities. *J. Rehabil. Med.* **2008**, *40*, 440–445. [CrossRef]
11. Bhatia, S.; Piasecki, D.P.; Nho, S.J.; Romeo, A.A.; Cole, B.J.; Nicholson, G.P.; Boniquit, N.; Verma, N.N. Early return to work in workers' compensation patients after arthroscopic full-thickness rotator cuff repair. *Arthroscopy* **2010**, *26*, 1027–1034. [CrossRef]
12. Brinker, M.R.; Savory, C.G.; Weeden, S.H.; Aucoin, H.C.; Curd, D.T. The results of total knee arthroplasty in workers' compensation patients. *Bull. Hosp. Jt. Dis.* **1998**, *57*, 80–83. [PubMed]
13. Denard, P.J.; Ladermann, A.; Burkhart, S.S. Long-term outcome after arthroscopic repair of type II SLAP lesions: Results according to age and workers' compensation status. *Arthroscopy* **2012**, *28*, 451–457. [CrossRef] [PubMed]
14. Holtby, R.; Razmjou, H. Impact of work-related compensation claims on surgical outcome of patients with rotator cuff related pathologies: A matched case-control study. *J. Shoulder Elbow Surg.* **2010**, *19*, 452–460. [CrossRef]
15. Bartys, S.; Frederiksen, P.; Bendix, T.; Burton, K. System influences on work disability due to low back pain: An international evidence synthesis. *Health Policy* **2017**, *121*, 903–912. [CrossRef]
16. Gum, J.L.; Glassman, S.D.; Carreon, L.Y. Is type of compensation a predictor of outcome after lumbar fusion? *Spine* **2013**, *38*, 443–448. [CrossRef]
17. Siambanes, D.; Miz, G.S. Treatment of symptomatic anterior cervical nonunion using the Rogers interspinous wiring technique. *Am. J. Orthop.* **1998**, *27*, 792–796.
18. Madan, S.S.; Boeree, N.R. Comparison of instrumented anterior interbody fusion with instrumented circumferential lumbar fusion. *Eur. Spine J.* **2003**, *12*, 567–575. [CrossRef]
19. Kohl, C.; McIntosh, E.J.; Unger, S.; Haddaway, N.R.; Kecke, S.; Schiemann, J.; Wilhelm, R. Online tools supporting the conduct and reporting of systematic reviews and systematic maps: A case study on CADIMA and review of existing tools. *Environ. Evid.* **2018**, *7*, 8. [CrossRef]
20. Moher, D.; Liberati, A.; Tetzlaff, J.; Altman, D.G.; Group, P. Preferred reporting items for systematic reviews and meta-analyses: The PRISMA statement. *PLoS Med.* **2009**, *6*, e1000097. [CrossRef]
21. NASS. Levels of Evidence for Primary Research Question. Available online: https://www.spine.org/Portals/0/Assets/Downloads/ResearchClinicalCare/LevelsofEvidence.pdf (accessed on 26 April 2021).

22. Sterne, J.A.; Hernan, M.A.; Reeves, B.C.; Savovic, J.; Berkman, N.D.; Viswanathan, M.; Henry, D.; Altman, D.G.; Ansari, M.T.; Boutron, I.; et al. ROBINS-I: A tool for assessing risk of bias in non-randomised studies of interventions. *BMJ* **2016**, *355*, i4919. [CrossRef] [PubMed]
23. Agazzi, S.; Reverdin, A.; May, D. Posterior lumbar interbody fusion with cages: An independent review of 71 cases. *J. Neurosurg.* **1999**, *91*, 186–192. [CrossRef]
24. Albert, T.J.; Pinto, M.; Denis, F. Management of symptomatic lumbar pseudarthrosis with anteroposterior fusion. A functional and radiographic outcome study. *Spine* **2000**, *25*, 123–129. [CrossRef]
25. Asch, H.L.; Lewis, P.J.; Moreland, D.B.; Egnatchik, J.G.; Yu, Y.J.; Clabeaux, D.E.; Hyland, A.H. Prospective multiple outcomes study of outpatient lumbar microdiscectomy: Should 75 to 80% success rates be the norm? *J. Neurosurg.* **2002**, *96*, 34–44. [CrossRef] [PubMed]
26. Atlas, S.J.; Chang, Y.; Kammann, E.; Keller, R.B.; Deyo, R.A.; Singer, D.E. Long-term disability and return to work among patients who have a herniated lumbar disc: The effect of disability compensation. *J. Bone Jt. Surg. Am.* **2000**, *82*, 4–15. [CrossRef] [PubMed]
27. Carreon, L.Y.; Glassman, S.D.; Kantamneni, N.R.; Mugavin, M.O.; Djurasovic, M. Clinical outcomes after posterolateral lumbar fusion in workers' compensation patients: A case-control study. *Spine* **2010**, *35*, 1812–1817. [CrossRef] [PubMed]
28. Deutsch, H.; Musacchio, M.J., Jr. Minimally invasive transforaminal lumbar interbody fusion with unilateral pedicle screw fixation. *Neurosurg. Focus* **2006**, *20*, E10. [CrossRef]
29. Greenough, C.G.; Taylor, L.J.; Fraser, R.D. Anterior lumbar fusion: Results, assessment techniques and prognostic factors. *Eur. Spine J.* **1994**, *3*, 225–230. [CrossRef]
30. Greenough, C.G.; Peterson, M.D.; Hadlow, S.; Fraser, R.D. Instrumented posterolateral lumbar fusion. Results and comparison with anterior interbody fusion. *Spine* **1998**, *23*, 479–486. [CrossRef]
31. Herron, L.D.; Turner, J.A.; Novell, L.A.; Kreif, S.L. Patient selection for lumbar discectomy with a revised objective rating system. *Clin. Orthop. Relat. Res.* **1996**, 148–155. [CrossRef]
32. Klekamp, J.; McCarty, E.; Spengler, D.M. Results of elective lumbar discectomy for patients involved in the workers' compensation system. *J. Spinal Disord.* **1998**, *11*, 277–282. [CrossRef]
33. Lew, S.M.; Mehalic, T.F.; Fagone, K.L. Transforaminal percutaneous endoscopic discectomy in the treatment of far-lateral and foraminal lumbar disc herniations. *J. Neurosurg.* **2001**, *94*, 216–220. [CrossRef]
34. MacKay, M.A.; Fischgrund, J.S.; Herkowitz, H.N.; Kurz, L.T.; Hecht, B.; Schwartz, M. The effect of interposition membrane on the outcome of lumbar laminectomy and discectomy. *Spine* **1995**, *20*, 1793–1796. [CrossRef]
35. Marks, R.A. Transcutaneous lumbar diskectomy for internal disk derangement: A new indication. *South. Med. J.* **2000**, *93*, 885–890. [PubMed]
36. Montgomery, A.S.; Cunningham, J.E.; Robertson, P.A. The Influence of No Fault Compensation on Functional Outcomes After Lumbar Spine Fusion. *Spine* **2015**, *40*, 1140–1147. [CrossRef]
37. Parker, L.M.; Murrell, S.E.; Boden, S.D.; Horton, W.C. The outcome of posterolateral fusion in highly selected patients with discogenic low back pain. *Spine* **1996**, *21*, 1909–1916. [CrossRef]
38. Pelton, M.A.; Phillips, F.M.; Singh, K. A comparison of perioperative costs and outcomes in patients with and without workers' compensation claims treated with minimally invasive or open transforaminal lumbar interbody fusion. *Spine* **2012**, *37*, 1914–1919. [CrossRef] [PubMed]
39. Penta, M.; Fraser, R.D. Anterior lumbar interbody fusion. A minimum 10-year follow-up. *Spine* **1997**, *22*, 2429–2434. [CrossRef]
40. Phan, K.; Davies, S.; Rao, P.J.; Mobbs, R.J. Worker's Compensation Status and Outcomes Following Anterior Lumbar Interbody Fusion: Prospective Observational Study. *World Neurosurg.* **2017**, *103*, 680–685. [CrossRef] [PubMed]
41. Rouben, D.; Casnellie, M.; Ferguson, M. Long-term durability of minimal invasive posterior transforaminal lumbar interbody fusion: A clinical and radiographic follow-up. *J. Spinal Disord. Tech.* **2011**, *24*, 288–296. [CrossRef] [PubMed]
42. Sanderson, P.L.; Fraser, R.D.; Hall, D.J.; Cain, C.M.; Osti, O.L.; Potter, G.R. Short segment fixation of thoracolumbar burst fractures without fusion. *Eur. Spine J.* **1999**, *8*, 495–500. [CrossRef]
43. Schnee, C.L.; Freese, A.; Ansell, L.V. Outcome analysis for adults with spondylolisthesis treated with posterolateral fusion and transpedicular screw fixation. *J. Neurosurg.* **1997**, *86*, 56–63. [CrossRef] [PubMed]
44. Slosar, P.J.; Reynolds, J.B.; Schofferman, J.; Goldthwaite, N.; White, A.H.; Keaney, D. Patient satisfaction after circumferential lumbar fusion. *Spine* **2000**, *25*, 722–726. [CrossRef]
45. Taylor, V.M.; Deyo, R.A.; Ciol, M.; Farrar, E.L.; Lawrence, M.S.; Shonnard, N.H.; Leek, K.M.; McNeney, B.; Goldberg, H.I. Patient-oriented outcomes from low back surgery: A community-based study. *Spine* **2000**, *25*, 2445–2452. [CrossRef] [PubMed]
46. Vaccaro, A.R.; Ring, D.; Scuderi, G.; Cohen, D.S.; Garfin, S.R. Predictors of outcome in patients with chronic back pain and low-grade spondylolisthesis. *Spine* **1997**, *22*, 2030–2034, discussion 2035. [CrossRef]
47. Cheriyan, T.; Harris, B.; Cheriyan, J.; Lafage, V.; Spivak, J.M.; Bendo, J.A.; Errico, T.J.; Goldstein, J.A. Association between compensation status and outcomes in spine surgery: A meta-analysis of 31 studies. *Spine J.* **2015**, *15*, 2564–2573. [CrossRef]
48. de Moraes, V.Y.; Godin, K.; Tamaoki, M.J.; Faloppa, F.; Bhandari, M.; Belloti, J.C. Workers' compensation status: Does it affect orthopaedic surgery outcomes? A meta-analysis. *PLoS ONE* **2012**, *7*, e50251. [CrossRef]
49. Anderson, J.T.; Haas, A.R.; Percy, R.; Woods, S.T.; Ahn, U.M.; Ahn, N.U. Return to Work After Diskogenic Fusion in Workers' Compensation Subjects. *Orthopedics* **2015**, *38*, e1065–e1072. [CrossRef]

50. Anderson, J.T.; Haas, A.R.; Percy, R.; Woods, S.T.; Ahn, U.M.; Ahn, N.U. Workers' Compensation, Return to Work, and Lumbar Fusion for Spondylolisthesis. *Orthopedics* **2016**, *39*, e1–e8. [CrossRef] [PubMed]
51. Deyo, R.A.; Mirza, S.K.; Martin, B.I.; Kreuter, W.; Goodman, D.C.; Jarvik, J.G. Trends, major medical complications, and charges associated with surgery for lumbar spinal stenosis in older adults. *JAMA* **2010**, *303*, 1259–1265. [CrossRef] [PubMed]
52. Lavin, R.A.; Tao, X.; Yuspeh, L.; Bernacki, E.J. Temporal relationship between lumbar spine surgeries, return to work, and workers' compensation costs in a cohort of injured workers. *J. Occup. Environ. Med.* **2013**, *55*, 539–543. [CrossRef]
53. Hedlund, R.; Johansson, C.; Hägg, O.; Fritzell, P.; Tullberg, T. Swedish Lumbar Spine Study Group. The long-term outcome of lumbar fusion in the Swedish lumbar spine study. *Spine J.* **2016**, *16*, 579–587. [CrossRef] [PubMed]
54. Rubinstein, S.M.; van Middelkoop, M.; Ostelo, R.; Verhagen, A.; Koes, B.; Peul, W.C. Surgery versus conservative management of sciatica due to a lumbar herniated disc: A systematic review. *Eur. Spine J.* **2011**, *20*, 513–522. [CrossRef]
55. Lequin, M.B.; Verbaan, D.; Jacobs, W.C.; Brand, R.; Bouma, G.J.; Vandertop, W.P.; Peul, W.C.; Leiden-The Hague Spine Intervention Prognostic Study Group; Wilco, C.P.; Bart, W.K.; et al. Surgery versus prolonged conservative treatment for sciatica: 5-year results of a randomised controlled trial. *BMJ Open* **2013**, *3*, e002534. [CrossRef] [PubMed]
56. Todd, N.V. The surgical treatment of non-specific low back pain. *Bone Jt. J.* **2017**, *99*, 1003–1005. [CrossRef]
57. Sears, J.M.; Blanar, L.; Bowman, S.M.; Adams, D.; Silverstein, B.A. Predicting work-related disability and medical cost outcomes: Estimating injury severity scores from workers' compensation data. *J. Occup. Rehabil.* **2013**, *23*, 19–31. [CrossRef]
58. Sears, J.M.; Bowman, S.M.; Rotert, M.; Hogg-Johnson, S. A New Method to Classify Injury Severity by Diagnosis: Validation Using Workers' Compensation and Trauma Registry Data. *J. Occup. Rehabil.* **2015**, *25*, 742–751. [CrossRef] [PubMed]
59. Daniels, A.H.; Kuris, E.O.; Kleinhenz, D.T.; Palumbo, M.A. Spine Surgery Outcomes in Workers' Compensation Patients. *J. Am. Acad. Orthop. Surg.* **2017**, *25*, e225–e234. [CrossRef] [PubMed]
60. Tabaraee, E.; Ahn, J.; Bohl, D.D.; Elboghdady, I.M.; Aboushaala, K.; Singh, K. The Impact of Worker's Compensation Claims on Outcomes and Costs Following an Anterior Cervical Discectomy and Fusion. *Spine* **2015**, *40*, 948–953. [CrossRef] [PubMed]
61. Anderson, J.T.; Haas, A.R.; Percy, R.; Woods, S.T.; Ahn, U.M.; Ahn, N.U. Chronic Opioid Therapy After Lumbar Fusion Surgery for Degenerative Disc Disease in a Workers' Compensation Setting. *Spine* **2015**, *40*, 1775–1784. [CrossRef] [PubMed]
62. Kukreja, S.; Kalakoti, P.; Ahmed, O.; Nanda, A. Predictors of reoperation-free survival following decompression-alone lumbar spine surgery for on-the-job injuries. *Clin. Neurol. Neurosurg.* **2015**, *135*, 41–45. [CrossRef]
63. DeBerard, M.S.; Wheeler, A.J.; Gundy, J.M.; Stein, D.M.; Colledge, A.L. Presurgical biopsychosocial variables predict medical, compensation, and aggregate costs of lumbar discectomy in Utah workers' compensation patients. *Spine J.* **2011**, *11*, 395–401. [CrossRef] [PubMed]
64. Anderson, J.T.; Haas, A.R.; Percy, R.; Woods, S.T.; Ahn, U.M.; Ahn, N.U. Clinical depression is a strong predictor of poor lumbar fusion outcomes among workers' compensation subjects. *Spine* **2015**, *40*, 748–756. [CrossRef]
65. Taylor, V.M.; Deyo, R.A.; Ciol, M.; Kreuter, W. Surgical treatment of patients with back problems covered by workers compensation versus those with other sources of payment. *Spine* **1996**, *21*, 2255–2259. [CrossRef] [PubMed]
66. Rainville, J.; Sobel, J.B.; Hartigan, C.; Wright, A. The effect of compensation involvement on the reporting of pain and disability by patients referred for rehabilitation of chronic low back pain. *Spine* **1997**, *22*, 2016–2024. [CrossRef] [PubMed]
67. Andreshak, T.G.; An, H.S.; Hall, J.; Stein, B. Lumbar spine surgery in the obese patient. *J. Spinal Disord.* **1997**, *10*, 376–379. [CrossRef] [PubMed]
68. Atlas, S.J.; Tosteson, T.D.; Hanscom, B.; Blood, E.A.; Pransky, G.S.; Abdu, W.A.; Andersson, G.B.; Weinstein, J.N. What is different about workers' compensation patients? Socioeconomic predictors of baseline disability status among patients with lumbar radiculopathy. *Spine* **2007**, *32*, 2019–2026. [CrossRef]
69. Hadler, N.M.; Carey, T.S.; Garrett, J. The influence of indemnification by workers' compensation insurance on recovery from acute backache. North Carolina Back Pain Project. *Spine* **1995**, *20*, 2710–2715. [CrossRef]
70. Volinn, E.; Van Koevering, D.; Loeser, J.D. Back sprain in industry. The role of socioeconomic factors in chronicity. *Spine* **1991**, *16*, 542–548. [CrossRef] [PubMed]
71. Schatman, M. Workers' Compensation and Its Potential for Perpetuation of Disability. In *Handbook of Occupational Health and Wellness*; Gatchel, R.J., Schultz, I.Z., Eds.; Springer: Boston, MA, USA, 2012; pp. 341–361. [CrossRef]
72. Khor, S.; Lavallee, D.; Cizik, A.M.; Bellabarba, C.; Chapman, J.R.; Howe, C.R.; Lu, D.; Mohit, A.A.; Oskouian, R.J.; Roh, J.R.; et al. Development and Validation of a Prediction Model for Pain and Functional Outcomes After Lumbar Spine Surgery. *JAMA Surg.* **2018**, *153*, 634–642. [CrossRef]
73. Rudbeck, M.; Johansen, J.P.; Omland, Ø. A Follow-Up Study on Return to Work in the Year After Reporting an Occupational Injury Stratified by Outcome of the Workers' Compensation System. *J. Occup. Environ. Med.* **2018**, *60*, 542–547. [CrossRef]
74. Gallagher, R.M.; Williams, R.A.; Skelly, J.; Haugh, L.D.; Rauh, V.; Milhous, R.; Frymoyer, J. Workers' Compensation and return-to-work in low back pain. *Pain* **1995**, *61*, 299–307. [CrossRef]
75. Kenny, D.T. Returning to work after workplace injury: Impact of worker and workplace factors. *J. Appl. Rehabil. Couns.* **1998**, *29*, 13–19. [CrossRef]
76. Glass, D. *Investigation into the Management of Complex Workers Compensation Claims and WorkSafe Oversight*; Victorian Ombudsman: Melbourne, Australia, 2016.
77. Berecki-Gisolf, J.; Clay, F.J.; Collie, A.; McClure, R.J. The impact of aging on work disability and return to work: Insights from workers' compensation claim records. *J. Occup. Environ. Med.* **2012**, *54*, 318–327. [CrossRef]

78. ILO. GEIP: Contributing to Decent Work and the Social Protection Floor Guarantee in the Workplace. Available online: https://www.ilo.org/wcmsp5/groups/public/---ed_emp/---emp_ent/documents/publication/wcms_573083.pdf (accessed on 1 April 2021).
79. Dworsky, M.; Broten, N. *How Can Workers' Compensation Systems Promote Occupational Safety and Health? Stakeholder Views on Policy and Research Priorities*; RAND Corporation: Santa Monica, CA, USA, 2018. [CrossRef]

International Journal of
Environmental Research and Public Health

MDPI

Systematic Review

The Effects of Workplace Interventions on Low Back Pain in Workers: A Systematic Review and Meta-Analysis

Fabrizio Russo [1,*], Giuseppe Francesco Papalia [1,*], Gianluca Vadalà [1], Luca Fontana [2], Sergio Iavicoli [3], Rocco Papalia [1] and Vincenzo Denaro [1]

1 Department of Orthopaedic and Trauma Surgery, Campus Bio-Medico University of Rome, 00128 Rome, Italy; g.vadala@unicampus.it (G.V.); r.papalia@unicampus.it (R.P.); denaro@unicampus.it (V.D.)
2 Department of Public Health, Section of Occupational Medicine, University of Naples Federico II, 80121 Naples, Italy; Luca.fontana@unina.it
3 Department of Occupational and Environmental Medicine, Epidemiology and Hygiene, Italian Workers' Compensation Authority (INAIL), 00078 Rome, Italy; s.iavicoli@inail.it
* Correspondence: fabrizio.russo@unicampus.it (F.R.); g.papalia@unicampus.it (G.F.P.)

Abstract: This systematic review and meta-analysis aimed to analyze the effects of workplace interventions (WI) on clinical outcomes related to low back pain (LBP) in a worker population, and to assess socio-economic parameters as participants on sick leave, days of sick leave, and return to work following WI. A systematic literature search was performed to select randomized clinical trials that investigated the effectiveness of WI on return to work, sick leave, and working capacity of workers affected by nonspecific LBP. Fourteen articles were included in the review and meta-analysis. The meta-analysis showed improvements in pain (p = 0.004), disability (p = 0.0008), fear-avoidance for psychical activity (p = 0.004), and quality of life (p = 0.001 for physical scale and p = 0.03 for mental scale) for patients who underwent WI compared to controls. Moreover, the pain reduction following WI was statistically significant in the healthcare workers' group (p = 0.005), but not in the other workers' group. The participants on sick leave and the number of days of sick leave decreased in the WI group without statistical significance (p = 0.85 and p = 0.10, respectively). Finally, LBP recurrence was significantly reduced in the WI group (p = 0.006). WI led to a significant improvement of clinical outcomes in a workers' population affected by LBP.

Keywords: workplace interventions; low back pain; workers; work ability; systematic review; meta-analysis

Citation: Russo, F.; Papalia, G.F.; Vadalà, G.; Fontana, L.; Iavicoli, S.; Papalia, R.; Denaro, V. The Effects of Workplace Interventions on Low Back Pain in Workers: A Systematic Review and Meta-Analysis. *IJERPH* **2021**, *18*, 12614. https://doi.org/10.3390/ijerph182312614

Academic Editor: Paul B. Tchounwou

Received: 15 October 2021
Accepted: 23 November 2021
Published: 30 November 2021

1. Introduction

Non-specific low back pain (LBP) is a common worldwide disorder with a significant impact on productivity, work ability, and quality of life [1]. Indeed, LBP is characterized by persisting pain, muscle weakness, reduction of physical activity [2], and sleep disorders, which have serious consequences on a person's quality of life by limiting daily life and work activities [3]. The etiopathogenesis of nonspecific LBP is multifactorial including lifestyle risk factors (i.e., excess weight) [4] but also, according to the type of job, several occupational risk factors such as manual handling of heavy loads, awkward and prolonged postures (i.e., sustained sedentary work), whole-body mechanical vibrations, and work-related stress (i.e., psychosocial factors) [5]. The lifetime prevalence of LBP in the general population is high and was estimated at about 60–70% in industrialized countries [6]. Notably, in the literature, there is evidence that the prevalence of this disorder in particular working populations and/or industrial/productive fields such as construction, forestry or fishing, agriculture, and healthcare sectors is significantly higher than in the general population [7]. In particular, healthcare workers represent a job category easily susceptible to LBP biomechanical risk factors [8] with an annual prevalence of 40–50% [9], while the prevalence of LBP is estimated at about 34% in office workers [10]. LBP frequently causes sick leave and persistent or recurrent disability, representing an important socioeconomic burden [11].

Therefore, the prevention of work absenteeism due to LBP recurrence has become a public and occupational health priority worldwide [12]. The treatments for non-specific LBP usually consisted of non-surgical procedures, such as physical exercise, cognitive behavioral therapy, and pharmacological treatment [13–15]. Physical exercise at the workplace is considered an activity able to prevent occupational musculoskeletal disorders being able to enhance the physical capacity of workers. However, previous studies regarding occupational interventions showed contrasting results about the reduction of LBP symptoms following only physical exercise at the workplace [16–18]. This should not be surprising since, considering the numerous and different variables in workplaces that can play an important role in the onset of this disorder, it is likely that its prevention needs a multidisciplinary approach that exploits the simultaneous adoption of technical, organizational, procedural, and training measures. In this regard, several studies developed, applied, and evaluated this type of preventive strategies in different work environments consisting of workplace interventions (WI) that include workplace assessment, educational programs with ergonomic posture training sessions, physical activity at the workplace, and cognitive-behavioral therapy for the treatment of physical, psychological, occupational, and ergonomic risk factors. WI aim to prevent and/or manage LBP, reduce disabilities and fears for work and psychical activity, promote personalized action plans, and improve outcomes regarding work ability and quality of life.

Nevertheless, our current knowledge of the real effectiveness of WI remains rather fragmented and, at the same time, the understanding of the key factors or best combination of WI for achieving significant prevention or reduction of work-related LBP is equally limited. For this reason, in this context, we performed this systematic review and meta-analysis in order to critically evaluate the effects of WI on LBP in workers. In detail, the primary aim was to analyze the effects on clinical and occupational outcomes related to LBP in workers after the implementation of specific WI programs. The secondary endpoint was to assess the impact of WI on socio-economic parameters as participants on sick leave, days of sick leave, and return to work.

2. Materials and Methods

This study was carried out according to the Preferred Reporting Items for Systematic Reviews and Meta-Analysis (PRISMA) (Table S1) guidelines [19]. In this systematic review, we included randomized clinical trials (RCTs) that evaluated the effects of workplace interventions for workers suffering from LBP.

2.1. Inclusion Criteria

The inclusion criteria were RCTs in the English language published in the last twenty years, which investigated the effectiveness of workplace interventions on return to work, sick leave, and the working capacity of workers affected by nonspecific LBP. The workplace interventions include technical interventions, physical exercise programs, behavioral training, educational programs, and participatory ergonomics. Exclusion criteria were represented by studies that enrolled patients with neurodegenerative diseases, previous spinal or brain surgery, or following spinal cord infections or injuries. We excluded studies that analyzed only physical or psychosocial activities as an intervention, and that only evaluated reducing sitting time as an outcome or the impact of a sit–stand workstation.

2.2. Search Methods

A systematic literature search was executed using PubMed–Medline, Cochrane Central, Scopus, and Google Scholar. We used the following search strings: ("workplace" [MeSH Terms] or "workplace" [All Fields] or "workplaces" [All Fields] or "workplaces" [All Fields]) and ("interventions" [All Fields] or "interventions" [All Fields] or "interventive" [All Fields] or "methods" [MeSH Terms] or "methods" [All Fields] or "intervention"

[All Fields] or "interventional" [All Fields]) and ("low back pain" [MeSH Terms] or ("low" [All Fields] and "back" [All Fields] and "pain" [All Fields]) or "low back pain" [All Fields]). The reference lists of the included RCTs were detected to obtain further eligible studies. After removing duplicates, two independent investigators reviewers (G.P. and F.R.) checked the abstracts of potentially included studies. Any divergence was discussed with the third review author (G.V.). Finally, two review authors (G.P. and F.R.) read the full articles in order to select the included studies for this review and meta-analysis.

2.3. Data Collection, Analysis, and Outcomes

Two independent reviewers (G.P. and F.R.) conducted data extraction. The following data were extracted from the included studies: Authors, year of publication, type of study, level of evidence, numbers of participants in study and control groups, age and sex of participants, types of workers, intervention in the experimental and in the control group, follow-up, and results. LBP, disability, fear-avoidance beliefs, quality of life, and work ability were assessed as outcomes in the included studies. Finally, participants on sick leave, days of sick leave, LBP recurrence, and return to work were compared between workplace intervention and control groups.

2.4. Risk of Bias Assessment

The risk of bias of the included RCTs was evaluated by two authors (G.P. and F.R.) by the guideline for systematic reviews in the Cochrane Back and Neck Group [20]. This tool assesses the following types of biases: selection bias, performance bias, attrition bias, detection bias, and reporting bias. The trials were judged at low, unclear, or high risk of bias in relation to the risk of bias of the various domains.

2.5. Statistical Analysis

A meta-analysis was performed using the Review Manager (RevMan) software Version 5.4.1 (The Nordic Cochrane Center, The Cochrane Collaboration, 2014, Copenhagen, Denmark). Low back pain, disability, fear-avoidance beliefs, quality of life, and work ability between the experimental and the control groups were calculated as continuous outcomes. Instead, participants on sick leave, days of sick leave, LBP recurrence, and return to work were evaluated as dichotomous outcomes. The continuous data are presented as the standard mean difference (SMD) with 95% confidence intervals due to the adoption of diverse scores in the included studies. The outcomes expressed with negative mean values of SMD present a higher improvement with lower values. Dichotomous data are shown as odds ratio (OR) with 95% confidence intervals. For the calculation of the weight of the samples of the trials, for the days of sick leave, we used mean days of sick leave per participant as events and the total number of days of follow-up per participant as the total. A subgroup analysis was performed to evaluate LBP in healthcare workers and other kinds of workers. The heterogeneity was calculated using the I^2 test. A fixed-effect model was adopted for low heterogeneity ($I^2 < 55\%$); otherwise, a random-effect model was involved. The statistical significance of the results was set at $p < 0.05$.

2.6. Quality Assessment

The quality of the evidence and strength of recommendation of the selected outcomes were analyzed by the GRADE (Grading of Recommendations Assessment, Development, and Evaluation) assessment. Five elements (risk of bias, inconsistency, indirectness, imprecision, and publication bias) were assessed for each result and were categorized as not serious, serious, or very serious. The outcomes for RCTs received an initial ranking of high quality of evidence, which could be downgraded to moderate, low, or very low concerning the valuation of the five items.

3. Results

3.1. Results of the Search

The literature search generated 691 articles. After removing duplicates, the reviewers screened titles and abstracts of 673 papers, and chose 41 eligible articles that were read in full. Afterwards, 27 studies were excluded for the following reasons: Not reporting selected outcomes (n = 9), not evaluating workplace interventions (n = 6), not specific for LBP (n = 4), patients with mental disorders (n = 3), validation of work rehabilitation program (n = 2), subgroup analysis of previous study (n = 2), and protocols of RCT (n = 2). Finally, 14 articles were included in the review and meta-analysis (Figure 1).

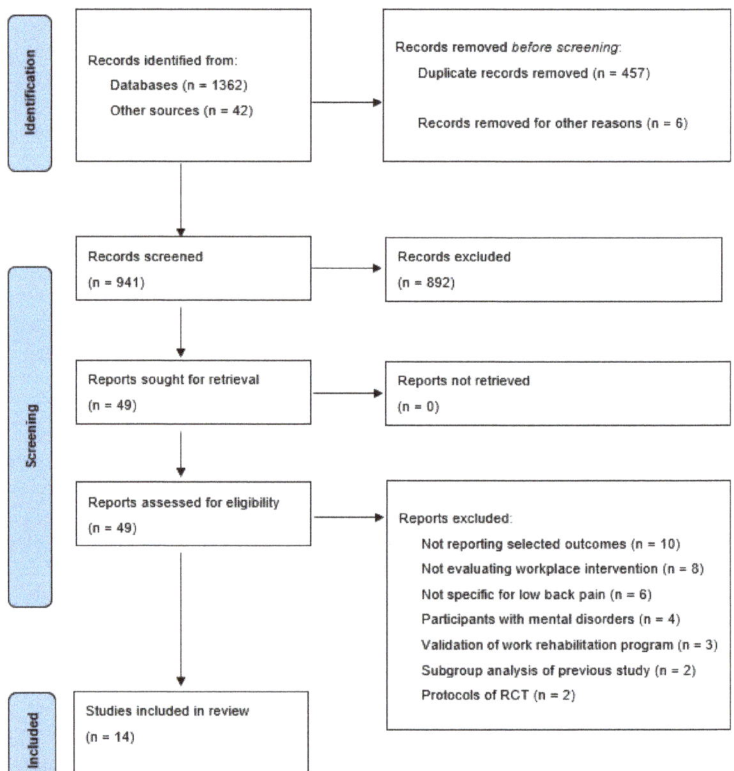

Figure 1. Preferred Reporting Items for Systematic Review and Meta-Analysis (PRISMA) flow diagram.

3.2. Demographic Data

The total number of participants in all the studies was 3197, divided into 1837 in the study group and 1360 in the control group (Table 1). Patients' ages ranged from 29.6 to 52 years in the study groups, and from 26.6 to 51 in the control groups. The percentages of men in the studies ranged from 99% to 0% in the intervention groups and from 98.4% to 0% in the control groups. Therefore, important heterogeneity in the gender of the participants of the included studies was reported. The workers analyzed were distributed as follows: Nursing assistants or healthcare workers in six studies (43%), office workers in two studies (14%), employees in the automotive industry in one study (7%), workers at a manufacturing company in one study (7%), physically demanding workers in one study (7%), and workers (without specification of the type of job) in three studies (22%).

3.3. Workplace Intervention Program

Workplace intervention protocols including multidisciplinary interventions consisted of a combination of the following programs: Work-related evaluation and a workplace assessment with work modifications (four studies); an educational program and ergonomic posture training sessions (six studies); a supervised intervention of exercise sessions of muscle strengthening, flexibility, segmental stabilization, and endurance training on the workplace (six studies); and behavioral counseling and cognitive-behavioral therapy for LBP or stress self-management (two studies). The mean follow-up was 11.3 months and ranged from 3 to 24 months.

3.4. Clinical Outcome Data

Clinical outcomes were diversified in the included studies. LBP was assessed in 10 studies, using the Visual Analogue scale (VAS) [21–25], Numeric Rating Scale (NRS) [26,27], Multidimensional Pain Inventory (MPI-D) [28], and Cornell musculoskeletal discomfort questionnaire [29]. Disability was evaluated in five trials, by the Quebec Disability Scale [23,24,27] and Oswestry Disability Index (ODI) [25,30]. The work subscale of Fear-Avoidance Beliefs Questionnaire (FABQ-W) was assessed in four studies [22,24,26,28], while the physical activity subscale (FABQ-P) was assessed in three studies [24,26,28]. Work ability was reported in two studies, using the Work Ability Index (WAI) [22] and subjective working capacity [30]. Quality of life was assessed in three studies [24,26,28] using physical and mental scales of the Short-Form-36 (SF-36) or Short-Form-12 (SF-12) Health Survey.

3.5. Methodological Evaluation

Through the guideline for systematic reviews in the Cochrane Back and Neck Group, four studies (28.6%) were at low risk of bias (A), eight studies (57.1%) were judged at unclear risk of bias (B), and only two studies (14.3%) were at a high risk of bias (C) (Table 2). Precisely, random sequence generation was adequate in all the studies (100%). Allocation concealment was graded as adequate in all the studies except one (93%). Blinding for patients and care providers was not adequate in all studies, due to the modality of the interventions. Blinding for outcome assessment resulted in being adequate in all the studies (100%). Selective reporting was evaluated as adequate in six studies (43%). Other sources of bias were adequate in four trials (28.6%).

3.6. Effect of Intervention

The meta-analysis demonstrated the effectiveness of workplace interventions on LBP, disability, the Fear-Avoidance Beliefs Questionnaire for work and psychical activity, work ability, and quality of life compared to controls (Figure 2). Pain decreased significantly in the intervention group in comparison with the control group (SMD −0.16, 95% CI −0.26 to −0.05, $p = 0.004$). Disability scores showed significant improvements for workplace interventions compared to controls (SMD −0.28, 95% CI −0.45 to −0.12, $p = 0.0008$). FABQ-W demonstrated lower fear-avoidance beliefs about work in workers who underwent WI compared to controls (SMD −0.07, 95% CI −0.21 to 0.07), but no significant differences ($p = 0.32$). FABQ-P showed a significant reduction of fear-avoidance beliefs about physical activity in the experimental group (SMD −0.21, 95% CI −0.35 to −0.07, $p = 0.004$). Work ability presented improvements in favor of the intervention group (SMD −0.17, 95% CI −0.52 to 0.17), without significant differences ($p = 0.31$). Short-Form Health Survey results showed statistically significant improvements in quality of life for both the scales for the participants in the workplace intervention group (SMD −0.23, 95% CI −0.38 to −0.09, $p = 0.001$ for physical scale, and SMD −0.16, 95% CI −0.30 to −0.01, $p = 0.03$ for mental scale, respectively). Finally, evaluating the clinical outcomes in totality, a significant difference was reported in favor of the workplace intervention group compared to the controls ($p < 0.00001$).

Table 1. Main characteristics and clinical results of the included studies.

Characteristics of Working Population		WI		Follow-Up	Results	References
Study Group	Control Group	Study Group	Control Group			
37 (29.7% M, 70.3% F) nursing staff working in the operating room (age 31.45 ± 8.19)	37 (18.9% M, 81.1% F) nursing staff working in the operating room (age 26.64 ± 5.83)	Ergonomics educational program	No intervention	3 months	The IG reported a reduction in the prevalence of musculoskeletal disorders, in particular of LBP ($p = 0.000$).	Abdollahi et al. (2020) [31]
96 (53% M, 47% F) unspecified workers (age 44 ± 8.6)	100 (33% M, 67% F) unspecified workers (age 41.2 ± 10.7)	Workplace intervention: workplace assessment, work modifications, and case management	Usual care	12 months	Time until return to work for workers with WI was 77 versus 104 days for workers without this intervention ($p = 0.02$). Functional status and pain intensity improved more in workers who received a WI, than in workers without this intervention.	Anema et al. (2007) [21]
13 (15% M, 85% F) office workers (age 52 ± 9)	14 (29% M, 71% F) office workers (age 51 ± 13)	Behavioural counselling, sit-stand desk attachment and cognitive behavioral therapy for LBP self-management	No intervention	6 months	The relative decrease in ODI from baseline was 50% in the IG and 14% in the CG ($p = 0.042$). LBP was not significantly reduced in IG versus CG, though small-to-moderate effect sizes favoring the IG were observed.	Barone Gibbs et al. (2018) [16]
171 (23% M, 77% F) healthcare workers from hospitals (age 47.1 ± 8.5)	171 (22% M, 78% F) healthcare workers from hospitals (age 47.3 ± 8.5)	Exercise training sessions in the workplace, and a home-based self-managed EP	Usual care	24 months	35 workers in the IG and 31 workers in the CG had at least one LBP recurrence with sick leave. The intervention was effective in reducing fear avoidance with a mean reduction of −3.6 points in the IG compared with −1.3 points in the CG ($p < 0.05$).	Chaléat-Valaye et al. (2016) [15]
28 (50% M, 50% F) employed patients (age 41.46 ± 11.93)	23 (43.5% M, 56.5% F) employed patients (age 48.30 ± 10.14)	Individually targeted vocational sessions in conjunction with group rehabilitation for LBP	Group Rehabilitation	6 months	The IG had a better outcome for disability or pain and fear-avoidance	Coole et al. (2012) [13]

Table 1. *Cont.*

Characteristics of Working Population		WI		Follow-Up	Results	References
Study Group	Control Group	Study Group	Control Group			
92 (8.7% M, 91.3%F) nurses (age 37.9 ± 11.6)	91 (6.6% M, 93.4%F) nurses (age 41.1 ± 10.8)	Psychological units, segmental stabilization exercises units, and ergonomic and workplace-specific units (plus General Physical EP)	General Physical EP	12 months	For the primary study end point of pain interference, the effect size at 12 months after intervention was 0.58 in the MP and 0.47 in the EP.	Ewert et al. (2009) [19]
153 (68% M, 32% F) physically demanding workers (age 45.3 ± 10.1)	152 (67.1% M, 32.9% F) physically demanding workers (age 45.7 ± 10.5)	Occupational medicine consultations, a work-related evaluation and workplace intervention plan, an optional workplace visit, and a physical activity program	No intervention	6 months	Both groups showed improvements in average pain score, disability, fear-avoidance beliefs for physical activities and work; no statistically significant difference was found between the groups.	Hansen et al. (2019) [17]
59 (100% F) healthcare and social care professionals at healthcare centers (age 46 ± 7.9)	61 (100% F) healthcare and social care professionals at healthcare centers (age 46.5 ± 7)	Physical training, relaxation training, and cognitive-behavioral stress management methods	Physical exercise and passive treatment	24 months	In the MR group, statistically significant differences (at least $p < 0.05$) were found during the follow-up in ODI, subjective working ability and beliefs in future working ability.	Kaapa et al. (2006) [21]
301 (99% M, 1% F) manufacturing company workers (age 35.4)	315 (98.4% M, 1.6% F) manufacturing company workers (age 36.5)	Training sessions of participatory workplace improvement-based provision of ergonomic training and ergonomic action checklists on workplace improvement activities	Usual care	12 months	In the IG the incident rate ratio of participatory workplace improvements for the LBP category was significantly elevated after the training sessions, but decreased during the 10-month follow-up period.	Kajiki et al. (2017) [23]

Table 1. *Cont.*

Characteristics of Working Population		WI		Follow-Up	Results	References
Study Group	Control Group	Study Group	Control Group			
107 (42% M, 58% F) employees at primary health care centers (age 44)	57 (40% M, 60% F) employees at primary health care centers (age 43)	Exercises for improving the function of the deep abdominal muscles and establishing symmetric use of the back (plus a worksite visit)	Usual care	24 months	There were no differences between the three treatment arms regarding the intensity of pain and the perceived disability. The average number of days on sick leave was lower in the IGs than in the CG ($p = 0.03$).	Karjalainen et al. (2004) [24]
37 (70.3% M, 29.7% F) employees working in assembly positions in the automotive industry (age 45.1 ± 9.11)	38 (44.8% M, 55.2% F) employees working in assembly positions in the automotive industry (age 45.34 ± 8.80)	Supervised WI of muscle strengthening, flexibility, and endurance training	No intervention	6 months	Significant beneficial effect ($p < 0.025$) for the IG at 2 and 6 months in pain parameters, specific flexibility, and in back functions.	Nassif et al. (2011) [18]
646 (gender not available) employees in two municipalities (age not available)	211 (gender not available) employees in two municipalities (age not available)	Educational meetings, peer support and access to an outpatient clinic	Usual care	12 months	The IG had significantly fewer days of sick leave at the three-month (4.9 days, $p = 0.001$) and six-month (4.4 days, $p = 0.016$) follow-ups compared with the CG.	Ree et al. (2016) [25]
34 (gender not available) white collars (age 29.64 ± 0.90)	28 (gender not available) white collars (age 28.74 ± 0.82)	Office-based stretching exercises mechanisms to rise the range and flexibility of motion in the muscles of the back plus "total workplace Occupational Safety and Health and ergonomic intervention"	No intervention	6 months	Significant differences were seen in pain scores for lower back (MD −6.87; 95% CI −10 to −3.74) between the combined exercise and ergonomic modification and CGs.	Shariat et al. (2017) [20]

Table 1. *Cont.*

Characteristics of Working Population		WI		Follow-Up	Results	References
Study Group	Control Group	Study Group	Control Group			
63 (82.5% M, 17.5% F) nursing assistants (age not available)	62 (75.8% M, 24.2% F) nursing assistants (age not available)	Multidisciplinary intervention consisted of an educational program and ergonomic posture training	Usual care	6 months	The comparison tests showed significant change from baseline in reduction of work-related LBP intensity following the multidisciplinary program, with scores of 5.01 ± 1.97 to 3.42 ± 2.53 after 6 months on the visual analog scale in the IG ($p < 0.001$) and no significant change in CGs.	Shojaei et al. (2017) [14]

All the studies reported in the table were randomized clinical trials and have a level of evidence = I; F: Female; M: Male; IG: Intervention Group; LBP: Low Back Pain; WI: Workplace Intervention; EP: Exercise Program; CG: Control Group; MP: Multimodal Program; ODI: Oswestry Disability Index; MR: Multidisciplinary Rehabilitation.

Table 2. Guideline for systematic reviews in the Cochrane back and neck group.

Study	Randomization	Allocation	Patient Blinded	Care Provider Blinded	Outcome Assessor Blinded	Drop-Out Rate	All Randomized Participants Analyzed in the Group	Free of Selective Reporting	Groups Similar at Baseline	Cointerventions Avoided	Compliance	Timing of Outcome Assessment	Other Sources of Bias	Risk of Bias
Abdollahi	Y	Y	N	N	Y	Y	Y	U	Y	Y	Y	Y	U	B
Anema	Y	Y	N	N	Y	Y	Y	U	Y	Y	N	Y	U	B
Barone Gibbs	Y	Y	N	N	Y	Y	Y	Y	Y	Y	Y	Y	U	A
Chaléat-Valaye	Y	Y	N	N	Y	U	Y	Y	Y	Y	Y	Y	Y	A
Coole	Y	U	N	N	Y	N	Y	U	Y	Y	N	Y	U	C

Table 2. *Cont.*

Study	Randomization	Allocation	Patient Blinded	Care Provider Blinded	Outcome Assessor Blinded	Drop-Out Rate	All Randomized Participants Analyzed in the Group	Free of Selective Reporting	Groups Similar at Baseline	Cointerventions Avoided	Compliance	Timing of Outcome Assessment	Other Sources of Bias	Risk of Bias
Ewert	Y	Y	N	N	Y	U	Y	Y	Y	Y	Y	Y	U	B
Hansen	Y	Y	N	N	Y	Y	Y	Y	Y	Y	Y	Y	Y	A
Kaapa	Y	Y	N	N	Y	U	Y	Y	Y	Y	Y	Y	U	B
Kajiki	Y	Y	N	N	Y	Y	Y	U	Y	Y	Y	Y	Y	A
Karjalainen	Y	Y	N	N	Y	Y	Y	U	Y	Y	Y	Y	U	B
Nassif	Y	Y	N	N	Y	N	Y	Y	Y	Y	U	Y	U	B
Ree	Y	Y	N	N	Y	U	Y	U	Y	Y	U	Y	U	C
Shariat	Y	Y	N	N	Y	N	Y	U	Y	Y	U	Y	Y	B
Shojaei	Y	Y	N	N	Y	Y	Y	U	Y	Y	Y	Y	U	B

Y: Yes; N: No; U: Unsure.

Study or Subgroup	Workplace Mean	Workplace SD	Workplace Total	Control Mean	Control SD	Control Total	Weight	Std. Mean Difference IV, Fixed, 95% CI
2.1.1 Pain								
Anema 2007	3.2	2.6	96	3.4	2.7	100	3.9%	−0.08 [−0.36, 0.21]
Chaléat-Valaye 2016	36.2	21	139	36.7	21	141	5.5%	−0.02 [−0.26, 0.21]
Coole 2012	5.19	1.87	19	5.33	2.17	19	0.7%	−0.07 [−0.70, 0.57]
Ewert 2009	1.08	0.93	83	1.3	0.98	86	3.3%	−0.23 [−0.53, 0.07]
Gibbs 2018	1.9	1.4	13	2.2	2	14	0.5%	−0.17 [−0.92, 0.59]
Hansen 2019	3.86	5.84	153	4	5.2	152	6.0%	−0.03 [−0.25, 0.20]
Kaapa 2006	3.5	2.6	49	4	2.9	46	1.9%	−0.18 [−0.58, 0.22]
Nassif 2011	3.15	2.3	29	3.53	2.47	23	1.0%	−0.16 [−0.71, 0.39]
Shariat 2017	2.07	12.4	34	8.63	13	28	1.2%	−0.51 [−1.02, −0.00]
Shojaei 2017	3.42	2.53	63	5.03	2.13	62	2.3%	−0.68 [−1.04, −0.32]
Subtotal (95% CI)			**678**			**671**	**26.3%**	**−0.16 [−0.26, −0.05]**
Heterogeneity: $Chi^2 = 13.24$, df = 9 (P = 0.15); $I^2 = 32\%$								
Test for overall effect: Z = 2.88 (P = 0.004)								
2.1.2 Disability								
Chaléat-Valaye 2016	19.8	15.8	139	24.4	16.4	141	5.5%	−0.28 [−0.52, −0.05]
Gibbs 2018	12.3	9	13	20.1	14.2	14	0.5%	−0.63 [−1.41, 0.15]
Kaapa 2006	19.7	14.3	49	19.3	13.1	46	1.9%	0.03 [−0.37, 0.43]
Nassif 2011	27.15	13.78	29	30.21	17.26	23	1.0%	−0.20 [−0.74, 0.35]
Shojaei 2017	21.33	19.37	63	30.33	18.03	62	2.4%	−0.48 [−0.83, −0.12]
Subtotal (95% CI)			**293**			**286**	**11.2%**	**−0.28 [−0.45, −0.12]**
Heterogeneity: $Chi^2 = 4.33$, df = 4 (P = 0.36); $I^2 = 8\%$								
Test for overall effect: Z = 3.36 (P = 0.0008)								
2.1.3 FABQ-W								
Chaléat-Valaye 2016	14.4	10.2	139	15.7	9.9	141	5.5%	−0.13 [−0.36, 0.11]
Coole 2012	18.61	10.8	19	20.67	9.7	19	0.7%	−0.20 [−0.83, 0.44]
Ewert 2009	2.2	1.36	83	2.29	1.31	86	3.3%	−0.07 [−0.37, 0.23]
Hansen 2019	21.36	20.5	153	21.43	18.6	152	6.0%	−0.00 [−0.23, 0.22]
Subtotal (95% CI)			**394**			**398**	**15.6%**	**−0.07 [−0.21, 0.07]**
Heterogeneity: $Chi^2 = 0.73$, df = 3 (P = 0.87); $I^2 = 0\%$								
Test for overall effect: Z = 0.99 (P = 0.32)								
2.1.4 FABQ-P								
Chaléat-Valaye 2016	8.5	5.8	139	11.7	6.1	141	5.3%	−0.54 [−0.77, −0.30]
Ewert 2009	2.59	1.15	83	2.66	1.22	86	3.3%	−0.06 [−0.36, 0.24]
Hansen 2019	21.36	20.5	153	21.43	18.6	152	6.0%	−0.00 [−0.23, 0.22]
Subtotal (95% CI)			**375**			**379**	**14.7%**	**−0.21 [−0.35, −0.07]**
Heterogeneity: $Chi^2 = 11.40$, df = 2 (P = 0.003); $I^2 = 82\%$								
Test for overall effect: Z = 2.86 (P = 0.004)								
2.1.5 Work ability								
Coole 2012	−7	2.8	19	−6.78	2.5	19	0.7%	−0.08 [−0.72, 0.56]
Kaapa 2006	2.9	2.8	49	3.5	2.8	46	1.9%	−0.21 [−0.62, 0.19]
Subtotal (95% CI)			**68**			**65**	**2.6%**	**−0.17 [−0.52, 0.17]**
Heterogeneity: $Chi^2 = 0.12$, df = 1 (P = 0.73); $I^2 = 0\%$								
Test for overall effect: Z = 1.01 (P = 0.31)								
2.1.6 SF physical								
Chaléat-Valaye 2016	−47.4	7.7	139	−45.1	8.3	141	5.5%	−0.29 [−0.52, −0.05]
Ewert 2009	−51.74	6.37	83	−49.73	7.89	86	3.3%	−0.28 [−0.58, 0.02]
Hansen 2019	−45.64	19.75	153	−42.11	24.4	152	6.0%	−0.16 [−0.38, 0.07]
Subtotal (95% CI)			**375**			**379**	**14.8%**	**−0.23 [−0.38, −0.09]**
Heterogeneity: $Chi^2 = 0.70$, df = 2 (P = 0.70); $I^2 = 0\%$								
Test for overall effect: Z = 3.18 (P = 0.001)								
2.1.7 SF mental								
Chaléat-Valaye 2016	−45.8	11.7	139	−44	11	141	5.5%	−0.16 [−0.39, 0.08]
Ewert 2009	−49.45	9.45	83	−49.06	8.67	86	3.3%	−0.04 [−0.34, 0.26]
Hansen 2019	−51.84	27.5	153	−45.35	32.6	152	6.0%	−0.21 [−0.44, 0.01]
Subtotal (95% CI)			**375**			**379**	**14.8%**	**−0.16 [−0.30, −0.01]**
Heterogeneity: $Chi^2 = 0.80$, df = 2 (P = 0.67); $I^2 = 0\%$								
Test for overall effect: Z = 2.12 (P = 0.03)								
Total (95% CI)			**2558**			**2557**	**100.0%**	**−0.18 [−0.23, −0.12]**
Heterogeneity: $Chi^2 = 36.10$, df = 29 (P = 0.17); $I^2 = 20\%$								
Test for overall effect: Z = 6.29 (P < 0.00001)								
Test for subgroup differences: $Chi^2 = 4.78$, df = 6 (P = 0.57), $I^2 = 0\%$								

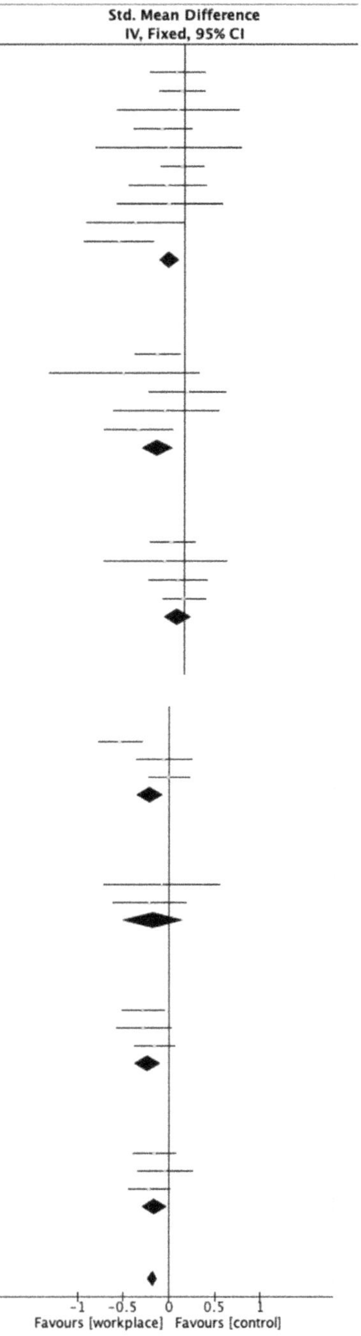

Figure 2. Outcome measurements.

The analysis of the participants on sick leave showed a reduction for patients who underwent intervention programs (OR 0.98, 95% CI 0.76 to 1.26) but no significant differences ($p = 0.85$) (Figure 3). The number of total days of sick leave decreased in the WI group (OR 0.80, 95% CI 0.62 to 1.04) without statistical significance ($p = 0.10$) (Figure 4). Return to work was analyzed in only one study, which reported a better result in favor of the study group (OR 0.77, 95% CI 0.57 to 1.04, $p = 0.08$) (Figure 5). Finally, LBP recurrence was significantly reduced in the intervention group compared to controls (OR 0.38, 95% CI 0.19 to 0.76, $p = 0.006$) (Figure 6).

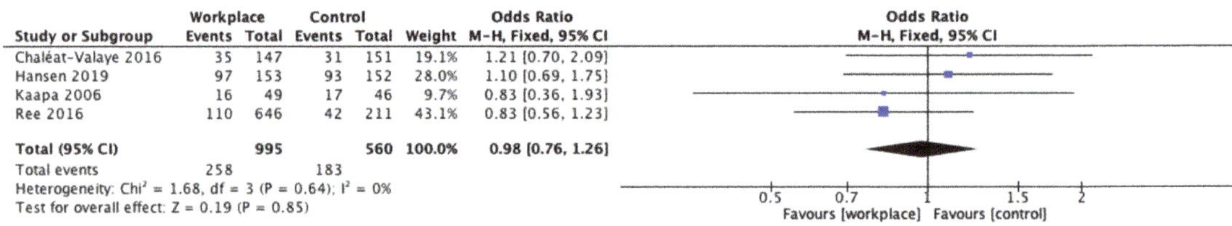

Figure 3. Participants on sick leave.

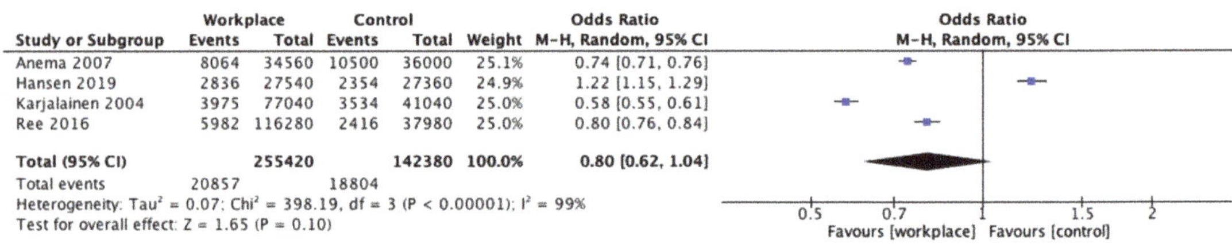

Figure 4. Days of sick leave.

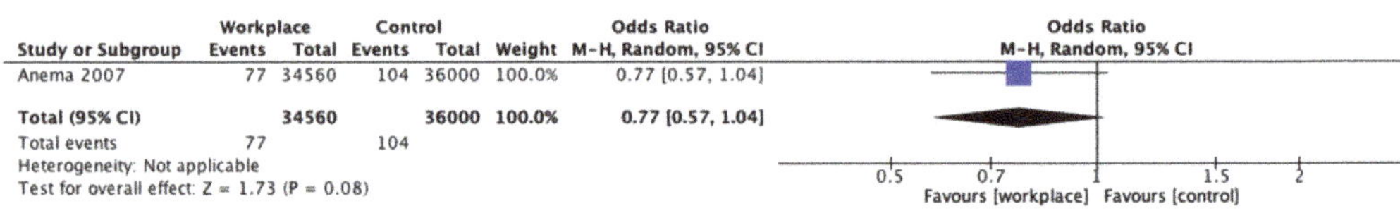

Figure 5. Return to work.

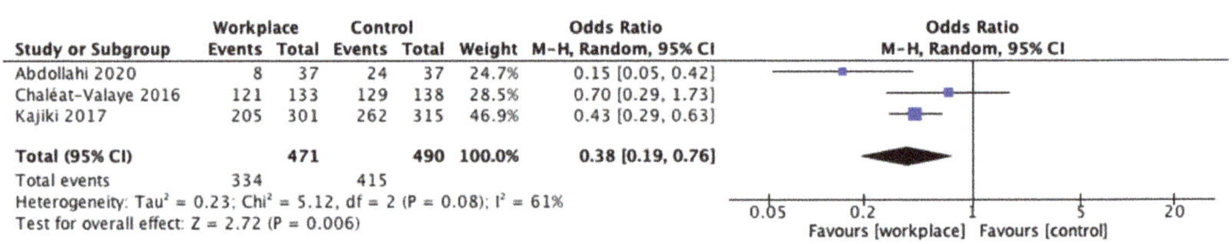

Figure 6. Low back pain recurrence.

In conclusion, the test for subgroup differences showed no statistically significant subgroup effect for LBP ($p = 0.29$). However, the pain reduction after WI was statistically significant in the healthcare workers' group ($p = 0.005$), but no difference was reported in the other workers' group (Figure 7).

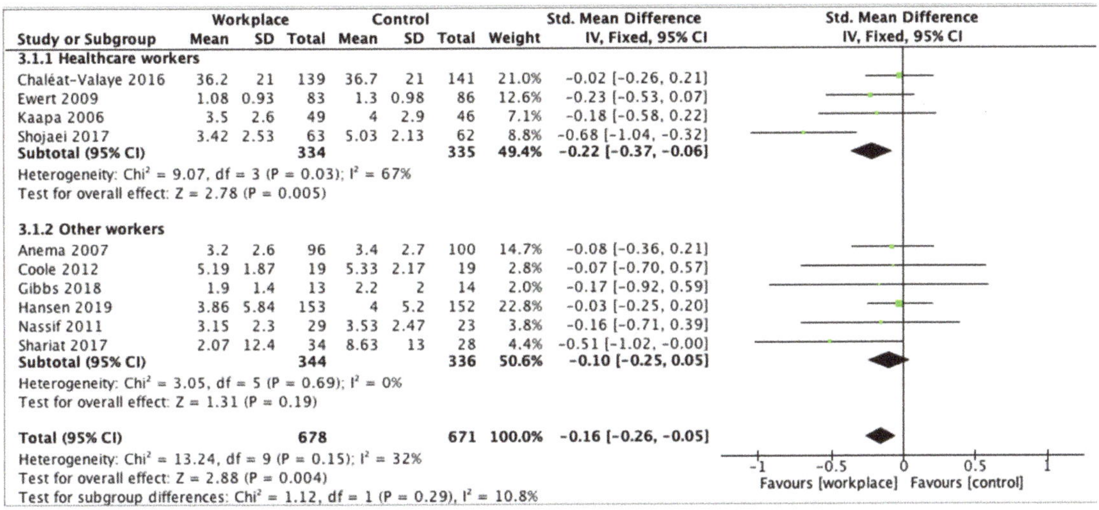

Figure 7. Subgroup analysis for low back pain.

3.7. Quality Assessment

GRADE was applied to evaluate the quality of the evidence given in the included RCTs (Table 3). It produced seven comparisons for continuous data and three for dichotomous data. Regarding clinical outcomes, disability, SF physical, and SF mental maintained a high quality of evidence, while pain obtained high-quality evidence because it received an upgrade due to the large effect. FABQ-W and FABQ-P were downgraded by one level for risk of bias and inconsistency, thus reporting a moderate quality of evidence. Finally, work ability presented a low quality of evidence, due to risk of bias and imprecision. In contrast, the outcomes of participants on sick leave and days of sick leave achieved low quality of evidence, while LBP recurrence reported a very low quality of evidence.

Table 3. GRADE.

Outcomes	N. of Participants (Studies)	Risk of Bias	Inconsistency	Indirectness	Imprecision	Other Considerations	Quality
Pain	1349 (10 RCT)	serious	not serious	not serious	not serious	not serious	⊕⊕⊕⊕ high *
Disability	579 (5 RCT)	not serious	not serious	not serious	not serious	not serious	⊕⊕⊕⊕ high
FABQ-W	792 (4 RCT)	serious	not serious	not serious	not serious	not serious	⊕⊕⊕◯ moderate
FABQ-P	754 (3 RCT)	not serious	serious	not serious	not serious	not serious	⊕⊕⊕◯ moderate
Work ability	133 (2 RCT)	serious	not serious	not serious	serious	not serious	⊕⊕◯◯ low
SF physical	754 (3 RCT)	not serious	not serious	not serious	not serious	not serious	⊕⊕⊕⊕ high
SF mental	754 (3 RCT)	not serious	not serious	not serious	not serious	not serious	⊕⊕⊕⊕ high
Participants on sick leave	1555 (4 RCT)	serious	not serious	not serious	serious	not serious	⊕⊕◯◯ low
Days of sick leave	1526 (4 RCT)	serious	serious	not serious	not serious	not serious	⊕⊕◯◯ low
LBP recurrence	961 (3 RCT)	serious	serious	not serious	serious	not serious	⊕◯◯◯ very low

N.: Number; FABQ-W: Fear-Avoidance Beliefs Questionnaire Work subscale; FABQ-P: Fear-Avoidance Beliefs Questionnaire Physical activity subscale; SF: Short Form; LBP: Low Back Pain; RCT: Randomized clinical trial. * Upgrade due to large effect.

4. Discussion

The primary aim of this systematic review and meta-analysis was to assess the effects of WI on workers in terms of clinical outcomes. The secondary endpoint was the interpretation of socio-economic parameters as participants on sick leave, days of sick leave, LBP recurrence, and return to work following specific workplace programs. Studies that analyzed only physical or psychosocial activities at the workplace or evaluated the effectiveness of a sit–stand workstation were not included since WI must be analyzed in its entirety in order to provide the best support to workers and obtain the best benefits in terms of LBP, work ability, and return to work.

Employees who underwent WI experienced improvements in LBP, disability, fear-avoidance beliefs, quality of life, and work ability compared to controls, with a significant increase for all the reported scores. The meta-analysis proved that the scores regarding LBP, disability, FABQ-P, and SF physical and mental obtained the most statistical significance compared to controls, showing the best improvements after WI. On the other hand, FABQ-W and work ability outcomes did not show significant differences compared to control groups. Therefore, it has been shown that WI led to excellent results in symptom reduction, daily living activities, and quality of life, but it remains a subjective limitation for workers to perform their job activities. Moreover, the meta-analysis showed non-significant improvements in participants on sick leave, days of sick leave, and return to work after WI. This could further demonstrate that workers did not feel able and ready to undergo workloads, although they have experienced a significant reduction in LBP and a global increase in quality of life. The subgroup analysis for LBP, even if in absence of a significant subgroup effect, showed a greater reduction of pain after WI in healthcare workers compared to other workers. Therefore, workplace interventions seem to ensure greater benefits for a population of nurses and healthcare workers, but further and more specific trials are needed to demonstrate these results. However, for the other clinical outcomes, it was not possible to observe a difference between the different kinds of works.

Furthermore, all the clinical outcomes showed high (pain, disability, SF physical, and SF mental) or moderate (FABQ-W and FABQ-P) quality of evidence and strength of recommendation at GRADE—except for work ability, which had a low quality of evidence—justifying a recommendation of workplace interventions in workers with LBP. On the other hand, GRADE reported a low quality of evidence for participants on sick leave and days of sick leave and very low quality for LBP recurrence. Finally, it should be noted that by the guideline for systematic reviews in the Cochrane Back and Neck Group, only two studies have been judged at a high risk of bias, showing an acceptable overall quality of the included studies. Almost all the studies showed a low risk of bias for random sequence generation and allocation concealment. However, in all studies, the risk of bias was high for blinding for patients and care providers, due to the impossibility to blind patients to the interventions that they were receiving. Instead, in all the trials, the outcomes were reported by assessors blinded to the group allocation.

In their review, Gobbo et al. [16] showed that exercise programs in the workplace reduce LBP symptoms, improve muscle strength and flexibility, and increase the quality of life in office workers. Contrarily, the meta-analysis performed by Maciel et al. [17] showed that physical exercise at the workplace did not reduce the occurrence of LBP ($p < 0.4$). Sowah et al. [18] evaluated occupational interventions as treatments for the prevention of LBP and demonstrated that exercise interventions, with or without educational interventions in the workplace, have the potential to prevent LBP. More specifically, Roman-Liu et al. [32] proved strong differences in effects among intervention strategies. In fact, they showed that technical modifications of the workstand and education based on practical training represent more effective strategies for LBP prevention than behavioral and physical training. Finally, in a meta-analysis conducted by Parry et al. [33], they did not show evidence that interventions to increase standing or walking in the workplace reduce musculoskeletal symptoms in sedentary workers.

Certain limitations may hinder the interpretation of data. The limitations of this study are related to the heterogeneity in the population of workers of the included RCTs. Indeed, three studies [21,22,25] did not specify the job of the participants, six studies involved nursing assistants or healthcare workers [23,24,28,30,31,34], while the patients in the remaining five trials [26,27,29,35,36] practiced other kinds of jobs. Other heterogeneities concerned the age and sex of the participants. In fact, the mean age, ranging from 26.6 to 52 years, correlated with different grades of LBP, which may need diverse treatments. Moreover, two studies showed a clear predominance of women [28,30], one study enrolled almost all men [35], and two studies did not report the gender distribution of the participants [29,36]. Due to the multiple works analyzed and the differences in the population groups, the workplace interventions performed in the various trials were very miscellaneous and not homogeneous. Moreover, two studies showed a clear predominance of women [26,28], one study enrolled almost all men [33] and two studies did not report the gender distribution of the participants [27,34]. Due to the multiple works analyzed and to the differences in the population groups, the follow-up was one year or less in 11 studies (78.5%), not allowing the comprehension of long-term effects of WI on LBP, disability, quality of life, and work ability.

The evaluation of findings provided by the studies included in this review clearly showed that different types of WI determine a beneficial effect both on clinical outcomes and socio-economic parameters related to LBP in workers. However, in consideration of the considerable heterogeneity (in terms of working population, socio-demographic characteristics, and diversification of WI) of the studies, trying to establish which is the best approach in terms of effectiveness in preventing LBP (and therefore in reducing its multiple negative effects) is a rather challenging task that can also lead to drawing conclusions that are not entirely correct. Indeed, in this regard, it should be taken into account that the degree of effectiveness of the different WI strategies may be affected by numerous factors such as the socio-demographic characteristics of the working population (e.g., gender, age, education level, presence of chronic degenerative diseases), the working activities carried out by employees, which determine a greater or lesser exposure to occupational risk factors for LBP (e.g., manual handling of heavy loads, awkward and prolonged postures, whole-body mechanical vibrations, work-related stress), and the number and type of WI (e.g., technical interventions, procedural measures, organizational tools, educational programs) [5]. Consequently, even if the literature data suggest, for example, that an engineering redesign of workstations is more effective than participatory ergonomics or that a tailored physical exercise achieves better results when coupled with cognitive and behavioral training or even that strength exercise is more beneficial than cardiorespiratory exercise, it is not obvious that a WI strategy based on the aforementioned indications will achieve the same level of effectiveness in all workplaces or working populations. Therefore, in our opinion, a prevention program based on WI to be truly decisive in reducing the negative effects of LBP in workers cannot be limited to replicating the same intervention strategy in all workplaces. In this regard, we believe that the design of an adequate WI approach must be based on a flexible decision-making process, which, starting from the occupational risk assessment and taking into account the characteristics of the working population, identifies, on the basis of the evidence of the literature, the best possible combination of the use of the different WI. Indeed, WI should be targeted for a specific work, with the simultaneous and combined presence of all the programs, such as a technical intervention, physical exercise, behavioral training, and educational and participatory ergonomics, in order to treat and prevent the LBP in the totality of its manifestations at workplace.

5. Conclusions

This systematic review demonstrated that workplace interventions led to a significant improvement of clinical outcomes in a worker population affected by LBP. The meta-analysis showed strong evidence that WI improved LBP, disability, and quality of life in workers. However, a statistical increase in purely working parameters has not been described, testifying to the fact that despite the pain decreased, workers were still afraid to fully return to work. WI should be practiced in order to prevent and treat musculoskeletal symptoms, which could reduce the work ability and increase the number of sick leave days for the workers. However, workplace interventions standardized for specific works are needed, and the follow-up should be longer to evaluate the long-term effects of WI on clinical and working outcomes.

Supplementary Materials: The following are available online at https://www.mdpi.com/article/10.3390/ijerph182312614/s1, Table S1: PRISMA Checklist.

Author Contributions: Conceptualization, F.R. and S.I.; methodology, F.R. and G.F.P.; formal analysis, F.R., G.F.P. and L.F.; data curation, F.R., G.F.P., G.V. and L.F.; writing—original draft preparation, F.R., G.F.P. and L.F.; writing—review and editing, F.R., G.F.P., G.V., L.F., S.I., R.P. and V.D.; supervision, F.R., G.V., S.I., R.P. and V.D.; funding acquisition, F.R., G.V., S.I. and V.D. All authors have read and agreed to the published version of the manuscript.

Funding: This research was funded by the Research Grant (BRiC-2018 ID3) of the Italian Workers' Compensation Authority (INAIL).

Institutional Review Board Statement: Not applicable.

Informed Consent Statement: Not applicable.

Conflicts of Interest: The authors declare no conflict of interest.

References

1. Fan, X.; Straube, S. Reporting on Work-Related Low Back Pain: Data Sources, Discrepancies and the Art of Discovering Truths. *Pain Manag.* **2016**, *6*, 553–559. [CrossRef] [PubMed]
2. Vadalà, G.; Russo, F.; De Salvatore, S.; Cortina, G.; Albo, E.; Papalia, R.; Denaro, V. Physical Activity for the Treatment of Chronic Low Back Pain in Elderly Patients: A Systematic Review. *J. Clin. Med.* **2020**, *9*, 1023. [CrossRef] [PubMed]
3. Cieza, A.; Weigl, M.; Disler, P.; van der Linden, S.; Kostanjsek, N.; de Bie, R.; Stucki, G.; Jäckel, W. ICF Core Sets for Low Back Pain. *J. Rehabil. Med.* **2004**, *36*, 69–74. [CrossRef] [PubMed]
4. Russo, F.; Ambrosio, L.; Ngo, K.; Vadalà, G.; Denaro, V.; Fan, Y.; Sowa, G.; Kang, J.D.; Vo, N. The Role of Type I Diabetes in Intervertebral Disc Degeneration. *Spine* **2019**, *44*, 1177–1185. [CrossRef]
5. Russo, F.; Di Tecco, C.; Fontana, L.; Adamo, G.; Papale, A.; Denaro, V.; Iavicoli, S. Prevalence of Work Related Musculoskeletal Disorders in Italian Workers: Is There an Underestimation of the Related Occupational Risk Factors? *BMC Musculoskelet. Disord.* **2020**, *21*, 738. [CrossRef]
6. Duthey, B. *Update on 2004 Background Paper, BP 6.24 Low Back Pain*; World Health Organization: Geneva, Switzerland, 2013.
7. European Agency for Safety and Health at Work; IKEI; Panteia. *Work-Related Musculoskeletal Disorders: Prevalence, Costs and Demographics in the EU*; Publications Office: Luxembourg, 2019.
8. Andersen, L.L.; Burdorf, A.; Fallentin, N.; Persson, R.; Jakobsen, M.D.; Mortensen, O.S.; Clausen, T.; Holtermann, A. Patient Transfers and Assistive Devices: Prospective Cohort Study on the Risk for Occupational Back Injury among Healthcare Workers. *Scand. J. Work. Environ. Health* **2014**, *40*, 74–81. [CrossRef]
9. Hignett, S. Work-Related Back Pain in Nurses. *J. Adv. Nurs.* **1996**, *23*, 1238–1246. [CrossRef]
10. Janwantanakul, P.; Pensri, P.; Jiamjarasrangsri, V.; Sinsongsook, T. Prevalence of Self-Reported Musculoskeletal Symptoms among Office Workers. *Occup. Med.* **2008**, *58*, 436–438. [CrossRef]
11. Lambeek, L.C.; van Tulder, M.W.; Swinkels, I.C.S.; Koppes, L.L.J.; Anema, J.R.; van Mechelen, W. The Trend in Total Cost of Back Pain in The Netherlands in the Period 2002 to 2007. *Spine* **2011**, *36*, 1050–1058. [CrossRef]
12. Russo, F.; De Salvatore, S.; Ambrosio, L.; Vadalà, G.; Fontana, L.; Papalia, R.; Rantanen, J.; Iavicoli, S.; Denaro, V. Does Workers' Compensation Status Affect Outcomes after Lumbar Spine Surgery? A Systematic Review and Meta-Analysis. *Int. J. Environ. Res. Public. Health* **2021**, *18*, 6165. [CrossRef]
13. Majchrzycki, M.; Kocur, P.; Kotwicki, T. Deep Tissue Massage and Nonsteroidal Anti-Inflammatory Drugs for Low Back Pain: A Prospective Randomized Trial. *Sci. World J.* **2014**, *2014*, 287597. [CrossRef]
14. Vadalà, G.; Russo, F.; Ambrosio, L.; Papalia, R.; Denaro, V. Mesenchymal Stem Cells for Intervertebral Disc Regeneration. *J. Biol. Regul. Homeost. Agents* **2016**, *30*, 173–179. [PubMed]
15. Vadala', G.; Russo, F.; Ambrosio, L.; Di Martino, A.; Papalia, R.; Denaro, V. Biotechnologies and biomaterials in spine surgery. *J. Biol. Regul. Homeost. Agents* **2015**, *29*, 137–147. [PubMed]
16. Gobbo, S.; Bullo, V.; Bergamo, M.; Duregon, F.; Vendramin, B.; Battista, F.; Roma, E.; Bocalini, D.S.; Rica, R.L.; Alberton, C.L.; et al. Physical Exercise Is Confirmed to Reduce Low Back Pain Symptoms in Office Workers: A Systematic Review of the Evidence to Improve Best Practices in the Workplace. *J. Funct. Morphol. Kinesiol.* **2019**, *4*, 43. [CrossRef] [PubMed]

17. Maciel, R.R.B.T.; Dos Santos, N.C.; Portella, D.D.A.; Alves, P.G.J.M.; Martinez, B.P. Effects of physical exercise at the workplace for treatment of low back pain: A systematic review with meta-analysis. *Rev. Bras. Med. Trab.* **2018**, *16*, 225–235. [CrossRef] [PubMed]
18. Sowah, D.; Boyko, R.; Antle, D.; Miller, L.; Zakhary, M.; Straube, S. Occupational Interventions for the Prevention of Back Pain: Overview of Systematic Reviews. *J. Saf. Res.* **2018**, *66*, 39–59. [CrossRef]
19. Page, M.J.; McKenzie, J.E.; Bossuyt, P.M.; Boutron, I.; Hoffmann, T.C.; Mulrow, C.D.; Shamseer, L.; Tetzlaff, J.M.; Akl, E.A.; Brennan, S.E.; et al. The PRISMA 2020 Statement: An Updated Guideline for Reporting Systematic Reviews. *BMJ* **2021**, *372*, n71. [CrossRef]
20. Furlan, A.D.; Malmivaara, A.; Chou, R.; Maher, C.G.; Deyo, R.A.; Schoene, M.; Bronfort, G.; van Tulder, M.W. 2015 Updated Method Guideline for Systematic Reviews in the Cochrane Back and Neck Group. *Spine* **2015**, *40*, 1660–1673. [CrossRef]
21. Anema, J.R.; Steenstra, I.A.; Bongers, P.M.; de Vet, H.C.W.; Knol, D.L.; Loisel, P.; van Mechelen, W. Multidisciplinary Rehabilitation for Subacute Low Back Pain: Graded Activity or Workplace Intervention or Both? A Randomized Controlled Trial. *Spine* **2007**, *32*, 291–298. [CrossRef] [PubMed]
22. Coole, C.; Drummond, A.; Watson, P.J. Individual Work Support for Employed Patients with Low Back Pain: A Randomized Controlled Pilot Trial. *Clin. Rehabil.* **2013**, *27*, 40–50. [CrossRef]
23. Shojaei, S.; Tavafian, S.S.; Jamshidi, A.R.; Wagner, J. A Multidisciplinary Workplace Intervention for Chronic Low Back Pain among Nursing Assistants in Iran. *Asian Spine J.* **2017**, *11*, 419–426. [CrossRef]
24. Chaléat-Valayer, E.; Denis, A.; Abelin-Genevois, K.; Zelmar, A.; Siani-Trebern, F.; Touzet, S.; Bergeret, A.; Colin, C.; Fassier, J.-B. Long-Term Effectiveness of an Educational and Physical Intervention for Preventing Low-Back Pain Recurrence: A Randomized Controlled Trial. *Scand. J. Work. Environ. Health* **2016**, *42*, 510–519. [CrossRef] [PubMed]
25. Barone Gibbs, B.; Hergenroeder, A.L.; Perdomo, S.J.; Kowalsky, R.J.; Delitto, A.; Jakicic, J.M. Reducing Sedentary Behaviour to Decrease Chronic Low Back Pain: The Stand Back Randomised Trial. *Occup. Environ. Med.* **2018**, *75*, 321–327. [CrossRef] [PubMed]
26. Hansen, B.B.; Kirkeskov, L.; Begtrup, L.M.; Boesen, M.; Bliddal, H.; Christensen, R.; Andreasen, D.L.; Kristensen, L.E.; Flachs, E.M.; Kryger, A.I. Early Occupational Intervention for People with Low Back Pain in Physically Demanding Jobs: A Randomized Clinical Trial. *PLoS Med.* **2019**, *16*, e1002898. [CrossRef] [PubMed]
27. Nassif, H.; Brosset, N.; Guillaume, M.; Delore-Milles, E.; Tafflet, M.; Buchholz, F.; Toussaint, J.-F. Evaluation of a Randomized Controlled Trial in the Management of Chronic Lower Back Pain in a French Automotive Industry: An Observational Study. *Arch. Phys. Med. Rehabil.* **2011**, *92*, 1927–1936. [CrossRef]
28. Ewert, T.; Limm, H.; Wessels, T.; Rackwitz, B.; von Garnier, K.; Freumuth, R.; Stucki, G. The Comparative Effectiveness of a Multimodal Program Versus Exercise Alone for the Secondary Prevention of Chronic Low Back Pain and Disability. *PM&R* **2009**, *1*, 798–808. [CrossRef]
29. Shariat, A.; Cleland, J.A.; Danaee, M.; Kargarfard, M.; Sangelaji, B.; Tamrin, S.B.M. Effects of Stretching Exercise Training and Ergonomic Modifications on Musculoskeletal Discomforts of Office Workers: A Randomized Controlled Trial. *Braz. J. Phys. Ther.* **2018**, *22*, 144–153. [CrossRef]
30. Kääpä, E.H.; Frantsi, K.; Sarna, S.; Malmivaara, A. Multidisciplinary Group Rehabilitation Versus Individual Physiotherapy for Chronic Nonspecific Low Back Pain: A Randomized Trial. *Spine* **2006**, *31*, 371–376. [CrossRef]
31. Abdollahi, T.; Pedram Razi, S.; Pahlevan, D.; Yekaninejad, M.S.; Amaniyan, S.; Leibold Sieloff, C.; Vaismoradi, M. Effect of an Ergonomics Educational Program on Musculoskeletal Disorders in Nursing Staff Working in the Operating Room: A Quasi-Randomized Controlled Clinical Trial. *Int. J. Environ. Res. Public. Health* **2020**, *17*, 7333. [CrossRef]
32. Roman-Liu, D.; Kamińska, J.; Tokarski, T. Effectiveness of Workplace Intervention Strategies in Lower Back Pain Prevention: A Review. *Ind. Health* **2020**, *58*, 503–519. [CrossRef]
33. Parry, S.P.; Coenen, P.; Shrestha, N.; O'Sullivan, P.B.; Maher, C.G.; Straker, L.M. Workplace Interventions for Increasing Standing or Walking for Decreasing Musculoskeletal Symptoms in Sedentary Workers. *Cochrane Database Syst. Rev.* **2019**. [CrossRef] [PubMed]
34. Karjalainen, K.; Malmivaara, A.; Mutanen, P.; Roine, R.; Hurri, H.; Pohjolainen, T. Mini-Intervention for Subacute Low Back Pain. *Spine* **2004**, *29*, 1069–1076. [CrossRef] [PubMed]
35. Kajiki, S.; Izumi, H.; Hayashida, K.; Kusumoto, A.; Nagata, T.; Mori, K. A randomized controlled trial of the effect of participatory ergonomic low back pain training on workplace improvement. *J. Occup. Health* **2017**, *59*, 256–266. [CrossRef] [PubMed]
36. Ree, E.; Lie, S.A.; Eriksen, H.R.; Malterud, K.; Indahl, A.; Samdal, O.; Harris, A. Reduction in Sick Leave by a Workplace Educational Low Back Pain Intervention: A Cluster Randomized Controlled Trial. *Scand. J. Public Health* **2016**, *44*, 571–579. [CrossRef] [PubMed]

International Journal of
Environmental Research and Public Health

MDPI

Review

Artificial Intelligence in Bone Metastases: An MRI and CT Imaging Review

Eliodoro Faiella [1], Domiziana Santucci [1], Alessandro Calabrese [2,*], Fabrizio Russo [3], Gianluca Vadalà [3], Bruno Beomonte Zobel [1], Paolo Soda [4], Giulio Iannello [4], Carlo de Felice [2] and Vincenzo Denaro [3]

1 Department of Radiology, University of Rome "Campus Bio-Medico", Via Alvaro del Portillo, 00128 Roma, Italy; e.faiella@unicampus.it (E.F.); d.santucci@unicampus.it (D.S.); b.zobel@unicampus.it (B.B.Z.)
2 Department of Radiology, University of Rome "Sapienza", Viale del Policlinico, 00161 Roma, Italy; c.df@uniroma1.it
3 Department of Orthopaedic and Trauma Surgery, University of Rome "Campus Bio-Medico", Via Alvaro del Portillo, 00128 Roma, Italy; f.russo@unicampus.it (F.R.); g.vadala@unicampus.it (G.V.); v.denaro@unicampus.it (V.D.)
4 Unit of Computer Systems and Bioinformatics, Department of Engineering, University of Rome "Campus Bio-Medico", Via Alvaro del Portillo, 00128 Roma, Italy; p.soda@unicampus.it (P.S.); g.iannello@unicampus.it (G.I.)
* Correspondence: alessandro.calabrese@uniroma1.it

Abstract: (1) Background: The purpose of this review is to study the role of radiomics as a supporting tool in predicting bone disease status, differentiating benign from malignant bone lesions, and characterizing malignant bone lesions. (2) Methods: Two reviewers conducted the literature search independently. Thirteen articles on radiomics as a decision support tool for bone lesions were selected. The quality of the methodology was evaluated according to the radiomics quality score (RQS). (3) Results: All studies were published between 2018 and 2021 and were retrospective in design. Eleven (85%) studies were MRI-based, and two (15%) were CT-based. The sample size was <200 patients for all studies. There is significant heterogeneity in the literature, as evidenced by the relatively low RQS value (average score = 22.6%). There is not a homogeneous protocol used for MRI sequences among the different studies, although the highest predictive ability was always obtained in T2W-FS. Six articles (46%) reported on the potential application of the model in a clinical setting with a decision curve analysis (DCA). (4) Conclusions: Despite the variability in the radiomics method application, the similarity of results and conclusions observed is encouraging. Substantial limits were found; prospective and multicentric studies are needed to affirm the role of radiomics as a supporting tool.

Keywords: MRI; CT; bone metastasis; bone cancer; lung cancer; prostate cancer; machine learning; radiomics; signature

Citation: Faiella, E.; Santucci, D.; Calabrese, A.; Russo, F.; Vadalà, G.; Zobel, B.B.; Soda, P.; Iannello, G.; de Felice, C.; Denaro, V. Artificial Intelligence in Bone Metastases: An MRI and CT Imaging Review. *IJERPH* **2022**, *19*, 1880. https://doi.org/10.3390/ijerph19031880

Academic Editors: Vincenzo Denaro, Sergio Iavicoli, Fabrizio Russo and Gianluca Vadalà

Received: 27 December 2021
Accepted: 5 February 2022
Published: 8 February 2022

1. Introduction

Bone is the third most frequent site for metastatic localization, after lung and liver [1], with breast and prostate cancer accounting for almost 70% of primary tumors [2]. In most cases, bone metastases influence a patient's short-term prognosis, as bone lesions can rarely be completely eradicated. Patients with bone metastases have the option of undergoing palliative care to reduce the size of the lesions, slow their growth, or allow for improvement in symptoms. Bone metastases lead to a sharp reduction in life expectancy: average survival in patients with bone metastases from melanoma is 6 months; from breast cancer, 19–25 months; and from prostate cancer, 53 months [3].

The improvement of therapeutic strategies to deal with the various forms of cancer has led to an increase in life expectancy and, consequently, a lengthening of the time a patient can coexist with metastatic disease [4]. The most frequent site of bone metastasis is the axial skeleton because of its high red marrow content [1,2,5,6], which is therefore frequently responsible for the increased morbidity and decreased quality of life of patients.

The spectrum of clinical manifestations is very heterogeneous, ranging from complete absence of symptoms to severe pain, reduced mobility, pathologic fractures, spinal cord compression, bone marrow aplasia, and hypercalcemia. Hypercalcemia is in turn responsible for constipation, polyuria, polydipsia, and fatigue [2,7]. In the final stages, hypercalcemia may lead to cardiac arrhythmias and acute renal failure [1].

Therefore, to identify a proper course of treatment, it is essential to differentiate metastatic lesions from any primary or benign lesions of the bone. In order to assess the patient's prognosis and choose the most appropriate medical treatment according to their life expectancy, bone metastases should be diagnosed at the time of the diagnosis of the primary tumor: the aim is to reduce the incidence of complications and improve the quality of life.

Bone scintigraphy, computed tomography (CT), magnetic resonance imaging (MRI), and positron emission tomography (PET) are all capable of assessing the presence of bone metastases [8]. The sensitivity and specificity of bone scintigraphy are 78% and 48%, respectively, but despite its relatively low specificity which may require further imaging examinations, it is still the most widely available technique and the most suggested by the guidelines for the study of bone disease. The CT exam, with a sensitivity and specificity of 74% and 56%, respectively, can be used as a guide during interventional diagnostic procedures. In addition, CT allows simultaneous evaluation of bone and systemic staging, reducing the burden of imaging for patients. MRI shows a sensitivity and specificity of 95% and 90%, respectively. It is a radiation-free technique and is considered the imaging modality of choice for assessing metastatic spread in the bone marrow. 18F FDG-PET (fluorodeoxyglucose) has a sensitivity and specificity of 98% and 56%, respectively: the sensitivity may vary among different histologies, as some well-differentiated tumors can go undetected because of their low metabolism [9].

Radiomics is an emerging branch of artificial intelligence (AI) that involves converting digital medical images that contain information related to tumor pathophysiology, also known as features, into measurable and quantifiable data. These data, combined with clinical and qualitative imaging-derived data, can improve medical decision making [10].

The field of radiomics is constantly and rapidly evolving. The purpose of AI is to aid the physician in the assessment of lesions beyond subjective visual interpretation in order to obtain additional information about tumor behavior and pathophysiology that is otherwise not inferable by the human eye with currently used techniques.

As a topic of relatively recent emergence and application, there is considerable variability in the workflow that determines the results of radiomics-related studies. For traditional radiomics approaches, the workflow is divided into specific steps: data selection, medical imaging evaluation/segmentation, feature extraction, exploratory analysis, and modeling. The acquisition technical specifications and medical image reading modalities, the software and how the segmentation of the regions of interest (ROIs) is produced, the feature extraction, and the algorithm of the predictive model are all subject to numerous factors, making the research, and therefore the literature on it, highly heterogeneous. The radiomics quality score (RQS) was introduced in order to evaluate the past and future radiomics studies by achieving homogeneity in study reporting [11].

The purpose of this review is to investigate the potential role of radiomics as a decision-supporting tool in predicting bone disease status, differentiating benign from malignant bone lesions, and characterizing malignant lesions at the genetic level.

2. Materials and Methods

MEDLINE databases, such as PubMed and Web of Science, were employed for the research, using the following strings: (("radiomics" OR "machine learning") AND (metastases OR metastasis) AND ("bone" OR "spine" OR "spinal")).

No limitations were applied to the search strategy. The following criteria were used for the inclusion of the studies: (a) imaging analysis involved only CT and MRI modalities; (b) the studies addressed the ability of radiomics to predict, diagnose, or characterize bone lesions; (c) the studies involved humans only; (d) the articles were accessible through our institution; and (e) the publications were in English.

Case studies, abstracts, reviews, letters to editors, editorials, and commentaries were excluded. We completed the search by manually reviewing the bibliography of all selected articles.

Two reviewers conducted the search, selected the studies, and extracted data from each study independently. From a total of 100 articles, 13 research articles were considered suitable and then collected.

The quality of the methodology was assessed according to the RQS as described by Lambin et al. [11].

Each of the 16 criteria, covering individual aspects of the radiomics workflow, was assigned a different maximum score in relation to its importance. The absence of feature selection and validation results in a reduction in the final score by −3 and −5 points, respectively. The two reviewers assigned, in agreement, the RQS to the selected studies in absolute and percentage values (maximum value of 36, representing 100%).

The following data were extracted from each study: title, authors, year and journal of publication, study objective, study design (retrospective or prospective), number of patients, CT and MRI technical information, software used for segmentation and feature selection, number and type of radiomics features considered, algorithms used for classification, summary of results, and RQS.

3. Results

Our search found 13 publications on radiomics as a decision support tool regarding bone lesions. All studies were published between 2018 and 2021 and were retrospective in design. Study characteristics, as recorded by the reviewers, are shown in Table 1.

Eleven (85%) studies were MRI-based, whereas two (15%) were CT-based. Four (30%) studies were focused on whether radiomics could predict epidermal growth factor receptor (EGFR) mutation in spinal metastases of primary lung adenocarcinoma. Three (23%) studied bone metastases from prostate cancer: two aimed to predict the presence of bone metastases from prostate cancer, one studied the prognostic role in terms of overall survival (OS) and cause-specific survival (CSS) of radiomics in prostate cancer patients with bone metastases. Four (30%) studies aimed to differentiate bone metastases from other pathological conditions: two studies evaluated the ability of radiomics to differentiate bone metastases from benign vertebral bone disease, and two studies evaluated the ability of radiomics to differentiate bone metastases from other pathological bone lesions. One (7%) study aimed to differentiate between metastatic and nonmetastatic vertebral bodies, and one aimed to differentiate between metastatic lesions in the spine originating from lung cancer and other nonpulmonary cancers.

3.1. EGFR Mutation Prediction in Spinal Metastasis from Primary Lung Adenocarcinoma

Jiang et al. [12] analyzed MRI-based multiparametric radiomics for EGFR mutation prediction on T2-weighted (T2W), T2-weighted fat-saturated (T2W-FS), and T1-weighted (T1W) images: both traditional handcrafted and deep learning-based features were derived from each MRI sequence. For each of the two types of approach, radiomics models showed better results using combined features from all the MRI sequences than those with features extracted from each individual sequence. A fusion model created by integrating traditional handcrafted and deep learning-based features from the three sequences achieved the best prediction performance. A radiomics nomogram was obtained by integrating the best performing radiomics features: a decision curve analysis (DCA) confirmed the potential clinical utility of the radiomics nomogram.

Table 1. Characteristics of the selected radiomics studies.

Authors	Publication Year	Objective	Journal	Number of Patients	Imaging Modality	Segmentation	Technique Used for Feature Selection	Validation	Classification	Features	Best Results	Calibration Statistics	Decision Curve Analysis	RQS
Jiang X et al. [12]	2021	Detect EGFR mutation in spinal metastasis in patients with primary lung adenocarcinoma	*Journal of Magnetic Resonance Imaging*	97	3T MRI T1W, T2W, T2W-FS	Manual, ITK-SNAP	Mann–Whitney U-test, LASSO, 10-fold cross-validation	Y	Logistic regression models	Handcrafted features: first-order, shape- and size- based, texture, filtered features	Fusion features: AUC = 0.771, ACC = 0.550, SEN = 0.750, SPE = 0.875	Y	Y	10/36 = 27.7%
Ren M et al. [13]	2021	Detect EGFR mutation in spinal metastasis in patients with primary lung adenocarcinoma	*Medical Physics*	110	3T MRI T1W, T2W, T2W-FS	Manual, ITK-SNAP	Intraclass correlation coefficient (ICC) analysis, Mann–Whitney U, LASSO, 10-fold cross-validation	Y	Logistic regression, random forest, neural network, and support vector machine	First-order, shape-based, and texture (1967)	Fusion features: AUC = 0.803 (0.682–0.924), SEN = 0.700, SPE = 0.818; nomogram, AUC = 0.882 (0.695–0.974), ACC = 0.808, SEN = 0.846, SPE = 0.846	Y	Y	11/36 = 30.5%
Fan Y et al. [14]	2021	Detect EGFR mutation in spinal metastasis in patients with primary lung adenocarcinoma	*Physics in Medicine & Biology*	94	3T MRI, T1W, T2W-FS	Manual, ITK-SNAP	Mann–Whitney U-test, LASSO, 10-fold cross-validation	Y	Logistic regression models	First-order, shape- and size- based, texture, high-dimensional features (1595)	Multiregional radiomics signature: AUC = 0.777 (0.612–0.967), ACC = 0.688, SEN = 0.615, SPE = 0.947	N	N	8/36 = 22.2%
Ran C et al. [15]	2020	Detect EGFR mutation subtypes in exons 19 and 21 in spinal metastasis in patients with primary lung adenocarcinoma	*Academic Radiology*	76	3T MRI, T1W, T2W-FS	Manual	Mann–Whitney U, LASSO, 10-fold cross-validation	Y	Logistic regression models	First-order, shape-based, and texture (1967)	T1W: AUC = 0.728 (0.526 0.903), ACC = 0.692, SEN = 0.692, SPE = 0.769; T2W-FS: AUC = 0.852 (0.706 0.998), ACC = 0.731, SEN = 0.846, SPE = 0.769; nomogram, AUC = 0.821 (0.692–0.929), SEN = 0.667, SPE = 0.909	Y	Y	12/36 = 33.3%
Wang Y et al. [16]	2019	Pretreatment prediction of bone metastasis in patients with prostate cancer	*Magnetic Resonance Imaging*	176	3T MRI T2W, T1W DCE	Manual, IBEX	Linear regression, ridge regression, logistic regression models	Y	Linear regression, ridge regression, logistics regression models	Shape, intensity, intensity histogram, GLCM, gray-level run (976)	Combined T2W and DCE: AUC = 0.898 (0.833–0.937), ACC = 0.821, SEN = 0.647, SPE = 0.782	Y	N	8/36 = 22.2%
Hayakawa T et al. [17]	2020	Investigate the potential prognostic value of clinical risk factors, image features, and radiomics of pelvic bone metastasis in newly diagnosed prostate cancer patients	*Japanese Journal of Radiology*	69	CT	Manual, 3D Slicer	N	N	Not available	Shape-based, first-order statistics, texture (105)	Maximum 2D diameter and least axis were detected as risk factors for OS (HR 1.007 and 1.013, respectively)	N	N	0/36 = 0%
Zhang W et al. [18]	2020	Pretreatment prediction of bone metastasis in patients with prostate cancer	*BMC Medical Imaging*	116	3T MRI, T2W-FS, DWI, DCE T1W	Manual, AK software	ANOVA	Y	Logistic regression models	Not available (204)	AUC = 0.84	Y	Y	14/36 = 38.8%
Sun W et al. [19]	2021	Distinguish between benign and malignant bone tumors	*Cancer Imaging*	206	CT	Manual, ITK-SNAP	LASSO	Y	Logistic regression models	Shape, statistical, texture, wavelet (1130)	Radiomic model, AUC = 0.781 (0.643–0.918); nomogram, AUC = 0.917	Y	Y	12/36 = 33.3%

Table 1. *Cont.*

Authors	Publication Year	Objective	Journal	Number of Patients	Imaging Modality	Segmentation	Technique Used for Feature Selection	Validation	Classification	Features	Best Results	Calibration Statistics	Decision Curve Analysis	RQS
Xiong X et al. [20]	2021	Differentiating between multiple myeloma and different tumor metastasis lesions of the lumbar vertebra	*Frontiers in Oncology*	107	3T MRI, T1W, T2W-FS	Manual	LASSO, 10-fold cross-validation	Y	Support vector machine, k-nearest neighbor, random forest, artificial neural networks (ANNs), and naïve Bayes	Histogram features, GLCM, GRLM, and an autoregressive model (282)	Differentiating myeloma and metastasis, ANN T2W-FS: AUC = 0.815, SEN = 0.879, SPE = 0.790; differentiating myeloma and metastasis subtypes, ANN T2W-FS: AUC = 0.648, SEN = 0.714, SPE = 0.775	N	N	8/36 = 22.2%
Yin P et al. [21]	2018	Differentiation between primary sacral chordoma, sacral giant cell tumor, and sacral metastatic tumor	*Journal of Magnetic Resonance Imaging*	167	3T MRI, T2W-FS, T1W CE	Manual, ITK-SNAP	ANOVA, LASSO, Pearson correlation, random forest	Y	Random forest	Histogram features, form factor features, Haralick, GLCM features, RLM (385).	Combined T2W and T1W CE: AUC = 0.773, ACC = 0.711; T2W, AUC = 0.678, ACC = 0.541; T1W CE, AUC = 0.592, ACC = 0.568	Y	N	9/36 = 25%
Zhong X et al. [22]	2020	Differentiating of cervical spine osteoradionecrosis from metastasis after radiotherapy in nasopharyngeal carcinoma	*BMC Medical Imaging*	123	1.5 MRI, T1W CE	Manual, MaZda	Intraclass correlation coefficient (ICC) analysis, combination feature selection algorithm (combination of Fisher coefficient, classification error probability combined with average correlation coefficients, and mutual information), LASSO, 10-fold cross-validation	Y	Logistic regression models	Histogram, gray-level co-occurrence matrix, run-length matrix, absolute gradient, autoregressive model, and wavelet (279)	Nomogram: AUC = 0.720 (0.573–0.867), ACC = 0.723, SEN = 0.800, SPE = 0.640	Y	Y	11/36 = 30.5%
Filograna L et al. [23]	2019	Differentiate between metastatic and nonmetastatic vertebral bodies in patients with bone marrow metastatic disease	*La Radiologia Medica*	8	1.5 MRI, T1W, T2W-FS	Not available	Wilcoxon test	N	Logistic regression models	Statistical/histogram, morphological, and textural features (89)	T1W: AUC = 0.814 (0.685–0.942); T2W: AUC = 0.911 (0.829–0.993)	N	N	2/36 = 5.5%
Lang N et al. [24]	2019	Differentiate metastatic cancer in the spine originated from lung cancer and other nonlung tumors	*Magnetic Resonance Imaging*	61	3T MRI DCE	Manual, automatic	Random forest algorithm	N	Logistic regression models	Texture, histogram (33 × 3 maps)	3 features, histogram + texture: ACC = 0.68; 5 features, histogram + texture: ACC = 0.71;	N	N	1/36 = 2.7%

Ren et al. [13] produced a nomogram using an MRI-based radiomics signature and smoking status to classify patients with EGFR mutation and wild-type EGFR through analysis of spinal metastases on T2W, T2W-FS, and T1W images. In addition to the radiomics model, a deep learning approach was considered: the combined signature generated higher AUCs than either feature type alone. Four different machine learning classifiers were developed and compared, with logistic regression outperforming the others. The nomogram achieved an AUC of 0.821 (SEN = 0.667, SPE = 0.909): DCA showed that the nomogram had a higher net benefit than all treatment and nontreatment strategies when the threshold was greater than 0.013.

Fan et al. [14] proposed a predictive model that could determine the presence of EGFR mutation in spinal metastasis subregions. Spinal metastases were divided into subregions based on patient- and population-level clustering: marginal, fragmentary, and internal subregions and the total tumor region. Radiomics features were extracted from the subregions' T2W-FS and T1W images. For both sequences, the radiomics signature derived from the inner subregions outperformed other subregions or the entire tumor regions in terms of AUC: the multiregion radiomics signature derived from merging the inner subregion from T1W and T2W-FS MRI achieved the best detection capabilities. The results suggest that the inner region is biologically more aggressive than the others.

Ran et al. [15] further investigated the predictive ability of the EGFR mutation in spinal metastases by constructing a radiomics model that could identify the mutation subtype in exon 19 and exon 21. The radiomics signature derived from the T2W-FS MRI consistently outperformed the T1W-derived signature in terms of AUC, ACC, sensitivity, and specificity. A nomogram model was constructed by incorporating the combined radiomic signature, age, and CEA level, achieving an AUC of 0.881 in the validation set: a decision curve analysis (DCA) confirmed that the model potentially guides individual treatments for patients with lung adenocarcinoma.

3.2. Bone Metastasis from Prostate Cancer

Wang et al. [16] determined that multiparametric prostate MRI predicted the presence/absence of bone metastasis in prostate cancer patients using radiomics features alone and combined with free PSA level and Gleason score. The combined MRI features derived from T2W and DCE showed higher prognostic performance than features derived from the single sequence and Gleason score. The radiomics MRI model combined with clinicopathological features (free PSA level, age, and Gleason score) yielded the highest AUC (AUC = 0.916), further improving predictive performance.

Hayakawa et al. [17] investigated the potential prognostic value of clinical risk factors (anamnestic and laboratory data and histological prostate cancer characteristics), imaging features, and radiomics of pelvic bone metastases in patients with newly diagnosed prostate cancer: patients were studied for OS and CSS. Only shape-based features were detected as risk factors for OS, and "maximum 2D diameter", defined as the largest tumor surface dimension in the axial plane, was detected as a risk factor for OS after multivariate analysis (HR = 1.007). None of the radiomics features were detected as a risk factor for CSS in the uni- and multivariate analysis. After multivariate analysis, LDH, hemoglobin, and "maximum 2D diameter" were detected as risk factors for OS, whereas total Gleason score, LDH, and maximum 2D diameter were detected as a risk factors for CSS. Radiation therapy to the prostate gland and bone metastases did not significantly improve both OS and CSS.

Zhang et al. [18] established and validated a radiomics model that combined prostate multiparametric MRI-based radiomics signature and clinical risk factors to predict bone metastasis in patients with prostate cancer before treatment. The radiomics signature constructed from features extracted from DWI, T2W-FS, and DCE images showed good predictive efficiency. The nomogram, which incorporated the radiomics signature based on MRI and clinical risk factors, had an AUC of 0.92 in the validation set. DCA also demonstrated the clinical use of the radiomics model, which had better discriminatory efficiency than t-PSA or radiomics signature alone.

3.3. Differentiation of Bone Metastases from Other Bone Diseases

Sun et al. [19] proposed a CT-based nomogram able to distinguish between benign and malignant bone tumors. The nomogram, obtained by combining the radiomics signature and clinical model (consisting of demographics and CT characteristics), had higher diagnostic performance than the clinical model, but there was no statistical difference compared with the radiomics signature (AUC = 0.823 in the validation set). The DCA showed that the nomogram had higher diagnostic performance than the clinical model and achieved greater net clinical benefits than the clinical and radiomics signature models when considered alone.

Xiong et al. [20] evaluated the discrimination ability in T1W and T2W-FS MRI sequences between bone lesions from multiple myeloma and metastasis through several machine learning models: support vector machine (SVM), k-nearest neighbor (KNN), random forest (RF), artificial neural networks (ANNs), and naïve Bayes (NB). The ANN classifier from T2W images showed the best performance, both in differentiating myeloma from metastases and for classifying metastasis subtypes.

Yin et al. [21] developed and validated a multiparametric prostate MRI-based radiomics model to differentiate primary sacral chordoma, giant cell sacral tumor, and metastatic sacral tumor. Radiomics features extracted from the combined T2W-FS and CE T1W images exceeded those from the T2W-FS or T1W images alone, but T2W-FS outperformed T1W images. The highest radiomics model AUC was achieved when clinical and imaging data were combined.

Zhong et al. [22] proposed an MRI-based radiomics nomogram to differentiate cervical spine osteoradionecrosis from metastasis in patients with nasopharyngeal carcinoma after radiotherapy. The nomogram model demonstrated good calibration and discrimination, and DCA indicated that, if the threshold probability of a lesion for diagnosis as osteoradionecrosis is >12%, the radiomics nomogram adds net benefit when compared to either the treat-all-patients scheme or the treat-none scheme.

3.4. Other Studies

The study of Filograna et al. [23] is the only study that demonstrated the ability of radiomics-based MRI to differentiate between metastatic and nonmetastatic vertebral bodies in non-radiotherapy-treated cancer patients with metastatic bone marrow disease from primary tumors of different nature (three lung cancers, one prostate cancer, one esophageal cancer, one nasopharyngeal cancer, one hepatocarcinoma, and one breast cancer). Internal cross-validation showed an AUC of 0.814 for T1W images and 0.911 for T2W images. One histogram feature (minimum gray level) and one textural feature (joint variance of the gray level co-occurrence matrix) were found to be the best-fitting features in T1W and T2W images, respectively.

Lang et al. [24] aimed to differentiate metastatic spine cancer derived from primary lung cancer and other nonpulmonary cancers (breast, thyroid, prostate, liver, kidney) using an ROI-based model, radiomics, and deep learning. The accuracy of the radiomics model when histogram and texture features were combined was higher than that when histogram and texture features were evaluated alone. By increasing the number of features from three to five, the accuracy showed slightly higher values (from 0.68 to 0.71 in the histogram + texture model). The accuracy of the radiomics model was worse than that of the hot-spot ROI-based (ACC = 0.79) and deep learning (ACC = 0.71 ± 0.043 and 0.81 ± 0.034) methods.

3.5. RQS Assessment and Study Limitations

The average recorded RQS was 22.6% (0–38.8%). This low score confirms what has been reported in other reviews in the field of radiomics, representing a relatively low quality of research methodology [25–30]. None of the reviewed studies were prospective in design, no external validation on a dataset from another institution was performed, no cost-effectiveness of the clinical application of the radiomic models was reported, and no datasets were made publicly available (although four authors allowed access to the datasets upon request). Two articles (15%) did not perform any validation of their results. In only four (30%) of the articles, multiple segmentations were performed to assess the robustness of features to segmentation variabilities. The majority of articles (12/13, 92%) performed a feature reduction to decrease the risk of overfitting. Eight (61%) studies reported discrimination statistics (such as ROC curve and/or AUC), and six (46%) studies reported calibration statistics. Six articles (46%) reported on the potential application of the model in a clinical setting with a DCA.

4. Discussion

The application of radiomics in the diagnosis and characterization of bone lesions is recent and constantly evolving, as is the entire field of radiomics. The articles identified by our two researchers are few in number and were all published within the period between 2018 and 2021, with approximately 70% in the period immediately after 2020. Reflecting the relative freshness of this area of research, all studies are retrospective, performed at a single center, and with a small study population, ranging from 8 to 176 patients.

Radiomics can not only predict the presence of bone metastases and differentiate skeletal regions without lesions from those containing metastases, but its application is able to determine the primary tumor, differentiate metastases from other bone lesions (both benign and malignant), and predict mutation status (such as EGFR). Apart from MRI and FDG-PET, which have high predictive values, the other imaging methods have relatively low sensitivity and specificity values, although they are easily accessible and widespread [8,9]. Despite the predictive capabilities of the traditional imaging methods, there is some clinical information regarding bone metastases, including the genetic status or the primary tumor, that the naked eye is not able to perceive, due to similar clinical and imaging manifestations. Complete pathological confirmation and histological analysis are currently only possible by sampling through bone biopsy, which is associated with relatively high procedural risks (such as vertebral artery or spinal cord damage) [31]. Radiomics models, by inferring quantifiable data from the features, allow obtaining information that, once applied in the clinical setting, can be decisive for the specific therapeutic treatment choice. Because data are extracted from noninvasive methods, and in most cases radiation-free methods, radiomics is a further step towards the reduction in a patient diagnostic burden, and at the same time towards a patient-centered medicine. Some studies have also constructed nomograms in order to graphically represent the mathematical relationship between radiomics features and other prognostic factors, both clinical and diagnostic, in order to improve the clinical applicability of a field still difficult for nonexperts to interpret.

All articles included among their limitations the relatively small sample size (<200 patients), the single-center nature of the study, and the selection bias introduced by the retrospective design. Even in studies in which validation was performed on an internal dataset, the absence of external validation leads to reduced evidence of the possible clinical application of the research: multicenter studies are necessary to validate radiomics models and nomograms. Some articles complained about the tediousness of manual segmentation, which, in addition to being time-consuming, is not free of human error despite the option of multiple segmentation: the hope is that the spread of automatic, or semiautomatic, segmentation will speed up the process and further reduce the margin of error.

Our review confirms the considerable heterogeneity in current radiomics research, as evidenced by the relatively low RQS value obtained when analyzing the reviewed studies (22.6%). There is not a homogeneous protocol used for MRI sequences among the different studies, although the highest predictive ability was always obtained in T2W-FS. Wide variability also exists in the software used for image segmentation and feature extraction; the number and the type of features explored, with and without feature selection method application; and even the models used to classify the final features. All of these elements contribute to the reduced reproducibility of the results, even if none of them are considered integral to the RQS assessment.

As described above, the most critical limitation concerns the small sample size, which leads to selection bias. A possible way to overcome this important limit is to increase the number of patients under investigation or to extend the research and results validation to other centers. In fact, it is well known that, after a first validation of a radiomic model, a subsequent path of validation through multicenter studies is necessary to allow radiomics to get closer and closer to widespread clinical applicability, even and especially through prospective studies.

This review has some important limitations. To our knowledge, no other review has exclusively investigated the role of radiomics in the analysis and prediction of bone metastases, particularly the spine localization. Even within the field of radiomics, this is a niche subfield, as is evident from the low number of studies analyzed. This novelty, in addition to the high variability of the included studies, both in methodology and in objectives, prevented us from pursuing a robust meta-analysis. We expect that as radiomics evolves and becomes more widespread, there will be an increase in the number of patients included and more extensive validation of existing datasets. Another critical issue at this early stage of research is the ability to share data across public datasets that have already been validated, as currently none of the papers publicly released their data.

In addition, we have deliberately eliminated from the research the studies based on scintigraphy and PET (we have not detected studies that have used ultrasound) and papers in non-English language or not accessible from our institution, reducing the number of the articles included. Due to an implicit publication bias, most articles on this topic focus on the use of MRI. This implies that many other methods, on which there are no current studies, do not result in a significant contribution to research in the radiomics field, a phenomenon that introduces further bias into our review. Furthermore, at the time of publication, it is safe to assume that there are additional feature extraction software and classification models currently in development that we are unaware of in the literature, which are therefore protected from our review.

5. Conclusions

In spite of the variability in the radiomics method application, the similarity of results and conclusions observed is encouraging. Furthermore, all six studies that have measured the possible application of the radiomics model in the clinical setting through DCA have shown a net benefit compared to the use of the other strategies alone, confirming the promising role of radiomics in guiding the choice of treatments for individual cancer patients.

Author Contributions: Conceptualization, E.F. and V.D.; methodology, A.C.; software, A.C. and P.S.; validation, D.S., E.F. and G.V.; formal analysis, P.S.; investigation, D.S.; resources, A.C. and F.R.; data curation, A.C.; writing—original draft preparation, A.C. and D.S.; writing—review and editing, D.S. and G.V.; visualization, E.F.; supervision, G.I., B.B.Z. and C.d.F.; project administration, E.F. and G.V.; funding acquisition, V.D. All authors have read and agreed to the published version of the manuscript.

Funding: This research was funded by the Research Grants (BRiC-2018 ID3 and BRIC-2021 ID4) of the Italian Workers' Compensation Authority (INAIL).

Institutional Review Board Statement: Not applicable.

Informed Consent Statement: Not applicable.

Data Availability Statement: The study did not report any data.

Conflicts of Interest: The authors declare no conflict of interest.

References

1. Coleman, R.E. Metastatic Bone Disease: Clinical Features, Pathophysiology and Treatment Strategies. *Cancer Treat. Rev.* **2001**, *27*, 165–176. [CrossRef] [PubMed]
2. Cecchini, M.G.; Wetterwald, A.; van der Pluijm, G.; Thalmann, G.N. Molecular and Biological Mechanisms of Bone Metastasis. *EAU Update Ser.* **2005**, *3*, 214–226. [CrossRef]
3. Selvaggi, G.; Scagliotti, G.V. Management of Bone Metastases in Cancer: A Review. *Crit. Rev. Oncol./Hematol.* **2005**, *56*, 365–378. [CrossRef]
4. Yu, H.-H.M.; Tsai, Y.-Y.; Hoffe, S.E. Overview of Diagnosis and Management of Metastatic Disease to Bone. *Cancer Control J. Moffitt Cancer Cent.* **2012**, *19*, 84–91. [CrossRef]
5. Bussard, K.M.; Gay, C.V.; Mastro, A.M. The Bone Microenvironment in Metastasis; What Is Special about Bone? *Cancer Metastasis Rev.* **2007**, *27*, 41–55. [CrossRef]
6. Coleman, R. The Role of Zoledronic Acid in Cancer: Clinical Studies in the Treatment and Prevention of Bone Metastases. *Semin. Oncol.* **2001**, *28*, 11–16. [CrossRef]
7. Cuccurullo, V.; Lucio Cascini, G.; Tamburrini, O.; Rotondo, A.; Mansi, L. Bone Metastases Radiopharmaceuticals: An Overview. *Curr. Radiopharm.* **2013**, *6*, 41–47. [CrossRef]
8. O'Sullivan, G.J. Imaging of Bone Metastasis: An Update. *World J. Radiol.* **2015**, *7*, 202. [CrossRef]
9. Talbot, J.N.; Paycha, F.; Balogova, S. Diagnosis of bone metastasis: Recent comparative studies of imaging modalities. *Q. J. Nucl. Med. Mol. Imaging* **2011**, *55*, 374–410.
10. Lambin, P.; Rios-Velazquez, E.; Leijenaar, R.; Carvalho, S.; van Stiphout, R.G.P.M.; Granton, P.; Zegers, C.M.L.; Gillies, R.; Boellard, R.; Dekker, A.; et al. Radiomics: Extracting More Information from Medical Images Using Advanced Feature Analysis. *Eur. J. Cancer* **2012**, *48*, 441–446. [CrossRef]
11. Lambin, P.; Leijenaar, R.T.H.; Deist, T.M.; Peerlings, J.; de Jong, E.E.C.; van Timmeren, J.; Sanduleanu, S.; Larue, R.T.H.M.; Even, A.J.G.; Jochems, A.; et al. Radiomics: The Bridge between Medical Imaging and Personalized Medicine. *Nat. Rev. Clin. Oncol.* **2017**, *14*, 749–762. [CrossRef] [PubMed]
12. Jiang, X.; Ren, M.; Shuang, X.; Yang, H.; Shi, D.; Lai, Q.; Dong, Y. Multiparametric MRI -Based Radiomics Approaches for Preoperative Prediction of EGFR Mutation Status in Spinal Bone Metastases in Patients with Lung Adenocarcinoma. *J. Magn. Reson. Imaging* **2021**, *54*, 497–507. [CrossRef] [PubMed]
13. Ren, M.; Yang, H.; Lai, Q.; Shi, D.; Liu, G.; Shuang, X.; Su, J.; Xie, L.; Dong, Y.; Jiang, X. MRI-Based Radiomics Analysis for Predicting the EGFR Mutation Based on Thoracic Spinal Metastases in Lung Adenocarcinoma Patients. *Med. Phys.* **2021**, *48*, 5142–5151. [CrossRef]
14. Fan, Y.; Dong, Y.; Yang, H.; Chen, H.; Yu, Y.; Wang, X.; Wang, X.; Yu, T.; Luo, Y.; Jiang, X. Subregional Radiomics Analysis for the Detection of the EGFR Mutation on Thoracic Spinal Metastases from Lung Cancer. *Phys. Med. Biol.* **2021**, *66*, 215008. [CrossRef] [PubMed]
15. Cao, R.; Dong, Y.; Wang, X.; Ren, M.; Wang, X.; Zhao, N.; Yu, T.; Zhang, L.; Luo, Y.; Cui, E.N.; et al. MRI-Based Radiomics Nomogram as a Potential Biomarker to Predict the EGFR Mutations in Exon 19 and 21 Based on Thoracic Spinal Metastases in Lung Adenocarcinoma. *Acad. Radiol.* **2021**, *29*, e9–e17. [CrossRef] [PubMed]
16. Wang, Y.; Yu, B.; Zhong, F.; Guo, Q.; Li, K.; Hou, Y.; Lin, N. MRI-Based Texture Analysis of the Primary Tumor for Pre-Treatment Prediction of Bone Metastases in Prostate Cancer. *Magn. Reson. Imaging* **2019**, *60*, 76–84. [CrossRef]
17. Hayakawa, T.; Tabata, K.; Tsumura, H.; Kawakami, S.; Katakura, T.; Hashimoto, M.; Watanabe, Y.; Iwamura, M.; Hasegawa, T.; Ishiyama, H. Size of Pelvic Bone Metastasis as a Significant Prognostic Factor for Metastatic Prostate Cancer Patients. *Jpn. J. Radiol.* **2020**, *38*, 993–996. [CrossRef]
18. Zhang, W.; Mao, N.; Wang, Y.; Xie, H.; Duan, S.; Zhang, X.; Wang, B. A Radiomics Nomogram for Predicting Bone Metastasis in Newly Diagnosed Prostate Cancer Patients. *Eur. J. Radiol.* **2020**, *128*, 109020. [CrossRef]
19. Sun, W.; Liu, S.; Guo, J.; Liu, S.; Hao, D.; Hou, F.; Wang, H.; Xu, W. A CT-Based Radiomics Nomogram for Distinguishing between Benign and Malignant Bone Tumours. *Cancer Imaging* **2021**, *21*, 20. [CrossRef]
20. Xiong, X.; Wang, J.; Hu, S.; Dai, Y.; Zhang, Y.; Hu, C. Differentiating between Multiple Myeloma and Metastasis Subtypes of Lumbar Vertebra Lesions Using Machine Learning–Based Radiomics. *Front. Oncol.* **2021**, *11*, 601699. [CrossRef]
21. Yin, P.; Mao, N.; Zhao, C.; Wu, J.; Chen, L.; Hong, N. A Triple-Classification Radiomics Model for the Differentiation of Primary Chordoma, Giant Cell Tumor, and Metastatic Tumor of Sacrum Based on T2-Weighted and Contrast-Enhanced T1-Weighted MRI. *J. Magn. Reson. Imaging* **2018**, *49*, 752–759. [CrossRef] [PubMed]
22. Zhong, X.; Li, L.; Jiang, H.; Yin, J.; Lu, B.; Han, W.; Li, J.; Zhang, J. Cervical Spine Osteoradionecrosis or Bone Metastasis after Radiotherapy for Nasopharyngeal Carcinoma? The MRI-Based Radiomics for Characterization. *BMC Med. Imaging* **2020**, *20*, 104. [CrossRef]
23. Filograna, L.; Lenkowicz, J.; Cellini, F.; Dinapoli, N.; Manfrida, S.; Magarelli, N.; Leone, A.; Colosimo, C.; Valentini, V. Identification of the Most Significant Magnetic Resonance Imaging (MRI) Radiomic Features in Oncological Patients with Vertebral Bone Marrow Metastatic Disease: A Feasibility Study. *La Radiol. Med.* **2018**, *124*, 50–57. [CrossRef] [PubMed]
24. Lang, N.; Zhang, Y.; Zhang, E.; Zhang, J.; Chow, D.; Chang, P.; Yu, H.J.; Yuan, H.; Su, M.-Y. Differentiation of Spinal Metastases Originated from Lung and Other Cancers Using Radiomics and Deep Learning Based on DCE-MRI. *Magn. Reson. Imaging* **2019**, *64*, 4–12. [CrossRef] [PubMed]

25. Fiz, F.; Viganò, L.; Gennaro, N.; Costa, G.; La Bella, L.; Boichuk, A.; Cavinato, L.; Sollini, M.; Politi, L.S.; Chiti, A.; et al. Radiomics of Liver Metastases: A Systematic Review. *Cancers* **2020**, *12*, 2881. [CrossRef] [PubMed]
26. Stanzione, A.; Gambardella, M.; Cuocolo, R.; Ponsiglione, A.; Romeo, V.; Imbriaco, M. Prostate MRI Radiomics: A Systematic Review and Radiomic Quality Score Assessment. *Eur. J. Radiol.* **2020**, *129*, 109095. [CrossRef] [PubMed]
27. Chetan, M.R.; Gleeson, F.V. Radiomics in Predicting Treatment Response in Non-Small-Cell Lung Cancer: Current Status, Challenges and Future Perspectives. *Eur. Radiol.* **2020**, *31*, 1049–1058. [CrossRef] [PubMed]
28. Valdora, F.; Houssami, N.; Rossi, F.; Calabrese, M.; Tagliafico, A.S. Rapid Review: Radiomics and Breast Cancer. *Breast Cancer Res. Treat.* **2018**, *169*, 217–229. [CrossRef]
29. Calabrese, A.; Santucci, D.; Landi, R.; Beomonte Zobel, B.; Faiella, E.; de Felice, C. Radiomics MRI for Lymph Node Status Prediction in Breast Cancer Patients: The State of Art. *J. Cancer Res. Clin. Oncol.* **2021**, *147*, 1587–1597. [CrossRef]
30. Kendrick, J.; Francis, R.; Hassan, G.M.; Rowshanfarzad, P.; Jeraj, R.; Kasisi, C.; Rusanov, B.; Ebert, M. Radiomics for Identification and Prediction in Metastatic Prostate Cancer: A Review of Studies. *Front. Oncol.* **2021**, *11*, 4489. [CrossRef]
31. Peh, W. CT-Guided Percutaneous Biopsy of Spinal Lesions. *Biomed. Imaging Interv. J.* **2006**, *2*, e25. [CrossRef] [PubMed]

International Journal of
Environmental Research and Public Health

MDPI

Review

Artificial Intelligence and Computer Vision in Low Back Pain: A Systematic Review

Federico D'Antoni [1], Fabrizio Russo [2,*], Luca Ambrosio [2], Luca Vollero [1], Gianluca Vadalà [2], Mario Merone [1,*], Rocco Papalia [2] and Vincenzo Denaro [2]

1 Unit of Computer Systems and Bioinformatics, Università Campus Bio-Medico di Roma, Via Alvaro Del Portillo 21, 00128 Rome, Italy; f.dantoni@unicampus.it (F.D.); l.vollero@unicampus.it (L.V.)

2 Department of Orthopaedic Surgery, Università Campus Bio-Medico di Roma, Via Alvaro Del Portillo 200, 00128 Rome, Italy; l.ambrosio@unicampus.it (L.A.); g.vadala@unicampus.it (G.V.); r.papalia@unicampus.it (R.P.); v.denaro@unicampus.it (V.D.)

* Correspondence: fabrizio.russo@unicampus.it (F.R.); m.merone@unicampus.it (M.M.)

Abstract: Chronic Low Back Pain (LBP) is a symptom that may be caused by several diseases, and it is currently the leading cause of disability worldwide. The increased amount of digital images in orthopaedics has led to the development of methods related to artificial intelligence, and to computer vision in particular, which aim to improve diagnosis and treatment of LBP. In this manuscript, we have systematically reviewed the available literature on the use of computer vision in the diagnosis and treatment of LBP. A systematic research of PubMed electronic database was performed. The search strategy was set as the combinations of the following keywords: "Artificial Intelligence", "Feature Extraction", "Segmentation", "Computer Vision", "Machine Learning", "Deep Learning", "Neural Network", "Low Back Pain", "Lumbar". Results: The search returned a total of 558 articles. After careful evaluation of the abstracts, 358 were excluded, whereas 124 papers were excluded after full-text examination, taking the number of eligible articles to 76. The main applications of computer vision in LBP include feature extraction and segmentation, which are usually followed by further tasks. Most recent methods use deep learning models rather than digital image processing techniques. The best performing methods for segmentation of vertebrae, intervertebral discs, spinal canal and lumbar muscles achieve Sørensen–Dice scores greater than 90%, whereas studies focusing on localization and identification of structures collectively showed an accuracy greater than 80%. Future advances in artificial intelligence are expected to increase systems' autonomy and reliability, thus providing even more effective tools for the diagnosis and treatment of LBP.

Keywords: low back pain; orthopaedics; artificial intelligence; computer vision; digital image processing; deep learning; decision support systems; computer aided diagnosis

Citation: D'Antoni, F.; Russo, F.; Ambrosio, L.; Vollero, L.; Vadalà, G.; Merone, M.; Papalia, R.; Denaro, V. Artificial Intelligence and Computer Vision in Low Back Pain: A Systematic Review. *IJERPH* **2021**, *18*, 10909. https://doi.org/10.3390/ijerph182010909

Academic Editor: Oliver Faust

Received: 7 September 2021
Accepted: 9 October 2021
Published: 17 October 2021

Publisher's Note: MDPI stays neutral with regard to jurisdictional claims in published maps and institutional affiliations.

1. Introduction

In the last decade, a significant increase in the use of Artificial Intelligence (AI) has been experienced in the most disparate fields, ranging from vocal assistants commonly employed during our daily life to self-driving cars. Thanks to the unique ability of intelligent machines to be trained and automatically acquire new tasks based on previous experience or provided data, the use of AI is being increasingly investigated for applications in medical research [1]. Indeed, AI-based computers have already shown to potentially revolutionize drug design and discovery [2,3], automatic segmentation and relevant data extraction from radiological datasets [4] as well as the formulation of diagnosis, outcome prediction and treatment planning in different medical fields [5–7]. The adoption of this ground-breaking technology is being explored in spine surgery as well [1]. Indeed, thanks to its interdisciplinary nature and the wide utilization of radiological images to inspect the anatomical structures of the spine, the use of AI may be of particular value in determining, for example, which are the pathological discs [8], classifying a scoliotic curve [9] and predict its progression [10]. In this study, we have systematically reviewed the available literature on the use of AI, and more specifically computer vision, in the prevention, diagnosis, and treatment of chronic Low Back Pain (LBP).

LBP is mainly caused by intervertebral disc degeneration, and it is currently the leading cause of disability worldwide, as well as the most common reason for workers' compensation claims [11]. AI has improved the clinical practice with regards to the treatment, prevention and outcome prediction of subjects suffering from LBP. This is mainly due to the ever-growing amount of clinical data available to practitioners, which allow to train and develop increasingly sophisticated AI methodologies. With particular regards to LBP, a huge amount of digital clinical images are gathered daily in order to detect signs of disease in the spinal structures. For this reason, several machine learning algorithms have been developed in recent years in order to speed-up the diagnostic process and to optimize patients' recovery. The latest AI improvements were accompanied by the outbreak of deep learning and by an increase of computing capacity, which allow to develop models that are getting more and more autonomous and accurate. In particular, computer vision techniques applied to clinical images allow to detect some image features that are invisible to the human eye. The importance of computer vision in relation to LBP is multifaceted: it allows to perform a plethora of tasks that may improve the clinical practice, such as automatically localizing and detecting lumbar structures with segmentation. Moreover, it allows to extract a set of features from the image that can be used as an input for further machine learning algorithms in order to provide a decision support to the physician or, in other cases, directly suggest the most appropriate diagnosis. For this reason, we have systematically reviewed the available literature on the application of computer vision on the diagnosis and treatment of LBP in order to describe the state of the art of such technology and its potential applications.

2. Materials and Methods

In order to perform an exhaustive research of AI articles related to LBP, we performed a query research on PubMed (Query research used: (((Artificial intelligence [Title/Abstract]) OR ((feature extraction[Title/Abstract]) OR ((segmentation[Title/Abstract]) OR (Computer Vision[Title/Abstract]) OR (Machine learning[Title/Abstract])) OR (deep learning[Title/Abstract]) OR (neural network[Title/Abstract]))) AND ((Low Back Pain [Title/Abstract]) OR (lumbar[Title/Abstract]))). All the search words had to be included in the title or in the abstract of the articles: the terms "low back pain" and "lumbar" were considered for the pathological part, and the terms "artificial intelligence", "feature extraction", "segmentation", "computer vision", "machine learning", "deep learning" and "neural network" were considered for the AI part. We selected all the articles that included at least one term of the pathological part and at least one term of the artificial intelligence part in their title or abstract.

2.1. Inclusion and Exclusion Criteria

The aim of this work was to gather all the works concerning the utilization of AI, and particularly of computer vision, in the diagnosis, prevention, and treatment of chronic LBP and related diseases. Straightforwardly, all the selected articles had to meet all the following inclusion criteria:

- Chronic LBP or lumbar diseases must have been among the main topics of the article. We included works on the prevention, diagnosis or treatment of chronic LBP and treating at least one of the structures involved in LBP (i.e., vertebrae, discs, muscles);
- AI must have been used in the work with application to clinical images. We included articles exploiting AI methods falling in the areas of computer vision, machine learning and artificial Neural Networks (NNs);
- Subjects of the study: all the articles must have been based on studies of human low back and related pathology, regardless of the age or employment of the subjects included in the study;

- Language: all articles must have been written in English.

Conversely, articles that were excluded did not meet the inclusion criteria for one of the following reasons:

- A different medical problem was considered: we excluded articles which did not consider chronic LBP and its related physical structures and medical data. For example, we excluded studies that considered only cervical or thoracic vertebrae, or that focused on osteoporosis, metastases, traumatic LBP, and other causes of non-discogenic LBP;
- AI was not considered: some articles in the search results proposed definitions and practice for LBP based only on medical observation without utilization of AI;
- Computer vision and clinical images were not considered in the study, regardless of whether AI was utilized for developing diagnosis or support systems;
- Animal studies: we excluded studies based on vertebral structures of animals;
- Embryonal studies: we excluded studies performed on embryos and concerning the embryogenesis of spinal structures.

A preliminary screening of the article selection allowed us to define three main categories in which the utilization of AI in LBP might be split, namely computer vision, computer aided diagnosis, and decision support systems (DSSs) (Figure 1). Computer vision is the field of AI that deals with how computers can gain high-level understanding from digital images or videos. With regards to LBP, its main applications concern feature extraction and image segmentation. Feature extraction is a dimensionality reduction process which is applied to images obtained using Magnetic Resonance Imaging (MRI), ultrasound, X-rays, and Computed Tomography (CT). The main goal of feature extraction is to retrieve a restricted number of relevant features from an image without losing important information, in order to facilitate subsequent tasks such as classification or regression.

Image segmentation is the task of dividing an image into subregions corresponding to different elements of the image. More in depth, the goal of image segmentation is the labeling of each pixel of an image with a corresponding class, e.g., foreground or background, in order to detect the relevant elements of an image. It mainly resorts to two principal techniques: deep learning, in which the image is directly given as input to an artificial NN which is trained on other images to automatically identify subregions, and digital image processing (DIP) techniques, which process digital images to find the edges of different regions based on semantic characteristics, exploiting methods such as gradient thresholding or statistical shape models.

Computer aided diagnosis is a group of techniques which help medical practitioners in identifying a pathology or in quantifying the grade of a disease. It can be split into classification and regression, in which machine or deep learning models are used to assign a predefined label or to generate a numeric output, respectively. In practice, classification is used to identify or categorize a pathology, whereas regression is used to produce a quantitative evaluation of some measure.

Decision support systems (DSSs) are software systems that allow medical practitioners to enhance the decision making and improve the outcome of patients suffering from a specific disease. The goal of the vast majority of DSSs is the outcome prediction, i.e., the prediction of the improvement that a patient would experience after exposure to a defined therapy. By predicting the extent to which a patient would benefit from a specific treatment, DSSs provide the physician with practical tools to assess whether or not surgery may be preferable to conservative treatment. Finally, DSSs can be used for prevention, e.g., by providing the user with recommendations or correct practice for preventing the onset of a disease. It is worth noting that computer vision techniques can be used as preprocessing for developing a DSS, as well as a for computer aided diagnosis.

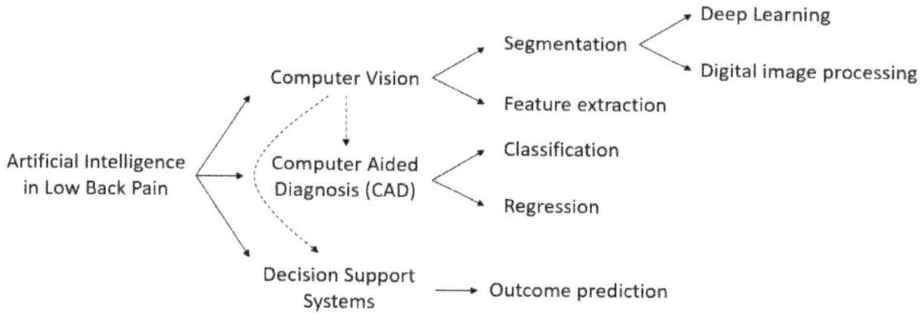

Figure 1. Schematic partitioning of the works concerning the application of AI in LBP.

2.2. Evaluation Metrics

Different tasks use different metrics to evaluate the performance of AI systems. However, considering the large amount of works reported in this review, different metrics were also considered within the same task. With regards to the feature extraction task, no specific evaluation metric was considered. This is because, in most cases, feature extraction is exploited as a preliminary step for further tasks such as classification and regression, and most papers only report the performance for the latter.

With regards to the classification task, we reported the results in terms of accuracy (Acc), where available. For brevity purposes, let us consider a binary classification task, e.g., positive vs. negative. Given a test set composed of N samples, defined the True Positives (TP) as the number of positive samples correctly classified, and the True Negatives (TN) as the number of negative samples correctly classified, then accuracy is defined as:

$$Acc\% = \frac{TP + TN}{N} \times 100 \tag{1}$$

Thus, greater values correspond to a better performance. For each class, recall and precision can be computed as well. Defined the False Positives (FP) and False Negatives (FN) as the number of misclassified positive/negative samples, then recall and precision are computed as:

$$Recall = \frac{TP}{TP + FN} \qquad Precision = \frac{TP}{TP + FP} \tag{2}$$

In binary problems, recall is also called True Positive Rate and corresponds to sensitivity, whereas the True Negative Rate is also called specificity. In the case of multi-class problems, accuracy is computed by considering the TP for each class, and recall and precision per class can be computed. For imbalanced datasets, the F1-Score can be computed for each class. The F1-Score for class c is defined as:

$$F1\text{-}Score_c = \frac{2 \cdot Recall_c \cdot Precision_c}{Recall_c + Precision_c} \tag{3}$$

and takes into account both recall and precision of the class. Another widely used evaluation metric is the Area Under the Curve (AUC), which corresponds to the area under the Receiver Operating Characteristic (ROC) curve showing the performance of a classification model at all classification thresholds, which is plotted considering the True Positive Rate against the False Positive Rate. Its values range from 0 to 1 (the closer to 1, the better the performance).

With regards to the regression task, let us consider a sequence of original values $x(t)$ and a sequence of predicted values $\tilde{x}(t)$. The Mean Absolute Error (MAE) for a sequence of N timestamps is defined as:

$$MAE = \sum_{t=1}^{N} \frac{|x(t) - \tilde{x}(t)|}{N} \tag{4}$$

Thus, the closer to 0 the value, the better the performance. In some cases, percentage error values are used to evaluate performance, the meaning of which varies with the investigated task.

With regards to the segmentation task, two main percent performance indices are used which evaluate to what extent the segmentation result is close to the desired segmentation. As stated, segmentation consists in labeling each pixel of an image. Given two sets of data A and B, corresponding to the desired and the effective segmented areas, the Sørensen–Dice coefficient (DICE) is defined as:

$$DICE(A,B) = \frac{2 \cdot |A \cap B|}{|A| + |B|} \tag{5}$$

where $|A|$ and $|B|$ are the cardinalities of the two sets. It divides the number of common elements of the two sets by the total number of elements of the two sets. When applied to binary data, it is equivalent to the *F1-Score*. Differently, the Jaccard index is defined as:

$$Jaccard(A,B) = \frac{|A \cap B|}{|A \cup B|} \tag{6}$$

and is also known as Intersection Over Union. For both indices, the closer to 100% the value, the better the performance. It is worth noting that $DICE(A,B) \geq Jaccard(A,B)$ for any couple of sets (A,B), and the relation $Jaccard = DICE/(2 - DICE)$ exists to compute one value from the other.

3. Quality of Evidence

The methodological quality of the included studies was graded independently by two reviewers (L.A. and F.R.), and any disagreement was resolved by the intervention of a third reviewer (G.V.) The risks of bias and applicability of the included studies were assessed by using customized assessment criteria based on the Quality Assessment of Diagnostic Accuracy Studies (QUADAS-2) [12]. This tool is based on 4 domains: patient selection, index test, reference standard, and flow and timing. Each domain is evaluated in terms of risk of bias, and the first 3 domains are also assessed in terms of concerns regarding applicability. Sixty-eight studies were rated on a 3-point scale, reflecting concerns about risk of bias and applicability as low, unclear or high, as shown in Figure 2 (the details of analysis are presented in Tables S1 and S2).

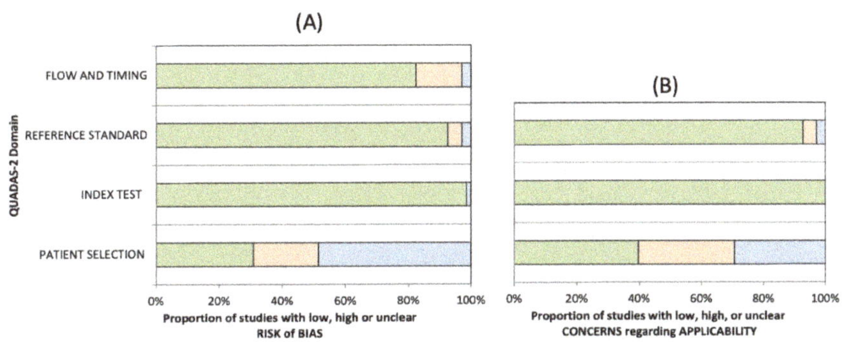

Figure 2. Summary of the methodological quality of included studies regarding the 4 domains assessing the risk of bias (**A**) and the 3 domains assessing applicability concerns (**B**) of the QUADAS-2 score. The portion of studies with a low risk of bias are highlighted in green, studies with an unclear risk of bias are depicted in blue and studies with a high risk of bias are represented in orange.

4. Results

The search was performed on 18 March 2021, and resulted in 558 articles. Nonetheless, many of these articles focused on a different topic from that of this review, so after a first screening based on the article titles and abstracts we reduced the number of eligible articles to 200. A second screening phase was performed after having read the full text of each article, which led the total amount of included articles to 76. We created a flow-chart diagram according to the PRISMA protocol that shows the selection process of the studies (Figure 3). The articles were screened by two independent reviewers and, in the event of discrepancies regarding the inclusion or exclusion of an article, they discussed together until consensus was reached.

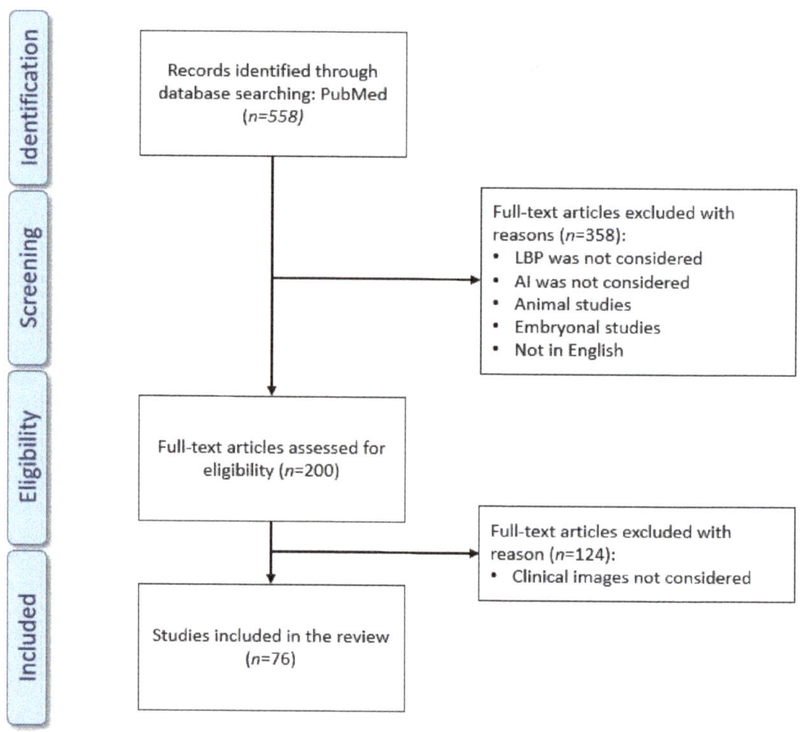

Figure 3. Preferred Reporting Items for Systematic reviews and Meta-Analyses (PRISMA) flow diagram.

It is worth noting how the amount of published work is increasing year by year, and that the number of articles published in 2020 is almost double that of 2019. This may be due to two main reasons: first, the ever-increasing amount of clinical images and data available to researchers and, secondly, the improvement of computing capacity observed in recent years. The final results of the search also include five reviews. One of them, published in 2020 by Tagliaferri et al. [13], is specifically focused on LBP, but considers only the diagnosis and prognosis capability of AI in comparison with the McKenzie and the STarT Back methods, and without taking into account works that exploit clinical images. The other four reviews do not focus specifically on LBP. In detail, in 2019 Tack [14] focused on musculoskeletal medicine in general, and determined in which fields AI had reached human prediction levels; in 2020, Azimi et al. [15] focused on the use of NNs for the treatment of the whole spine; in 2019, Galbusera et al. [1] described the application of AI to problems related to the whole spine; finally, in 2016 Yao et al. [16] performed a multi-center milestone comparative study for vertebral segmentation methods based on CT images. Two articles presenting databases were also found: LUMINOUS, which is a database of ultrasound images from 109 patients for multifidus muscle segmentation [17], and MyoSegmentum, which includes MRI images of 54 patients for the segmentation of lumbar muscles and vertebral bodies [18].

The remainder of this section reports the results of the search that include works concerning computer vision. In particular, we have listed manuscripts that performed a feature extraction task or that performed semantic segmentation, and we have described papers that used DIP/NN approaches in two different subsections.

4.1. Feature Extraction

Feature extraction is a dimensionality reduction process aimed at identifying a restricted set of relevant features in order to improve the predictive capability of a system. In this review, we identified a total of 8 papers, whose main characteristics are reported in Table 1, aiming to extract relevant features from several types of LBP-related images. In detail, we included:

- six articles on MRI (1 of which considers 3D MRI);
- one article on 3D images of the back surface;
- one article on X-ray imaging.

Table 1. Feature extraction. For each work, it is reported whether or not other tasks are performed following feature extraction. The reported results are related to the task following feature extraction. Abbreviations are used for Magnetic Resonance Imaging (MRI), Low Back Pain (LBP), Accuracy (Acc), Mean Absolute Error (MAE), Machine Learning (ML), Support Vector Machine (SVM).

Author/Year	Main Task	Data Type	# Patients	Structures Involved	Results	Model
Adankon, 2012 [19]	Feature Extraction and Classification	3D image of the back surface	165	Vertebrae	Acc = 95%	Local Geometric Descriptors and SVM
Castro-Mateos, 2014 [20]	Feature Extraction and Segmentation	3D MRI	59	Discs	DICE = 88.4%	Statistical shape model space and B-Spline space
Raudner, 2020 [21]	Feature Extraction	MRI	58	Discs	/	GRAPPATINI
Abdollah, 2020 [22]	Feature Extraction	MRI	28	Discs, Vertebrae	/	Random Forest and texture analysis
Yang 2020 [8]	Feature Extraction and Classification	MRI	109	Discs	Acc = 88.3%	Gabor wavelet transformation and KLT feature tracker
Ruiz-España, 2015 [23]	Feature Extraction and Classification	MRI	67	Discs	Acc > 90%	Gradient Vector Flow, several ML models
Ketola, 2020 [24]	Feature Extraction and Classification	MRI	518	LBP	Acc = 83%	Texture feature extraction and Logistic Regression
Garcia-Cano, 2018 [10]	Feature Extraction and Regression	X-rays	150	Vertebrae	Cobb angle MAE = 4.79°	Independent component analysis and Random Forest

Intervertebral discs (IVDs) are the most investigated lumbar structures (five papers), followed by vertebrae (three papers), whereas one paper evaluated LBP without focusing on a specific structure. It is worth noting that only two out of eight articles have exclusively focused on feature extraction, i.e., the work of Raudner et al. [21] in which the GRAPPATINI method is presented for IVD feature extraction from MRI, and the work of Abdollah et al. [22] in which a Random Forest and a Texture analysis are exploited on MRI for feature extraction from IVDs and vertebrae, respectively. The remaining six articles described the performance of further tasks after feature extraction. In detail, four of them performed classification, one performed regression, and one performed segmentation tasks.

All the works that performed further tasks following feature extraction exploited machine learning techniques rather than deep learning: this is one of the advantages of feature extraction, as it allows to achieve results using much faster and less computationally-expansive methods. With regards to classification, Adankon et al. [19] were the only ones to use 3D images of the surface of the human back: they extracted features for 165 patients using local geometric descriptors, and fed them to a least-squares Support Vector Machine (SVM) for the classification of scoliosis curve types, achieving 95% accuracy. Yang et al. [8] used a Gabor wavelet transform to extract features from MRI of 109 subjects, and a Kanade–Lucas–Tomasi (KLT) feature tracker to identify lumbar degenerative changes with an accuracy of 88.3%. Ruiz-España et al. [23] extracted features from MRI of 67 patients using Gradient Vector Flow, and tested several machine learning models to classify degenerated IVDs achieving accuracies greater than 90%. Ketola et al. [24] performed texture feature extraction from 518 MRI and used Logistic Regression to discriminate between symptomatic and asymptomatic LBP with an accuracy of 83%.

With regards to the regression task, Garcia-Cano et al. [10] extracted features from X-ray images of 150 patients through the medium of Independent Component Analysis, and used Random Forest Regression to predict the spinal curve progression in adolescents with idiopathic scoliosis, achieving a MAE of 4.79° for the Cobb angle.

With regards to the segmentation task, Castro-Mateos et al. [20] extracted features from 3D MRI of 59 subjects and performed IVDs segmentation using statistical shape model space and B-Spline space, achieving an average DICE score of 88.4%.

4.2. Segmentation

Image segmentation is the task of dividing an image into sub-regions corresponding to different elements of the image, with the aim of accurately identifying the borders of different elements in the image. This approach usually exploits manually-segmented images to train an AI model. Several manuscripts included in the reviewed performed a segmentation task, and some used segmentation as a preliminary step for further tasks. For this reason, in the next sections we report, where applicable, not only the segmentation results, but also those of the successive tasks for which segmentation is used with the aim of localizing and/or identifying structures. In this review, we refer to the task of detecting specific components (e.g., vertebrae) as "localization", whereas we refer to the task of assigning a label to specific components (e.g., L1, L2, etc.) as "identification". Moreover, we have differentiated included papers based on whether they exploited DIP techniques or NNs. In this review, we identified 38 manuscripts using DIP techniques, and 23 using NNs. However, it is worth noting how most recent research efforts are moving towards deep learning techniques: taking into account the articles published in the last 5 years (2016-2021), this review includes 16 papers using DIP, and 23 using NNs.

4.2.1. Digital Image Processing

DIP segmentation techniques process digital images to find the edges of different regions based on semantic characteristics, exploiting methods such as gradient thresholding or statistical shape models. In this review, we identified a total of 38 papers that performed DIP segmentation on different types of images (Table 2):

- 15 articles on MRI (2 of which considered 3D MRI);
- 15 articles on CT images;
- 1 articles on both MRI and CT images;
- 3 articles on fluoroscopic images;
- 2 articles on ultrasound images;
- 2 articles on X-ray images.

Vertebrae are the most investigated lumbar structures (26 papers), followed by IVDs (10 papers) and muscles (6 papers). It is worth noting that only one [25] out of the 21 works using CT, X-ray or fluoroscopic images did not involve segmentation of vertebral structures. In total, 20 articles focused only on segmentation without further tasks. Among the others, 12 performed successive structure localization, 6 conducted successive structure identification (4 of which performed both localization and identification), whereas regression, tracking, and 3D reconstruction were investigated by 1 manuscript for each task, respectively.

Table 2. Segmentation—Digital Image Processing. For each work, the main task is reported, whether it concerns only the segmentation of lumbar components, or if it aims to localize specific parts (e.g., the center of mass) of the components, or if it aims to identify each component (e.g., differentiating vertebrae between each other). If more structures are investigated, the correspondent results are reported in the same order in which structures are presented in the column "Structures involved". Abbreviations are used for Magnetic Resonance Imaging (MRI), Computed Tomography (CT), Accuracy (Acc), Sensitivity (Sen), Specificity (Spec), Area Under the Curve (AUC), Gradient Vector Flow (GVF), Support Vector Machine (SVM).

Author/Year	Main Task	Data Type	# Patients	Structures Involved	Results	Model
Haq, 2015 [26]	Segmentation	3D MRI	21	Discs	DICE = 91.7%	Shape-aware models
Neubert, 2012 [27]	Segmentation and Identification	3D MRI	28	Discs and Vertebrae	DICE = 89 and 91%, Sen = 100%, Spec = 98%	Statistical shape model
Haq, 2020 [25]	Segmentation	CT images	18 SpineWeb	Discs	DICE = from 91,7 to 95,4%	Shape statistics deformable model
Li, 2018 [28]	Segmentation	CT images	115 (Microsoft R.+ SpineWeb)	Vertebrae	DICE = 92.1%	Gaussian Mixture Model + threshold
Ibragimov, 2017 [29]	Segmentation	CT images	30 vertebrae	Vertebrae	DICE = 84.7%	Landmark detection and deformable models
Yu, 2018 [30]	Segmentation	CT images	21 images	Vertebrae	DICE = 93.9%	Bone-sheetness assisted grid cut
Korez, 2015 [31]	Segmentation	CT images	220	Vertebrae	DICE = 94.6%	Shape-constrained deformable model
Al-Helo, 2011 [32]	Segmentation	CT images	50	Vertebrae	Visual evaluation	Active shape models and GVF-snake
Ruiz-España, 2015 [33]	Segmentation	CT images	10	Vertebrae	DICE = 95%	Selective Binary Gaussian Filtering Regularized Level Set
Huang, 2013 [34]	Segmentation	CT images	56	Vertebrae	DICE = 94%	Otsu thresholding, edge- and region-based level set
Mahdy, 2018 [35]	Segmentation and Localization	CT images	10	Vertebrae	Visual evaluation	Threshold and adaptive K-Means
Courbot, 2016 [36]	Localization	CT images	15	Vertebrae	Visual evaluation, Acc = 89.4%	Hidden Markov Chain segmentation
Rasoulian, 2013 [37]	Localization	CT images	32	Vertebrae	Visual evaluation, Center of mass MAE = 2mm	Multi-object shape model
Mastmeyer, 2006 [38]	Segmentation	CT images	41	Vertebrae	DICE > 98.6%	Volume growing and morphological operations
Jimenez-Pastor, 2020 [39]	Localization and Identification	CT images	272 images	Vertebrae	Localization error = 13.7mm, Acc = 74,8%	Decision forest + morphological image processing
Lee, 2011 [40]	Localization and Identification	CT images	19	Vertebrae	Localization error = 0.14mm, Acc = 93.2%	Threshold and thinning-based integrated cost
Klinder, 2009 [41]	Localization and Identification	CT images	64	Vertebrae	Localization error = 1.1mm, Acc = 92%	Triangulated shape models
Štern, 2009 [42]	Localization	MRI and CT images	13 and 29 images	Discs and Vertebrae	Localization error = 2.8 and 1.8 mm	Analysis of the geometry of spinal structures
Wong, 2008 [43]	Segmentation and Tracking	Fluoroscopic images	2 videos	Vertebrae	Visual evaluation	Wavelet and shape-active contour based
Zheng, 2011 [44]	Segmentation and 3D reconstruction	Fluoroscopic images	4	Vertebrae	Mean reconstruction error<1.6mm	Statistical shape models
Michopoulou, 2009 [45]	Segmentation	MRI	34	Discs	DICE = 90%	Atlas-robust-fuzzy C-Means
Fallah, 2018 [46]	Segmentation	MRI	50	Discs and Vertebrae	DICE = 92.5 and 91.4%	Hierarchical conditional random field and Random Forest
Ghosh, 2014 [47]	Segmentation	MRI	212	Discs and Vertebrae	DICE = 87 and 84%	Random Forest and context features
Kim, 2018 [48]	Segmentation	MRI	19	Vertebrae	DICE = 90%	Graph-based and line-based segmentation algorithms
Gaonkar, 2017 [49]	Segmentation	MRI	63	Vertebrae	DICE = 83%	Multi-parametric ensemble
Gawel, 2018 [50]	Segmentation	MRI	50	Vertebrae	DICE = 91.4%	Cascade classifier and Active Appearance Model
Engstrom, 2011 [51]	Segmentation	MRI	20	Muscles	DICE = 87%	Statistical shape model
Baum, 2018 [52]	Segmentation	MRI	10	Muscles	DICE = 83%	Average shape model and dual feature model

Table 2. *Cont.*

Author/Year	Main Task	Data Type	# Patients	Structures Involved	Results	Model
Zheng, 2004 [53]	Segmentation	Fluoroscopic images	1	Vertebrae	Visual evaluation	Hough transform and Fourier descriptors
Jurcak, 2008 [54]	Segmentation	MRI	20	Muscles	DICE = 77%	Probabilistic atlases and geodesic active contours
Fortin, 2017 [55]	Segmentation and Regression	MRI	30	Muscles	Reliability coefficient = 97-99%	Threshold
Neubert, 2013 [56]	Segmentation and Localization	MRI	44	Discs	DICE = 92.3%, AUC = 0.98	Active shape model, Linear Discriminant Analysis, SVM
Oktay, 2011 [57]	Localization and Identification	MRI	40	Discs	Localization rate = 95.4%, Acc = 97%	Probabilistic model and SVM
Castro-Mateos, 2016 [58]	Identification	MRI	48	Discs	Sensitivity = 87%	Active contour model and Feedforward NN
Kim, 2020 [59]	Localization	Ultrasound	50	Muscles	2mm discrepancy	Fuzzy C-Means Clustering
Lui, 2014 [60]	Localization	Ultrasound	10	Muscles	F1-Score = 90.9%	Decoupled Active Contour
Ribeiro, 2010 [61]	Segmentation	X-rays	41	Vertebrae	DICE = 91.7%	Gabor Filters and NN
Sa, 2016 [62]	Localization	X-rays	30	Vertebrae	True Positive Rate = 75%	GVF-snake and SVM

With regards to the papers that focused exclusively on segmentation, Haq et al. [26] used shape-aware models on 3D MRI of 21 patients for the segmentation of IVDs, achieving an average DICE of 91.7%. In addition, in a successive article Haq et al. [25] utilized a shape statistical deformable model for the segmentation of IVDs on CT images of 18 subjects from the SpineWeb dataset, achieving DICE scores ranging from 91.7 to 95.4%. Li et al. [28] applied a threshold to the results of a Gaussian Mixture Model for segmenting vertebrae on a total of 115 CT images from the SpineWeb and the Microsoft Research datasets, with an average DICE of 92.1%. Ibragimov et al. [29] used landmark detection and deformable models for segmenting 30 vertebrae on CT images, with a DICE of 84.7%. Yu et al. [30] utilized bone-sheet assisted grid cut to segment vertebrae from 21 CT images, achieving an average DICE of 93.9%. Korez et al. [31] applied a shape-constrained deformable model for vertebrae segmentation from CT images of 220 patients, with a DICE of 94.6%. Al-Helo et al. [32] combined Active-shape models and GVF-snake for the segmentation of vertebrae from CT images of 50 subjects, assessing the segmentation quality by visual evaluation. Ruiz-España et al. [33] used a Selective Binary Gaussian Filtering Regularized Level Set to segment vertebrae on CT images of 10 subjects, achieving an average DICE of 95%. Huang et al. [34] exploited Otsu thresholding, Edge- and Region-based level sets to segment vertebrae on CT images of 56 subjects, with a 94% DICE. Mastmeyer et al. [38] utilized volume growing and morphological operations to segment vertebrae on CT images of 41 subjects, achieving DICE scores greater than 98.6%. Zhang et al. [53] applied Hough transform and Fourier descriptors for vertebrae segmentation on one fluoroscopic image, assessing the segmentation quality by visual evaluation. Michopoulou et al. [45] used an Atlas-robust-fuzzy C-Means for segmenting IVDs on MRI of 34 subjects, achieving a 90% DICE. Fallah et al. [46] exploited Hierarchical Conditional Random Fields and a Random Forest for the segmentation of IVDs and vertebrae, respectively, on MRI of 34 subjects, achieving a DICE of 92.5 and 91.4%, respectively. Ghosh et al. [47] combined Random Forest and context features for the segmentation of IVDs and vertebrae, respectively, on MRI of 212 subjects, achieving a DICE of 87 and 84%, respectively. Kim et al. [48] used graph-based and line-based segmentation algorithms for segmenting vertebrae on MRI of 19 patients, achieving a 90% DICE. Gaonkar et al. [49] applied a multi-parametric ensemble to segment vertebrae on MRI of 63 subjects, with an average DICE of 83%. Gawel et al. [50] combined a cascade classifier and an Active Appearance Model to segment vertebrae on 50 MRI, achieving a DICE of 91.4%. Engstrom et al. [51] used a Statistical Shape model for the segmentation of the quadratus lumborum muscle on MRI of 20 patients, achieving a DICE of 87%. Baum et al. [52] exploited an Average Shape model and a Dual Feature model for paraspinal muscle segmentation on MRI of 10 subjects, with a DICE of 83%. Jurcak et al. [54] applied Probabilistic atlases and Geodesic Active Contours for the segmentation of quadratus lumborum muscle on MRI of 20 subjects with a 77% DICE. Ribeiro et al. [61] used Gabor Filters and an ANN to segment vertebrae on X-ray images of 41 patients, achieving a DICE of 91.7%.

With regards to the articles that performed localization following segmentation, Mahdy et al. [35] used a threshold method followed by an adaptive K-Means for the segmentation and localization of lumbar vertebrae on CT images of 10 subjects in order to identify degenerated IVDs, and evaluated the performance by visual evaluation. Courbot et al. [36] exploited a Hidden Markov Chain for semi-automated segmentation of vertebrae on CT images of 15 subjects, achieving a localization accuracy of 89.4%. Rasoulian et al. [37] developed a multi-object shape model for vertebrae localization on 32 CT images, correctly localizing the centers of mass with a MAE of 2 mm with the aim of identifying the optimal location for spinal needle injection. Štern et al. [42] performed an analysis of the geometry of the spinal structures to localize the centers of IVDs and vertebrae on 13 MRI and 29 CT images, respectively, with a localization error of 2.8 and 1.8 mm, respectively. Neubert et al. [56] used an Active Shape model to segment IVDs on MRI of 44 subjects achieving a DICE of 92.3%, and an AUC of 0.98 for localization of degenerated IVDs using Linear Discriminant Analysis and SVM. Kim et al. [59] exploited Fuzzy C-Means Clustering for the localization of lumbar multifidus muscle on ultrasound images of 50 subjects, with a 2 mm localization discrepancy. Lui et al. [60] utilized Decoupled Active Contour for the localization of lumbar multifidus muscle on ultrasound images of 10 subjects, achieving an F1-Score of 90.9%. Sa et al. [62] used Gradient Vector Flow Snake and SVM for the localization of vertebrae on X-ray images of 30 subjects, achieving a True Positive Rate of 75%.

With regards to the papers that performed identification following segmentation, Neubert et al. [27] used a Statistical Shape model on 3D MRI of 28 subjects to segment and identify IVDs and vertebrae, achieving segmentation DICE of 89 and 91%, respectively, and 98.3% specificity and 100% sensitivity for the identification of degenerated IVDs. Castro-Mateos et al. [58] described an Active Contour Model for the segmentation and a Feedforward NN for the identification and classification of IVDs on MRI of 48 subjects, achieving 87% Sensitivity.

With regards to the papers that performed both localization and identification, Jimenez-Pastor et al. [39] used a Decision Forest and morphological image processing to localize and identify vertebrae on 272 CT images, achieving a localization error of 13.7 mm and an accuracy of 74.8%. Lee et al. [40] exploited threshold and thinning-based integrated cost on CT images of 19 subjects, for the localization and identification of lumbar pedicles in order to increase accuracy and safety during transpedicular screw placement, with a localization error of 0.14 mm and 93.2% accuracy. Klinder et al. [41] used a Triangulated Shape model on CT images of 64 subjects, achieving a vertebrae localization error of 1.1 mm and 92% accuracy. Oktay et al. [57] combined a Probabilistic model with an SVM to localize and detect IVDs on MRI of 40 subjects, achieving a localization rate of 95.4% and an accuracy of 97%.

In addition, Wong et al. [43] used Wavelets and a Shape-Active Contour-Based model for vertebrae segmentation and Tracking on 2 videos of fluoroscopic images, evaluating the performance by visual evaluation. Zheng et al. [44] utilized Statistical Shape models for vertebrae segmentation and 3D reconstruction on 4 fluoroscopic images, achieving a mean reconstruction error of less than 1.6 mm. Finally, Fortin et al. [55] used a threshold algorithm for segmentation and quantification of paraspinal muscle composition with a reliability coefficient ranging between 97 and 99%.

4.2.2. Deep Learning

Deep learning is a class of AI algorithms based on Artificial Neural Networks. More in detail, an NN is said to be "deep" if it is composed of more than 2 hidden layers. Deep learning techniques for segmentation take as an input the whole original image, and perform feature extraction, feature selection, segmentation and any further step (e.g., classification, regression) in one single model. In this review, we identified a total of 23 papers that performed deep learning segmentation, and their main characteristics are reported in Table 3. In detail:

- 13 articles on MRI (2 of which considered 3D MRI and 1 with the addition of clinical notes);
- 5 articles on CT images;
- 4 articles on X-ray images (1 of which in combination with Moire images);
- 1 article on ultrasound images.

Vertebrae were the most investigated lumbar structures (16 papers), followed by IVDs (11 papers), spinal canal (7 papers), and muscles (5 papers). In total, 9 articles focused exclusively on segmentation without further tasks. Among the others, 5 manuscripts performed successive structure identification, 3 carried out a regression task, 3 performed successive structure reconstruction, 1 work performed classification, 1 performed structure localization, and 1 carried out both structure localization and identification. It is worth noting that the vast majority of the works included in this section exploited Convolutional Neural Networks (CNNs) or models that derive from them.

Table 3. Segmentation—Deep Learning. For each work, the main task is reported, whether it concerns only the segmentation of lumbar components, or if it aims to localize specific parts (e.g., the center of mass) of the components, or if it aims to identify each component (e.g., differentiating vertebrae between each other). If more structures are investigated, the correspondent results are reported following the same order by which structures are presented in the "Structures involved" column. Abbreviations are used for Magnetic Resonance Imaging (MRI), Computed Tomography (CT), Mean Absolute Error (MAE), Accuracy (Acc), Convolutional Neural Network (CNN), Support Vector Machine (SVM), Regression Trees (RT).

Author/Year	Main Task	Data Type	# Patients	Structures Involved	Results	Model
Iriondo, 2020 [63]	Segmentation	3D MRI	31	Discs	DICE > 85%	Coarse-to-fine context memory NN
Staartjes, 2021 [64]	Segmentation and Reconstruction	3D MRI	3	All structures	Visual evaluation	CNN
Lee, 2020 [65]	Segmentation and Reconstruction	CT images	280 images	All structures	MAE = 21 pixels	Generative Adversarial Networks
Fan, 2020 [66]	Segmentation and Reconstruction	CT images	108	All structures	Kambin triangle = 161 mm^2	U-net
Malinda, 2020[67]	Segmentation	CT images	120	Vertebrae	DICE = 94.2%	Generative Adversarial Networks
Siemionow, 2020 [68]	Identification	CT images	45	Vertebrae	Acc = 96 to 99%	CNN
Netherton, 2020 [69]	Localization and Identification	CT images	330 images	Vertebrae	Localization error = 2.2 mm, Acc = 94%	X-net ensemble
Watanabe 2019 [70]	Regression	Moire images + X-rays	1996	Vertebrae	Cobb angle MAE = 3.42°	CNN
Kim, 2018 [71]	Segmentation	MRI	SpineWeb 20	Discs	DICE = 89.4%	CNN (BSU-net)
Shen, 2021 [72]	Segmentation	MRI	120	Discs, Spinal canal and Muscles	Jaccard: 87, 82 and 85%	Feedforward NN
Gaonkar, 2019 [73]	Segmentation	MRI	39295	Discs and Spinal canal	DICE = 88 and 87%	Discs: U-net, Canal: SVM and RT
Huang, 2020 [74]	Segmentation	MRI	100	Discs and Vertebrae	Jaccard = 92.6 and 94.7%	U-net
Li, 2021 [75]	Segmentation	MRI	120	Vertebrae and Spinal canal	DICE = 92.5%	CNN
Li, 2019 [76]	Segmentation	MRI	120	Muscles	DICE > 91.3%	Deformed U-net
Zhou, 2020 [77]	Segmentation	MRI	57	Vertebrae	DICE = 84.9%	U-net
Jamaludin, 2017 [78]	Classification	MRI	2009	Discs and Vertebrae	Acc = 95.6%	CNN
Natalia, 2020 [79]	Regression	MRI	515	Discs and Spinal canal	Mean error: 0.9 mm	SegNet and Contour Evolution Algorithm
Zhou, 2019 [80]	Identification	MRI	1318	Vertebrae	Acc = 98.9%	CNN
Forsberg, 2017 [81]	Identification	MRI with clinical notes	475	Vertebrae	Acc = 97%	CNN and parts-based graphical models
Baka, 2017 [82]	Identification	Ultrasound	19 data sets	Vertebrae	Acc = 92%	CNN and matching strategy
Cho, 2020 [83]	Segmentation and Regression	X-rays	629	Vertebrae	DICE = 82.1%, MAE = 8,055°	U-net
Li, 2016 [84]	Identification	X-rays	110	Vertebrae	Acc = 80.4%	CNN
Sa, 2017 [85]	Localization	X-rays	1081 images	Discs	Precision = 90.5%	Faster R-CNN

With regards to the articles that focused exclusively on segmentation, Iriondo et al. [63] used a Coarse-to-fine context memory NN to segment IVDs on 3D MRI of 31 subjects, achieving a DICE greater than 85%. Malinda et al. [67] utilized Generative Adversarial Networks (GANs) for vertebrae segmentation on CT images of 120 subjects, achieving a DICE of 94.2%. Kim et al. [71] exploited a BSU-net for IVDs segmentation on 20 MRI from the SpineWeb dataset, achieving a DICE of 89.4%. Shen et al. [72] used a Feedforward NN on MRI of 120 subjects, achieving a Jaccard index for the segmentation of IVDs, spinal canal and muscles of 87, 82 and 85%, respectively. Gaonkar et al. [73] applied a U-net to segment IVDs on 39295 MRI images, achieving an 88% DICE; they also combined an SVM with a Regression Tree to segment the spinal canal with a DICE of 87%. Huang et al. [74] used a U-net to segment IVDs and vertebrae on 100 MRI achieving a Jaccard index of 92.6 and 94.7%, respectively. Li et al. [75] utilized a CNN to segment vertebrae and spinal canal on MRI of 120 patients achieving an overall DICE of 92.5%. Moreover, they used a deformed U-net [76] for the segmentation of paraspinal muscles on 120 MRI achieving an overall DICE greater than 91.3%. Zhou et al. [77] utilized a U-net for vertebrae segmentation on MRI of 57 subjects, achieving a DICE of 84.9%.

With regards to the papers that performed structure identification following segmentation, Siemionow et al. [68] used a CNN to identify vertebrae on CT images of 45 subjects, with an overall accuracy ranging from 96 to 99%. Zhou et al. [80] combined a CNN and similarity with a beforehand lumbar image for vertebrae identification on MRI images of 1318 healthy and unhealthy subjects, achieving an accuracy of 98.9%. Forsberg et al. [81] combined a CNN and graph-based graphical models on MRI enriched with clinical notes to identify vertebrae of 475 patients, achieving an accuracy of 97%. Baka et al. [82] utilized a CNN and a matching strategy for vertebrae identification on ultrasound images from 19 datasets, achieving an accuracy of 92%. Li et al. [84] were the only to perform vertebrae identification on X-ray images. They applied a CNN on 110 images, achieving an 80.4% accuracy.

With regards to the articles that performed a regression task, Watanabe et al. [70] used a CNN to estimate spinal alignment on 1996 Moire images, with a Cobb angle MAE of 3.42°. Natalia et al. [79] combined a SegNet and a Contour Evolution Algorithm to measure anteroposterior diameter and foraminal widths on MRI of 515 patients suffering from lumbar spinal stenosis with a mean error of 0.9 mm. Cho et al. [83] used a U-net for the automated segmentation and measurement of lumbar lordosis on X-ray images of 629 patients, achieving a DICE of 82.1% and a MAE of 8.06°.

With regards to the articles performing a Reconstruction task, Staartjes et al. [64] developed a CNN to segment and reconstruct the lumbar structures from 3D MRI of 3 patients, evaluating the performance by visual evaluation. Lee et al. [65] used GANs to generate synthetic spine lumbar structures MRI from 280 CT images, with a MAE of 21 pixels. Fan et al. [66] exploited a U-net to reconstruct lumbar structures from CT images of 108 subjects, with a Kambin triangle of 161 mm^2.

With regards to the articles performing a classification task, Jamuladin et al. [78] used a CNN for classification of IVDs and vertebrae on MRI of 2009 subjects achieving an accuracy of 95.6%.

In addition, Sa et al. [85] fine-tuned a Faster Region-based CNN (R-CNN) for IVD localization on 1081 X-ray images with a 90.5% precision. Finally, Netherton et al. [69] used an X-net ensemble to localize and identify vertebrae on 330 CT images, achieving a localization error of 2.2 mm and an accuracy of 94%.

5. Discussion

Due to the extensive use of advanced imaging modalities and the complexity of anatomical structures involved in the development of LBP and its sequelae, a vast body of research has been investigating the utilization of AI in the elaboration of digital images for different purposes. The vast majority of the works in the literature exploit MRI or CT imaging, whereas a minority of works exploit X-ray, fluoroscopic or ultrasound imaging.

It is worth noting that vertebral structures are the main focus of articles performing segmentation, both with DIP and deep learning techniques; conversely, articles performing feature extraction are mainly focused on IVDs.

With regards to feature extraction, which is the capacity of a system to recognize a specific set of relevant features, all included studies collectively showed an accuracy > 80% in identifying the location of vertebrae [24] and IVDs [8,20–24], with the ability to even detect annular tears and lumbar disc herniation [21–23]. Although the majority of the studies were conducted on MRI images [8,20–24], one study utilized X-ray imaging [10] and another study built a 3D model of patients' backs using a noninvasive surface acquisition technology [19]. Moreover, some of these studies also reported the capacity of the described systems to perform classification and regression tasks on extracted data, such as estimating the degree of IVD degeneration [8,22–24], scoliosis curve type classification [19] and prediction of curve progression [10], the presence of spinal stenosis [23] and to explore the correlation between degenerative changes and the presence of LBP [24].

However, most studies focused on segmentation, which is the differentiation of specific subregions of an image based on distinct parameters. Traditionally, segmentation tasks have been performed by DIP systems via subdivision of elements within an image based on gradient thresholding or statistical shape models, which fall under the definition of semantic segmentation [86]. However, recent research has been exploring the use of deep learning-based AI systems which are able to perform multiple tasks at the basic and advanced level in a single model [1]. Vertebrae are by far the most investigated structure, with AI systems reaching > 90% DICE and > 90% accuracy in the majority of studies included in our review, both using DIP [28–41,43,44,48–50,53,61,62] and deep learning models [67,69,77,80–84]. In particular, a study from Lee et al. [40] proposed a model to obtain an automated segmentation of lumbar pedicles from CT images in order to increase accuracy and safety during transpedicular screw placement. On the other hand, a study from Watanabe and colleagues [70] described a CNN able to estimate spinal alignment, vertebral rotation and Cobb angle with a mean absolute error of 3.6 pixels for vertebral position, 2.9° for vertebral rotation and 3.42° with regards to the estimated Cobb angle. Similarly, Cho et al. [83] presented a CNN capable of segmenting lumbar vertebrae and subsequently calculate lumbar lordosis, with a mean absolute error of 8.055°. In this manuscript, Several AI systems for automated segmentation of IVDs have been described as well [25,26,45,48,56–58,63] with a reported DICE > 90% in nearly all studies. Besides, performance of systems developed for the segmentation of paraspinal muscles have reported a higher variability compared to other structures [51,52,54,55,60,71], with higher DICE values for systems based on deep learning models [76]. In addition, some studies evaluated the simultaneous segmentation of multiple structures, in particular IVDs and vertebrae [27,42,46,56,74,78], with a DICE > 90% in DIP-based systems [27,42,46,56] and a reported accuracy > 95% in most deep learning-based systems [68,71,74,78,85]. Furthermore, some of the latter have been used in order to synthesize CT images from MRI and vice versa. For example, Staartjes et al. [64] introduced a CNN-based system able to generate synthetic CT images from spine MRI, so as to acquire more precise information about osseous structures compared to traditional MRI without the need to expose patients to additional radiation. On the other hand, Lee and colleagues [65] presented a model based on GANs capable of producing a synthetic MRI from spine CT scans, which resulted in a mean overall similarity with real MRI scans of 80.2%. This study demonstrated the possibility to extract accurate information about soft tissues from spine CT without the necessity to order an MRI, which is often expensive and time-consuming. Other studies have also shown the possibility to automatically calculate the spinal canal area [73] as well as segmenting and reconstructing multiple structures at the same time [47,66,72,75,79] with an elevate degree of accuracy.

Figure 4 shows a boxplot that summarizes the results for the segmentation of IVDs, vertebrae and lumbar muscles, and the identification accuracy for different lumbar structures. With regards to the segmentation of IVDs and vertebrae, it is worth noting that DIP and deep learning techniques achieve very similar results, with DIP methods performing slightly better. This is mainly due to the regular and homogeneous surface of such structures, whose well-defined edges can be effectively identified using DIP techniques such as threshold and region-growing methods. Conversely, lumbar muscle segmentation performance of deep learning techniques is sensitively better than that of DIP methods. Indeed, the structure of muscles is irregular and more challenging to detect properly, and deep NNs provide a better tool for such a task. With regards to the identification accuracy, deep learning provides generally better results; nonetheless, DIP methods followed by machine learning techniques are typically faster and less computationally expensive, and, in some cases, provide similar performance.

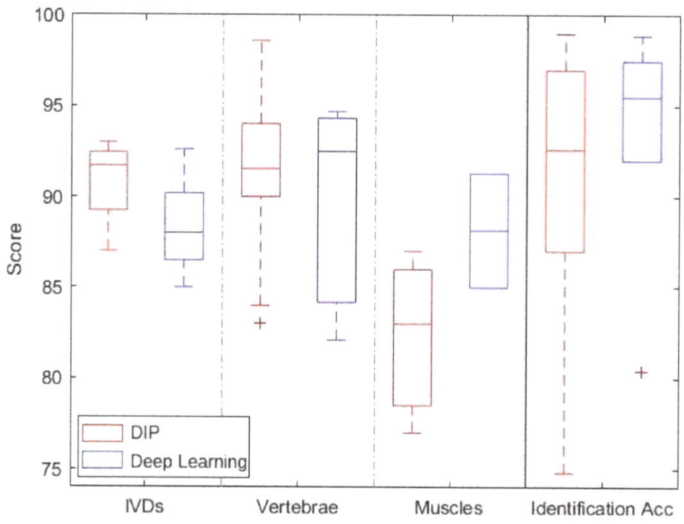

Figure 4. Boxplot summarizing the results for different structures and tasks. The three left columns refer to the DICE scores for the segmentation of IVDs, vertebrae and muscles; the right column refers to the identification accuracy for different structures.

Although the application of computer vision to the elaboration of radiological images of the spine is continuously increasing, some concerns still exist. Indeed, system validation still largely depends on multiple user interventions and cannot replace the human counterpart for obvious reasons, from both clinical and ethical perspectives. Furthermore, the best performing methods are based on the application of NNs, which usually require a large amount of images and computational capacity for training, which are not available to all researchers. However, some DIP techniques provide equal or better performance in the segmentation of regular-shaped structures such as vertebrae and IVDs, while requiring a smaller amount of data for training and limiting the computational burden. Moreover, some methods already exist for the automatic detection and grading of conditions such as spondylolisthesis, disc herniation and scoliosis.

6. Conclusions

In the last decade, the utilization of AI has increased considerably in all fields, and medical research made no exception. Indeed, AI-based computers have already shown the potential to revolutionize the medical field, including spine surgery. In this study, we have systematically reviewed the available literature on the use of AI, and more specifically computer vision, in the prevention, diagnosis, and treatment of LBP. In conclusion, computer vision techniques bear promises for effectively improving clinical practice in coming years, thanks to the availability of public datasets and to the natural upcoming increase of the computational capacity. Furthermore, steps are being taken towards the interpretability of AI and, in particular, of deep learning models. Such improvements will lead to the development of systems that will not require multiple user interventions, thus providing a valid assessment tool for physicians. LBP diagnosis and treatment often require the utilization and integration of advanced imaging modalities. In addition, several structural alterations, often subtle and nonunivocal to interpret, concur to define the clinical scenario. In this picture, the use of AI and computer vision may effectively assist and implement the diagnostic process, thus possibly improving clinical outcomes and diagnostic accuracy.

Supplementary Materials: The following are available online at https://www.mdpi.com/article/10.3390/ijerph182010909/s1, Table S1: Summary of the methodological quality of included studies regarding the 4 domains assessing the risk of bias of the QUADAS-2 score; Table S2: Summary of the methodological quality of included studies regarding the 3 domains assessing applicability concerns of the QUADAS-2 score.

Author Contributions: Conceptualization, F.D., F.R., M.M.; methodology, F.D., F.R., L.A., L.V., G.V., M.M.; writing—original draft preparation, F.D., F.R., L.A., M.M.;investigation, F.D.; writing—review and editing, F.D., F.R., L.A., L.V., G.V., M.M., R.P., V.D.; visualization, F.D.; supervision, F.R., L.V., G.V., M.M., R.P., V.D.; funding acquisition, F.R., G.V., R.P., V.D. All authors have read and agreed to the published version of the manuscript.

Funding: This research was funded by the Research Grant (BRiC-2018 ID3) of the Italian Workers' Compensation Authority (INAIL).

Institutional Review Board Statement: Not applicable.

Informed Consent Statement: Not applicable.

Data Availability Statement: No new data were created or analyzed in this study. Data sharing is not applicable to this article.

Conflicts of Interest: The authors declare no conflict of interest.

References

1. Galbusera, F.; Casaroli, G.; Bassani, T. Artificial intelligence and machine learning in spine research. *JOR Spine* **2019**, *2*, e1044. [CrossRef] [PubMed]
2. Chan, H.S.; Shan, H.; Dahoun, T.; Vogel, H.; Yuan, S. Advancing drug discovery via artificial intelligence. *Trends Pharmacol. Sci.* **2019**, *40*, 592–604. [CrossRef] [PubMed]
3. Rashid, M.B.M.A.; Chow, E.K.H. Artificial intelligence-driven designer drug combinations: From drug development to personalized medicine. *SLAS Technol. Transl. Life Sci. Innov.* **2019**, *24*, 124–125. [CrossRef] [PubMed]
4. Hosny, A.; Parmar, C.; Quackenbush, J.; Schwartz, L.H.; Aerts, H.J. Artificial intelligence in radiology. *Nat. Rev. Cancer* **2018**, *18*, 500–510. [CrossRef]
5. Bur, A.M.; Shew, M.; New, J. Artificial intelligence for the otolaryngologist: A state of the art review. *Otolaryngol.-Neck Surg.* **2019**, *160*, 603–611. [CrossRef]
6. Boon, I.S.; Au Yong, T.; Boon, C.S. Assessing the role of artificial intelligence (AI) in clinical oncology: Utility of machine learning in radiotherapy target volume delineation. *Medicines* **2018**, *5*, 131. [CrossRef]
7. Loftus, T.J.; Tighe, P.J.; Filiberto, A.C.; Efron, P.A.; Brakenridge, S.C.; Mohr, A.M.; Rashidi, P.; Upchurch, G.R.; Bihorac, A. Artificial intelligence and surgical decision-making. *JAMA Surg.* **2020**, *155*, 148–158. [CrossRef]
8. Yang, T.; Li, R.; Liang, N.; Li, J.; Yang, Y.; Huang, Q.; Li, Y.; Cao, W.; Wang, Q.; Zhang, H. The application of key feature extraction algorithm based on Gabor wavelet transformation in the diagnosis of lumbar intervertebral disc degenerative changes. *PLoS ONE* **2020**, *15*, e0227894. [CrossRef]
9. Thong, W.; Parent, S.; Wu, J.; Aubin, C.E.; Labelle, H.; Kadoury, S. Three-dimensional morphology study of surgical adolescent idiopathic scoliosis patient from encoded geometric models. *Eur. Spine J.* **2016**, *25*, 3104–3113. [CrossRef]
10. Garcia-Cano, E.; Cosío, F.A.; Duong, L.; Bellefleur, C.; Roy-Beaudry, M.; Joncas, J.; Parent, S.; Labelle, H. Prediction of spinal curve progression in adolescent idiopathic scoliosis using random forest regression. *Comput. Biol. Med.* **2018**, *103*, 34–43. [CrossRef]
11. Franklin, G.M.; Wickizer, T.M.; Coe, N.B.; Fulton-Kehoe, D. Workers' compensation: Poor quality health care and the growing disability problem in the United States. *Am. J. Ind. Med.* **2015**, *58*, 245–251. [CrossRef]
12. Whiting, P.F.; Rutjes, A.W.; Westwood, M.E.; Mallett, S.; Deeks, J.J.; Reitsma, J.B.; Leeflang, M.M.; Sterne, J.A.; Bossuyt, P.M. QUADAS-2: A revised tool for the quality assessment of diagnostic accuracy studies. *Ann. Intern. Med.* **2011**, *155*, 529–536. [CrossRef]

13. Tagliaferri, S.D.; Angelova, M.; Zhao, X.; Owen, P.J.; Miller, C.T.; Wilkin, T.; Belavy, D.L. Artificial intelligence to improve back pain outcomes and lessons learnt from clinical classification approaches: Three systematic reviews. *NPJ Digit. Med.* **2020**, *3*, 1–16. [CrossRef]
14. Tack, C. Artificial intelligence and machine learning | applications in musculoskeletal physiotherapy. *Musculoskelet. Sci. Pract.* **2019**, *39*, 164–169. [CrossRef] [PubMed]
15. Azimi, P.; Yazdanian, T.; Benzel, E.C.; Aghaei, H.N.; Azhari, S.; Sadeghi, S.; Montazeri, A. A Review on the Use of Artificial Intelligence in Spinal Diseases. *Asian Spine J.* **2020**, *14*, 543. [CrossRef] [PubMed]
16. Yao, J.; Burns, J.E.; Forsberg, D.; Seitel, A.; Rasoulian, A.; Abolmaesumi, P.; Hammernik, K.; Urschler, M.; Ibragimov, B.; Korez, R.; et al. A multi-center milestone study of clinical vertebral CT segmentation. *Comput. Med. Imaging Graph.* **2016**, *49*, 16–28. [CrossRef] [PubMed]
17. Belasso, C.J.; Behboodi, B.; Benali, H.; Boily, M.; Rivaz, H.; Fortin, M. LUMINOUS database: Lumbar multifidus muscle segmentation from ultrasound images. *BMC Musculoskelet. Disord.* **2020**, *21*, 1–11. [CrossRef]
18. Burian, E.; Rohrmeier, A.; Schlaeger, S.; Dieckmeyer, M.; Diefenbach, M.N.; Syväri, J.; Klupp, E.; Weidlich, D.; Zimmer, C.; Rummeny, E.J.; et al. Lumbar muscle and vertebral bodies segmentation of chemical shift encoding-based water-fat MRI: The reference database myosegmentum spine. *BMC Musculoskelet. Disord.* **2019**, *20*, 1–7. [CrossRef]
19. Adankon, M.M.; Dansereau, J.; Labelle, H.; Cheriet, F. Non invasive classification system of scoliosis curve types using least-squares support vector machines. *Artif. Intell. Med.* **2012**, *56*, 99–107. [CrossRef]
20. Castro-Mateos, I.; Pozo, J.M.; Eltes, P.E.; Del Rio, L.; Lazary, A.; Frangi, A.F. 3D segmentation of annulus fibrosus and nucleus pulposus from T2-weighted magnetic resonance images. *Phys. Med. Biol.* **2014**, *59*, 7847. [CrossRef]
21. Raudner, M.; Schreiner, M.M.; Hilbert, T.; Kober, T.; Weber, M.; Szelényi, A.; Windhager, R.; Juras, V.; Trattnig, S. Clinical implementation of accelerated T 2 mapping: Quantitative magnetic resonance imaging as a biomarker for annular tear and lumbar disc herniation. *Eur. Radiol.* **2020**, *31*, 3590–3599. [CrossRef] [PubMed]
22. Abdollah, V.; Parent, E.C.; Dolatabadi, S.; Marr, E.; Croutze, R.; Wachowicz, K.; Kawchuk, G. Texture Analysis in the Classification of T2 Weighted Magnetic Resonance Images in Persons with and without Low Back Pain. *J. Orthop. Res.®* **2020**, *39*, 2187–2196. [CrossRef] [PubMed]
23. Ruiz-España, S.; Arana, E.; Moratal, D. Semiautomatic computer-aided classification of degenerative lumbar spine disease in magnetic resonance imaging. *Comput. Biol. Med.* **2015**, *62*, 196–205. [CrossRef]
24. Ketola, J.H.; Inkinen, S.I.; Karppinen, J.; Niinimäki, J.; Tervonen, O.; Nieminen, M.T. T 2-weighted magnetic resonance imaging texture as predictor of low back pain: A texture analysis-based classification pipeline to symptomatic and asymptomatic cases. *J. Orthop. Res.®* **2020**. [CrossRef]
25. Haq, R.; Schmid, J.; Borgie, R.; Cates, J.; Audette, M.A. Deformable multisurface segmentation of the spine for orthopedic surgery planning and simulation. *J. Med. Imaging* **2020**, *7*, 015002. [CrossRef]
26. Haq, R.; Aras, R.; Besachio, D.A.; Borgie, R.C.; Audette, M.A. 3D lumbar spine intervertebral disc segmentation and compression simulation from MRI using shape-aware models. *Int. J. Comput. Assist. Radiol. Surg.* **2015**, *10*, 45–54. [CrossRef]
27. Neubert, A.; Fripp, J.; Engstrom, C.; Schwarz, R.; Lauer, L.; Salvado, O.; Crozier, S. Automated detection, 3D segmentation and analysis of high resolution spine MR images using statistical shape models. *Phys. Med. Biol.* **2012**, *57*, 8357. [CrossRef]
28. Li, Y.; Liang, W.; Zhang, Y.; Tan, J. Automatic global level set approach for lumbar vertebrae CT image segmentation. *BioMed Res. Int.* **2018**, *2018*, 6319879. [CrossRef]
29. Ibragimov, B.; Korez, R.; Likar, B.; Pernuš, F.; Xing, L.; Vrtovec, T. Segmentation of pathological structures by landmark-assisted deformable models. *IEEE Trans. Med. Imaging* **2017**, *36*, 1457–1469. [CrossRef]
30. Yu, W.; Liu, W.; Tan, L.; Zhang, S.; Zheng, G. Multi-object Model-Based Multi-atlas Segmentation Constrained Grid Cut for Automatic Segmentation of Lumbar Vertebrae from CT Images. In *Intelligent Orthopaedics*; Springer: Singapore, 2018; pp. 65–71._5. [CrossRef]
31. Korez, R.; Ibragimov, B.; Likar, B.; Pernuš, F.; Vrtovec, T. A framework for automated spine and vertebrae interpolation-based detection and model-based segmentation. *IEEE Trans. Med. Imaging* **2015**, *34*, 1649–1662. [CrossRef] [PubMed]
32. Al-Helo, S.; Raja'S, A.; Chaudhary, V.; Al-Zoubi, M. Segmentation of lumbar vertebrae from clinical CT using active shape models and GVF-snake. In Proceedings of the 2011 Annual International Conference of the IEEE Engineering in Medicine and Biology Society, Boston, MA, USA, 30 August–3 September 2011; pp. 8033–8036. [CrossRef]
33. Ruiz-España, S.; Díaz-Parra, A.; Arana, E.; Moratal, D. A fully automated level-set based segmentation method of thoracic and lumbar vertebral bodies in Computed Tomography images. In Proceedings of the 2015 37th Annual International Conference of the IEEE Engineering in Medicine and Biology Society (EMBC), Milan, Italy, 25–29 August 2015; pp. 3049–3052. [CrossRef]
34. Huang, J.; Jian, F.; Wu, H.; Li, H. An improved level set method for vertebra CT image segmentation. *Biomed. Eng. Online* **2013**, *12*, 1–16. [CrossRef]
35. Mahdy, L.N.; Ezzat, K.A.; Hassanien, A.E. Automatic detection System for Degenerative Disk and simulation for artificial disc replacement surgery in the spine. *ISA Trans.* **2018**, *81*, 244–258. [CrossRef]
36. Courbot, J.B.; Rust, E.; Monfrini, E.; Collet, C. Vertebra segmentation based on two-step refinement. *J. Comput. Surg.* **2016**, *4*, 1–22. [CrossRef]
37. Rasoulian, A.; Rohling, R.; Abolmaesumi, P. Lumbar spine segmentation using a statistical multi-vertebrae anatomical shape+ pose model. *IEEE Trans. Med. Imaging* **2013**, *32*, 1890–1900. [CrossRef]

38. Mastmeyer, A.; Engelke, K.; Fuchs, C.; Kalender, W.A. A hierarchical 3D segmentation method and the definition of vertebral body coordinate systems for QCT of the lumbar spine. *Med. Image Anal.* **2006**, *10*, 560–577. [CrossRef]
39. Jimenez-Pastor, A.; Alberich-Bayarri, A.; Fos-Guarinos, B.; Garcia-Castro, F.; Garcia-Juan, D.; Glocker, B.; Marti-Bonmati, L. Automated vertebrae localization and identification by decision forests and image-based refinement on real-world CT data. *La Radiol. Medica* **2020**, *125*, 48–56. [CrossRef] [PubMed]
40. Lee, J.; Kim, S.; Kim, Y.S.; Chung, W.K. Automated segmentation of the lumbar pedicle in CT images for spinal fusion surgery. *IEEE Trans. Biomed. Eng.* **2011**, *58*, 2051–2063. [CrossRef]
41. Klinder, T.; Ostermann, J.; Ehm, M.; Franz, A.; Kneser, R.; Lorenz, C. Automated model-based vertebra detection, identification, and segmentation in CT images. *Med. Image Anal.* **2009**, *13*, 471–482. [CrossRef] [PubMed]
42. Štern, D.; Likar, B.; Pernuš, F.; Vrtovec, T. Automated detection of spinal centrelines, vertebral bodies and intervertebral discs in CT and MR images of lumbar spine. *Phys. Med. Biol.* **2009**, *55*, 247. [CrossRef] [PubMed]
43. Wong, A.; Mishra, A.; Fieguth, P.; Clausi, D.; Dunk, N.M.; Callaghan, J.P. Shape-guided active contour based segmentation and tracking of lumbar vertebrae in video fluoroscopy using complex wavelets. In Proceedings of the 2008 30th Annual International Conference of the IEEE Engineering in Medicine and Biology Society, Vancouver, BC, Canada, 20–25 August 2008; pp. 863–866. [CrossRef]
44. Zheng, G.; Nolte, L.P.; Ferguson, S.J. Scaled, patient-specific 3D vertebral model reconstruction based on 2D lateral fluoroscopy. *Int. J. Comput. Assist. Radiol. Surg.* **2011**, *6*, 351–366. [CrossRef]
45. Michopoulou, S.K.; Costaridou, L.; Panagiotopoulos, E.; Speller, R.; Panayiotakis, G.; Todd-Pokropek, A. Atlas-based segmentation of degenerated lumbar intervertebral discs from MR images of the spine. *IEEE Trans. Biomed. Eng.* **2009**, *56*, 2225–2231. [CrossRef] [PubMed]
46. Fallah, F.; Walter, S.S.; Bamberg, F.; Yang, B. Simultaneous volumetric segmentation of vertebral bodies and intervertebral discs on fat-water MR images. *IEEE J. Biomed. Health Inform.* **2018**, *23*, 1692–1701. [CrossRef] [PubMed]
47. Ghosh, S.; Chaudhary, V. Supervised methods for detection and segmentation of tissues in clinical lumbar MRI. *Comput. Med. Imaging Graph.* **2014**, *38*, 639–649. [CrossRef]
48. Kim, S.; Bae, W.C.; Masuda, K.; Chung, C.B.; Hwang, D. Semi-automatic segmentation of vertebral bodies in MR images of human lumbar spines. *Appl. Sci.* **2018**, *8*, 1586. [CrossRef]
49. Gaonkar, B.; Xia, Y.; Villaroman, D.S.; Ko, A.; Attiah, M.; Beckett, J.S.; Macyszyn, L. Multi-parameter ensemble learning for automated vertebral body segmentation in heterogeneously acquired clinical MR images. *IEEE J. Transl. Eng. Health Med.* **2017**, *5*, 1–12. [CrossRef]
50. Gaweł, D.; Główka, P.; Kotwicki, T.; Nowak, M. Automatic spine tissue segmentation from MRI data based on cascade of boosted classifiers and active appearance model. *BioMed Res. Int.* **2018**, *2018*, 7952946. [CrossRef]
51. Engstrom, C.M.; Fripp, J.; Jurcak, V.; Walker, D.G.; Salvado, O.; Crozier, S. Segmentation of the quadratus lumborum muscle using statistical shape modeling. *J. Magn. Reson. Imaging* **2011**, *33*, 1422–1429. [CrossRef] [PubMed]
52. Baum, T.; Lorenz, C.; Buerger, C.; Freitag, F.; Dieckmeyer, M.; Eggers, H.; Zimmer, C.; Karampinos, D.C.; Kirschke, J.S. Automated assessment of paraspinal muscle fat composition based on the segmentation of chemical shift encoding-based water/fat-separated images. *Eur. Radiol. Exp.* **2018**, *2*, 1–5. [CrossRef]
53. Zheng, Y.; Nixon, M.S.; Allen, R. Automated segmentation of lumbar vertebrae in digital videofluoroscopic images. *IEEE Trans. Med. Imaging* **2004**, *23*, 45–52. [CrossRef]
54. Jurcak, V.; Fripp, J.; Engstrom, C.; Walker, D.; Salvado, O.; Ourselin, S.; Crozier, S. Automated segmentation of the quadratus lumborum muscle from magnetic resonance images using a hybrid atlas based-geodesic active contour scheme. In Proceedings of the 2008 30th Annual International Conference of the IEEE Engineering in Medicine and Biology Society, Vancouver, BC, Canada, 20–25 August 2008; pp. 867–870. [CrossRef]
55. Fortin, M.; Omidyeganeh, M.; Battié, M.C.; Ahmad, O.; Rivaz, H. Evaluation of an automated thresholding algorithm for the quantification of paraspinal muscle composition from MRI images. *Biomed. Eng. Online* **2017**, *16*, 1–12. [CrossRef]
56. Neubert, A.; Fripp, J.; Engstrom, C.; Walker, D.; Weber, M.; Schwarz, R.; Crozier, S. Three-dimensional morphological and signal intensity features for detection of intervertebral disc degeneration from magnetic resonance images. *J. Am. Med. Inform. Assoc.* **2013**, *20*, 1082–1090. [CrossRef]
57. Oktay, A.B.; Akgul, Y.S. Localization of the Lumbar Discs Using Machine Learning and Exact Probabilistic Inference. In *Medical Image Computing and Computer-Assisted Intervention—MICCAI 2011*; Fichtinger, G., Martel, A., Peters, T., Eds.; Springer: Berlin/Heidelberg, Germany, 2011; pp. 158–165._20. [CrossRef]
58. Castro-Mateos, I.; Hua, R.; Pozo, J.M.; Lazary, A.; Frangi, A.F. Intervertebral disc classification by its degree of degeneration from T2-weighted magnetic resonance images. *Eur. Spine J.* **2016**, *25*, 2721–2727. [CrossRef]
59. Kim, K.B.; Park, H.J.; Song, D.H. Automatic Characterizations of Lumbar Multifidus Muscle and Intramuscular Fat with Fuzzy C-means based Quantization from Ultrasound Images. *Curr. Med. Imaging* **2020**, *16*, 592–600. [CrossRef]
60. Lui, D.; Scharfenberger, C.; De Carvalho, D.E.; Callaghan, J.P.; Wong, A. Semi-automatic Fisher-Tippett guided active contour for lumbar multifidus muscle segmentation. In Proceedings of the 2014 36th Annual International Conference of the IEEE Engineering in Medicine and Biology Society, Chicago, IL, USA, 26–30 August 2014; pp. 5530–5533. [CrossRef]

61. Ribeiro, E.A.; Nogueira-Barbosa, M.H.; Rangayyan, R.M.; Azevedo-Marques, P.M. Detection of vertebral plateaus in lateral lumbar spinal X-ray images with Gabor filters. In Proceedings of the 2010 Annual International Conference of the IEEE Engineering in Medicine and Biology, Buenos Aires, Argentina, 31 August–4 September 2010; pp. 4052–4055. [CrossRef]
62. Sa, R.; Owens, W.; Wiegand, R.; Chaudhary, V. Fast scale-invariant lateral lumbar vertebrae detection and segmentation in X-ray images. In Proceedings of the 2016 38th Annual International Conference of the IEEE Engineering in Medicine and Biology Society (EMBC), Orlando, FL, USA, 16–20 August 2016; pp. 1054–1057. [CrossRef]
63. Iriondo, C.; Pedoia, V.; Majumdar, S. Lumbar intervertebral disc characterization through quantitative MRI analysis: An automatic voxel-based relaxometry approach. *Magn. Reson. Med.* **2020**, *84*, 1376–1390. [CrossRef] [PubMed]
64. Staartjes, V.E.; Seevinck, P.R.; Vandertop, W.P.; van Stralen, M.; Schröder, M.L. Magnetic resonance imaging–based synthetic computed tomography of the lumbar spine for surgical planning: A clinical proof-of-concept. *Neurosurg. Focus* **2021**, *50*, E13. [CrossRef] [PubMed]
65. Lee, J.H.; Han, I.H.; Kim, D.H.; Yu, S.; Lee, I.S.; Song, Y.S.; Joo, S.; Jin, C.B.; Kim, H. Spine computed tomography to magnetic resonance image synthesis using generative adversarial networks: A preliminary study. *J. Korean Neurosurg. Soc.* **2020**, *63*, 386. [CrossRef] [PubMed]
66. Fan, G.; Liu, H.; Wang, D.; Feng, C.; Li, Y.; Yin, B.; Zhou, Z.; Gu, X.; Zhang, H.; Lu, Y.; et al. Deep learning-based lumbosacral reconstruction for difficulty prediction of percutaneous endoscopic transforaminal discectomy at L5/S1 level: A retrospective cohort study. *Int. J. Surg.* **2020**, *82*, 162–169. [CrossRef] [PubMed]
67. Malinda, V.; Lee, D. Lumbar Vertebrae Synthetic Segmentation in Computed Tomography Images Using Hybrid Deep Generative Adversarial Networks. In Proceedings of the 2020 42nd Annual International Conference of the IEEE Engineering in Medicine & Biology Society (EMBC), Montreal, QC, Canada, 20–24 July 2020; pp. 1327–1330. [CrossRef]
68. Siemionow, K.; Luciano, C.; Forsthoefel, C.; Aydogmus, S. Autonomous image segmentation and identification of anatomical landmarks from lumbar spine intraoperative computed tomography scans using machine learning: A validation study. *J. Craniovertebral Junction Spine* **2020**, *11*, 99._37_20. [CrossRef] [PubMed]
69. Netherton, T.J.; Rhee, D.J.; Cardenas, C.E.; Chung, C.; Klopp, A.H.; Peterson, C.B.; Howell, R.M.; Balter, P.A.; Court, L.E. Evaluation of a multiview architecture for automatic vertebral labeling of palliative radiotherapy simulation CT images. *Med. Phys.* **2020**, *47*, 5592. [CrossRef]
70. Watanabe, K.; Aoki, Y.; Matsumoto, M. An application of artificial intelligence to diagnostic imaging of spine disease: Estimating spinal alignment from moire images. *Neurospine* **2019**, *16*, 697. [CrossRef]
71. Kim, S.; Bae, W.C.; Masuda, K.; Chung, C.B.; Hwang, D. Fine-grain segmentation of the intervertebral discs from MR spine images using deep convolutional neural networks: BSU-Net. *Appl. Sci.* **2018**, *8*, 1656. [CrossRef] [PubMed]
72. Shen, H.; Huang, J.; Zheng, Q.; Zhu, Z.; Lv, X.; Liu, Y.; Wang, Y. A Deep-Learning–Based, Fully Automated Program to Segment and Quantify Major Spinal Components on Axial Lumbar Spine Magnetic Resonance Imaging. *Phys. Ther.* **2021**, *101*, pzab041. [CrossRef]
73. Gaonkar, B.; Villaroman, D.; Beckett, J.; Ahn, C.; Attiah, M.; Babayan, D.; Villablanca, J.; Salamon, N.; Bui, A.; Macyszyn, L. Quantitative Analysis of Spinal Canal Areas in the Lumbar Spine: An Imaging Informatics and Machine Learning Study. *Am. J. Neuroradiol.* **2019**, *40*, 1586–1591. [CrossRef] [PubMed]
74. Huang, J.; Shen, H.; Wu, J.; Hu, X.; Zhu, Z.; Lv, X.; Liu, Y.; Wang, Y. Spine Explorer: A deep learning based fully automated program for efficient and reliable quantifications of the vertebrae and discs on sagittal lumbar spine MR images. *Spine J.* **2020**, *20*, 590–599. [CrossRef]
75. Li, H.; Luo, H.; Huan, W.; Shi, Z.; Yan, C.; Wang, L.; Mu, Y.; Liu, Y. Automatic lumbar spinal MRI image segmentation with a multi-scale attention network. *Neural Comput. Appl.* **2021**, 1–14. [CrossRef]
76. Li, H.; Luo, H.; Liu, Y. Paraspinal muscle segmentation based on deep neural network. *Sensors* **2019**, *19*, 2650. [CrossRef] [PubMed]
77. Zhou, J.; Damasceno, P.F.; Chachad, R.; Cheung, J.R.; Ballatori, A.; Lotz, J.C.; Lazar, A.A.; Link, T.M.; Fields, A.J.; Krug, R. Automatic Vertebral Body Segmentation Based on Deep Learning of Dixon Images for Bone Marrow Fat Fraction Quantification. *Front. Endocrinol.* **2020**, *11*, 612. [CrossRef]
78. Jamaludin, A.; Lootus, M.; Kadir, T.; Zisserman, A.; Urban, J.; Battié, M.C.; Fairbank, J.; McCall, I. ISSLS PRIZE IN BIOENGINEERING SCIENCE 2017: Automation of reading of radiological features from magnetic resonance images (MRIs) of the lumbar spine without human intervention is comparable with an expert radiologist. *Eur. Spine J.* **2017**, *26*, 1374–1383. [CrossRef]
79. Natalia, F.; Meidia, H.; Afriliana, N.; Young, J.C.; Yunus, R.E.; Al-Jumaily, M.; Al-Kafri, A.; Sudirman, S. Automated measurement of anteroposterior diameter and foraminal widths in MRI images for lumbar spinal stenosis diagnosis. *PLoS ONE* **2020**, *15*, e0241309. [CrossRef]
80. Zhou, Y.; Liu, Y.; Chen, Q.; Gu, G.; Sui, X. Automatic lumbar MRI detection and identification based on deep learning. *J. Digit. Imaging* **2019**, *32*, 513–520. [CrossRef]
81. Forsberg, D.; Sjöblom, E.; Sunshine, J.L. Detection and labeling of vertebrae in MR images using deep learning with clinical annotations as training data. *J. Digit. Imaging* **2017**, *30*, 406–412. [CrossRef] [PubMed]
82. Baka, N.; Leenstra, S.; van Walsum, T. Ultrasound aided vertebral level localization for lumbar surgery. *IEEE Trans. Med. Imaging* **2017**, *36*, 2138–2147. [CrossRef] [PubMed]

83. Cho, B.H.; Kaji, D.; Cheung, Z.B.; Ye, I.B.; Tang, R.; Ahn, A.; Carrillo, O.; Schwartz, J.T.; Valliani, A.A.; Oermann, E.K.; et al. Automated measurement of lumbar lordosis on radiographs using machine learning and computer vision. *Glob. Spine J.* **2020**, *10*, 611–618. [CrossRef] [PubMed]
84. Li, Y.; Liang, W.; Zhang, Y.; An, H.; Tan, J. Automatic lumbar vertebrae detection based on feature fusion deep learning for partial occluded C-arm X-ray images. In Proceedings of the 2016 38th Annual International Conference of the IEEE Engineering in Medicine and Biology Society (EMBC), Orlando, FL, USA, 16–20 August 2016; pp. 647–650. [CrossRef]
85. Sa, R.; Owens, W.; Wiegand, R.; Studin, M.; Capoferri, D.; Barooha, K.; Greaux, A.; Rattray, R.; Hutton, A.; Cintineo, J.; et al. Intervertebral disc detection in X-ray images using faster R-CNN. In Proceedings of the 2017 39th Annual International Conference of the IEEE Engineering in Medicine and Biology Society (EMBC), Jeju, Korea, 11–15 July 2017; pp. 564–567. [CrossRef]
86. Liu, X.; Deng, Z.; Yang, Y. Recent progress in semantic image segmentation. *Artif. Intell. Rev.* **2019**, *52*, 1089–1106. [CrossRef]

International Journal of
Environmental Research and Public Health

MDPI

Review

Epidural Steroid Injections for Low Back Pain: A Narrative Review

Massimiliano Carassiti [1,*], Giuseppe Pascarella [1], Alessandro Strumia [1], Fabrizio Russo [2], Giuseppe Francesco Papalia [2], Rita Cataldo [1], Francesca Gargano [1], Fabio Costa [1], Michelangelo Pierri [3], Francesca De Tommasi [4], Carlo Massaroni [4], Emiliano Schena [4] and Felice Eugenio Agrò [1]

1 Unit of Anaesthesia, Intensive Care and Pain Management, Department of Medicine, Campus Bio-Medico University of Rome, 00128 Rome, Italy; g.pascarella@unicampus.it (G.P.); a.strumia@unicampus.it (A.S.); r.cataldo@unicampus.it (R.C.); f.gargano@unicampus.it (F.G.); f.costa@unicampus.it (F.C.); f.agro@unicampus.it (F.E.A.)

2 Department of Orthopaedic and Trauma Surgery, Campus Bio-Medico University of Rome, 00128 Rome, Italy; fabrizio.russo@unicampus.it (F.R.); g.papalia@unicampus.it (G.F.P.)

3 Integrated Sleep Surgery Team UCBM, Unit of Otolaryngology, Integrated Therapies in Otolaryngology, Campus Bio-Medico University of Rome, 00128 Rome, Italy; michelangelo.pierri@unicampus.it

4 Unit of Measurements and Biomedical Instrumentation, Department of Engineering, Campus Bio-Medico University of Rome, 00128 Rome, Italy; f.detommasi@unicampus.it (F.D.T.); c.massaroni@unicampus.it (C.M.); e.schena@unicampus.it (E.S.)

* Correspondence: m.carassiti@unicampus.it

Abstract: Low back pain represents a significant socioeconomic burden. Several nonsurgical medical treatments have been proposed for the treatment of this disabling condition. Epidural steroid injections (ESIs) are commonly used to treat lumbosacral radicular pain and to avoid surgery. Even though it is still not clear which type of conservative intervention is superior, several studies have proved that ESIs are able to increase patients' quality of life, relieve lumbosacral radicular pain and finally, reduce or delay more invasive interventions, such as spinal surgery. The aim of this narrative review is to analyze the mechanism of action of ESIs in patients affected by low back pain and investigate their current application in treating this widespread pathology.

Keywords: epidural steroid injections; low back pain; lumbosacral radicular pain; disk herniation; canal stenosis; review

Citation: Carassiti, M.; Pascarella, G.; Strumia, A.; Russo, F.; Papalia, G.F.; Cataldo, R.; Gargano, F.; Costa, F.; Pierri, M.; De Tommasi, F.; et al. Epidural Steroid Injections for Low Back Pain: A Narrative Review. *IJERPH* **2022**, *19*, 231. https://doi.org/10.3390/ijerph19010231

Academic Editor: Paul B. Tchounwou

Received: 3 December 2021
Accepted: 22 December 2021
Published: 26 December 2021

1. Introduction

Low back pain (LBP) and lumbosacral radicular pain are common causes of physical and mental morbidity and they are also a significant economic burden, causing an expenditure of more than USD 100 billion per year in the United States alone [1,2]. In the medical literature, low back pain is referred to as sciatica, lumbosacral radicular syndrome, lumbar radiculopathy, nerve root pain and nerve root entrapment/irritation, and is commonly described as a pain starting in the back and radiating to the legs. The etiological cause of low back pain is first represented by intervertebral disk disease. The pathophysiological changes involved in the intervertebral disk disease may lead to disk herniation or degenerative diseases, such as canal stenosis or chronic instability of the diseased segments. The most common cause of sciatica is the herniation of the nucleus pulposus, a component of the intervertebral disk in the lumbar region, which causes stenosis and inflammation [3,4]. Some estimate that sciatica caused by herniation of the lumbar disk has a prevalence of 9.8 out of 1000 [5], meaning that of all reported cases of sciatica it appears that 90% are caused by herniation of the lumbar disk [6].

Several nonsurgical medical treatments have been proposed for lumbosacral radicular pain, from lifestyle changes, exercise and physical therapy to analgesic local/oral drugs and epidural steroid injections (ESIs) [7,8]. The conservative management of LBP aims to delay or avoid surgery. As a matter of fact, LBP can improve spontaneously or with conservative treatment. Cases which do not respond to treatment are candidates for surgical intervention.

An ESI is a common and minimally invasive procedure, performed to successfully treat lumbosacral radicular pain, which has also proved its effectiveness in the treatment of back acute pain and leg symptoms. The injections are used to deliver steroids, and sometimes local anesthetics, to the epidural space, directly to the site that causes the pain using a caudal, interlaminar or transforaminal approach [9]. The epidural injection is a well-founded anesthetic and analgesic technique; moreover, nowadays, new technological devices can help anesthesiologists to learn and to administer it [10–16]. Even though it is still not clear which type of conservative intervention is superior, several studies have proved that an ESI is able to increase patients' quality of life, relieve lumbosacral radicular pain and finally, reduce or delay more invasive interventions, such as spinal surgery. Although ESIs should represent a treatment of choice in the case of acute LBP or leg pain, in our research we focused on the efficacy of ESIs in the treatment of chronic LBP.

The aim of this narrative review is to analyze the mechanism of action of ESIs in chronic lumbar pain patients and to understand their current use, application and success in treating this significant widespread pathology.

2. Materials and Methods

A literature review using online databases was carried out regarding the use of epidural steroid injections for lumbar canal stenosis and disk herniation. Articles were extracted from PubMed, Google Scholar, MEDLINE, UpToDate, Embase and Web of Science, combining the terms "spinal disease," "radicular pain," "spinal stenosis," "canal stenosis," "disk herniation" and "epidural steroid injection" as keywords for the research. Only papers in the English language and regarding human studies were taken into consideration. Non-English language studies were excluded. Scientific publications up to September 2021were included. Only papers focusing on epidural steroid injections for lumbar canal stenosis or disk herniation were included. All reference lists of the relevant studies were then screened to identify any missing publications. The search and the study selection were performed by two investigators (G.P.; A.S.) working independently. At the first level, the titles and abstracts of identified studies were screened. At the second level, the full texts were retrieved and assessed. Ethical approval and patient informed consent were not required because this was a review of previously published studies and did not involve direct contact with patients or alterations to patient care. Any discrepancies were resolved by a third author (M.C.) through consensus. The following data were extracted from each eligible study: first author's name; publication year; study design; intervention protocol type (the type and amount of steroid and local anesthetic used for the ESI and therapies or medication used for conservative treatment); outcome parameters including Visual Analogic Scale (VAS), Numeric Rating Scale (NRS), Oswestry Disability Index (ODI) and successful events; and the summary of findings.

3. Results

3.1. Mechanism of Lumbosacral Radicular Pain

Low back pain and radicular pain are caused by interrelated biomechanical and biochemical factors. With the advancement of age and the presence of chronic diseases such as diabetes, obesity, smoking and overload, a series of degenerative processes occur inside the intervertebral disk [17]. The intervertebral disk is approximately 7 to 10 mm thick and 4 cm in diameter and is formed of two different components: the nucleus pulposus, rich in water and glycopeptides, and the anulus fibrosus, constituted of a series of 15 to 25 rings, or lamellae, with collagen fibers parallel to the lamellae in addition to elastin fibers. A thin hyaline cartilage endplate is the interface between the disk and the superior and inferior vertebrae bodies. When the nucleus becomes less elastic and the anulus less continent due to aging, dehydration, inflammatory conditions and/or prolonged misusage of the back, a part of the nucleus can herniate, usually backward. This causes an inflammatory state in the epidural space and the increase in cytokines and other inflammation mediators. This condition, on the one hand favors the ulterior herniation of the nucleus pulposus and on the other, it compresses and stimulates the spinal nerve roots, resulting in back and radicular pain [18,19].

Generally, sciatica from lumbar disk herniation is a self-limiting condition that improves in weeks or months without medical intervention; in some cases, rest, analgesic drugs and a structured exercise program may be needed. Usually, the inflammatory state is more important than the mechanical compression in the pathogenesis and the chronicity of the disease, unless there are no neurological deficits [20]. However, in patients who are refractory to conservative treatment, surgery is usually recommended.

Lumbar spinal canal stenosis is a process that could be part of the aging process and can be related to herniation of an intervertebral disk. Other common causes of stenosis are: congenital deformities; spondylolisthesis; osteophytes; arthritic degeneration; synovial cysts; hypertrophy of the facet joints; hypertrophy of the ligamentum flavum; epidural lipomatosis; spondylosis of the intervertebral disk margins; previous surgery; and neoplastic diseases. All these factors could cause lumbar nerve root compression with microvascular ischemia, axonal injury, intraneural fibrosis and an inflammatory state, leading to chronic back pain [21].

3.2. Rationale of Epidural Steroid Injections

Epidural injections are performed using a Tuohy needle with the tip placed inside the epidural space, which is located between the ligamentum flavum and the dura mater. Usually, the epidural space is localized thanks to the loss of resistance (LOR) technique, where the needle is advanced between the spinal processes of the vertebras with the help of a syringe full of air or saline solution, which is used to continuously test the pressure on the piston of the syringe. The needle passes through the ligamentum flavum and, when the epidural space is reached, a loss of resistance is felt by the operator on the syringe piston. Moreover, epidural injections can also be performed rapidly under CT and navigation guidance (Figure 1). These techniques can be used to precisely guide needle placement, allowing for the visualization of the optimal needle path and identification of potential problems, such as narrow intralaminar spaces and spinal stenosis, before needle insertion (Figure 2).

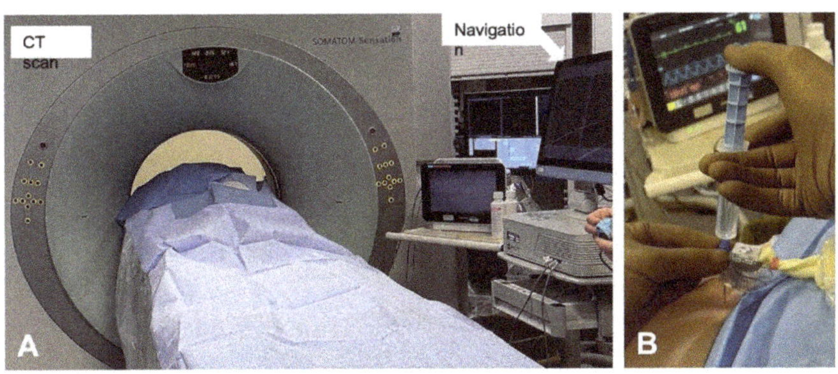

Figure 1. Representative images of (**A**) intraoperative setting for CT and navigation guided epidural injection and (**B**) navigated needle insertion.

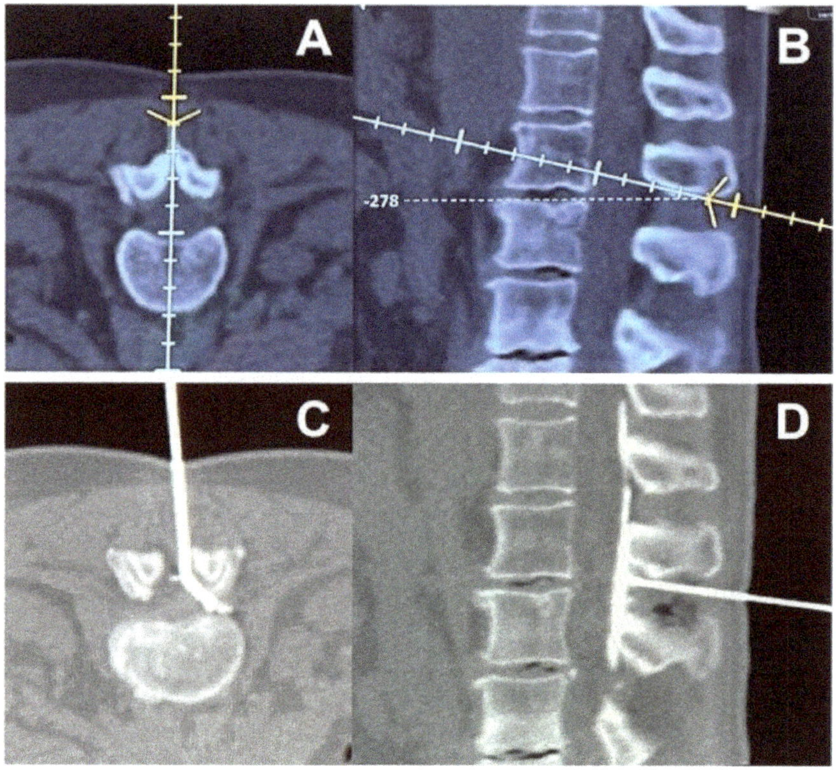

Figure 2. Representative images of (**A**,**B**) navigated planning of needle path to the epidural space and (**C**,**D**) evidence of injection in the epidural space by the use of a contrast agent.

Corticosteroids are widely used in regional anesthesia and chronic pain procedures, such as epidural injections, intraarticular injections and nerve blocks. Corticosteroids have a similar structure and activity to the endogenous produced hormone cortisol, which has an anti-inflammatory, immunosuppressive, vasoconstrictive and antiproliferative effect. They work by preventing the enzyme PLA2 from liberating arachidonic acid from the cells. This inhibits the cyclo-oxygenase and lipoxygenase production, which is responsible for the level of prostaglandins, thromboxanes and leukotrienes, before finally decreasing the inflammatory state [22]. They also inhibit the nerve transmission in nociceptive C fibers and reduce vasal permeability, which decreases intraneural and perineural oedema. Local anesthetics have been administered in the epidural space since 1901; however, the epidural use of corticosteroids has only been documented since 1952 [23]. Their efficacy when administrated via epidural injections has been demonstrated in various studies and a stronger effect has been proven in patients with a higher protein count in the cerebrospinal fluid, which is usually associated with an inflammatory state [24]. The prolonged use of corticosteroids at high doses has many systemic side effects and can also result in iatrogenic adrenal gland suppression; however, the epidural administration limits the systemic side effects because a smaller dose is necessary to reach the pharmacological target and its diffusion into systemic circulation is more difficult than in other types of administration [25]. In the Yang et al. [26] meta-analysis regarding lumbosacral radicular pain due to any cause, the use of ESIs resulted in the more effective in control of lumbosacral pain compared to pure conservative treatment, both in short and intermediate terms. However, two other recent meta-analyses have shown a similarity in efficacy and duration, in terms of pain reduction and functional gain, between local anesthetic alone or local anesthetic with a corticosteroid epidural injection [27,28].

3.3. Epidural Steroid Injections for Disk Herniation Lumbar Pain

The epidural administration of corticosteroids is one of the most common mini-invasive medical treatments for chronic spinal pain caused by disk herniation [29,30]. In fact, in the absence of chronic severe pain or neurological deficit, epidural steroid injections may be the treatment of choice for disk herniation. As mentioned before, it reduces the concentration of inflammatory mediators in the epidural space and vascular permeability [31]; it also reduces the damage of C fibers, which diminishes the pain [32,33]. In particular, the anesthetic effect of methylprednisolone over other steroids and non-steroid anti-inflammatory drugs has been proven when injected into the epidural space [34].

In clinical practice, both corticosteroids and local anesthetics are used [35,36]; the former are used to reduce the inflammation for a prolonged time, while the latter are used to mitigate the discomfort of the procedure and immediately decrease pain.

The difference in the use of a local anesthetic alone or local anesthetic with a corticosteroid in the epidural administration to treat disk herniation pain has been indagated in numerous papers without a clear result [37,38], although a meta-analysis written by Lee at al. evidenced a small difference between the epidural injection of lidocaine and lidocaine plus corticosteroids [39].

However, a good number of studies have described significant pain relief and improvement of functional status after an ESI, especially in short–medium terms [40–45] (Table 1). In fact, Kennedy and colleagues have found a high rate of success of ESIs at 6 months in their study, but there was also a recurrence of the symptoms during the 5 years follow-up after the injection [46]. In a similar way, Buchner et al. found a significant improvement in patients treated with epidural steroid injections for a very short period after the treatment but no improvement was seen after 6 weeks and 6 months, compared to the control group who did not receive the injection [47]. A response to the treatment after 1 h of having the procedure has been suggested as predictive for favorable medium-term success [48]. Interestingly, Buttermann et al. suggested that ESIs could be more effective in patients who presented magnetic resonance imaging of inflammatory endplate changes [49].

On the other hand, the Spine Patient Outcomes Research Trial (SPORT) [50], a prospective multi-center study of the operative versus nonoperative treatment of lumbar intervertebral disk herniation, found no improvement in short- or long-term outcomes in patients who received ESIs compared to patients who did not. However, it is important to say that an increased rate of surgical avoidance was observed in the group treated with ESIs; this could underline the role of conservative treatments, also considering the high incidence of the spontaneous reabsorption of lumbar disk herniation (66.66% according to Zhong et al.) [51].

Finally, Kreiner et al. [52], in their guideline for the diagnosis and treatment of disk herniation with radicular pain, stated that an ESI is indicated for a proportion of patients with lumbar disk herniation to provide symptom relief in the short term (2–4 weeks) with a grade A recommendation. Additionally, at the moment, no sufficient evidence exists to make a recommendation regarding the 12-month, or more, efficacy of ESIs.

Table 1. Epidural steroid injections for disk herniation studies.

Author, Year	Study Design	Study Protocol		Outcome Measures	Summary of Findings
		Steroid Injection	Control		
Sariyildiz MA, 2017 [40]	Prospective (repeated measures)	Transforaminal, 40 mg betamethasone + lidocaine 2%	Baseline	VAS, Oswestry Disability Index (ODI), hospital anxiety and depression scale, and Pittsburgh Sleep Quality Index (PSQI)	Compared to baseline measurements, there were significant improvements (> 50%) in radicular pain, ODI, depressive symptoms and PSQI scores at two weeks and 12 months after injection
Guclu B, 2020 [41]	Prospective (repeated measures)	Transforaminal 3 mL 0.33% lidocaine + 4 mg dexamethasone	Baseline	VAS scores at 12 weeks	Transforaminal epidural steroid injection is effective in relieving radicular pain, especially in paramedian lumbar disk herniation
Kennedy DJ, 2017 [46]	Prospective	Transforaminal epidural steroid injection	Baseline	Presence of recurrent or persistent pain, pain within the previous week, current opioid use for radicular symptoms, need for additional spinal injections, progression to surgery and unemployment due to pain	Despite a high success rate at 6 months, the majority of subjects experienced a recurrence of symptoms at some time during the subsequent 5 years. Few reported current symptoms and a small minority required additional injections, surgery or opioid pain medications
Manchikanti L, 2014 [29]	RCT, double-blind	Transforaminal 1% lidocaine, followed by 3 mg or 0.5 mL betamethasone	1.5 mL 1% lidocaine + 0.5 mL sodium chloride	Numeric Rating Scale (NRS), Oswestry Disability Index 2.0 (ODI), opioid intake	At 2 years, there was significant improvement in all participants, although there was a lack of evidence of the superiority of steroids compared to local anesthetic
Manchikanti L, 2014 [30]	RCT, double-blind	Interlaminar 0.5% lidocaine (6 mL) + 1 mL betamethasone	0.5% lidocaine (6 mL)	Numeric Rating Scale (NRS), Oswestry Disability Index 2.0 (ODI), opioid intake	Improvement in 70% of the steroid group and 60% of the control group at the end of 2 years.
Manchikanti L, 2013 [31]	RCT, double-blind	Interlaminar 0.5% lidocaine (5 mL) + 1 mL betamethasone	0.5% lidocaine (6 mL)	Pain relief and functional status improvement of $\geq 50\%$	Average relief of 33.7 ± 18.1 weeks in the local anesthetic group and 39.1 ± 12.2 weeks in the local anesthetic and steroid group
Buchner M, 2000 [47]	RCT	Interforaminal 100 mg methylprednisolone in 10 mL bupivacaine 0.25%	10 mL bupivacaine 0.25%	VAS, straight leg raising test and functional status	No significance on pain relief, improvement of straight leg raising and improvement of functional status at 6 weeks and 6 months

Table 1. *Cont.*

Author, Year	Study Design	Study Protocol		Outcome Measures	Summary of Findings
		Steroid Injection	Control		
Vad VB, 2002 [44]	RCT	Transforaminal epidural steroid injection	Saline trigger-point injection	VAS, patient satisfaction scale, Roland–Morris low back pain questionnaire	At 1.4 years, the group receiving transforaminal epidural steroid injections had a success rate of 84%, vs. 48% for the control group
Butterman GR, 2004 [49]	Prospective	Epidural steroid injection	Baseline	VAS, Oswestry Disability Index [ODI], pain diagram	At 2 years, it was beneficial for a small number of patients with advanced disk degeneration and chronic low back pain. It was more effective in discogenic inflammation
Manchikanti L, 2008 [45]	RCT	Caudal epidural injections with 9 mL 0.5% lidocaine mixed with 1 mL steroid (6 mg betamethasone or 40 mg methylprednisolone)	Caudal epidural injections with 0.5% lidocaine 9 mL	NRS, ODI, opiod intake	Comparable efficacy in both groups at 12 months
Radcliff K, 2012 [50]	Prospective	Epidural steroid injection	No epidural steroid injection	VAS, ODI, patient satisfaction	No improvement in short- or long-term outcomes (4 years) compared to patients who were not treated with ESIs
Şencan S, 2020 [48]	Retrospective	Transforaminal epidural steroid injection	Baseline	NRS	A decreased pain scores at 1 h is a predictor for a favorable 3-month response to an ESI

LDH = lumbar disk herniation; VAS = visual analogue scale; FU = follow-up; ESI = epidural steroid injections.

3.4. Epidural Steroid Injections for Canal Stenosis Lumbar Pain

The administration of steroids via epidural injection as a nonsurgical treatment for lumbar spinal stenosis (LSS) has been analyzed in various studies but, again, there is not a clear consensus regarding their efficacy in relieving the symptoms, especially in the long-term follow-ups. It is important to notice that a consistent number of studies have reported some degree of benefit, especially regarding short-term improvements [53–59] (Table 2). A more favorable response seems to be associated with relative youth, female sex and patients with single level stenosis, while BMI, MRI severity and the dimension of the spinal canal are probably not predictive [55,60,61]. Additionally, individual pain sensitivity does not seem to influence the outcome of an ESI in the patients affected by LSS [62]. Interestingly, Milburn and colleagues, in a randomized study, suggested that the response to the treatment is maximized when the ESI is performed at the intervertebral level of maximal stenosis [63], and their result was confirmed by the trial conducted a few years later by Bajpai et al. [64].

A randomized, double-blind controlled trial with a 2-year follow-up was conducted by Manchikanti et al., which compared the epidural injection of local anesthetic alone to local anesthetic plus steroids, and the authors found a significant relief of the symptoms in a convincing percentage of the patients treated, but without significant difference between the two groups [65]. Accordingly, another large randomized trial on 400 patients, conducted by Friedly and colleagues, found minimal or no short-term benefits in adding steroids to a local anesthetic epidural injection for the treatment of LSS [66].

Moreover, some other studies did not find any significant improvement in symptoms or quality of life after an ESI for the treatment of LSS [67–70]. Tran et al. wrote a review regarding the nonsurgical treatment of LSS and concluded that the literature could provide only limited evidence to formulate recommendations pertaining to the nonsurgical treatment of LSS [71].

Finally, Liu et al., in their systematic review and metanalysis, also concluded that there is minimal evidence to show that epidural steroids are better than local anesthetic alone in the treatment of LSS patients [72].

Table 2. Epidural steroid injections for canal stenosis studies.

Authors, Year	Study Design	Study Protocol		Outcome Measures	Summary of Findings
		Intervention	Control		
Sabbaghan S et al., 2020 [53]	Retrospective, single arm	Bupivacaine hydrochloride 0.5% (3 mL) + triamcinolone acetonide 80 mg (2 mL)	Baseline	VAS for lumbar pain, VAS for lower limb pain and ODI	Improvement in pain (both lumbar than lower limb) and ability
Park CH et al., 2014 [54]	Prospective, single arm	2 mg preservative-free ropivacaine + 1500 units hyaluronidase + 40 mg triamcinolone acetonide	Baseline	5-point patient satisfaction scale at 2 and 8 weeks post-treatment	The ESI seems to provide effective short-term pain relief from LSS. The grade of LSS appears to have no effect on the degree of pain relief
Cosgrove JL et al., 2011 [55]	Prospective, single arm	Steroids, not specified	Baseline	Self-reported activity level and measured walking ability using the SSSQ and SMWT. The results were correlated through demographic data, magnetic resonance imaging (MRI) characteristics and electrodiagnostic results	Significant improvement as measured by changes in SMWT and SSSQ. Relative youth and female sex are associated with a more favorable response
Farooque M et al., 2017 [56]	Case series	10 mg dexamethasone (1 mL) + an equal volume of 2% preservative-free lidocaine on each side (transforaminal)	Baseline	Pain score and Swiss Spinal Stenosis score at baseline, 1, 3 and 6 months	Reduction of ≥ 50% in numeric pain scale score in 30% of participants at 1 month, 53% at 3 months and 44% at 6 months. Swiss Spinal Stenosis subscale scores indicated a significant reduction in the proportion of participants reporting the presence of severe pain in the back, buttocks and legs during FU compared to baseline ($p < 0.05$)
Hammerich A et al., 2019 [57]	Randomized parallel-group trial	1.5 mL steroid (not specified) at each site injected. Reassess at 3–4 and 6–8 weeks for potential second and third injections	1.5 mL of steroid (not specified) at each site injected. Reassess at 3–4 and 6–8 weeks for potential second and third injections + physical therapy (PT)	Disability, pain, quality of life and global rating of change were collected at 10 weeks, 6 months and 1 year, and then analyzed using linear mixed model analysis	The ESI plus PT was not superior to ESI alone for reducing disability in individuals with LSS.

Table 2. *Cont.*

Authors, Year	Study Design	Study Protocol		Outcome Measures	Summary of Findings
		Intervention	Control		
Brown LL et al., 2012 [58]	Randomized controlled trial	80 mg triamcinolone acetate (40 mg for diabetic patients) mixed with 6 mL preservative-free saline injected in divided doses at the treated levels	MILD procedure: a minimally invasive posterior lumbar decompression performed fluoroscopically through a small 6-gauge port	Visual Analog Scale, Oswestry Disability Index and Zurich Claudication Questionnaire (ZCQ) for patient satisfaction	MILD procedure was superior compared to ESI in pain reduction and the improvement of functional mobility
Kim HJ et al., 2013 [62]	Prospective, single arm	40 mg (1 mL) triamcinolone acetonide suspension + 1 mL bupivacaine hydrochloride 0.5% + 1 mL of saline	Baseline	Pain sensitivity questionnaire (PSQ), Oswestry Disability Index (ODI), and Visual Analog Scale (VAS) for back and leg pain	Significant decreases in VAS for back/leg pain and ODI 2 months after ESI. Individual pain sensitivity does not influence the outcomes of ESI treatment in patients with LSS
Campbell MJ et al., 2007 [60]	Controlled clinical trial	Steroids (not specified) once a week for 3 weeks	Baseline	Spinal canal dimension, resolution of symptoms after ESI, necessity of surgery after ESI	Spinal canal dimension is not predictive of the success or failure of ESIs in patients with LSS
Milburn J et al., 2014 [63]	Randomized controlled trial	2 mL 40 mg/mL methylprednisolone + 2 mL bupivacaine 0.25% + 2 mL normal saline at the most stenotic level	2 mL 40 mg/mL methylprednisolone + 2 mL bupivacaine 0.25% + 2 mL normal saline at 2 intervertebral levels cephalad of the level of maximal stenosis	Analog pain scale from 0 to 10 during ambulation and at rest, Roland–Morris Disability Questionnaire (RDQ) at baseline, immediately after ESI and at 1, 4, and 12 weeks post-injection	Symptom improvement is optimized when the ESI is performed at the intervertebral level of maximal stenosis
Bajpai S et al., 2020 [64]	Randomized controlled trial	5 mL bupivacaine (0.25%) + 2 mL methylprednisolone acetate (40 mg/mL) + 1 mL normal saline at maximal stenotic intervertebral level	5 mL bupivacaine (0.25%) + 2 mL methylprednisolone acetate (40 mg/mL) + 1 mL normal saline 2 intervertebral levels cephalad to the level of maximal stenosis	Numeric Pain Rating Scale (NPRS) and Oswestry Disability Index (ODI) at 2, 6 and 12 weeks after the intervention	Pain relief improvement is optimized when the ESI is performed at the maximum stenotic intervertebral level

Table 2. *Cont.*

Authors, Year	Study Design	Study Protocol		Outcome Measures	Summary of Findings
		Intervention	Control		
Koc Z et al., 2009 [59]	Randomized controlled trial	Group 1: inpatient physical therapy program for 2 weeks Group 2: 60 mg triamcinolone acetonide + 15 mg 0.5% bupivacaine hydrochloride + 5.5 mL saline	Group 3: no intervention	Pain severity by Visual Analog Scale (VAS), Finger Floor Distance (FFD) (cm), Treadmill Walk Test, Sit-to-Stand Test (Seconds), Weight-Carrying (WC) Test (Seconds)	Both ESI and physical therapy groups demonstrated an improvement in symptoms and in outcomes measured without any significant differences
Manchikanti L et al., 2015 [65]	Randomized controlled trial	Local anesthetic (lidocaine 0.5%) 5 mL mixed + 6 mg betamethasone (1 mL)	6 mL local anesthetic (lidocaine 0.5%)	Numeric Pain Rating Rcale (NPRS) and Oswestry Disability Index (ODI) at 3, 6, 12, 18 and 24 months post-treatment	Epidural injections of local anesthetic with or without steroids provide relief in a significant proportion of patients with LSS
Shamov T et al., 2020 [61]	Prospective, two arms	Group 1: 10 mg dexamethasone in 3 cc 0.25% bupivacaine for patients with discogenic sciatica Group2: 10 mg dexamethasone in 3 cc 0.25% bupivacaine for patients with LSS		Pain intensity was assessed by VAS at baseline and on days 1, 15 and 30 after intervention	ESIs are more effective in patients with discogenic sciatica than in single level LSS. In multiple level LSS, results are disappointing
Friedly JL et al., 2014 [66]	Randomized controlled trial	1 to 3 mL 0.25% to 1% lidocaine followed by 1 to 3 mL triamcinolone (60 to 120 mg), betamethasone (6 to 12 mg), dexamethasone (8 to 10 mg) or methylprednisolone (60 to 120 mg)	1 to 3 mL 0.25% to 1% lidocaine alone	Roland–Morris Disability Questionnaire (RDQ) and the rating of leg pain intensity (on a scale from 0 to 10) at 6 weeks after injection	Epidural injections of glucocorticoids plus lidocaine offered minimal or no short-term benefits compared to epidural injections of lidocaine alone
Makris UE et al., 2017 [66]	Subsequent analysis of a randomized controlled trial (Friedly JL et al., 2014 [58])	1 to 3 mL 0.25% to 1% lidocaine followed by 1 to 3 mL triamcinolone (60 to 120 mg), betamethasone (6 to 12 mg), dexamethasone (8 to 10 mg) or methylprednisolone (60 to 120 mg)	1 to 3 mL 0.25% to 1% lidocaine alone	RDQ (Roland–Morris Disability Questionnaire), Sickness Impact Profile (SIP) weights assigned to the RDQ items and patient-prioritized RDQ items at 6 weeks after injection	No significant difference between groups for the RDQ or patient-prioritized RDQ, and while the difference between groups for RDQ using SIP weights was statistically significant, this was not clinically important

Table 2. *Cont.*

Authors, Year	Study Design	Study Protocol		Outcome Measures	Summary of Findings
		Intervention	Control		
Tomkins-Lane CC et al., 2012 [68]	Prospective, single arm	Steroids, not specified	Baseline	Total activity (performance) measured over 7 days and maximum continuous activity (capacity), walking capacity was also assessed with the Self-Paced Walking Test and subjects completed the ODI, SSSQ, Medical Outcomes Study 36-Item Short-Form Health Survey, Visual Analog Pain scales and body diagrams	At 1 week postinjection, 58.8% of the subjects demonstrated increased total activity and 53% had increased maximum continuous activity, although neither change was statistically significant. Patients perceived improvements in symptoms, but these were not reflected in significant changes in performance or capacity
Rivest C et al., 1998 [69]	Prospective, two arms	Group 1: 3 mL 0.5% lidocaine and 3 mL methylprednisolone acetate, followed by 3 mL saline for patients with LSS Group 2: 3 mL 0.5% lidocaine and 3 mL methylprednisolone acetate, followed by 3 mL saline for patients with discogenic sciatica		Pain was assessed at baseline and 2 weeks following a single ESI using a Visual Analog Scale	LSS patients have worse responses to ESIs than herniated disk patients
Fukusaki M et al., 1998 [70]	Randomized controlled trial	Group 3: 8 mL 1% mepivacaine and 40 mg methylprednisolone	Group 1: 8 mL saline Group 2: 8 mL 1% mepivacaine	Evaluation of improvement on pseudo-claudication associated with LSS as follows: excellent effect, > 100 m in walking distance; good effect, 20–100 m in walking distance; poor effect, < 20 m in walking distance	ESIs have no beneficial effects on walking ability associated with LSS compared to epidural injections with a LA alone

ESI = epidural steroid injection; LSS = lumbar spinal stenosis; LA = local anesthetic; FU = follow-up; VAS = Visual Analog Scale; ODI = Oswestry Disability Index; SSSQ = Swiss Spinal Stenosis Questionnaire; SMWT = 6-Minute Walk Test; MRI = magnetic resonance imaging; PT = physical therapy; MILD = minimally invasive lumbar decompression; ZCQ = Zurich Claudication Questionnaire; PSQ = Pain Sensitivity Questionnaire; FFD = Finger Floor Distance; SIP = Sickness Impact Profile.

3.5. Image-Guided Epidural Injections

Special consideration should be given to image-guided techniques that may help the clinician in performing epidural injections, although no significant difference has been shown regarding outcomes. The use of ultrasound in performing an interlaminar approach could help to estimate the distance from the skin to the epidural space and the optimal needle direction [73,74]. However, although it is a radiation-free technique, ultrasound is limited by the operator's experience and the real-time visualization of needle tip advancement could be challenging, especially in obese patients [75]. Fluoroscopy (x-rays) and computerized tomography (CT) have both proved to be effective and safe techniques for guiding transforaminal epidural injections, although the former provides less radiation exposure for patients [76,77]. Moreover, new generation CT devices may integrate neuronavigation systems that are able to perform a computerized analysis in order to best define the needle's path towards the epidural space. Its use has been described for spinal surgery but it may be expanded to transforaminal epidural injections as well, although the high cost of these devices should be considered [78]. Future studies are expected to determine the best technique in terms of efficacy and safety for both patients and clinicians.

4. Discussion

Chronic lumbar pain is a widespread problem, which affects a large part of population at some point of their life. Disk herniation and canal stenosis are the most common causes and they need to be treated due to the high impact of the symptoms on patients' quality of life, especially because they could affect walking and ability to work [18].

Surgical intervention has been proven to be effective but is not usually considered as the first option [7]. On the other hand, nonsurgical treatments, such as epidural steroid injections, do not have clear literature consensus. Fully understanding whether ESIs would be able to relieve symptoms and delay or prevent surgery could be a crucial step, especially because chronic back pain patients are typically elderly and multimorbid who could be more affected by the impact of surgery.

In this narrative review, we tried to analyze the existing literature regarding the use ESIs for these kinds of patients, considering randomized controlled studies as well as reviews, meta-analyses and guidelines.

Overall, ESIs seem to be safe and quite effective in relieving the main symptoms, especially in short-term follow-ups, and in delaying surgery, according to a consistent number of studies and to the guideline written by Kreiner and colleagues [52]. Moreover, ESIs could be more powerful in the case of patients with disk herniation than patients with canal stenosis [61]. Attention should also be paid to the technique and to the vertebral level of the injection, at least for spinal stenosis [63,64].

As mentioned before, literature consensus is still missing and numerous studies did not find significative improvements, especially in long-term follow-ups. In addition, it seems to be difficult to find significative differences between using local anesthetics alone or local anesthetics plus steroids in the injection.

Due to the anti-inflammatory action of steroids, patients with high local inflammatory status could probably benefit more from using steroids [49]. However, epidural steroid injections have also been associated with potential adverse effects, including acute neurological symptoms [79,80], in addition to other possible complications related to the epidural technique, i.e., inadvertent dural puncture, hematomas and infection. [81]

In this context, the identity of the patients who could benefit the most from this procedure has not been completely established yet and could be a crucial future goal.

This narrative review has some limitations. Firstly, the studies taken into consideration did have different epidural injection approaches. Secondly, the heterogeneity of the enrolled patients, analyzed parameters and data collection in the studies taken into consideration could be an important bias. Lastly, the heterogeneity of the purposes of the studies, such as the comparison of steroid injections versus nothing or local anesthetic injections versus local anesthetic plus steroids, could make global analysis difficult.

Surely, more randomized studies with larger numbers of patients are needed to fully understand the efficacy of ESIs and define which patients could benefit more from the procedure, especially in order to delay or prevent surgery.

5. Conclusions

According to the literature analyzed in this narrative review, there is no consensus on the use of ESIs for patients with chronic lumbar pain. ESIs seem to be effective in relieving symptoms in the short term and delaying surgery, while evidence of any long-terms benefits is still lacking. More studies are needed to better understand which patients could benefit more from epidural steroid injections.

Author Contributions: Conceptualization, M.C. and G.P.; methodology, G.F.P. and M.P.; software, C.M.; validation, A.S., G.P. and M.C.; formal analysis, F.D.T.; investigation, A.S. and E.S.; resources, R.C.; data curation, F.R.; writing—original draft preparation, A.S.; writing—review and editing, F.C. and G.F.P.; visualization, F.G. and F.R.; supervision, M.C.; project administration, F.E.A. All authors have read and agreed to the published version of the manuscript.

Funding: This research received no external funding.

Institutional Review Board Statement: Not applicable.

Informed Consent Statement: Not applicable.

Data Availability Statement: Not applicable.

Conflicts of Interest: The authors declare no conflict of interest.

References

1. Dieleman, J.L.; Cao, J.; Chapin, A.; Chen, C.; Li, Z.; Liu, A.; Horst, C.; Kaldjian, A.; Matyasz, T.; Scott, K.; et al. US Health Care Spending by Payer and Health Condition, 1996–2016. *JAMA* **2020**, *323*, 863–884. [CrossRef]
2. Mokdad, A.H.; Ballestros, K.; Echko, M.; Glenn, S.; Olsen, H.E.; Mullany, E.; Lee, A.; Khan, A.R.; Ahmadi, A.; Ferrari, A.J.; et al. The State of US Health, 1990–2016: Burden of Diseases, Injuries, and Risk Factors Among US States. *JAMA* **2018**, *319*, 1444–1472. [PubMed]
3. Kennedy, D.; Plastaras, C.; Casey, E.; Visco, C.J.; Rittenberg, J.; Conrad, B.; Sigler, J.; Dreyfuss, P. Comparative Effectiveness of Lumbar Transforaminal Epidural Steroid Injections with Particulate Versus Nonparticulate Corticosteroids for Lumbar Radicular Pain due to Intervertebral Disc Herniation: A Prospective, Randomized, Double-Blind Trial. *Pain Med.* **2014**, *15*, 548–555. [CrossRef] [PubMed]
4. Spijker-Huiges, A.; Winters, J.C.; Van Wijhe, M.; Groenier, K. Steroid injections added to the usual treatment of lumbar radicular syndrome: A pragmatic randomized controlled trial in general practice. *BMC Musculoskelet. Disord.* **2014**, *15*, 341. [CrossRef]
5. Zhai, J.; Zhang, L.; Li, M.; Tian, Y.; Zheng, W.; Chen, J.; Huang, T.; Li, X.; Tian, Z. Epidural injection with or without steroid in managing chronic low back and lower extremity pain: Ameta-analysis of ten randomized controlled trials. *Int. J. Clin. Exp. Med.* **2015**, *8*, 8304–8316. [CrossRef] [PubMed]
6. Lee, J.; Shin, J.-S.; Lee, Y.J.; Kim, M.-R.; Ahn, Y.-J.; Park, K.B.; Kropf, M.A.; Shin, B.-C.; Lee, M.S.; Ha, I.-H. Effects of Shinbaro pharmacopuncture in sciatic pain patients with lumbar disc herniation: Study protocol for a randomized controlled trial. *Trials* **2015**, *16*, 455. [CrossRef]
7. Chen, B.-L.; Guo, J.-B.; Zhang, H.-W.; Zhang, Y.-J.; Zhu, Y.; Zhang, J.; Hu, H.-Y.; Zheng, Y.-L.; Wang, X.-Q. Surgical versus non-operative treatment for lumbar disc herniation: A systematic review and meta-analysis. *Clin. Rehabilit.* **2018**, *32*, 146–160. [CrossRef]
8. Carassiti, M.; Di Martino, A.; Centonze, A.; Quattrocchi, C.C.; Caldaria, A.; Agrò, F.; Denaro, V. Failed back surgery syndrome: A new strategy by the epidural injection of MESNA. *Musculoskelet. Surg.* **2018**, *102*, 179–184. [CrossRef]
9. Kozlov, N.; Benzon, H.T.; Malik, K.M. Epidural steroid injections: Update on efficacy, safety, and newer medications for injection. *Minerva Anestesiol.* **2015**, *81*, 901–909.
10. Carassiti, M.; Cataldo, R.; Formica, D.; Massaroni, C.; De Filippis, A.; Palermo, P.; Di Tocco, J.; Setola, R.; Valenti, C.; Schena, E. A new pressure guided management tool for epidural space detection: Feasibility assessment in a clinical scenario. *Minerva Anestesiol.* **2020**, *86*, 736–741. [CrossRef] [PubMed]
11. Carassiti, M.; Mattei, A.; Quarta, R.; Massaroni, C.; Saccomandi, P.; Tesei, M.; Setola, R.; Schena, E. A New Pressure Guided Management Tool for Epidural Space Detection: Feasibility Assessment on a Simulator. *Artif. Organs* **2017**, *41*, E320–E325. [CrossRef]

12. Tesei, M.; Saccomandi, P.; Massaroni, C.; Quarta, R.; Carassiti, M.; Schena, E.; Setola, R. A cost-effective, non-invasive system for pressure monitoring during epidural needle insertion: Design, development and bench tests. In Proceedings of the 2016 38th Annual International Conference of the IEEE Engineering in Medicine and Biology Society (EMBC), Orlando, FL, USA, 16–20 August 2016; pp. 194–197.
13. Carassiti, M.; Quarta, R.; Mattei, A.; Tesei, M.; Saccomandi, P.; Massaroni, C.; Setola, R.; Schena, E. Ex vivo animal-model assessment of a non-invasive system for loss of resistance detection during epidural blockade. In Proceedings of the 2017 39th Annual International Conference of the IEEE Engineering in Medicine and Biology Society (EMBC), Jeju, Korea, 11–15 July 2017; pp. 759–762.
14. Carassiti, M.; Pascarella, G.; Strumia, A.; Cataldo, R.; Antinolfi, V.; Costa, F.; Agrò, F.E. Pressure monitoring devices may undetect epidural space: A report on the use of Compuflo®system for epidural injection. *J. Clin. Monit.* **2021**. [CrossRef]
15. de Tommasi, F.; Lo Presti, D.; Virgili, F.; Massaroni, C.; Schena, E.; Carassiti, M. Soft System Based on Fiber Bragg Grating Sensor for Loss of Resistance Detection during Epidural Procedures: In Silico and In Vivo Assessment. *Sensors* **2021**, *21*, 5329. [CrossRef] [PubMed]
16. de Tommasi, F.; Presti, D.L.; Massaroni, C.; Schena, E.; Carassiti, M. FBG-based System for Loss of Resistance Detection During Epidural Injections. In Proceedings of the 2021 IEEE International Workshop on Metrology for Industry 4.0 & IoT (MetroInd4.0&IoT), Rome, Italy, 7–9 June 2021; IEEE: Rome, Italy, 2021; pp. 172–176.
17. Huang, W.; Han, Z.; Liu, J.; Yu, L.; Yu, X. Risk Factors for Recurrent Lumbar Disc Herniation: A Systematic Review and Meta-Analysis. *Medicine* **2016**, *95*, e2378. [CrossRef] [PubMed]
18. Wheeler, A.H.; Murrey, D.B. Chronic lumbar spine and radicular pain: Pathophysiology and treatment. *Curr. Pain Headache Rep.* **2002**, *6*, 97–105. [CrossRef] [PubMed]
19. McCarron, R.F.; Wimpee, M.W.; Hudkins, P.G.; Laros, G.S. The inflammatory effect of nucleus pulposus. A possible element in the pathogenesis of low-back pain. *Spine* **1987**, *12*, 760–764. [CrossRef]
20. Deyo, R.A.; Mirza, S.K. Clinical Practice. Herniated Lumbar Intervertebral Disk. *N. Engl. J. Med.* **2016**, *374*, 1763–1772. [CrossRef]
21. Deer, T.; Sayed, D.; Michels, J.; Josephson, Y.; Li, S.; Calodney, A.K. A Review of Lumbar Spinal Stenosis with Intermittent Neurogenic Claudication: Disease and Diagnosis. *Pain Med.* **2019**, *20*, S32–S44. [CrossRef]
22. Coutinho, A.E.; Chapman, K.E. The anti-inflammatory and immunosuppressive effects of glucocorticoids, recent developments and mechanistic insights. *Mol. Cell. Endocrinol.* **2011**, *335*, 2–13. [CrossRef]
23. Manchikanti, L.; Knezevic, N.N.; Parr, A.; Kaye, A.D.; Sanapati, M.; Hirsch, J.A. Does Epidural Bupivacaine with or Without Steroids Provide Long-Term Relief? A Systematic Review and Meta-analysis. *Curr. Pain Headache Rep.* **2020**, *24*, 26. [CrossRef]
24. Bogduk, N. Epidural steroids. *Spine* **1995**, *20*, 845–848. [CrossRef] [PubMed]
25. Yasir, M.; Goyal, A.; Bansal, P.; Sonthalia, S. Corticosteroid Adverse Effects. In *StatPearls*; StatPearls Publishing: Treasure Island, FL, USA, 2021.
26. Yang, S.; Kim, W.; Kong, H.H.; Do, K.H.; Choi, K.H. Epidural steroid injection versus conservative treatment for patients with lumbosacral radicular pain: A meta-analysis of randomized controlled trials. *Medicine* **2020**, *99*, e21283. [CrossRef] [PubMed]
27. Knezevic, N.N.; Manchikanti, L.; Urits, I.; Orhurhu, V.; Vangala, B.P.; Vanaparthy, R.; Sanapati, M.R.; Shah, S.; Soin, A.; Mahajan, A.; et al. Lack of Superiority of Epidural Injections with Lidocaine with Steroids Compared to Without Steroids in Spinal Pain: A Systematic Review and Meta-Analysis. *Pain Physician* **2020**, *23*, 239.
28. Zhao, W.; Wang, Y.; Wu, J.; Gao, X.; Wei, Q.; Lai, X.; An, J. Long-Term Outcomes of Epidurals with Lidocaine With or Without Steroids for Lumbar Disc Herniation and Spinal Stenosis: A Meta-Analysis. *Pain Physician* **2020**, *23*, 365–374.
29. Manchikanti, L.; Soin, A.; Mann, D.P.; Bakshi, S.; Pampati, V.; Hirsch, J.A. Comparative Analysis of Utilization of Epidural Procedures in Managing Chronic Pain in the Medicare Population: Pre and Post Affordable Care Act. *Spine* **2019**, *44*, 220–232. [CrossRef] [PubMed]
30. Manchikanti, L.; Sanapati, M.R.; Pampati, V.; Soin, A.; Atluri, S.; Kaye, A.D.; Subramanian, J.; Hirsch, J.A. Update of Utilization Patterns of Facet Joint Interventions in Managing Spinal Pain from 2000 to 2018 in the US Fee-for-Service Medicare Population. *Pain Physician* **2020**, *23*, E133–E149. [CrossRef]
31. Manchikanti, L.; Gharibo, C.G.; Gilligan, C.J.; Hirsch, J.A.; Nampiaparampil, D.E.; Manchikanti, K.N.; Falco, F.E.; Singh, V.; Benyamin, R.M.; Kaye, A.D.; et al. Comparison of the efficacy of saline, local anesthetics, and steroids in epidural and facet joint injections for the management of spinal pain: A systematic review of randomized controlled trials. *Surg. Neurol. Int.* **2015**, *6*, 194–235. [CrossRef] [PubMed]
32. Lundin, A.; Magnuson, A.; Axelsson, K.; Nilsson, O.; Samuelsson, L. Corticosteroids Peroperatively Diminishes Damage to the C-Fibers in Microscopic Lumbar Disc Surgery. *Spine* **2005**, *30*, 2362–2367. [CrossRef]
33. Hayashi, N.; Weinstein, J.N.; Meller, S.T.; Lee, H.-M.; Spratt, K.F.; Gebhart, G.F. The Effect of Epidural Injection of Betamethasone or Bupivacaine in a Rat Model of Lumbar Radiculopathy. *Spine* **1998**, *23*, 877–885. [CrossRef]
34. Johansson, A.; Hao, J.; Sjölund, B. Local corticosteroid application blocks transmission in normal nociceptive C-fibres. *Acta Anaesthesiol. Scand.* **1990**, *34*, 335–338. [CrossRef]
35. Wong, S.H.; Wong, C.S.; Li, T.T. Steroids in regional analgesia. *Expert Opin. Pharmacother.* **2010**, *11*, 2839–2848. [CrossRef] [PubMed]

36. Shanthanna, H.; Busse, J.W.; Thabane, L.; Paul, J.; Couban, R.; Choudhary, H.; Kaushal, A.; Suzumura, E.; Kim, I.; Harsha, P. Local anesthetic injections with or without steroid for chronic non-cancer pain: A protocol for a systematic review and meta-analysis of randomized controlled trials. *Syst. Rev.* **2016**, *5*, 18. [CrossRef] [PubMed]
37. Manchikanti, L.; Cash, K.A.; Pampati, V.; Falco, F.J.E. Transforaminal epidural injections in chronic lumbar disc herniation: A randomized, double-blind, active-control trial. *Pain Physician* **2014**, *17*, 489–501. [CrossRef]
38. Manchikanti, L.; Singh, V.; Cash, K.A.; Pampati, V.; Falco, F.J.E. A randomized, double-blind, active-control trial of the effectiveness of lumbar interlaminar epidural injections in disc herniation. *Pain Physician* **2014**, *17*, 61–74. [CrossRef]
39. Lee, J.H.; Shin, K.-H.; Park, S.J.; Lee, G.J.; Lee, C.-H.; Kim, D.H.; Kim, D.H.; Yang, H.S. Comparison of Clinical Efficacy Between Transforaminal and Interlaminar Epidural Injections in Lumbosacral Disc Herniation: A Systematic Review and Meta-Analysis. *Pain Physician* **2018**, *21*, 433–448. [CrossRef]
40. Sariyildiz, M.A.; Batmaz, I.; Yazmalar, L.; Güneş, M.; Turan, Y. The effectiveness of transforaminal epidural steroid injections on radicular pain, functionality, psychological status and sleep quality in patients with lumbar disc herniation. *J. Back Musculoskelet. Rehabilitation* **2017**, *30*, 265–270. [CrossRef]
41. Guclu, B.; Deniz, L.; Yuce, Y.; Adilay, U.; Aytar, H.; Turkoglu, M.; Tiryaki, M.; Ozdek, R.; Boran, B.O. Transforaminal Epidural Steroid Injection in the Treatment of Pain in Foraminal and Paramedian Lumbar Disc Herniations. *Turk. Neurosurg.* **2020**, *30*, 394–399. [CrossRef]
42. Manchikanti, L.; Singh, V.; Cash, K.A.; Pampati, V.; Falco, F.J.E. The Role of Fluoroscopic Interlaminar Epidural Injections in Managing Chronic Pain of Lumbar Disc Herniation or Radiculitis: A Randomized, Double-Blind Trial. *Pain Pract.* **2012**, *13*, 547–558. [CrossRef]
43. Wang, J.C.; Lin, E.; Brodke, D.S.; Youssef, J.A. Epidural Injections for the Treatment of Symptomatic Lumbar Herniated Discs. *J. Spinal Disord. Tech.* **2002**, *15*, 269–272. [CrossRef]
44. Vad, V.B.; Bhat, A.L.; Lutz, G.E.; Cammisa, F. Transforaminal epidural steroid injections in lumbosacral radiculopathy: A prospective randomized study. *Spine* **2002**, *27*, 11–16. [CrossRef] [PubMed]
45. Manchikanti, L.; Singh, V.; Cash, K.A.; Pampati, V.; Damron, K.S.; Boswell, M.V. Preliminary results of a randomized, equivalence trial of fluoroscopic caudal epidural injections in managing chronic low back pain: Part 2—Disc herniation and radiculitis. *Pain Physician* **2008**, *11*, 801–815. [CrossRef] [PubMed]
46. Kennedy, D.; Zheng, P.Z.; Smuck, M.; McCormick, Z.L.; Huynh, L.; Schneider, B.J. A minimum of 5-year follow-up after lumbar transforaminal epidural steroid injections in patients with lumbar radicular pain due to intervertebral disc herniation. *Spine J.* **2018**, *18*, 29–35. [CrossRef]
47. Buchner, M.; Zeifang, F.; Brocai, D.R.; Schiltenwolf, M. Epidural Corticosteroid Injection in the Conservative Management of Sciatica. *Clin. Orthop. Relat. Res.* **2000**, *375*, 149–156. [CrossRef] [PubMed]
48. Şencan, S.; Çelenlioğlu, A.E.; Asadov, R.; Gündüz, O.H. Predictive factors for treatment success of transforaminal epidural steroid injection in lumbar disc herniation-induced sciatica. *Turk. J. Med Sci.* **2020**, *50*, 126–131. [CrossRef]
49. Buttermann, G.R. The effect of spinal steroid injections for degenerative disc disease. *Spine J.* **2004**, *4*, 495–505. [CrossRef] [PubMed]
50. Radcliff, K.; Hilibrand, A.; Lurie, J.D.; Tosteson, T.D.; Delasotta, L.; Rihn, J.; Zhao, W.; Vaccaro, A.; Albert, T.J.; Weinstein, J.N. The impact of epidural steroid injections on the outcomes of patients treated for lumbar disc herniation: A subgroup analysis of the SPORT trial. *J. Bone Jt. Surg. Am.* **2012**, *94*, 1353–1358. [CrossRef]
51. Zhong, M.; Liu, J.-T.; Jiang, H.; Mo, W.; Yu, P.-F.; Li, X.-C.; Xue, R.R. Incidence of Spontaneous Resorption of Lumbar Disc Herniation: A Meta-Analysis. *Pain Physician* **2017**, *20*, E45–E52.
52. Kreiner, D.S.; Hwang, S.W.; Easa, J.E.; Resnick, D.K.; Baisden, J.L.; Bess, S.; Cho, C.H.; DePalma, M.J.; Dougherty, P.; Fernand, R.; et al. An evidence-based clinical guideline for the diagnosis and treatment of lumbar disc herniation with radiculopathy. *Spine J.* **2014**, *14*, 180–191. [CrossRef]
53. Sabbaghan, S.; Mirzamohammadi, E.; Mahabadi, M.A.; Nikouei, F.; Rahbarian, F.; Ahmadichaboki, S.; Eftekhari, S.; Zamankhani, M.; Aghdam, A.A. Short-Term Efficacy of Epidural Injection of Triamcinolone Through Translaminar Approach for the Treatment of Lumbar Canal Stenosis. *Anesthesiol. Pain Med.* **2020**, *10*, e99764. [CrossRef]
54. Park, C.-H.; Lee, S.-H. Correlation between Severity of Lumbar Spinal Stenosis and Lumbar Epidural Steroid Injection. *Pain Med.* **2014**, *15*, 556–561. [CrossRef] [PubMed]
55. Cosgrove, J.L.; Bertolet, M.; Chase, S.L.; Cosgrove, G.K. Epidural Steroid Injections in the Treatment of Lumbar Spinal Stenosis Efficacy and Predictability of Successful Response. *Am. J. Phys. Med. Rehabilit.* **2011**, *90*, 1050–1055. [CrossRef] [PubMed]
56. Farooque, M.; Salzman, M.M.; Ye, Z. Effectiveness of Bilateral Transforaminal Epidural Steroid Injections in Degenerative Lumbar Spinal Stenosis Patients With Neurogenic Claudication: A Case Series. *PM&R* **2017**, *9*, 26–31. [CrossRef]
57. Hammerich, A.; Whitman, J.; Mintken, P.; Denninger, T.; Akuthota, V.; Sawyer, E.E.; Hofmann, M.; Childs, J.D.; Cleland, J. Effectiveness of Physical Therapy Combined With Epidural Steroid Injection for Individuals With Lumbar Spinal Stenosis: A Randomized Parallel-Group Trial. *Arch. Phys. Med. Rehabilit.* **2019**, *100*, 797–810. [CrossRef]
58. Brown, L.L. A Double-blind, Randomized, Prospective Study of Epidural Steroid Injection vs. The mild®Procedure in Patients with Symptomatic Lumbar Spinal Stenosis. *Pain Pract.* **2012**, *12*, 333–341. [CrossRef] [PubMed]
59. Koc, Z.; Ozcakir, S.; Sivrioglu, K.; Gurbet, A.; Kucukoglu, S. Effectiveness of Physical Therapy and Epidural Steroid Injections in Lumbar Spinal Stenosis. *Spine* **2009**, *34*, 985–989. [CrossRef] [PubMed]

60. Campbell, M.J.; Carreon, L.Y.; Glassman, S.D.; McGinnis, M.D.; Elmlinger, B.S. Correlation of Spinal Canal Dimensions to Efficacy of Epidural Steroid Injection in Spinal Stenosis. *J. Spinal Disord. Tech.* **2007**, *20*, 168–171. [CrossRef]
61. SShamov, T.; Al-Hashel, J.Y.; Rousseff, R.T. Fluoroscopic Epidural Steroid Injection: Pain Relief in Discogenic Sciatica Versus Lumbar Spinal Stenosis. A Study on Middle Eastern Patients. *Acta Med.* **2020**, *63*, 73–78. [CrossRef]
62. Kim, H.-J.; Yeom, J.S.; Lee, J.W.; Chang, B.-S.; Lee, C.-K.; Lee, G.-W.; Im, S.-B.; Kim, H.J. The Influence of Pain Sensitivity on the Treatment Outcome of Transforaminal Epidural Steroid Injection in Patients with Lumbar Spinal Stenosis. *Pain Pract.* **2014**, *14*, 405–412. [CrossRef] [PubMed]
63. Milburn, J.; Freeman, J.; Steven, A.; Altmeyer, W.; Kay, D. Interlaminar Epidural Steroid Injection for Degenerative Lumbar Spinal Canal Stenosis: Does the Intervertebral Level of Performance Matter? *Ochsner J.* **2014**, *14*, 62–66. [PubMed]
64. Bajpai, S.; Yelavarthi, R. Effect of intervertebral level on interlaminar epidural steroid injection in lumbar spinal canal stenosis: A randomized controlled trial. *Anesth. Essays Res.* **2020**, *14*, 100–103. [CrossRef]
65. Manchikanti, L.; Cash, K.A.; McManus, C.D.; Damron, K.S.; Pampati, V.; Falco, F.J.E. A randomized, double-blind controlled trial of lumbar interlaminar epidural injections in central spinal stenosis: 2-year follow-up. *Pain Physician* **2015**, *18*, 79–92. [CrossRef] [PubMed]
66. Friedly, J.L.; Comstock, B.A.; Turner, J.A.; Heagerty, P.J.; Deyo, R.A.; Sullivan, S.D.; Bauer, Z.; Bresnahan, B.W.; Avins, A.L.; Nedeljkovic, S.S.; et al. A Randomized Trial of Epidural Glucocorticoid Injections for Spinal Stenosis. *N. Engl. J. Med.* **2014**, *371*, 11–21. [CrossRef]
67. Makris, U.E.; Edwards, T.C.; Lavallee, D.C.; Bauer, Z.; Comstock, B.A.; Jarvik, J.G.; Patrick, D.L.; Lotfi, M.; Friedly, J.L. Patient Priority Weighting of the Roland Morris Disability Questionnaire Does Not Change Results of the Lumbar Epidural Steroid Injections for Spinal Stenosis Trial. *Spine* **2017**, *42*, 42–48. [CrossRef] [PubMed]
68. Tomkins-Lane, C.C.; Conway, J.; Hepler, C.; Haig, A.J. Changes in Objectively Measured Physical Activity (Performance) after Epidural Steroid Injection for Lumbar Spinal Stenosis. *Arch. Phys. Med. Rehabilit.* **2012**, *93*, 2008–2014. [CrossRef] [PubMed]
69. Rivest, C.; Katz, J.N.; Ferrante, F.M.; Jamison, R.N. Effects of epidural steroid injection on pain due to lumbar spinal stenosis or herniated disks: A prospective study. *Arthritis Care Res.* **1998**, *11*, 291–297. [CrossRef] [PubMed]
70. Fukusaki, M.; Kobayashi, I.; Hara, T.; Sumikawa, K. Symptoms of Spinal Stenosis Do Not Improve After Epidural Steroid Injection. *Clin. J. Pain* **1998**, *14*, 148–151. [CrossRef]
71. Tran, D.Q.H.; Duong, S.; Finlayson, R.J. Lumbar spinal stenosis: A brief review of the nonsurgical management. *Can. J. Anesth.* **2010**, *57*, 694–703. [CrossRef]
72. Wu, X.; Liu, K.; Liu, P.-C.; Liu, R.; Cai, M. Steroid for epidural injection in spinal stenosis: A systematic review and meta-analysis. *Drug Des. Dev. Ther.* **2015**, *9*, 707–716. [CrossRef]
73. Hashemi, M.; Dadkhah, P.; Taheri, M.; Abootorabi, S.M.H.S.; Naderi-Nabi, B.; Haji, S.M. Ultrasound-Guided Lumbar Transforaminal Epidural Injections; A Single Center Fluoroscopic Validation Study. *Bull. Emerg. Trauma* **2019**, *7*, 251–255. [CrossRef]
74. Park, K.D.; Kim, T.K.; Lee, W.Y.; Ahn, J.; Koh, S.H.; Park, Y. Ultrasound-Guided Versus Fluoroscopy-Guided Caudal Epidural Steroid Injection for the Treatment of Unilateral Lower Lumbar Radicular Pain. *Medicine* **2015**, *94*, e2261. [CrossRef]
75. Sahin, T.; Balaban, O. Lumbar Ultrasonography for Obstetric Neuraxial Blocks: Sonoanatomy and Literature Review. *Turk. J. Anesth. Reanim.* **2018**, *46*, 257–267. [CrossRef]
76. Bui, J.; Bogduk, N. A Systematic Review of the Effectiveness of CT-Guided, Lumbar Transforaminal Injection of Steroids. *Pain Med.* **2013**, *14*, 1860–1865. [CrossRef] [PubMed]
77. Dietrich, T.J.; Peterson, C.K.; Zeimpekis, K.G.; Bensler, S.; Sutter, R.; Pfirrmann, C.W.A. Fluoroscopy-guided versus CT-guided Lumbar Steroid Injections: Comparison of Radiation Exposure and Outcomes. *Radiology* **2019**, *290*, 752–759. [CrossRef] [PubMed]
78. Tehli, O.; Harman, F.; Temiz, C.; Kacar, Y.; Kazanci, B.; Daneyemez, M.; Solmaz, I.; Kural, C. The use of neuronavigation and intraoperative imaging in spinal stabilization surgery. *Turk. Neurosurg.* **2016**, *26*, 771–776. [CrossRef]
79. Viswanath, O.; Suthar, R.; Kannan, M.; Baskin, M. Post Procedural Complication following Cervical Epidural Local Anesthetic Injection: A Case Report. *Anesthesiol. Pain Med.* **2017**, *7*, e44636. [CrossRef]
80. Field, J.; Rathmell, J.P.; Stephenson, J.H.; Katz, N.P. Neuropathic Pain following Cervical Epidural Steroid Injection. *Anesthesiology* **2000**, *93*, 885–888. [CrossRef] [PubMed]
81. Kang, X.H.; Bao, F.P.; Xiong, X.X.; Li, M.; Jin, T.T.; Shao, J.; Zhu, S.M. Major complications of epidural anesthesia: A prospective study of 5083 cases at a single hospital. *Acta Anaesthesiol. Scand.* **2014**, *58*, 858–866. [CrossRef]

International Journal of
Environmental Research and Public Health

MDPI

Review

Psychological Approaches for the Integrative Care of Chronic Low Back Pain: A Systematic Review and Metanalysis

Giorgia Petrucci [1], Giuseppe Francesco Papalia [1], Fabrizio Russo [1,*], Gianluca Vadalà [1], Michela Piredda [2], Maria Grazia De Marinis [2], Rocco Papalia [1] and Vincenzo Denaro [1]

1 Department of Orthopaedic and Trauma Surgery, Campus Bio-Medico University of Rome, 00128 Rome, Italy; g.petrucci@unicampus.it (G.P.); g.papalia@unicampus.it (G.F.P.); g.vadala@unicampus.it (G.V.); r.papalia@unicampus.it (R.P.); denaro@unicampus.it (V.D.)

2 Research Unit Nursing Science, Campus Bio-Medico University of Rome, 00128 Rome, Italy; m.piredda@unicampus.it (M.P.); m.demarinis@unicampus.it (M.G.D.M.)

* Correspondence: fabrizio.russo@unicampus.it

Abstract: Chronic low back pain (CLBP) is the most common cause of disability worldwide, affecting about 12% to 30% of the adult population. Psychological factors play an important role in the experience of pain, and may be predictive of pain persistence, disability, and long-term sick leave. The aim of this meta-analysis was to identify and to describe the most common psychological approaches used to treat patients who suffer from CLBP. A systematic search was performed on PubMed/MEDLINE and Cochrane Central. Overall, 16 studies with a total of 1058 patients were included in the analysis. Our results suggest that cognitive behavioral therapy (CBT) and mindfulness-based stress reduction (MBSR) interventions are both associated with an improvement in terms of pain intensity and quality of life when singularly compared to usual care. Disability also improved in both groups when compared to usual care. Significant differences in fear-avoidance beliefs were noted in the CBT group compared to usual care. Therefore, psychological factors are related to and influence CLBP. It is crucial to develop curative approaches that take these variables into account. Our findings suggest that CBT and MBSR modify pain-related outcomes and that they could be implemented in clinical practice.

Keywords: low back pain; cognitive behavioral therapy; mindfulness-based stress reduction; depression; disability; fear-avoidance beliefs

Citation: Petrucci, G.; Papalia, G.F.; Russo, F.; Vadalà, G.; Piredda, M.; De Marinis, M.G.; Papalia, R.; Denaro, V. Psychological Approaches for the Integrative Care of Chronic Low Back Pain: A Systematic Review and Metanalysis. *IJERPH* **2022**, *19*, 60. https://doi.org/10.3390/ijerph19010060

Academic Editor: Paul B. Tchounwou

Received: 20 November 2021
Accepted: 14 December 2021
Published: 22 December 2021

1. Introduction

Chronic low back pain (CLBP) is the most common cause of disability worldwide [1,2], affecting about 12% to 30% of the adult population [3,4]. It is estimated that 50% to 80% of adults feel at least one episode of back pain during their lifetime [5]. Therefore, managing CLBP becomes crucial for both individuals and health care systems [1]. Chronic pain has a multidimensional nature and in addition to nociceptive and physiological aspects, it also includes aspects relating to the emotional and cognitive sphere [6]. Low back pain pathogenesis can also be diverse, including organic, non-specific etiology, and psychological causes [7,8]. Psychological factors play an important role in the experience of pain [9,10], as patients with CLBP who experience anxiety tend to exacerbate the painful sensation and increase illness behavior [11], catastrophizing pain [12–14]. These factors can make the pain experience, as well as the mechanical and physiological processes, last longer [15,16], causing physical and psychosocial disability [9]. In this regard, it has been shown that patients with CLBP suffering from depression experience higher levels of pain, functional disability, and lower levels of health-related quality of life (QoL) [16]. So, all psychological variables may be predictive of pain persistence, disability, long-term sick leave [11,15], significantly influencing the quality of life perceived by patients. Therefore, it is crucial to assess and address the psychological sphere as much as the other aspects, designing a holistic and integrative framework to treat patients affected by LBP [7,17]. The American Pain Society (APS) published specific evidence-based guidelines for an interdisciplinary treatment and rehabilitation (defined as an integrated intervention with rehabilitation plus a psychological and/or social/occupational component) as a treatment option for patients with chronic LBP [18]. With the advancements in health psychology, several approaches were implemented in the care of patients with chronic pain. To our knowledge, there exist different systematic reviews in literature [16,19–22] that analyze psychological approaches to treating patients who suffer from CLBP. These studies do not evaluate which approach is most used. Moreover, a comparison between different types of psychological approaches, in order to evaluate the effectiveness in terms of improvement of clinical outcomes, has not been performed. The objectives of this systematic review and meta-analysis are (1) to identify and to describe the most common psychological approaches used to treat patients who suffer from Chronic LBP, and (2) to study the effectiveness of these approaches in terms of reduction of pain, disability, fear-avoidance behaviors, anxiety, depression, and of increase in quality of life of patients with Chronic LBP.

2. Materials and Methods

This systematic review was performed in agreement with the Preferred Reporting Items for Systematic Reviews and Meta-Analysis (PRISMA) guidelines [23]. The protocol was previously registered on PROSPERO (registration number CRD42021255687). This review included only randomized clinical trials (RCTs) that assessed the effectiveness of the most common psychological approaches on quality of life (QoL), pain, disability and fear-avoidance behaviors in adult patients suffering from chronic low back pain (CLBP).

2.1. Inclusion Criteria

We included RCTs published in the last 25 years that included adult patients with CLBP; compared psychological interventions with either comparator (usual care such as health education, physical exercise, information package and waiting list); and assessed reduction of pain, disability, fear-avoidance behaviors, anxiety, depression, and increase in quality of life. Studies were excluded if they were not RCTs, if they analyzed acute or sub-acute low back pain and if they included back-surgery patients.

2.2. Search Methods

We performed a systematic literature search on the following databases: PubMed/MEDLINE, Scopus and Cochrane Central. No language restrictions were set. The search strategy was checked by three reviewers (G.P., G.F.P. and F.R.). We developed a specific question defining the intervention, the population, and outcomes to analyze (according with PICO method).

PICO methods. Definition of elements.

- Population: the reference population included not hospitalized patients suffering from chronic low back pain. The patients included should be at least 18 years old, and they did not have to undergo surgery.
- Interventions: Selected psychological approaches
- Comparison Intervention: usual care, education program, supportive care, physical exercise, physiotherapy and waiting list
- Outcomes: pain, disability, fear-avoidance, anxiety and depression reduction and the improvement of quality of life

The search string included the following keywords (both Mesh and free-terms in PubMed/MEDLINE): Low back pain OR "Low back pain *" OR lumbago OR "lower back pain" OR "lower back pain *" OR "Low Back Ache" OR "Low Back Ache *" OR "Low Backache" OR "Low Backache *"; cognitive behavioral therapy OR "behavioral treatment" OR "behavior treatment" OR "behavior therapy" OR "cognitive behavior treatment" OR "cognitive treatment" OR "cognitive therapy", Mindfulness OR Meditation OR "mindfulness meditation", "operant behavioral therapy", hypnotism OR hypnoanalysis

OR hypnotherapy, "acceptance and commitment therapy". The reference lists of the included RCTs were examined to choose additional studies for inclusion. After removing duplicates, two reviewers (G.P. and G.F.P.) independently analyzed the abstracts. Conflicts of opinion were solved discussing with a third reviewer (F.R.). In the end, the full texts were read and checked by two reviewers (G.P. and G.F.P.), choosing the studies to include in the review and meta-analysis.

2.3. Data Collection, Analysis, and Outcomes

Two authors (G.P. and G.F.P.) independently extracted the following data from the studies selected: authors, year of publication, country, sample size, patients' age and sex, intervention (s) in the experimental and in the control group, follow-up period, outcomes analyzed, tools used and conclusions.

2.4. Risk of Bias Assessment

Two independent reviewers (G.P. and G.F.P.) evaluated the risk of bias of the included RCTs using the Cochrane risk-of-bias tool [24]. Possible differences in the assessment were checked by a third reviewer (F.R.). Each item was classified with a low, unclear, or high risk of bias. Thus, the studies present low risk of bias in case of six or seven domains at low risk of bias, unclear risk of bias in presence of four or five domains at low risk of bias, and high risk of bias if fewer than four domains were reported at low risk of bias.

2.5. Statistical Analysis

A meta-analysis was produced by Review Manager (RevMan) software Version 5.4.1. Pain, disability, quality of life, depression and fear-avoidance beliefs were assessed between CBT, MBRS and control groups as continuous outcomes. In presence of different scores, the relative outcome was presented as standard mean difference (SMD) with 95% confidence intervals, while we adopted mean difference (MD) for the outcomes assessed by the same score. Instead, days without pain was calculated as a dichotomous outcome using odds ratio (OR) with 95% confidence intervals. The evaluation of the samples' weight for this outcome was assessed by the mean value of days without pain per number of patients as events and the number of patients per number of weeks of follow-up as total. The I^2 test was adopted to check the heterogeneity of studies included. In case of low heterogeneity ($I^2 < 55\%$), a fixed-effect model was used, otherwise, we adopted a random-effect model. The statistical significance of the results was set at $p < 0.05$.

3. Results

3.1. Results of the Literature Search

The literature search yielded 3277 articles. After removal of duplicates, the reading of titles and abstracts led to 48 eligible papers. All 48 full-texts were read. Afterwards, 32 studies were eliminated for these reasons: patients who suffered from acute pain ($n = 8$), patients who suffered from sub-acute low back pain ($n = 7$), not reporting selected outcomes ($n = 5$), back surgery patients ($n = 5$), inpatients ($n = 4$), pediatric patients ($n = 2$), and hypochondriacal patients ($n = 1$). At the end of selection, 16 RCTs were included in the systematic review and meta-analysis (Figure 1).

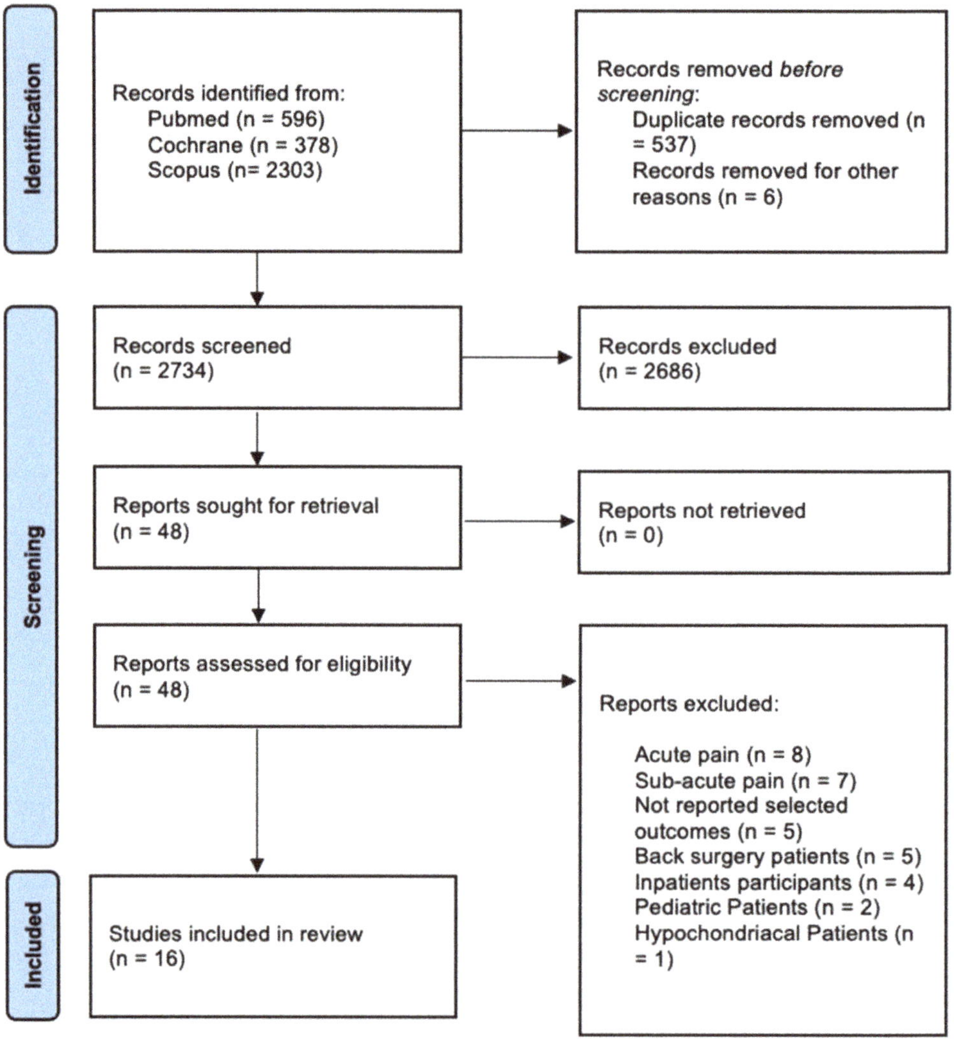

Figure 1. Preferred reporting items for systematic review and meta-analysis (PRISMA 2020).

3.2. Demographic Data

The total sample consisted of 2038 adults with CLBP reviewed—1058 were in the intervention group and 980 were in the control group. Most studies were published in the USA (*n* = 8; 50%), two studies were published in UK (12.5%) and Germany (12.5%), one study was published in Italy (6.25%), in The Netherlands (6.25%), in Pakistan (6.25%) and in Sweden (6.25%). The age of the patients ranged from 40.7 to 78 years in the experimental groups, and from 40.5 to 75.6 in the control groups. The percentage of women in the studies ranged from 13% to 80% in the intervention groups and from 6% to 87% in the control groups. In Table 1 the main characteristics of included studies and samples are reported.

Table 1. Main characteristics of the included studies and samples.

Author	Year	Country	Study Group N.	Study Group Age	Study Group Sex	Control Group N.	Control Group Age (Years)	Control Group Sex
Cherkin et al.	2016	USA	116	50 ± 11.9	71% F 29% M	113	48.9 ± 12.5	87% F 13% M
			112	49.1 ± 12.6	66% F 34% M			
Monticone et al.	2013	Italy	45	49 ± 8	60% F 40% M	45	49.7 ± 7	55% F 45% M
Johnson et al.	2007	UK	116	47.3 ± 10.9	61% F 39% M	118	48.5 ± 11.4	58% F 42% M
Smeets et al.	2006	The Netherlands	58	42.5 ± 9.7	58.6% F 41.4% M	53	42.7 ± 9.1	41.5% F 58.5% M
			61	40.7 ± 10.1	37.7% F 62.3% M	51	40.5 ± 11.2	37.3% F 62.7% M
Rutledge et al.	2018	USA	30	62.5 ± 11.3	13% F 87% M	31	64.3 ± 12.7	6% F 92% M
Rutledge et al.	2018	USA	33	54 ± 14.8	37.5% F 62.5% M	33	52.6 ± 12.5	39.4% F 60.6% M
Linden et al.	2014	Germany	53	50.4 ± 6.9	68% F 32% M	50	49.7 ± 7	68% F 32% M
Khan et al.	2016	Pakistan	27	39.61 ± 5.3	54% F 46% M	27	39.61 ± 5.3	54% F 46% M
Pincus et al.	2015	UK	45	43.7 ± 16.3	60% F 40% M	44	45.4 ± 15.8	38.6% F 61.4% M
Basler et al.	1997	Germany	36	49.3 ± 9.7	75.6% F 24.4% M	40	49.3 ± 9.7	75.6% F 24.4%M
Linton et al.	2000	Sweden	107	44	70% F 30% M	70	45	71% F 29% M
						66	44	74% F 26% M
Zgierska et al.	2016	USA	21	51.8 ± 9.7	80% F 20% M	14	51.8 ± 9.7	80% F 20% M
Morone et al.	.2008	USA	19	74.1 ± 6.1	53% F 47% M	18	75.6 ± 5	61% F 39% M
Morone et al.	2009	USA	16	78 ± 7.1	69% F 31% M	19	73 ± 6.2	58% F 42% M
Morone et al.	2016	USA	140	75 ± 7.2	66% F 34% M	142	74 ± 6.0	66% F 34% M
Day et al.	2019	USA	23	49.9 ± 11.9	61% F 39% M	23	48.1 ± 16.1	52% F 48% M
						23	54.3 ± 14.9	44% F 56% M

3.3. Type of Interventions

The psychological approaches most used are the cognitive behavioral therapy (CBT) and mindfulness-based stress reduction (MBSR). CBT was evaluated in eleven studies, while the remaining three studies [25–27] examined MBSR. Two studies evaluated both CBT and MBSR [9,28] versus usual care (Table 2). The mean follow-up was 7.8 months and ranged from 3 weeks to 15 months.

3.4. Clinical Outcome Data

The outcomes were analyzed by different tools. Disability was assessed using the Roland and Morris Disability Questionnaire (RMDQ) in 11 studies [9,13,26,27,29–34], the Oswestry Disability Index (ODI) in one study [35], the pain-related disability (PRD) in one study [36], the Dusseldorf disability scale in one study [37], and the PROMIS—physical function in one study [28]. Intensity of pain was assessed using the Numeric Rating Scale (NRS) in six studies [27–31,37], the Visual Analogue Scale (VAS) in four studies [12,31,35,37], the Brief Pain Inventory (BPI) in two studies [34,35], and the McGill pain Questionnaire Short Form and SF-36 pain scale in two studies [25,26]. Quality of life was assessed using the Short-Form Health Survey (SF-36) in five studies [25–27,29,34], the EQ-5D in two studies [32,34], and the Short Form Health Survey (SF-12) in one study [9]. Fear-avoidance behaviors were assessed using the Fear-Avoidance Belief Questionnaire (FABQ) in two studies [36,38], and the Tampa Scale of Kinesiophobia (TSK) in two studies [29,34]. Psychological disorders were assessed using the Generalized Anxiety Disorder 2-item (GAD-2) in three studies [8,33,37], the Personal Health Questionnaire Depression Scale (PHQ-8) in one study [9], the Hospital Anxiety and Depression Scale (HADS) in one study [34], the Beck's Depression Inventory (BDI) in two studies [13,31] and the PROMIS—depression in one study [28].

3.5. Methodological Evaluation

After the application of the Cochrane risk-of-bias tool, nine studies (56%) were at moderate risk of bias, four studies (25%) were at low risk of bias and three studies were determined to be at high risk of bias (Table 3).

Table 2. Clinical results of the included studies.

Study	Intervention (s)	Control	Follow-Up	Outcomes (Tool)	Conclusion
Cherkin et al., 2016	Mindfulness: body scan, yoga, meditation, for 8 weeks. CBT: education about chronic pain, relationships between thoughts and emotional and physical reactions, sleep hygiene, setting and working toward behavioral goals, relaxation skills, activity pacing, and pain-coping strategies, for 8 weeks	Usual care (whatever care participants received)	12 months	Disability (RMDQ) QoL (SF-12) Depression (PHQ-8) Anxiety (GAD-2)	Among adults with CLBP, treatment with MBSR or CBT, compared with usual care, resulted in greater improvement in back pain and functional limitations at 26 weeks, with no significant differences in outcomes between MBSR and CBT
Monticone et al., 2013	CBT: intervention to modify fear of movement beliefs, catastrophizing thinking, and negative feelings, and ensuring gradual reactions to illness behaviors, for 5 weeks	Active and passive mobilizations of the spine, and exercises aimed at stretching and strengthening muscles, and improving postural control, for 5 weeks	12 months	Disability (RMDQ) Pain (NRS) QOL (SF-36) Fear advoidance behaviours (TSK)	The long-lasting multidisciplinary program was superior to the exercise program in reducing disability, fear- avoidance beliefs and pain, and enhancing the quality of life of patients with chronic low back pain. The effects were clinically tangible and lasted for at least 1 year after the intervention ended.
Johnson et al., 2007	CBT: educational pack containing a booklet and audio-cassette + problem solving, pacing and regulation of activity, challenging distorted cognitions about activity and harm, for 6 week	Educational pack containing a booklet and audio-cassette + usual care for 6 weeks	15 months	Pain (VAS) Disability (RMDQ) QoL (EQ-5D)	CBT intervention program produces only modest effects in reducing LBP and disability over a 1-year period.

Table 2. *Cont.*

Study	Intervention (s)	Control	Follow-Up	Outcomes (Tool)	Conclusion
Smeets et al., 2006	CBT: operant behavioral graded activity training and problem solving training Active Physical Treatment (APT): aerobic training, and three dynamic static strengthening exercises for 4 weeks Combined Treatment (CT): CBT + APT	Waiting List (WL) for 10 weeks	12 months	Disability (RMDQ) Pain (VAS) Depression (BDI)	CBT are as effective in reducing the subjective experienced level of functioning
Rutledge et al., 2018	CBT: to provide core educational information, guide patients' learning and skills development, and structure self-monitoring exercises for the respective session, for 8 weeks	Supportive Care: - Education by distribution of a standard text - Active Listening by the therapist to participant's concerns - Supportive care following Rogerian principles	12 months	Disability (RMDQ) Pain (NRS) Depression (BDI)	No evidence of meaningful effect size differences between the treatments.
Rutledgeet al., 2018	CBT: managing pain, managing stress, thinking differently, assertive communication, setting goals for 8 weeks	Supportive Care: - Education by distribution of a standard text - Active Listening by the therapist to participant's concerns - Supportive care following Rogerian principles	12 months	Disability (RMDQ) Pain (NRS)	CBT versus SC therapy demonstrated statistically significant and comparable patterns of improved outcomes on measures of back pain disability, pain severity, and self- rated improvement.

Table 2. *Cont.*

Study	Intervention (s)	Control	Follow-Up	Outcomes (Tool)	Conclusion
Linden et al., 2014	general orthopedic inpatient treatment + therapy in reference to the GRIP and the pain and illness management program from Geissner at al. with additional cognitive behavior therapy interventions which aim at stress reduction and problem solving, self monitoring, pain management, change in dysfunctional cognitions, reduction of avoidance behavior, and wellbeing therapy for 3 weeks	General orthopedic inpatient treatment	3 weeks	Fear advoidance behaviours (FABQ) Pain (VAS) Pain related disability (PDI)	CBT can reduce back pain and increase functional coping, and that this is not mediated by an improvement in mental health and a reduction of depression, anxiety or somatization in general or by induc- tion of some general optimistic views. Pain is not identical with mental problems.
Khan et al., 2016	general exercise + CBT aimed to guide patients to achieve their daily life goals. CBT consisted of operant behavioural graded activity and problem solving training, for 12 weeks	General exercise at home 2 times per day and at least 5 times a week (for 12 weeks)	12 weeks	Disability (RMDQ) Pain (VAS)	This study found that both CBT with General exercises and General exercises alone significantly reduced pain intensity and disability in patients with chronic low back pain. Furthermore, subjects treated with CBT & Exercises showed an additional clinical benefit as compared to General Exercises only. Hence, CBT & Exercises could be a better option in clinical practice.
Pincus et al., 2015	Session content was not structured, and at the discretion of therapists, included any features of Contextual Cognitive-Behavioural Therapy (CCBT) they thought were appropriate at the point with that patient.	Physiotherapy, comprised back to fitness group exercises with at least 60% of content exercise-based.	3 months	Fear advoidance behaviours (TSK) pain (Brief Pain Inventory) disability (RMDQ) anxiety and depression (HADS) QoL (EQ-5D and SF-36)	CCBT is a credible and acceptable intervention for LBP patients who exhibit psychological obstacles to recovery.

Table 2. *Cont.*

Study	Intervention (s)	Control	Follow-Up	Outcomes (Tool)	Conclusion
Basler et al., 1997	medical treatment such as pain medication, nerve blocks, TENS, and physical therapy + CBT therapy: education, relaxation, Modifying thoughts and feelings, Pleasant activity scheduling, Training of posture	Medical treatment such as pain medication, nerve blocks, TENS, and physical therapy	6 months	Pain (NRS) Disability (Dusseldorf disability scale)	Experimental subjects reported less pain, more pleasurable activities and feelings, less avoidance and less catastrophizing, and disability was reduced. The results were maintained at follow-up. Patients who only received medical treatment showed little improvement. Data indicate that the program meets the needs of the patients and should be continued.
Linton et al., 2007	Sessions were organized to activate participants and promote coping. Each session began with a short review, in which homework was covered. The treatment lasts 6 weeks	1. pamphlet: straightforward advice about the best way to cope with back pain by remaining active and thinking positively. 2. Information package: advice and illustrations showing how the patient might cope with spinal pain or prevent it by such methods as lifting properly and maintaining good posture.	12 months	Pain (VAS) Depression and anxiety (HAD) Fear Advoidance (FABQ)	This study demonstrates that CBT group intervention can lower the risk of a long-term disability developing.
Zgierska et al., 2016	Usual care and opioid therapy management + manualized training in the meditation-CBT intervention 2 h per week for 8 weeks	Pharmacotherapy, opioid therapy management and physical therapy	26 weeks	Pain (Brief Pain Inventory) Disability (ODI)	Mindfulness meditation and CBT-based interventions have the potential to safely reduce pain severity and sensitivity in patients with opioid-treated CLBP
Morone et al., 2008	Mindfulness: body scan, sitting practice, walking meditation	Waiting List	3 months	Pain (McGill pain Questionnaire- Short Form and SF-36 pain scale) Disability (RMDQ) QoL (SF-36)	The mindfulness intervention sustained improvement in physical function and pain acceptance.

Table 2. *Cont.*

Study	Intervention (s)	Control	Follow-Up	Outcomes (Tool)	Conclusion
Morone et al., 2009	Mindfulness: body scan, sitting practice, walking meditation	Educational program (8 weeks), including lectures, group discussion, and homework assignments based on the health topics discussed	4 months	Disability (RMDQ) Pain (McGill pain Questionnaire- Short Form and SF-36 pain scale) QoL (SF-36)	A mindfulness meditation program and an education control group both showed improvement at program completion on measures of pain, and physical and psychological function.
Morone et al., 2016	Mindfulness: body scan, sitting practice, walking meditation for 8 weeks	Educational program on a successful aging curriculum known as the 10 Keys to Healthy Aging	6 months	Disability (RMDQ) Pain (NRS) QoL (SF-36)	A mind-body program for chronic LBP improved short-term function and long-term current and most severe pain. The functional improvement was not sustained.
Day et al., 2019	MBCT for pain protocol integrates cognitive and behavioral techniques with mindfulness-based strategies	CT techniques delivered: treatment involved traditional Beckian style column technique restructuring exercises Mindufulness: cognitive-behavioral and mindful movement components removed	6 months	Pain (NRS) Physical function (PROMIS) Depression (PROMIS)	The findings show that MBCT is a feasible, tolerable, acceptable, and potentially efficacious treatment option for CLBP. Further, MBCT, and possibly CT, could have sustained benefits that exceed MM on some important CLBP outcomes.

Table 3. Cochrane risk-of-bias tool for randomized controlled trials.

	Random Sequence Generation	Allocation Concealment	Blinding (Participants and Personnel)	Blinding (Outcome Assessment)	Incomplete Outcome Data	Selective Reporting	Other Bias	Risk of Bias
Cherkin et al., 2016	L	L	H	L	L	L	H	U
Monticone et al., 2013	L	L	H	L	L	L	L	L
Johnson et al., 2007	L	L	H	H	L	L	H	U
Smeets et al., 2006	L	L	H	L	L	L	H	U
Rutledge et al., 2018	L	L	H	L	L	L	H	U
Rutledge T et al., 2018	L	L	H	H	L	L	L	U
Linden et al., 2014	L	U	H	L	L	L	H	U
Khan et al., 2016	L	U	H	L	L	L	H	U
Pincus et al., 2015	L	L	H	U	L	L	H	U
Basler et al., 1997	L	L	H	L	L	L	H	U
Linton et al., 2000	L	U	H	U	L	L	H	H
Zgierska et al., 2016	L	U	H	H	L	L	H	H
Morone et al., 2008	L	U	U	H	L	L	H	H
Morone et al., 2009	L	L	U	L	H	L	L	L
Morone et al., 2016	L	L	U	L	L	U	L	L
Day et al., 2019	L	L	H	L	L	L	L	L

L: low; U: unclear; H: high.

3.6. *Effect of Intervention*

The meta-analysis analyzed the effectiveness of CBT and MBSR in terms of pain, disability, quality of life, depression and Fear-Avoidance Beliefs compared to controls.

3.6.1. Pain

Pain showed a significant decrease both in CBT and MBSR group compared with the control group, respectively SMD −0.73, 95% CI −1.20 to −0.26, $p = 0.002$ for CBT (Figure 2) and SMD −0.30, 95% CI −0.47 to −0.13, $p = 0.0005$ for MBSR (Figure 3). No significant pain reduction was demonstrated (MD −0.05, 95% CI −0.50 to 0.39, $p = 0.81$) when comparing MBSR and CBT (Figure 4).

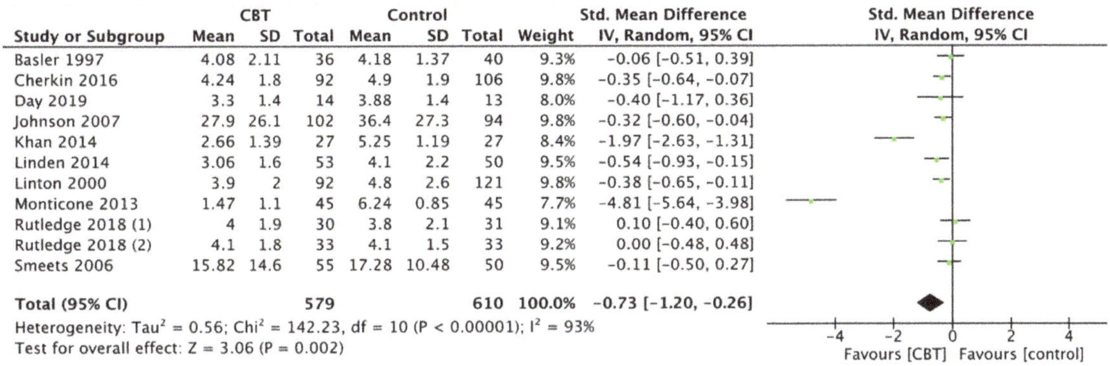

Figure 2. Pain: CBT versus control.

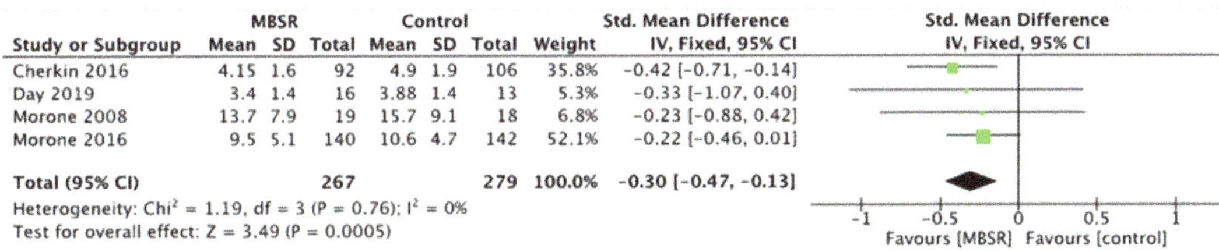

Figure 3. Pain: MBSR versus control.

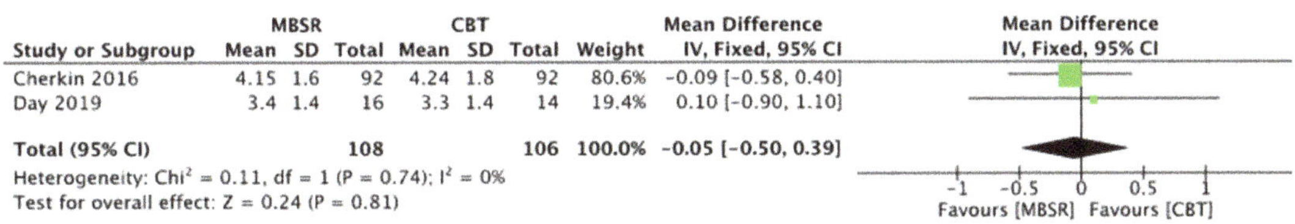

Figure 4. Pain: MBSR versus CBT.

3.6.2. Disability

Disability scores demonstrated significant improvements after CBT in comparison with controls (SMD −0.88, 95% CI −1.50 to −0.26, p = 0.005) (Figure 5). Instead, the reduction of disability after MBSR was not statistically significant compared to controls (MD −0.71, 95% CI −1.53 to −0.11, p = 0.09) (Figure 6).

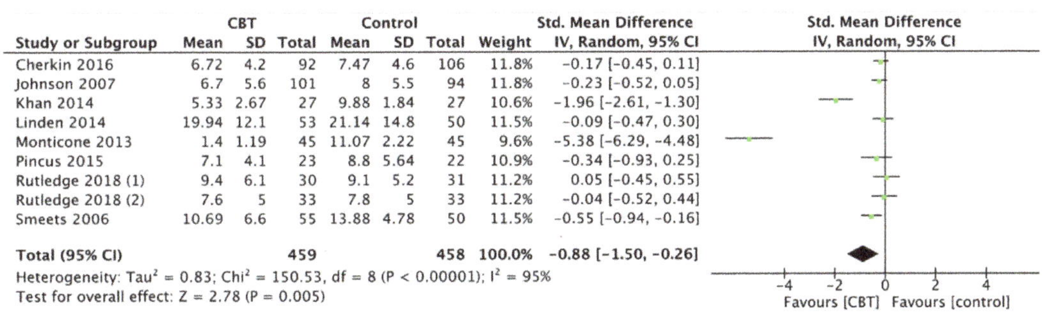

Figure 5. Disability: CBT versus control.

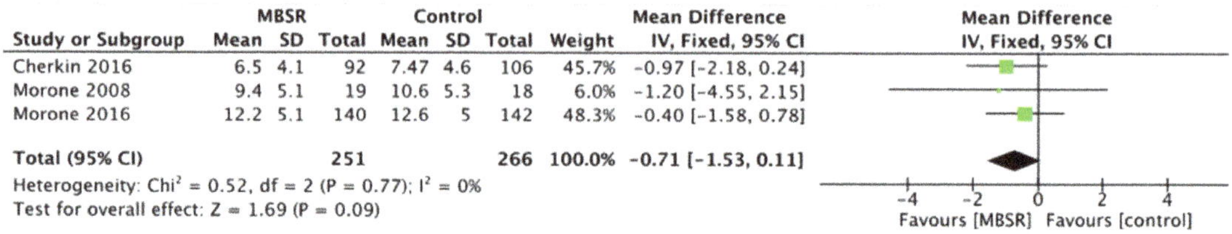

Figure 6. Disability: MBSR versus control.

3.6.3. Quality of Life

Quality of life showed significant improvement in CBT and MBSR group compared to controls, respectively SMD 0.69, 95% CI 0.00 to 1.37, $p = 0.05$ for CBT (Figure 7) and MD 2.84, 95% CI 0.31 to 5.37, $p = 0.03$ for MBSR (Figure 8). Moreover, comparing the two intervention groups, a significant difference in quality of life was shown in favor of MBSR (MD 2.54, 95% CI 0.84 to 4.24, $p = 0.003$) (Figure 9).

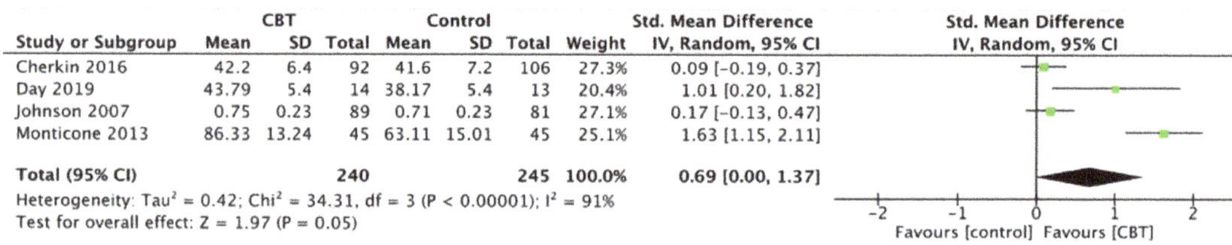

Figure 7. Quality of Life: CBT versus control.

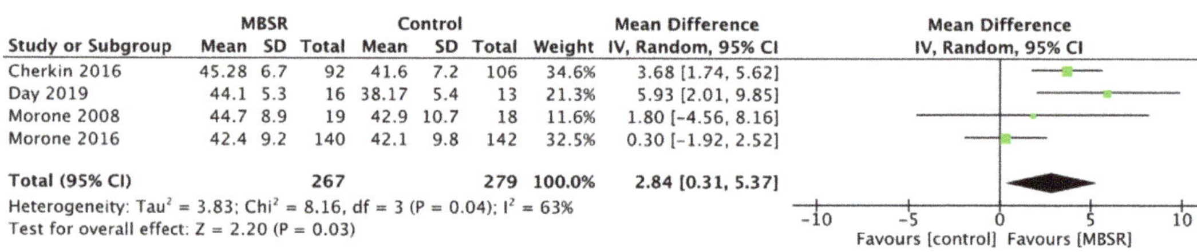

Figure 8. Quality of Life: MBSR versus control.

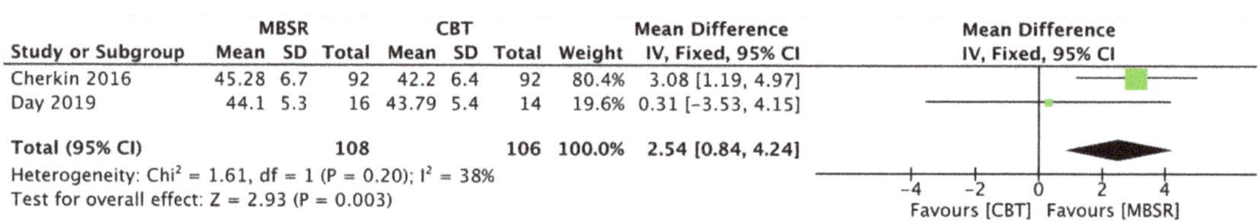

Figure 9. Quality of Life: MBSR versus CBT.

3.6.4. Depression

Depression scales did not show significant differences between the groups. More precisely, depression did not report statistical improvements between CBT and controls (SMD −0.26, 95% CI −0.72 to 0.19, $p = 0.26$) (Figure 10), MBRS and controls (SMD −1.55, 95% CI −4.53 to 1.43, $p = 0.31$) (Figure 11), and also MBRS and CBT (SMD 0.00, 95% CI −0.27 to 0.27, $p = 1.00$) (Figure 12).

Study or Subgroup	CBT Mean	CBT SD	CBT Total	Control Mean	Control SD	Control Total	Weight	Std. Mean Difference IV, Random, 95% CI
Cherkin 2016	3.98	2.6	92	4.42	2.9	106	23.7%	-0.16 [-0.44, 0.12]
Day 2019	46.14	4.1	14	57.54	4.3	13	10.7%	-2.63 [-3.71, -1.56]
Linton 2000	3.9	3.5	92	3.95	3	121	23.9%	-0.02 [-0.29, 0.26]
Rutledge 2018 (1)	15	13.6	30	10.3	9.9	31	19.7%	0.39 [-0.12, 0.90]
Smeets 2006	8.8	6.7	55	9.42	7.81	50	22.0%	-0.08 [-0.47, 0.30]
Total (95% CI)			**283**			**321**	**100.0%**	**-0.26 [-0.72, 0.19]**

Heterogeneity: Tau² = 0.21; Chi² = 25.64, df = 4 (P < 0.0001); I² = 84%
Test for overall effect: Z = 1.14 (P = 0.26)

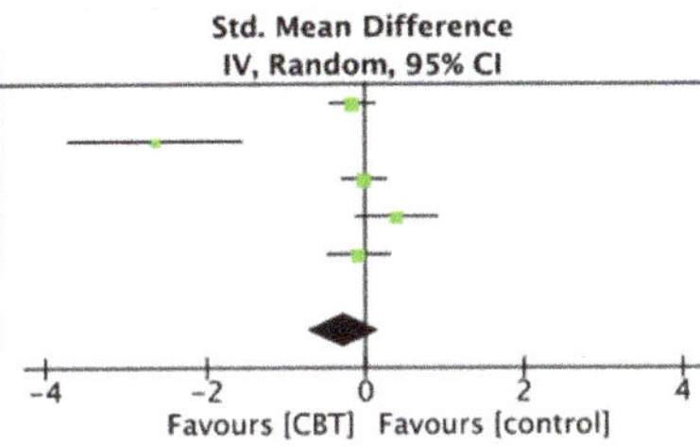

Figure 10. Depression: CBT versus control.

Study or Subgroup	MBSR Mean	MBSR SD	MBSR Total	Control Mean	Control SD	Control Total	Weight	Std. Mean Difference IV, Random, 95% CI
Day 2019	44.08	4.1	16	57.54	4.3	13	48.3%	-3.12 [-4.25, -1.99]
Cherkin 2016	4.19	2.8	92	4.42	2.9	106	51.7%	-0.08 [-0.36, 0.20]
Total (95% CI)			**108**			**119**	**100.0%**	**-1.55 [-4.53, 1.43]**

Heterogeneity: Tau² = 4.45; Chi² = 26.12, df = 1 (P < 0.00001); I² = 96%
Test for overall effect: Z = 1.02 (P = 0.31)

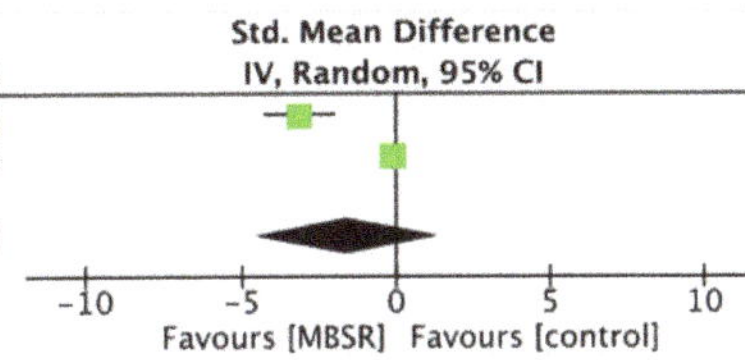

Figure 11. Depression: MBSR versus control.

Study or Subgroup	MBSR Mean	MBSR SD	MBSR Total	CBT Mean	CBT SD	CBT Total	Weight	Std. Mean Difference IV, Fixed, 95% CI
Cherkin 2016	4.19	2.8	92	3.98	2.6	92	86.4%	0.08 [-0.21, 0.37]
Day 2019	44.08	4.1	16	46.14	4.1	14	13.6%	-0.49 [-1.22, 0.24]
Total (95% CI)			**108**			**106**	**100.0%**	**0.00 [-0.27, 0.27]**

Heterogeneity: Chi² = 2.00, df = 1 (P = 0.16); I² = 50%
Test for overall effect: Z = 0.00 (P = 1.00)

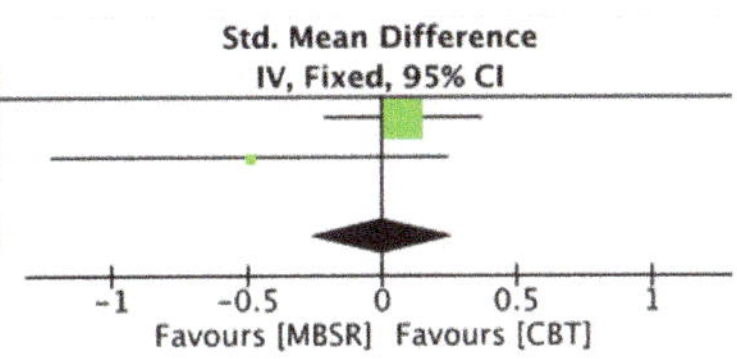

Figure 12. Depression: MBSR versus CBT.

3.6.5. Fear-Avoidance Beliefs

The meta-analysis demonstrated lower fear-avoidance beliefs in patients who underwent CBT compared to control group (SMD −2.17, 95% CI −4.22 to −012, $p = 0.04$) (Figure 13).

Study or Subgroup	CBT Mean	CBT SD	CBT Total	Control Mean	Control SD	Control Total	Weight	Std. Mean Difference IV, Random, 95% CI
Linden 2014	34.28	17.9	53	46.1	20.3	50	34.0%	-0.61 [-1.01, -0.22]
Linton 2000	8	4.7	92	8.7	5.8	121	34.2%	-0.13 [-0.40, 0.14]
Monticone 2013	17.67	1.62	45	40.96	5.17	45	31.8%	-6.03 [-7.02, -5.04]
Total (95% CI)			**190**			**216**	**100.0%**	**-2.17 [-4.22, -0.12]**

Heterogeneity: Tau² = 3.18; Chi² = 126.69, df = 2 (P < 0.00001); I² = 98%
Test for overall effect: Z = 2.08 (P = 0.04)

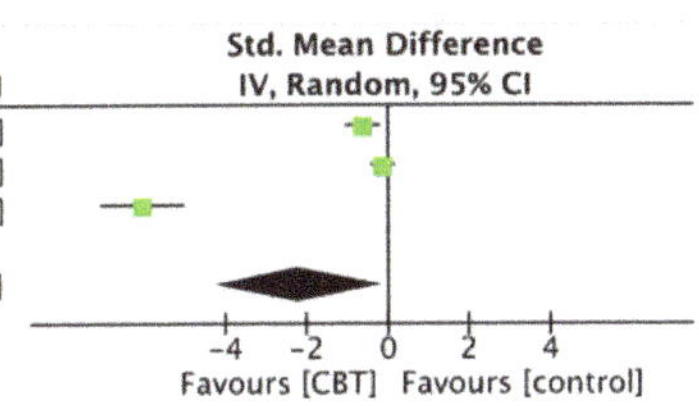

Figure 13. Fear-Avoidance beliefs: CBT versus control.

3.6.6. Days without Pain

Finally, the number of days without LBP increased in CBT group compared to controls, but without statistical significance (OR 1.38, 95% CI 0.73 to 2.61, $p = 0.32$) (Figure 14).

Study or Subgroup	CBT Events	CBT Total	Control Events	Control Total	Weight	Odds Ratio M-H, Random, 95% CI
Basler 1997	21	1872	11	2080	35.9%	2.13 [1.03, 4.44]
Linton 2000	193	4784	235	6292	64.1%	1.08 [0.89, 1.32]
Total (95% CI)		**6656**		**8372**	**100.0%**	**1.38 [0.73, 2.61]**
Total events	214		246			

Heterogeneity: $Tau^2 = 0.16$; $Chi^2 = 3.08$, df = 1 (P = 0.08); $I^2 = 67\%$
Test for overall effect: Z = 0.99 (P = 0.32)

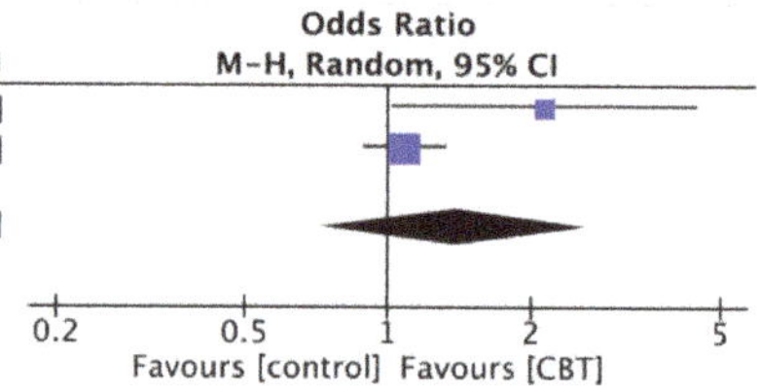

Figure 14. Days without pain: CBT versus control.

4. Discussion

The link between psychological factors and CLBP has been widely demonstrated in several studies. The aims of this Systematic Review and Meta-Analysis were (1) to identify and describe the most frequently used psychological approaches to treat patients affected by CLBP, and (2) to study the effectiveness of these approaches in terms of reduction of pain, disability, fear-avoidance behaviors, anxiety, depression, and of increase in quality of life. According to the literature [16,19,21], the most common psychological approaches used to treat CLBP are cognitive behavioral therapy and mindfulness-based intervention. CBT demonstrated its effectiveness for different chronic pain conditions [7]. This approach helps patients with maladaptive emotions, behaviors, and cognitions through a goal-oriented and systematic process. CBT was initially used to treat disorders like insomnia, anxiety, and depression, and was later implemented to manage chronic pain [39]. The CBT intervention consists in several sessions guided by a skilled therapist, with different frequency and duration. In these sessions, activities like pain education, relaxation training, managing of automatic thoughts, stress reduction, problem solving and sleep education [39] are performed. MBSR is also becoming increasingly popular and available in the United States [9]. With this treatment, patients are educated about the psychophysiology of stress and are provided opportunities to apply MBSR skills to specific situations [40]. This approach has several contemporary interpretations, based on formal and informal systematic meditation training, patient education, yoga exercises, and individual and group dialogue [41].

The findings of this systematic review and meta-analysis suggest that CBT and MBSR interventions were both associated with an improvement in terms of pain intensity and quality of life when singularly compared to usual care. Disability also improved in both groups when compared to usual care, although it was only statistically significant in patients treated with CBT, which may be due to the paucity of studies that analyzed MBSR intervention. Significant differences in fear-avoidance beliefs were noted in the CBT group compared to usual care. However, no studies analyzed this outcome for the MBSR approach. No meaningful results were noted for depression in both MBSR and CBT groups. Moreover, only two RCT compared CBT to MBSR, showing no significant improvements in pain intensity and depression along with a better quality of life for the MBSR intervention [9,28].

Another meta-analysis [20] studied several psychosocial interventions to treat patients affected by CLBP. This study demonstrated an improvement in pain, QoL and work-related disability in the intervention group towards the waiting list group.

Our results agree with those by Gotink et al. [22], who studied MBSR applied to chronic illness. Their review shows the large use of this treatment with patients affected by cancer, cardiovascular diseases, mental disorders, and non-specific chronic pain. Indeed, an improvement in depressive symptoms and physical health and a decrease in pain burden, intensity and disability are reported in patients affected by non-specific chronic pain.

Regarding the implementation of CBT, Richmond et al. [19] shows its effectiveness for non-specific low back pain, with improvement of pain, functional disability, and quality of life. In addition, Morley et al. [42] show the improvement of pain and functional disability in patients with chronic pain.

Psychological factors in people affected by LBP are associated with increased risk of developing disability [43]. For instance, the symptom of depression and the catastrophizing of pain predict poor low back pain-related outcomes [44,45]. Therefore, cognitive and emotional factors have a crucial impact on pain perception and, in line with the literature [46], it is fundamental to identify and take care of psychological factors, through a multidisciplinary approach in patients with chronic pain. Indeed this approach should be considered because each aspect requires specific interventions [47].

This review has several limitations. Firstly, there is some difference in heterogeneity between studies regarding CBT and MBSR. In particular, the studies involving CBT presented high heterogeneity due to the greater number of studies included, the different types of CBT performed, the different duration of interventions, and the tools used. Instead, the studies investigating MBSR used the same tools for the analyzed outcomes, resulting therefore in lower heterogeneity. Additionally, demographic characteristics of the participants were different in the included studies, with various gender and age distribution. However, this did not influence the statistical analysis of the studies. There are differences regarding the quality of the studies; indeed, the major number of the studies included were of moderate quality ($n = 9$), whereas four studies had a low risk of bias, and three studies had a high risk of bias. We also decided to include in the meta-analysis the studies with a high risk of bias, which is another limitation regarding the number of studies included. The participants of most studies, except for three studies [25–27], were not blind to treatment allocation, given the nature of the intervention. Another limitation of this study is the heterogeneity of the types of treatment in the control groups. Indeed, we found different types of treatments in the control groups such as physiotherapy, physical exercises, educational programs and drug treatments. In only two studies we found the waiting list, and we suppose it may be the most adequate control group for the reduction of bias, hence the need to develop RCTs with waiting lists as a control group for an appropriate analysis of the effectiveness of psychological interventions. To reduce the heterogeneity in the analysis of outcomes as pain, it would be appropriate to develop RCTs that use the same tools.

5. Conclusions

In the present study we analyzed the most used and effective psychological approaches to treat patients affected by chronic LBP. CBT and MBSR have proven their significant effectiveness to improve pain intensity and quality of life compared to controls. These approaches also demonstrated their efficacy in reducing disability and fear-avoidance, but without significant results. The importance of treating psychological aspects is widely proven, but the paucity and heterogeneity of the studies included cannot make us confident to affirm which is the most effective treatment. Further studies are needed to compare CBT and MBSR.

Author Contributions: Conceptualization, G.P., G.F.P., F.R. and G.V.; methodology, G.P., G.F.P., F.R. and G.V.; formal analysis, G.P., G.F.P. and F.R.; data curation, G.P., G.F.P. and F.R.; writing—original draft preparation, G.P., G.F.P., F.R., G.V., M.P. and M.G.D.M.; writing—review and editing, G.P., G.F.P., F.R., G.V., M.P., M.G.D.M., R.P. and V.D.; supervision, F.R., G.V., M.P., M.G.D.M., R.P. and V.D. All authors have read and agreed to the published version of the manuscript.

Funding: This research was funded by the Research Grant (BRiC-2018 ID3) of the Italian Workers' Compensation Authority (INAIL).

Institutional Review Board Statement: Not applicable.

Informed Consent Statement: Not applicable.

Acknowledgments: Not applicable.

Conflicts of Interest: The authors declare no conflict of interest.

References

1. Vos, T.; Barber, R.M.; Bell, B.; Bertozzi-Villa, A.; Biryukov, S.; Bolliger, I.; Charlson, F.; Davis, A.; Degenhardt, L.; Dicker, D.; et al. Global, regional, and national incidence, prevalence, and years lived with disability for 301 acute and chronic diseases and injuries in 188 countries, 1990–2013: A systematic analysis for the Global Burden of Disease Study 2013. *Lancet* **2015**, *386*, 743–800. [CrossRef]
2. Russo, F.; Papalia, G.F.; Vadalà, G.; Fontana, L.; Iavicoli, S.; Papalia, R.; Denaro, V. The Effects of Workplace Interventions on Low Back Pain in Workers: A Systematic Review and Meta-Analysis. *Int. J. Environ. Res. Public. Health* **2021**, *18*, 12614. [CrossRef] [PubMed]
3. Andersson, G.B. Epidemiological features of chronic low-back pain. *Lancet* **1999**, *354*, 581–585. [CrossRef]
4. Deyo, R.A.; Mirza, S.K.; Martin, B.I. Back Pain Prevalence and Visit Rates: Estimates from U.S. National Surveys, 2002. *Spine* **2006**, *31*, 2724–2727. [CrossRef] [PubMed]
5. Rubin, D.I. Epidemiology and Risk Factors for Spine Pain. *Neurol. Clin.* **2007**, *25*, 353–371. [CrossRef] [PubMed]
6. Neugebauer, V.; Galhardo, V.; Maione, S.; Mackey, S. Forebrain pain mechanisms. *Brain Res. Rev.* **2009**, *60*, 226–242. [CrossRef]
7. Williams, A.C.D.C.; Fisher, E.; Hearn, L.; Eccleston, C. Psychological therapies for the management of chronic pain (excluding headache) in adults. *Cochrane Database Syst. Rev.* **2020**, *2021*. [CrossRef]
8. Russo, F.; Ambrosio, L.; Ngo, K.; Vadalà, G.; Denaro, V.; Fan, Y.; Sowa, G.; Kang, J.D.; Vo, N. The Role of Type I Diabetes in Intervertebral Disc Degeneration. *Spine* **2019**, *44*, 1177–1185. [CrossRef]
9. Cherkin, D.C.; Sherman, K.J.; Balderson, B.H.; Cook, A.J.; Anderson, M.L.; Hawkes, R.J.; Hansen, K.E.; Turner, J.A. Effect of Mindfulness-Based Stress Reduction vs. Cognitive Behavioral Therapy or Usual Care on Back Pain and Functional Limitations in Adults With Chronic Low Back Pain: A Randomized Clinical Trial. *JAMA* **2016**, *315*, 1240. [CrossRef]
10. Russo, F.; De Salvatore, S.; Ambrosio, L.; Vadalà, G.; Fontana, L.; Papalia, R.; Rantanen, J.; Iavicoli, S.; Denaro, V. Does Workers' Compensation Status Affect Outcomes after Lumbar Spine Surgery? A Systematic Review and Meta-Analysis. *Int. J. Environ. Res. Public Health* **2021**, *18*, 6165. [CrossRef]
11. Gatchel, R.J.; Peng, Y.B.; Peters, M.; Fuchs, P.; Turk, D.C. The biopsychosocial approach to chronic pain: Scientific advances and future directions. *Psychol. Bull.* **2007**, *133*, 581–624. [CrossRef]
12. Keefe, F.J.; Rumble, M.E.; Scipio, C.D.; Giordano, L.A.; Perri, L.M. Psychological aspects of persistent pain: Current state of the science. *J. Pain* **2004**, *5*, 195–211. [CrossRef] [PubMed]
13. Smeets, R.J.; Vlaeyen, J.; Kester, A.D.; Knottnerus, J.A. Reduction of Pain Catastrophizing Mediates the Outcome of Both Physical and Cognitive-Behavioral Treatment in Chronic Low Back Pain. *J. Pain* **2006**, *7*, 261–271. [CrossRef]
14. Sullivan, M.J.L.; Thorn, B.; Haythornthwaite, J.A.; Keefe, F.; Martin, M.; Bradley, L.; Lefebvre, J.C. Theoretical Perspectives on the Relation Between Catastrophizing and Pain. *Clin. J. Pain* **2001**, *17*, 52–64. [CrossRef] [PubMed]
15. Jarvik, J.G.; Hollingworth, W.; Heagerty, P.J.; Haynor, D.R.; Boyko, E.J.; Deyo, R.A. Three-Year Incidence of Low Back Pain in an Initially Asymptomatic Cohort: Clinical and Imaging Risk Factors. *Spine* **2005**, *30*, 1541–1548. [CrossRef] [PubMed]
16. Hajihasani, A.; Rouhani, M.; Salavati, M.; Hedayati, R.; Kahlaee, A.H. The Influence of Cognitive Behavioral Therapy on Pain, Quality of Life, and Depression in Patients Receiving Physical Therapy for Chronic Low Back Pain: A Systematic Review. *PM&R* **2019**, *11*, 167–176. [CrossRef]
17. Vadalà, G.; Russo, F.; De Salvatore, S.; Cortina, G.; Albo, E.; Papalia, R.; Denaro, V. Physical Activity for the Treatment of Chronic Low Back Pain in Elderly Patients: A Systematic Review. *J. Clin. Med.* **2020**, *9*, 1023. [CrossRef]
18. Chou, R.; Loeser, J.D.; Owens, D.K.; Rosenquist, R.W.; Atlas, S.J.; Baisden, J.; Carragee, E.J.; Grabois, M.; Murphy, D.R.; Resnick, D.K.; et al. Interventional Therapies, Surgery, and Interdisciplinary Rehabilitation for Low Back Pain: An Evidence-Based Clinical Practice Guideline from the American Pain Society. *Spine* **2009**, *34*, 1066–1077. [CrossRef]
19. Richmond, H.; Hall, A.M.; Copsey, B.; Hansen, Z.; Williamson, E.; Hoxey-Thomas, N.; Cooper, Z.; Lamb, S.E. The Effectiveness of Cognitive Behavioural Treatment for Non-Specific Low Back Pain: A Systematic Review and Meta-Analysis. *PLoS ONE* **2015**, *10*, e0134192. [CrossRef] [PubMed]
20. Hoffman, B.M.; Papas, R.K.; Chatkoff, D.K.; Kerns, R.D. Meta-analysis of psychological interventions for chronic low back pain. *Health Psychol.* **2007**, *26*, 1–9. [CrossRef]
21. Anheyer, D.; Haller, H.; Barth, J.; Lauche, R.; Dobos, G.; Cramer, H. Mindfulness-Based Stress Reduction for Treating Low Back Pain: A Systematic Review and Meta-Analysis. *Ann. Intern. Med.* **2017**, *166*, 799. [CrossRef]
22. Gotink, R.A.; Chu, P.; Busschbach, J.J.V.; Benson, H.; Fricchione, G.L.; Hunink, M.G.M. Standardised Mindfulness-Based Interventions in Healthcare: An Overview of Systematic Reviews and Meta-Analyses of RCTs. *PLoS ONE* **2015**, *10*, e0124344. [CrossRef]
23. Page, M.J.; Moher, D. Evaluations of the uptake and impact of the Preferred Reporting Items for Systematic reviews and Meta-Analyses (PRISMA) Statement and extensions: A scoping review. *Syst. Rev.* **2017**, *6*, 263. [CrossRef] [PubMed]
24. Higgins, J.P.T.; Altman, D.G.; Gøtzsche, P.C.; Jüni, P.; Moher, D.; Oxman, A.D.; Savović, J.; Schulz, K.F.; Weeks, L.; Sterne, J.; et al. The Cochrane Collaboration's tool for assessing risk of bias in randomised trials. *BMJ* **2011**, *343*, d5928. [CrossRef] [PubMed]
25. Morone, N.E.; Greco, C.M.; Weiner, D.K. Mindfulness meditation for the treatment of chronic low back pain in older adults: A randomized controlled pilot study. *Pain* **2008**, *134*, 310–319. [CrossRef] [PubMed]
26. Morone, N.E.; Rollman, B.L.; Moore, C.G.; Li, Q.; Weiner, D.K. A Mind–Body Program for Older Adults with Chronic Low Back Pain: Results of a Pilot Study. *Pain Med.* **2009**, *10*, 1395–1407. [CrossRef]

27. Morone, N.E.; Greco, C.M.; Moore, C.G.; Rollman, B.L.; Lane, B.; Morrow, L.A.; Glynn, N.W.; Weiner, D.K. A Mind-Body Program for Older Adults with Chronic Low Back Pain: A Randomized Clinical Trial. *JAMA Intern. Med.* **2016**, *176*, 329. [CrossRef]
28. Day, M.A.; Ward, L.C.; Ehde, D.M.; Thorn, B.E.; Burns, J.; Barnier, A.; Mattingley, J.B.; Jensen, M.P. A Pilot Randomized Controlled Trial Comparing Mindfulness Meditation, Cognitive Therapy, and Mindfulness-Based Cognitive Therapy for Chronic Low Back Pain. *Pain Med.* **2019**, *20*, 2134–2148. [CrossRef]
29. Monticone, M.; Ferrante, S.; Rocca, B.; Baiardi, P.; Farra, F.D.; Foti, C. Effect of a Long-Lasting Multidisciplinary Program on Disability and Fear-Avoidance Behaviors in Patients With Chronic Low Back Pain: Results of a Randomized Controlled Trial. *Clin. J. Pain* **2013**, *29*, 929–938. [CrossRef]
30. Rutledge, T.; Atkinson, J.H.; Chircop-Rollick, T.; D'Andrea, J.; Garfin, S.; Patel, S.; Penzien, D.B.; Wallace, M.; Weickgenant, A.L.; Slater, M. Randomized Controlled Trial of Telephone-delivered Cognitive Behavioral Therapy Versus Supportive Care for Chronic Back Pain. *Clin. J. Pain* **2018**, *34*, 322–327. [CrossRef]
31. Rutledge, T.; Atkinson, J.H.; Holloway, R.; Chircop-Rollick, T.; D'Andrea, J.; Garfin, S.R.; Patel, S.; Penzien, D.B.; Wallace, M.; Weickgenant, A.L.; et al. Randomized Controlled Trial of Nurse-Delivered Cognitive-Behavioral Therapy Versus Supportive Psychotherapy Telehealth Interventions for Chronic Back Pain. *J. Pain* **2018**, *19*, 1033–1039. [CrossRef] [PubMed]
32. Johnson, R.E.; Jones, G.T.; Wiles, N.J.; Chaddock, C.; Potter, R.G.; Roberts, C.; Symmons, D.P.M.; Watson, P.J.; Torgerson, D.J.; Macfarlane, G.J. Active Exercise, Education, and Cognitive Behavioral Therapy for Persistent Disabling Low Back Pain: A Randomized Controlled Trial. *Spine* **2007**, *32*, 1578–1585. [CrossRef] [PubMed]
33. Khan, M.; Akhter, S.; Soomro, R.R.; Ali, S.S. The effectiveness of Cognitive Behavioral Therapy (CBT) with general exercises versus general exercises alone in the management of chronic low back pain. *Pak. J. Pharm. Sci.* **2014**, *27*, 1113–1116. [PubMed]
34. Pincus, T.; Anwar, S.; McCracken, L.M.; McGregor, A.; Graham, L.; Collinson, M.; McBeth, J.; Watson, P.; Morley, S.; on behalf of the OBI Trial Management Team; et al. Delivering an Optimised Behavioural Intervention (OBI) to People with Low Back Pain with High Psychological Risk; Results and Lessons Learnt from a Feasibility Randomised Controlled Trial of Contextual Cognitive Behavioural Therapy (CCBT) vs. Physiotherapy. *BMC Musculoskelet. Disord.* **2015**, *16*, 147. [CrossRef] [PubMed]
35. Zgierska, A.E.; Burzinski, C.A.; Cox, J.; Kloke, J.; Stegner, A.; Cook, D.B.; Singles, J.; Mirgain, S.; Coe, C.L.; Bačkonja, M. Mindfulness Meditation and Cognitive Behavioral Therapy Intervention Reduces Pain Severity and Sensitivity in Opioid-Treated Chronic Low Back Pain: Pilot Findings from a Randomized Controlled Trial. *Pain Med.* **2016**, *17*, 1865–1881. [CrossRef] [PubMed]
36. Linden, M.; Scherbe, S.; Cicholas, B. Randomized controlled trial on the effectiveness of cognitive behavior group therapy in chronic back pain patients. *J. Back Musculoskelet. Rehabil.* **2014**, *27*, 563–568. [CrossRef] [PubMed]
37. Basler, H.-D.; Jäkle, C.; Kröner-Herwig, B. Incorporation of cognitive-behavioral treatment into the medical care of chronic low back patients: A controlled randomized study in German pain treatment centers. *Patient Educ. Couns.* **1997**, *31*, 113–124. [CrossRef]
38. Linton, S.J.; Andersson, T. Can Chronic Disability Be Prevented?: A Randomized Trial of a Cognitive-Behavior Intervention and Two Forms of Information for Patients With Spinal Pain. *Spine* **2000**, *25*, 2825–2831. [CrossRef]
39. Murphy, J.L.; Cordova, M.J.; Dedert, E.A. Cognitive behavioral therapy for chronic pain in veterans: Evidence for clinical effectiveness in a model program. *Psychol. Serv.* **2020**. [CrossRef]
40. Proulx, K. Integrating Mindfulness-Based Stress Reduction. *Holist. Nurs. Pract.* **2003**, *17*, 201–208. [CrossRef]
41. Majeed, M.H.; Ali, A.A.; Sudak, D.M. Mindfulness-based interventions for chronic pain: Evidence and applications. *Asian J. Psychiatry* **2018**, *32*, 79–83. [CrossRef] [PubMed]
42. Morley, S.; Eccleston, C.; Williams, A. Systematic review and meta-analysis of randomized controlled trials of cognitive behaviour therapy and behaviour therapy for chronic pain in adults, excluding headache. *Pain* **1999**, *80*, 1–13. [CrossRef]
43. Hartvigsen, J.; Hancock, M.; Kongsted, A.; Louw, Q.; Ferreira, M.L.; Genevay, S.; Hoy, D.; Karppinen, J.; Pransky, G.; Sieper, J.; et al. What low back pain is and why we need to pay attention. *Lancet* **2018**, *391*, 2356–2367. [CrossRef]
44. Wertli, M.M.; Eugster, R.; Held, U.; Steurer, J.; Kofmehl, R.; Weiser, S. Catastrophizing—a prognostic factor for outcome in patients with low back pain: A systematic review. *Spine J.* **2014**, *14*, 2639–2657. [CrossRef] [PubMed]
45. Pinheiro, M.B.; Ferreira, M.L.; Refshauge, K.; Maher, C.; Ordoñana, J.R.; Andrade, T.B.; Tsathas, A.; Ferreira, P. Symptoms of depression as a prognostic factor for low back pain: A systematic review. *Spine J.* **2016**, *16*, 105–116. [CrossRef] [PubMed]
46. Crofford, L.J. Psychological aspects of chronic musculoskeletal pain. *Best Pract. Res. Clin. Rheumatol.* **2015**, *29*, 147–155. [CrossRef] [PubMed]
47. Tagliaferri, S.D.; Miller, C.T.; Owen, P.J.; Mitchell, U.H.; Brisby, H.; FitzGibbon, B.; Masse-Alarie, H.; Van Oosterwijck, J.; Belavy, D.L. Domains of Chronic Low Back Pain and Assessing Treatment Effectiveness: A Clinical Perspective. *Pain Pract.* **2020**, *20*, 211–225. [CrossRef]

International Journal of
Environmental Research and Public Health

MDPI

Article

Thirty Minutes Identified as the Threshold for Development of Pain in Low Back and Feet Regions, and Predictors of Intensity of Pain during 1-h Laboratory-Based Standing in Office Workers

Beatriz Rodríguez-Romero [1,*], Michelle D. Smith [2], Sonia Pértega-Díaz [3], Alejandro Quintela-del-Rio [4] and Venerina Johnston [2]

[1] Psychosocial Intervention and Functional Rehabilitation Research Group, Department of Physiotherapy, Medicine and Biomedical Sciences, Campus Oza, University of A Coruña, 15071 A Coruna, Spain
[2] School of Health and Rehabilitation Sciences, The University of Queensland, Brisbane 4072, Australia; m.smith5@uq.edu.au (M.D.S.); v.johnston@uq.edu.au (V.J.)
[3] Rheumatology and Health Research Group, Department of Health Sciences, Campus Esteiro, University of A Coruña, 15471 Ferrol, Spain; s.pertega@udc.es
[4] Department of Mathematics, University of A Coruña, 15071 A Coruna, Spain; alejandro.quintela@udc.es
* Correspondence: beatriz.romero@udc.es; Tel.: +34-981-167-000 (ext. 5841)

Citation: Rodríguez-Romero, B.; Smith, M.D.; Pértega-Díaz, S.; Quintela-del-Rio, A.; Johnston, V. Thirty Minutes Identified as the Threshold for Development of Pain in Low Back and Feet Regions, and Predictors of Intensity of Pain during 1-h Laboratory-Based Standing in Office Workers. *IJERPH* **2022**, *19*, 2221. https://doi.org/10.3390/ijerph19042221

Academic Editors: Sergio Iavicoli, Vincenzo Denaro, Fabrizio Russo and Gianluca Vadalà

Received: 29 December 2021
Accepted: 13 February 2022
Published: 16 February 2022

Publisher's Note: MDPI stays neutral with regard to jurisdictional claims in published maps and institutional affiliations.

Abstract: This study with 40 office workers investigated (a) the effect of time spent standing on low-back and lower limb pain during a 1-h laboratory-based task; (b) the standing time after which a significant increase in pain is likely; and (c) the individual, physical and psychosocial factors that predict pain. The primary outcome was bodily location of pain and pain intensity on a 100-mm Visual Analogue Scale recorded at baseline and every 15 min. Physical measures included trunk and hip motor control and endurance. Self-report history of pain, physical activity, psychosocial job characteristics, pain catastrophizing and general health status were collected. Univariate analysis and regression models were included. The prevalence of low-back pain increased from 15% to 40% after 30 min while feet pain increased to 25% from 0 at baseline. The intensity of low-back and lower limb pain also increased over time. A thirty-minute interval was identified as the threshold for the development and increase in low-back and feet pain. Modifiable factors were associated with low-back pain intensity—lower hip abductor muscle endurance and poorer physical health, and with feet symptoms—greater body mass index and less core stability.

Keywords: low-back pain; standing position; musculoskeletal pain; lower extremity

1. Introduction

Office workers are known to adopt sedentary behaviors at work [1]. Although the available evidence has not confirmed a consistent causal relationship between occupational sitting and musculoskeletal pain [2–4], a growing body of evidence suggests that prolonged sitting is a major concern for the development of several chronic diseases, such as cardiovascular disease, diabetes type 2 and premature mortality [5]. Thus, it is no surprise that there is heightened interest in workplace initiatives to reduce the amount of sitting time for office workers [6,7] with alternatives such as sit-stand workstations or breaking up seated-work with standing-work. However, some concerns have been expressed that substituting sitting with standing may expose workers to new hazards and/or other health consequences. Two systematic reviews with meta-analysis in laboratory [8] and occupational [9] settings suggested that prolonged standing is associated with the occurrence of low-back and lower extremity symptoms, although the conclusions are inconclusive for the association with lower extremity symptoms. Similarly, causality between occupational standing and LBP has not been resolved, and not all people who are exposed to prolonged standing will develop LBP [2,10].

Experimental laboratory studies which simulate occupational standing have used an induced pain paradigm to identify factors which could be associated with developing low-back and/or lower extremity pain [8]. Some factors suggested to predispose a person to the development of LBP during prolonged standing are: (i) fatigue of the trunk and hip muscles [11–13]; (ii) movement control dysfunction [13–16]; and (iii) postural stiffness through increased levels of coactivation of hip and trunk muscles [14,17]. Specific factors associated with an increase in LBP during prolonged standing are unknown. Discomfort experienced in the feet and lower limbs during standing is often attributed to reductions in venous return and muscular fatigue [18]. The flowmetry, leg circumference, skin temperature, force through feet, and lower limb and trunk muscle activity have been reported as the main outcomes to study the possible mechanisms for lower limb symptoms; although the underlying mechanisms require further investigation [8,9,18].

Evidence for determining thresholds of excessive standing has not been elucidated. In a 2-year prospective study, Andersen et al. [19] demonstrated that standing at work for 30 min or more every hour elevated the odds for LBP by a factor of 2.1, and for pain in the hip, knee or foot by a factor of 1.7. A meta-analysis [9] suggested a statistically significant association between 2 or 4 h/workday of occupational standing and the occurrence of low back/lower extremity symptoms, although the authors highlighted that conclusions on the dose–response association cannot be drawn.

The aims of this laboratory study in office workers were to determine: (i) the effect of time spent standing on pain status during a 1-h laboratory-based standing task; (ii) the point after which significant increases in pain are likely; and (iii) the individual (e.g., age, sex, history of LBP, self-rated health), physical (e.g., deficits in motor control, muscle endurance) and psychosocial (e.g., job demands) factors that are associated with higher levels of low-back and lower limb pain after a 1-h standing task. Given the findings of previous studies, it was hypothesized that there would be a significant effect of time spent standing on the prevalence and intensity of low-back and lower limb pain, with a common threshold time when significant increases occur. We expect that specific individual (health status, BMI), physical (deficits in muscle endurance) and psychosocial (e.g., low job control) factors would be associated with higher scores of low-back and lower limb pain after a one-hour standing task.

2. Materials and Methods

2.1. Participants

A convenience sample of forty office workers, aged $\geq$ 18 years, who performed mostly sedentary work for $\geq$30 h per week were recruited. Participants were excluded if they: (i) were pregnant or less than six months postpartum, (ii) had any major trauma or surgery to the spine or lower limb in the last 12 months or (iii) had a diagnosis of neurological or systemic pathology. The recruitment process and sample size are described elsewhere [13].

The University of Queensland Human Research Ethics Committee B approved this study (Approval Number: #2017000666) and all participants provided informed consent prior to study participation. This study was registered in the Protocol Registration and Results System (PRS) (NCT03678623).

2.2. Study Procedure

Participants completed self-reported measures, undertook a physical examination conducted by a trained physiotherapist, and then participated in a 1-h standing task. These self-report measures were administered via an online survey completed the day prior or the same day as the laboratory testing session.

2.3. Measurements

Self-reported measures included (i) demographics; (ii) history of LBP (7-day prevalence); (iii) location of any bodily pain assessed with the Nordic Musculoskeletal Questionnaire [20] and pain assessed with a 100-mm Visual Analogue Scale (VAS) anchored with "no pain" at 0 and "worst pain imaginable" at 100 mm for each body location [21]; (iv) total and occupational physical activity assessed with the International Physical Activity Questionnaire (IPAQ) (MET-min/week) [22] and the Occupational Sitting and Physical Activity Questionnaire (OSPAQ) (minutes) [23], respectively; (v) psychosocial job characteristics evaluated through the Job Content Questionnaire (JCQ) which includes four domains (job control, psychological job demands, social support and physical demands) (4-point Likert) [24]; (vi) propensity for pain catastrophizing assessed with the Pain Catastrophizing Scale (PCS-total) (scores ranged from 0 to 52 with greater scores indicate a greater degree of catastrophizing) [25]; and (vii) general health status evaluated with the SF-12 through the Physical (PCS) and Mental (MCS) Component Summary scores [26].

At the start of the laboratory session, participants were given 15 min of seated rest while the testing protocol was explained. Physical testing was then undertaken in the following order: (i) height and weight; (ii) three motor control impairment tests: the active hip abduction test (AHAbd) (ranging from 0–5 as rated by participants and 0–3 by the examiner, with lower scores indicating better motor control for both ratings) [27], and the active straight leg raise test (ASLR) (ranging from 0–10 as the summed score of participant and examiner ratings, with lower scores indicating better motor control) [28]; (iii) endurance tests of the following trunk and hip muscles (measured as seconds able to holding a static position): abdominal endurance [29], supine bridge [30], isometric hip abduction [31] and Biering–Sorensen test [29]. The specific methodology on how each of these tests was applied is explained elsewhere [13].

The standing paradigm consisted of participants standing for an hour while performing their usual computer-based work. Participants stood within a rectangular floor space (122 × 61 cm) with their body fist-width away from the edge of a height-adjustable workstation. The workstation was standardized to each participant so that the desk height was 5–6 cm below the lateral epicondyle, the computer monitor was at arm's length from the body, and the top of the computer monitor was at eye level. Participants were allowed to shift their weight as often as desired but were asked to keep both feet on the ground the majority of the time. The participant was not allowed to lean on the workstation with their arms, legs or trunk [13].

The primary outcome was pain status (yes/no) and severity of pain (VAS, 0–100 mm) in the low back and lower extremity (Figure 1). The workers self-reported the location of their pain on the body map and indicated pain intensity on the VAS at baseline, every 15 min during, and at the end of the standing test. The investigator verbally asked the participants to rate their pain. Participants were not given access to their previous scores.

2.4. Statistical Analysis

Descriptive statistics were computed for pain ratings over the 1-h task (0, 15, 30, 45 and 60 min). At each timepoint, prevalence of pain for each location was computed, and VAS scores were summarized as mean ± standard deviation and median.

For hip-thigh and knee-calf regions, the number of subjects who developed pain was small, and only descriptive analyses were performed. Therefore, only low-back and ankle-feet regions were analyzed in more detail. For those locations, an increment of ≥10 mm in VAS pain at any time between start and end of the test was considered to classify participants as Pain Developers (PD) or Non-Pain Developers (NPD) [9,32], and both groups were analyzed independently.

Since repeated pain evaluations were obtained for each subject at different timepoints, appropriate tests for paired data were employed in bivariate analysis. McNemar's test for paired data was used to analyze the significance of changes in the prevalence of pain for each location during the 1-h task. In the same way, Wilcoxon's signed-rank test was employed to compare pain scores between different timepoints.

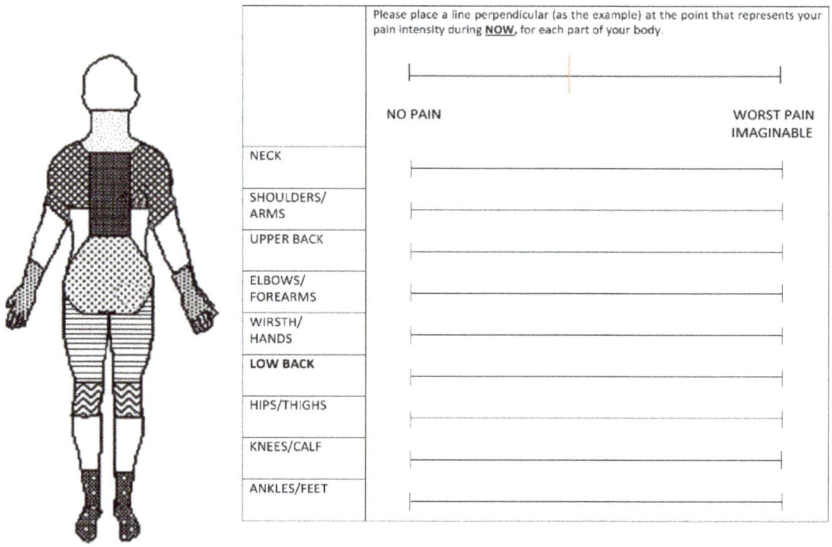

Figure 1. Body Pain Diagram and Visual Analogue Scale used every 15 min for reporting location and intensity of pain during the standing task.

Association of the different individual, physical and psychosocial factors recorded with pain ratings during the 1-h task was explored. First, the maximum change in VAS score from baseline and during the task was considered as the outcome using univariate and multivariate analyses. Spearman's correlations were used to determine the strength of the relationship between each of the individual, physical and psychosocial factors included and the increase in low-back and ankle-feet pain during the task. Then, a stepwise multivariate linear regression model was adjusted, including as covariates those with the highest associated correlations in the univariate analysis, both for the whole group of workers and for the PD group. Finally, in order to compare the consistency of the results, univariate and multivariate linear mixed-effects random-slope repeated measures models were also adjusted [33]. This type of model assumes that time effects (changes in pain rating over time) are random among individuals, considering the correlation among repeated measures in the same subject. Regression coefficients were estimated for the interaction between each of the covariates and time, allowing the rate of change to vary for different baseline characteristics.

Statistical analyses were performed using software SPSS version 25.0 (SPSS Inc., Chicago, IL, USA) and R version 4.0.5 (R Foundation for Statistical Computing, Vienna, Austria), with a bilateral significance level set at $p < 0.05$.

3. Results

Forty office workers (22 females; mean age: 37.4 ± 6.6 years; BMI: 26.3 ± 5.7 and 58% considered within healthy weight range) were included in the study. All participants completed the laboratory testing (physical testing and 1-h standing task) with no adverse events reported.

3.1. Standing-Time Effect on Pain Status: Any Reported Pain

The number of workers who reported pain in the low back and lower limb, throughout the task, increased over time (Table 1). At the beginning of the standing task (0 min), 15% of the participants reported some degree of LBP, increasing up to 30% at 15 min ($p = 0.070$) and reaching 40% ($p = 0.006$) and 42.5% ($p = 0.003$) at 30 and 45 min, respectively (Figure 2a). None of the participants had ankle-feet pain at the beginning of the task. Prevalence of ankle-feet pain was 10% after 15 min, increasing to 25% at 30 min ($p = 0.031$) and reaching 35.0% both at 45 min ($p = 0.006$) and 60 min ($p = 0.002$) (Figure 2b). Low-back and ankle-feet pain prevalence did not significantly increase after 30 min. While there was an increase in the number of participants who reported lower limb pain between baseline and 60 min (2.5% to 15% for the hip-thigh region and 5% to 27.5% for the knee-calf), this did not reach statistical significance.

Table 1. Prevalence of any low-back and lower extremity pain at 0, 15, 30, 45 and 60 min throughout the 1-h standing task.

				Low Back						
Pain	0 min		15 min		30 min		45 min		60 min	
	n	%	*n*	%	*n*	%	*n*	%	*n*	%
No	34	85.0	28	70.0	24	60.0	23	57.5	23	57.5
Yes	6	15.0	12	30.0	16	40.0	17	42.5	17	42.5
Total	40	100.0	40	100.0	40	100.0	40	100.0	40	100.0
				Hip-Thigh						
Pain	0 min		15 min		30 min		45 min		60 min	
	n	%	*n*	%	*n*		*n*	%	*n*	%
No	39	97.5	39	97.5	35	87.5	34	85.0	34	85.0
Yes	1	2.5	1	2.5	5	12.5	6	15.0	6	15.0
Total	40	100.0	40	100.0	40	100.0	40	100.0	40	100.0
				Knee-Calf						
Pain	0 min		15 min		30 min		45 min		60 min	
	n	%	*n*	%	*n*		*n*	%	*n*	%
No	38	95.0	36	90.0	33	82.5	30	75.0	29	72.5
Yes	2	5.0	4	10.0	7	17.5	10	25.0	11	27.5
Total	40	100.0	40	100.0	40	100.0	40	100.0	40	100.0
				Ankle-feet						
Pain	0 min		15 min		30 min		45 min		60 min	
	n	%	*n*	%	*n*	%	*n*	%	*n*	%
No	40	100	36	90.0	30	75.0	26	65.0	26	65.0
Yes	-	-	4	10.0	10	25.0	14	35.0	14	35.0
Total	40	100.0	40	100.0	40	100.0	40	100.0	40	100.0

3.2. Standing-Time Effect on Pain Status: Intensity of Pain

Of the 40 participants, 14 office workers were considered low-back pain developers and 9 were ankle-feet-pain developers (abbreviated as ankle-feet-PD onwards). For the hip-thigh and knee-calf regions, the number of participants who reported having a change of $\geq$10 on the VAS was small (3 and 6, respectively) preventing analysis.

The raw VAS score and VAS score increased over time for the total sample, PD and NPD groups, both for low-back and ankle-feet regions, are shown in Table 2 and Supplementary Material Figure S1 There was a significant standing-time effect, with individuals identified as PD showing increased levels of pain over time and the NPD group remaining at a very low level. The low-back-PD group averaged a mean VAS score of 30.8 $\pm$ 20.5 mm and the ankle-feet-PD group averaged a mean VAS score of 22.6 $\pm$ 9.7 mm at the end of standing. In addition, these results show that after 30 min of standing, significant differences in pain scores appear from the baseline.

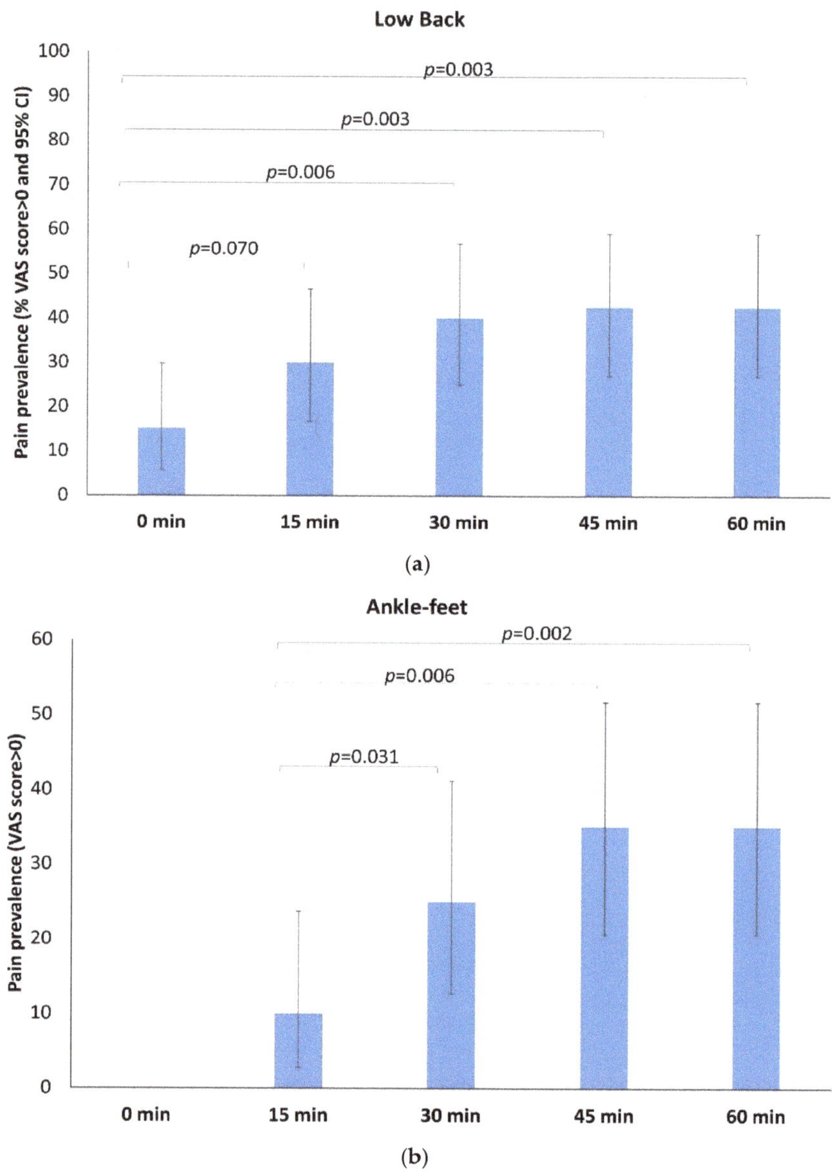

Figure 2. Prevalence and time-based changes of any reported pain (VAS scores > 0) for each area: (**a**) for low back; (**b**) for ankle-feet. *p*: McNemar test.

Table 2. VAS scores (0–100) and change in VAS scores from baseline at the low-back and ankle-feet regions at 0, 15, 30, 45 and 60 min of the 1 h standing task for all study participants, PD and NPD groups.

	Low Back				
	VAS Scores (0–100 mm)		VAS Scores Increase from Baseline (0–100 mm) #		
	Mean ± SD	Median	Mean ± SD	Median	*p* *
Total (*n* = 40)					
0 min	1.63 ± 4.5	0	-	-	
15 min	3.57 ± 8.7	0	1.94 ± 6.4	0	0.050
30 min	7.54 ± 13.2	0	5.90 ± 11.8	0	0.004
45 min	10.01 ± 17.6	0	8.38 ± 16.4	0	0.002
60 min	11.64 ± 19.0	0	9.97 ± 17.5	0	0.001
PD (*n* = 14)					
0 min	2.9 ± 5.9	0.0	-	-	-
15 min	8.7 ± 13.3	4.0	5.9 ± 8.1	3.0	0.050
30 min	20.5 ± 15.5	19.0	17.6 ± 12.7	19.0	0.004
45 min	27.0 ± 21.1	18.0	24.1 ± 19.0	18.0	0.002
60 min	30.8 ± 20.5	23.5	27.9 ± 18.1	23.5	0.001
NPD (*n* = 26)					
0 min	1.0 ± 3.5	0.0	-	-	-
15 min	0.8 ± 2.1	0.0	−0.2 ± 4.0	0.0	0.999
30 min	0.6 ± 1.6	0.0	−0.4 ± 3.9	0.0	0.684
45 min	0.9 ± 2.1	0.0	−0.1 ± 4.1	0.0	0.916
60 min	0.9 ± 3.1	0.0	−0.1 ± 3.6	0.0	0.715
	Ankle-feet				
Total (*n* = 40)					
0 min	0.0 ± 0.0	0	-	-	
15 min	1.5 ± 5.4	0	1.5 ± 5.4	0.0	0.068
30 min	3.2 ± 6.3	0	3.2 ± 6.3	0.0	0.005
45 min	5.0 ± 7.9	0	5.0 ± 7.9	0.0	0.001
60 min	5.8 ± 70.4	0	5.8 ± 10.4	0.0	0.001
PD (*n* = 9)					
0 min	0.0 ± 0.0	0.0	-	-	-
15 min	6.8 ± 10.1	0.0	6.8 ± 10.1	0.0	0.068
30 min	11.6 ± 7.2	12.0	11.6 ± 7.2	12.0	0.012
45 min	15.7 ± 4.6	17.0	15.7 ± 4.6	17.0	0.008
60 min	22.6 ± 9.7	21.0	22.6 ± 9.7	21.0	0.008
NPD (*n* = 31)					
0 min	0.0 ± 0.0	0.0	-	-	-
15 min	0.0 ± 0.0	0.0	0.0 ± 0.0	0.0	0.999
30 min	0.8 ± 3.3	0.0	0.8 ± 3.3	0.0	0.180
45 min	1.8 ± 5.5	0.0	1.8 ± 5.5	0.0	0.027
60 min	0.9 ± 2.3	0.0	0.9 ± 2.3	0.0	0.026

* *p*-values from two-sided Wilcoxon's signed-rank test comparing VAS score at each moment relative to baseline VAS pain. # Positive scores indicate an increase in VAS; PD: pain developers; NPD: non-pain developers.

3.3. Association of Individual, Physical and Psychosocial Factors with Pain Ratings over Time

Table 3 shows the Spearman's rho correlation coefficients between the maximum increment of VAS scores for the low back and ankle-feet during the task and different individual, physical and psychosocial factors in the whole sample. Correlations for PD and NPD groups are shown in Supplementary Material Table S1.

Table 3. Spearman correlation coefficients between the maximum increment at VAS scores at low-back and ankle-feet regions throughout of the standing task and variables for the entire sample (n = 40).

	Low Back		Ankle-Feet	
	Total n = 40		Total n = 40	
	Rho	p	Rho	p
Age (years)	0.162	0.318	−0.008	0.961
BMI (kg/m^2)	−0.098	0.546	0.379 *	0.016
IPAQ, during de last 7 days, how much time did you usually spend sitting on a weekend day (minutes)	−0.002	0.988	0.284	0.076
IPAQ, during de last 7 days, how much time did you usually spend sitting on a weekday (minutes)	0.142	0.381	0.025	0.878
IPAQ, MET min/week	0.005	0.975	0.067	0.690
LBP severity, last 7 days (0–100)	0.538 *	0.000	−0.046	0.777
OSPAQ, minutes sitting at work per week	−0.268	0.094	0.283	0.077
OSPAQ, minutes standing at work per week	0.028	0.862	−0.009	0.956
OSPAQ, minutes walking at work per week	−0.083	0.612	−0.208	0.198
JCQ, Job Control	−0.085	0.601	−0.068	0.679
JCQ, Psychological Job Demands	−0.140	0.390	−0.220	0.172
JCQ, Social Support	0.009	0.957	0.104	0.523
JCQ, Physical Demands	0.158	0.329	0.010	0.953
PCS, Rumination	0.222	0.168	−0.121	0.455
PCS, Magnification	0.141	0.386	−0.224	0.164
PCS, Helplessness	0.263	0.101	0.041	0.801
PCS-total	0.249	0.122	−0.121	0.458
SF-12, Mental Component Summary	0.278	0.083	−0.387 *	0.014
SF-12, Physical Component Summary	−0.345 *	0.029	0.168	0.299
ASLR, total examiner-score (0–10)	0.210	0.193	0.051	0.755
ASLR, total participant-score (0–10)	0.346 *	0.029	−0.058	0.724
AHAbd, right side, examiner-score (0–3)	0.038	0.816	0.025	0.881
AHAbd, left side, examiner-score (0–3)	0.129	0.426	0.033	0.839
AHAbd, right side, participant-score (0–5)	0.226	0.162	0.115	0.479
AHAbd, left side, participant-score (0–5)	0.325 *	0.041	0.008	0.960
Abdominal (s)	−0.269	0.094	−0.156	0.337
Side Bridge right side (s)	−0.254	0.114	−0.097	0.550
Side Bridge left side (s)	−0.246	0.126	−0.069	0.671
Supine Bridge (s)	−0.298	0.062	−0.327 *	0.040
Isometric hip abduction (right leg) (s)	−0.472 *	0.002	−0.003	0.985
Isometric hip abduction (left leg) (s)	−0.484 *	0.002	0.045	0.782
Sorensen (s)	−0.290	0.070	−0.315 *	0.048

* $p \leq 0.05$; BMI, Body Mass Index; MET, Metabolic Equivalent of Task (computed as the sum of walking, moderate-intensity, and vigorous-intensity physical activity); OSPAQ, Occupational Sitting and Physical Activity Questionnaire; JCQ, Job Content Questionnaire; ASLR, Active Straight Leg Raise; AHAbd, Active Hip Abduction; s, seconds.

Considering the whole sample, the maximum increase in VAS scores at the low back throughout of the 1-h standing task was significantly correlated with (i) history of LBP measured as the severity of LBP in the last 7 days (Rho = 0.54, p = 0.000); (ii) SF-12 (physical component summary) (Rho = −0.35, p = 0.029); (iii) ASLR, total participant-score (Rho =0.35, p = 0.029); (iv) AHAbd left side, participant-score (Rho = 0.33, p = 0.04); (v) isometric hip abduction (right leg) (Rho = −0.47, p = 0.002); (vi) isometric hip abduction (left leg) (Rho = −0.48, p = 0.002). In the feet area, the maximum increase in VAS scores was significantly correlated with (i) BMI (Rho = 0.38, p = 0.016); (ii) SF-12 (mental component summary) (Rho = 0.39, p = 0.014); (iii) Supine Bridge (Rho = −0.33, p = 0.04); and (iv) Biering–Sorensen test (Rho = −0.32, p = 0.048).

3.4. Regression Analysis-Predicting the Magnitude of Low-Back Pain throughout of 1-h Standing Task

A multivariate linear regression model was adjusted with the maximum increment in VAS scores throughout of standing as the dependent variable and the magnitude of the highest and significant correlated variables (determined using Spearman's correlation) as covariates (Table 4). Lower hip abductor muscle endurance (B = −0.23, p = 0.007) and lower physical health (PCS-SF-12) (B = −0.86, p = 0.008) predicted higher level of LBP throughout of the 1-h standing task for the entire sample. These two variables explained 41% of the variability in the increment of the pain (R^2 = 0.41).

Table 4. Multivariate lineal and mixed regression models for the prediction of low-back pain throughout of the 1-h standing task.

Lineal Regression Analysis					
Model	**B**	**SE**	***p***	**95% CI**	
				Lower	**Upper**
Intercept	74.661	15.375	0.000	43.507	105.814
Isometric Hip Abduction endurance test	−0.233	0.081	0.007	−0.397	−0.069
Physical Component Summary (SF-12)	−0.864	0.309	0.008	−1.489	−0.239
Mixed Regression Analysis					
	B	SE	*p*		
Fixed effects					
Intercept	−8.07	6.31	0.203		
Time	0.79	0.42	0.068		
Age	0.076	0.076	0.322		
Time × Age	0.009	0.005	0.078		
Isometric Hip Abduction endurance test	−0.002	0.025	0.939		
Time*Isometric Hip Abduction endurance test	−0.004	0.002	0.022		
Physical Component Summary (SF-12)	0.132	0.098	0.177		
Time*Physical Component Summary (SF-12)	−0.012	0.006	0.062		
	Estimate	SE			
Random effects					
Linear slope (time)	0.073	0.270			
Residual	32.30	5.68			

SE: Standard error; CI: Confidence Interval.

The mixed regression model reinforced these results, with the interaction time × minimum isometric hip abduction and time × PCS-SF-12 being significant predictors of the highest level of LBP during the task. These results indicate that workers with lower hip abductor muscle endurance and lower physical health experience a higher rate of increase of LBP during the task.

4. Discussion

This study showed significant time-based changes for the reported prevalence and severity of pain in the low back and ankle-feet regions of the body, with 30 min identified as the threshold for observing these differences. The regression models demonstrated that less hip abductor muscle endurance and less physical health (as measured with the SF-12) predicted a greater increase of LBP at the end of a 1-h standing task. The correlation analysis suggested that the maximum increase in VAS score in the ankle-feet area was associated with higher BMI, less back and hip muscle endurance (Supine Bridge and Biering–Sorensen test) and mental health (SF-12).

There was a significant effect of standing at a workstation for 1-h on the presence and intensity of pain in the low-back and ankle-feet regions. This has consistently been reported for the low back [9,11,14,17,34], but has been less commonly investigated in the ankle-feet region [9,18,35]. The percentage of office workers who developed LBP at the end of the 1-h standing task (42.5%) was similar to the average of 44% reported in the systematic review of Coenen et al. [8]; but lower than other studies which reported rates of 81% [34], 71% [11] and 65% [17]. The VAS level of pain reported by low-back PD in this study was within the range of what has been reported previously (19 mm [34] to 32 mm [11]), despite most previous work using populations of students without a history of LBP and using a longer standing task. The average increase in pain scores from baseline in our study was 27.9 ± 18.1 mm which is lower than clinical low back populations but higher than the absolute cut-off value of 15 points for the minimal important change (MIC) on the VAS for LBP patients [36].

Lower limb pain was commonly reported during the standing task by our participants, with 15% reporting hip-thigh pain, 27.5% reporting pain in the knee-calf region and 35% developing ankle-feet pain at the end of standing. It should be noted that no participants had ankle-feet pain at the beginning of the standing task. The increase in ankle-feet-PD pain was 22.6 ± 21 mm. Although this level of pain could be interpreted as low, it is higher than the MIC (9.3 mm) reported for the clinical interpretation of results in patients with foot or ankle pathologies [37]. This pain intensity for the ankle-feet region is similar to that reported by Antle and Cote [18] after 34 min of standing (3.5 out of 10) and Smith et al. [35] after 2 h of standing (1.8 out of 10).

Our results demonstrate that 30 min of standing affects the prevalence and severity of pain at the low-back and ankle-feet regions in office workers. After 30 min, the PD group reported pain scores which exceeded the MIC [36,37]. It has been demonstrated that in PD low back usually begins to develop within 30–60 min of standing [11,14,17,35]. Prolonged periods of occupational standing (greater than 30 min each hour), was one of the strongest predictors of LPB in employees [19] and low-back symptoms were reached after 71 min with this reduced to 42 min in those considered PD, after pooled data from laboratory studies [8]. Dose–response associations for lower extremity symptoms are more heterogeneous in the literature [8], with one study showing lower limb discomfort after 34 min of standing [18]. While a significant interaction of time on pain development during standing has been illustrated [11,14,17], most studies did not determine the point of time when the significant increase in pain occurred. Consequently, based on our results, we recommend that office workers restrict their maximum standing time to no more than 30 min which is less than the suggested time of 40 min [8].

Few studies have considered the predictors of pain intensity during standing. Multivariate regression modelling demonstrated that lower hip abductor muscle endurance and lower physical health were independently associated with a higher level of LBP, explaining 41% of the variability in pain in our study. In turn, mixed regression modelling demonstrated a significant interaction for time × isometric hip abduction and time × SF-12 (physical component summary), reinforcing the importance of these two factors as predictors of the highest-level LBP during the 1-h standing task. Hip abductor muscle endurance has previously suggested to be associated with LBP development during standing. Viggiani and Callaghan [12] determined that hip abductor fatigability (measured using isometric hip abduction) differentiated those PD from NPD, with PD having lower hip abductor endurance. Marshall et al. [11] showed that low-back PD had lower gluteus medius endurance (measured with a side-bridge test), there was no association between the side bridge test and pain levels in linear regression analysis. Finally, Hwang et al. [38] identified that hip abductor muscle strength (measured in the similar way as this study) was the variable that most contributed to VAS scores in workers with LBP who performed occupational standing. While causality between LBP and hip abductor muscle function cannot be confirmed from existing research, the negative slope in our analysis indicates that low-back VAS scores decrease as endurance of the hip abductors muscles increases. Together, data suggest that hip abduction muscle weakness and fatiguability may be important to consider in preventing or managing standing-induced LBP.

To our knowledge, this is the first study to investigate self-perceived physical health, measured through Physical Component Summary (SF-12), as predictor of higher pain intensity during a prolonged standing task. In the calculation of the PCS summary score, the highest weights were given to four domains: physical functioning, role physical, bodily pain, and general health. Bodily Pain was the domain with the lowest scores for our sample (mean: 49.6, IQR: 46.9–57.5, data not shown previously) and for the LBP-PD group (47.4 ± 10.5). This is consistent with previous reports stating that the presence of pain (in the low back and/or other sites) may be associated with the incidence and prevalence of LBP [39,40].

Factors associated with ankle-feet pain have not previously been investigated in relation to occupational standing in office workers. While regression analysis was not undertaken for the maximum VAS for ankle-feet pain due to small proportion of participants with this pain, ankle-feet region VAS was moderately correlated with greater BMI, lower trunk/hip muscle function (lower supine bridge and Biering–Sorensen test ability) and mental health. There is a strong association between high BMI and non-specific foot pain in the general population [41] and in people with plantar heel pain [42]. Mechanical loading with increased body mass is a possible mechanism for this relationship, as well as metabolic and psychological factors [41]. Trunk/hip stability is thought to be important for the production, transfer and control of force throughout the entire kinetic chain [43]. Gluteus maximus and medius muscles weakness has been identified in people with ankle injuries [44,45], and it has been hypothesized that deficits in core stability may increase risk of lower extremity injury [46,47]. Altered loading throughout the lower kinetic due to hip/trunk muscle deficits may contribute to ankle-feet pain during prolonged standing at a standing workstation. The association between better mental health and ankle-feet pain could be due to a spurious finding and a consequence of the small sample size.

There are limitations to this study that must be considered. First, the study has a relatively small sample size (n = 40), with only a small proportion of subjects classified as low-back or ankle-feet PD. Although previous studies have had similar sample sizes [12,14], this sample size may have led to two important issues: (i) failing to detect a real effect on pain development for any of the variables studied, and (ii) finding effects that seem supported by the data but are spurious. The use of alternative statistical methods would have been preferred [48] but were difficult due to the small sample. The results of this study should be confirmed in future studies involving more individuals and more analysis techniques. Second, this study did not include outcomes to identify potential vascular mechanisms that may be associated with the development of musculoskeletal ankle-feet symptoms [8,18]. We recommend that future research include such measures.

This study has shown that there is an impact of 1 h-standing exposure on several aspects of pain status in office workers, determined the significant dose–response relationship for standing, and clarified the factors associated with the intensity of low-back and ankle-feet pain. Based on the findings of our study, practitioners and clinicians should advise office workers to avoid standing for more than 30 min in light of the dose–response relationship for standing and pain. Due to the relationship between hip abductor muscle endurance, physical health status and intensity of LBP, future research is needed to determine if improving these factors decreases LBP intensity during standing. Similarly, further investigation is needed to understand the relationship between BMI, trunk/hip muscle function and mental health on ankle-feet pain during standing.

5. Conclusions

This study in office workers demonstrated that the prevalence and intensity of low-back and ankle-feet pain increased during a 1-h laboratory-based standing task, with 30 min identified as the threshold for the development/provocation of pain. Lower hip abductor muscle endurance and physical health predicted the low-back pain intensity. In the ankle-feet area, results suggest that the increase in pain scores was correlated with greater BMI and lower trunk/hip muscle function.

Supplementary Materials: The following supporting information can be downloaded at: https://www.mdpi.com/article/10.3390/ijerph19042221/s1, Figure S1: Pain development during 1 h-standing for PD and NPD groups: (a) at low-back; (b) at ankle-feet areas; Table S1: Spearman correlation coefficients between the maximum increment at VAS scores at low-back and ankle-feet regions throughout of the standing task and measured variables for PD groups.

Author Contributions: Conceptualization, V.J., B.R.-R. and M.D.S.; methodology, V.J., B.R.-R., M.D.S. and S.P.-D.; formal analysis, S.P.-D. and A.Q.-d.-R.; investigation, B.R.-R., V.J. and M.D.S.; resources, V.J. and B.R.-R.; writing—original draft preparation, B.R.-R. and S.P.-D.; writing—review and editing, V.J., M.D.S., B.R.-R. and S.P.-D.; visualization, B.R.-R., S.P.-D., M.D.S., A.Q.-d.-R. and V.J.; supervision, V.J.; project administration, V.J. All authors have read and agreed to the published version of the manuscript.

Funding: This research received no external funding.

Institutional Review Board Statement: The study was conducted in accordance with the Declaration of Helsinki and approved by the Institutional Review Board (or Ethics Committee) of The University of Queensland Human Research Ethics Committee B (Approval Number: #2017000666; 5 March 2018).

Informed Consent Statement: Informed consent was obtained from all subjects involved in the study.

Data Availability Statement: https://zenodo.org/record/5947650#.YfwuJviCFhF.

Acknowledgments: The authors thank all the participants who took part in the study.

Conflicts of Interest: The authors declare no conflict of interest.

References

1. Dutheil, F.; Duclos, M.; Esquirol, Y. Editorial: Sedentary Behaviors at Work. *Front. Public Health* **2020**, *8*, 57. [CrossRef] [PubMed]
2. Roffey, D.M.; Wai, E.K.; Bishop, P.; Kwon, B.K.; Dagenais, S. Causal Assessment of Occupational Standing or Walking and Low Back Pain: Results of a Systematic Review. *Spine J.* **2010**, *10*, 262–272. [CrossRef] [PubMed]
3. De Carvalho, D.E.; de Luca, K.; Funabashi, M.; Breen, A.; Wong, A.Y.L.; Johansson, M.S.; Ferreira, M.L.; Swab, M.; Neil Kawchuk, G.; Adams, J.; et al. Association of Exposures to Seated Postures With Immediate Increases in Back Pain: A Systematic Review of Studies With Objectively Measured Sitting Time. *J. Manip. Physiol. Ther.* **2020**, *43*, 1–12. [CrossRef]
4. Korshøj, M.; Hallman, D.M.; Mathiassen, S.E.; Aadahl, M.; Holtermann, A.; Jørgensen, M.B. Is Objectively Measured Sitting at Work Associated with Low-Back Pain? A Cross Sectional Study in the DPhacto Cohort. *Scand. J. Work. Environ. Health* **2018**, *44*, 96–105. [CrossRef] [PubMed]
5. Zhao, R.; Bu, W.; Chen, Y.; Chen, X. The Dose-Response Associations of Sedentary Time with Chronic Diseases and the Risk for All-Cause Mortality Affected by Different Health Status: A Systematic Review and Meta-Analysis. *J. Nutr. Health Aging* **2020**, *24*, 63–70. [CrossRef]
6. Buckley, J.P.; Hedge, A.; Yates, T.; Copeland, R.J.; Loosemore, M.; Hamer, M.; Bradley, G.; Dunstan, D.W. The Sedentary Office: An Expert Statement on the Growing Case for Change towards Better Health and Productivity. *Br. J. Sport. Med.* **2015**, *49*, 1357–1362. [CrossRef]
7. Shrestha, N.; Kukkonen-Harjula, K.T.; Verbeek, J.H.; Ijaz, S.; Hermans, V.; Pedisic, Z. Workplace Interventions for Reducing Sitting at Work. *Cochrane Database Syst. Rev.* **2018**, *6*, CD010912.
8. Coenen, P.; Parry, S.; Willenberg, L.; Shi, J.W.; Romero, L.; Blackwood, D.M.; Healy, G.N.; Dunstan, D.W.; Straker, L.M. Associations of Prolonged Standing with Musculoskeletal Symptoms-A Systematic Review of Laboratory Studies. *Gait Posture* **2017**, *58*, 310–318. [CrossRef]
9. Coenen, P.; Willenberg, L.; Parry, S.; Shi, J.W.; Romero, L.; Blackwood, D.M.; Maher, C.G.; Healy, G.N.; Dunstan, D.W.; Straker, L.M. Associations of Occupational Standing with Musculoskeletal Symptoms: A Systematic Review with Meta-Analysis. *Br. J. Sports Med.* **2018**, *52*, 174–181. [CrossRef]
10. Takala, E.-P. Lack of "Statistically Significant" Association Does Not Exclude Causality. *Spine J.* **2010**, *10*, 944. [CrossRef]
11. Marshall, P.W.M.; Patel, H.; Callaghan, J.P. Gluteus Medius Strength, Endurance, and Co-Activation in the Development of Low Back Pain during Prolonged Standing. *Hum. Mov. Sci.* **2011**, *30*, 63–73. [CrossRef] [PubMed]
12. Viggiani, D.; Callaghan, J.P. Hip Abductor Fatigability and Recovery Are Related to the Development of Low Back Pain During Prolonged Standing. *J. Appl. Biomech.* **2018**, *34*, 39–46. [CrossRef]
13. Rodríguez-Romero, B.; Smith, M.D.; Quintela-Del-Rio, A.; Johnston, V. What Psychosocial and Physical Characteristics Differentiate Office Workers Who Develop Standing-Induced Low Back Pain? A Cross-Sectional Study. *Int. J. Environ. Res. Public Health* **2020**, *17*, 7104. [CrossRef] [PubMed]
14. Nelson-Wong, E.; Callaghan, J.P. Is Muscle Co-Activation a Predisposing Factor for Low Back Pain Development during Standing? A Multifactorial Approach for Early Identification of at-Risk Individuals. *J. Electromyogr. Kinesiol.* **2010**, *20*, 256–263. [CrossRef] [PubMed]

15. Sorensen, C.J.; Johnson, M.B.; Norton, B.J.; Callaghan, J.P.; Van Dillen, L.R. Asymmetry of Lumbopelvic Movement Patterns during Active Hip Abduction Is a Risk Factor for Low Back Pain Development during Standing. *Hum. Mov. Sci.* **2016**, *50*, 38–46. [CrossRef] [PubMed]
16. Nelson Wong, E.; Flynn, T.; Callaghan, J.P. Development of Active Hip Abduction Screening Test for Identifying Occupational Low Back Pain. *J. Orthop. Sports Phys. Ther.* **2009**, *39*, 649–657. [CrossRef]
17. Nelson-Wong, E.; Gregory, D.E.; Winter, D.A.; Callaghan, J.P. Gluteus Medius Muscle Activation Patterns as a Predictor of Low Back Pain during Standing. *Clin. Biomech.* **2008**, *23*, 545–553. [CrossRef]
18. Antle, D.M.; Côté, J.N. Relationships between Lower Limb and Trunk Discomfort and Vascular, Muscular and Kinetic Outcomes during Stationary Standing Work. *Gait Posture* **2013**, *37*, 615–619. [CrossRef]
19. Andersen, J.H.; Haahr, J.P.; Frost, P. Risk Factors for More Severe Regional Musculoskeletal Symptoms: A Two-Year Prospective Study of a General Working Population. *Arthritis Rheum.* **2007**, *56*, 1355–1364. [CrossRef]
20. Kuorinka, I.; Jonsson, B.; Kilbom, A.; Vinterberg, H.; Biering-Sørensen, F.; Andersson, G.; Jørgensen, K. Standardised Nordic Questionnaires for the Analysis of Musculoskeletal Symptoms. *Appl. Ergon.* **1987**, *18*, 233–237. [CrossRef]
21. Von Korff, M.; Jensen, M.P.; Karoly, P. Assessing Global Pain Severity by Self-Report in Clinical and Health Services Research. *Spine* **2000**, *25*, 3140–3151. [CrossRef] [PubMed]
22. Silsbury, Z.; Goldsmith, R.; Rushton, A. Systematic Review of the Measurement Properties of Self-Report Physical Activity Questionnaires in Healthy Adult Populations. *BMJ Open* **2015**, *5*, e008430. [CrossRef] [PubMed]
23. Jancey, J.; Tye, M.; McGann, S.; Blackford, K.; Lee, A.H. Application of the Occupational Sitting and Physical Activity Questionnaire (OSPAQ) to Office Based Workers. *BMC Public Health* **2014**, *14*, 762. [CrossRef] [PubMed]
24. Karasek, R.; Brisson, C.; Kawakami, N.; Houtman, I.; Bongers, P.; Amick, B. The Job Content Questionnaire (JCQ): An Instrument for Internationally Comparative Assessments of Psychosocial Job Characteristics. *J. Occup Health Psychol.* **1998**, *3*, 322–355. [CrossRef]
25. Sullivan, M.J.; Bishop, S.R.; Pivik, J. The Pain Catastrophizing Scale: Development and Validation. *Psychol. Assess.* **1995**, *74*, 524–532. [CrossRef]
26. Gandek, B.; Ware, J.E.; Aaronson, N.K.; Apolone, G.; Bjorner, J.B.; Brazier, J.E.; Bullinger, M.; Kaasa, S.; Leplege, A.; Prieto, L.; et al. Cross-Validation of Item Selection and Scoring for the SF-12 Health Survey in Nine Countries: Results from the IQOLA Project. International Quality of Life Assessment. *J. Clin. Epidemiol.* **1998**, *51*, 1171–1178. [CrossRef]
27. Davis, A.M.; Bridge, P.; Miller, J.; Nelson-Wong, E. Interrater and Intrarater Reliability of the Active Hip Abduction Test. *J. Orthop. Sport. Phys. Ther.* **2011**, *41*, 953–960. [CrossRef]
28. Mens, J.M.; Vleeming, A.; Snijders, C.J.; Koes, B.W.; Stam, H.J. Reliability and Validity of the Active Straight Leg Raise Test in Posterior Pelvic Pain since Pregnancy. *Spine* **2001**, *26*, 1167–1171. [CrossRef]
29. McGill, S.M.; Childs, A.; Liebenson, C. Endurance Times for Low Back Stabilization Exercises: Clinical Targets for Testing and Training from a Normal Database. *Arch. Phys. Med. Rehabil.* **1999**, *80*, 941–944. [CrossRef]
30. Schellenberg, K.L.; Lang, J.M.; Chan, K.M.; Burnham, R.S. A Clinical Tool for Office Assessment of Lumbar Spine Stabilization Endurance: Prone and Supine Bridge Maneuvers. *Am. J. Phys. Med. Rehabil.* **2007**, *86*, 380–386. [CrossRef]
31. Van Cant, J.; Dumont, G.; Pitance, L.; Demoulin, C.; Feipel, V. Test-Retest Reliability of Two Clinical Tests for the Assessment of Hip Abductor Endurance in Healthy Females. *Int. J. Sports Phys. Ther.* **2016**, *11*, 24–33.
32. Sorensen, C.J.; Johnson, M.B.; Callaghan, J.P.; George, S.Z.; Van Dillen, L.R. Validity of a Paradigm for Low Back Pain Symptom Development During Prolonged Standing. *Clin. J. Pain* **2015**, *31*, 652–659. [CrossRef] [PubMed]
33. Brown, H.; Prescott, R. *Applied Mixed Models in Medicine.*; John Wiley & Sons: New York, NY, USA, 1999.
34. Gregory, D.E.; Callaghan, J.P. Prolonged Standing as a Precursor for the Development of Low Back Discomfort: An Investigation of Possible Mechanisms. *Gait Posture* **2008**, *28*, 86–92. [CrossRef] [PubMed]
35. Smith, M.D.; Vicentino, B.; Brown, W.; Gilson, N.; Gane, E.M.; Johnston, V. Characteristics of Symptoms Experienced by Individuals during a Prolonged Standing Task: An Office-Based Study. *Brazilian J. Phys. Ther.* **2022**, in press.
36. Ostelo, R.W.; Deyo, R.A.; Stratford, P.; Waddell, G.; Croft, P.; Von Korff, M.; Bouter, L.M.; de Vet, H.C. Interpreting Change Scores for Pain and Functional Status in Low Back Pain: Towards International Consensus Regarding Minimal Important Change. *Spine* **2008**, *33*, 90–94. [CrossRef]
37. Saarinen, A.J.; Uimonen, M.M.; Sandelin, H.; Toom, A.; Richter, M.; Repo, J.P. Minimal Important Change for the Visual Analogue Scale Foot and Ankle (VAS-FA). *Foot Ankle Surg. Off. J. Eur. Soc. Foot Ankle Surg.* **2021**, *27*, 196–200. [CrossRef]
38. Hwang, U.-J.; Kwon, O.-Y.; Jung, S.-H.; Ahn, S.-H.; Kim, H.-A. Predictors of Pain Intensity and Oswestry Disability Index in Prolonged Standing Service Workers with Nonspecific Chronic Low Back Pain Subclassified as Active Extension Pattern. *Musculoskelet. Sci. Pract.* **2019**, *40*, 58–64. [CrossRef]
39. Øverås, C.K.; Johansson, M.S.; de Campos, T.F.; Ferreira, M.L.; Natvig, B.; Mork, P.J.; Hartvigsen, J. Distribution and Prevalence of Musculoskeletal Pain Co-Occurring with Persistent Low Back Pain: A Systematic Review. *BMC Musculoskelet. Disord.* **2021**, *22*, 91. [CrossRef]
40. Taylor, J.B.; Goode, A.P.; George, S.Z.; Cook, C.E. Incidence and Risk Factors for First-Time Incident Low Back Pain: A Systematic Review and Meta-Analysis. *Spine J.* **2014**, *14*, 2299–2319. [CrossRef]
41. Butterworth, P.A.; Landorf, K.B.; Smith, S.E.; Menz, H.B. The Association between Body Mass Index and Musculoskeletal Foot Disorders: A Systematic Review. *Obes. Rev. Off. J. Int. Assoc. Study Obes.* **2012**, *13*, 630–642. [CrossRef]

42. Sullivan, J.; Pappas, E.; Burns, J. Role of Mechanical Factors in the Clinical Presentation of Plantar Heel Pain: Implications for Management. *Foot* **2020**, *42*, 101636. [CrossRef] [PubMed]
43. Neumann, D.A. Kinesiology of the Hip: A Focus on Muscular Actions. *J. Orthop. Sports Phys. Ther.* **2010**, *40*, 82–94. [CrossRef] [PubMed]
44. McCann, R.S.; Crossett, I.D.; Terada, M.; Kosik, K.B.; Bolding, B.A.; Gribble, P.A. Hip Strength and Star Excursion Balance Test Deficits of Patients with Chronic Ankle Instability. *J. Sci. Med. Sport* **2017**, *20*, 992–996. [CrossRef]
45. Khalaj, N.; Vicenzino, B.; Smith, M.D. Hip and Knee Muscle Torque and Its Relationship with Dynamic Balance in Chronic Ankle Instability, Copers and Controls. *J. Sci. Med. Sport* **2021**, *24*, 647–652. [CrossRef]
46. Willson, J.D.; Dougherty, C.P.; Ireland, M.L.; Davis, I.M. Core Stability and Its Relationship to Lower Extremity Function and Injury. *J. Am. Acad. Orthop. Surg.* **2005**, *13*, 316–325. [CrossRef] [PubMed]
47. De Blaiser, C.; De Ridder, R.; Willems, T.; Vanden Bossche, L.; Danneels, L.; Roosen, P. Impaired Core Stability as a Risk Factor for the Development of Lower Extremity Overuse Injuries: A Prospective Cohort Study. *Am. J. Sports Med.* **2019**, *47*, 1713–1721. [CrossRef]
48. Goulet, J.L.; Buta, E.; Bathulapalli, H.; Gueorguieva, R.; Brandt, C.A. Statistical Models for the Analysis of Zero-Inflated Pain Intensity Numeric Rating Scale Data. *J. Pain* **2017**, *18*, 340–348. [CrossRef]

International Journal of
Environmental Research and Public Health

MDPI

Article

The Effect of Core Stabilization Exercise with the Abdominal Drawing-in Maneuver Technique on Stature Change during Prolonged Sitting in Sedentary Workers with Chronic Low Back Pain

Pongsatorn Saiklang [1], Rungthip Puntumetakul [2,3,*] and Thiwaphon Chatprem [2]

[1] Division of Physical Therapy, Faculty of Physical Therapy, Srinakharinwiroj University, Nakhonnayok 26120, Thailand; Pongsatornsa@g.swu.ac.th
[2] Research Center of Back, Neck, Other Joint Pain and Human Performance (BNOJPH), Khon Kaen University, Khon Kaen 40002, Thailand; Thiwaphon.ao@gmail.com
[3] Department of Physical Therapy, Faculty of Associated Medical Sciences, Khon Kaen University, Khon Kaen 40002, Thailand
* Correspondence: Rungthiprt@gmail.com

Abstract: To enhance stature recovery, lumbar spine stabilization by stimulating the deep trunk muscle activation for compensation forces originating from the upper body was introduced. The abdominal drawing-in maneuver (ADIM) technique has been found mainly to activate deep trunk muscles. The purpose of the current study was to determine whether 5 weeks of training of deep trunk muscles using the ADIM technique could improve stature recovery, delay trunk muscle fatigue, and decrease pain intensity during prolonged sitting. Thirty participants with chronic low back pain (CLBP) conducted a core stabilization exercise (CSE) with the ADIM technique for 5 weeks. Participants were required to sit for 41 min before and after the exercise intervention. Stature change was measured using a seated stadiometer with a resolution of ±0.006 mm. During sitting, the stature change, pain intensity, and trunk muscle fatigue were recorded. A comparison between measurements at baseline and after 5 weeks of training demonstrated: (i) stature recovery and pain intensity significantly improved throughout the 41 min sitting condition; (ii) the bilaterally trunk muscle showed significantly decreased fatigue. The CSE with the ADIM technique was shown to provide a protective effect on detrimental reductions in stature change and trunk muscle fatigue during prolonged sitting in young participants under controlled conditions in a laboratory. This information may help to prevent the risk of LBP from prolonged sitting activities in real life situations.

Keywords: spinal load; core stability; ergonomics

Citation: Saiklang, P.; Puntumetakul, R.; Chatprem, T. The Effect of Core Stabilization Exercise with the Abdominal Drawing-in Maneuver Technique on Stature Change during Prolonged Sitting in Sedentary Workers with Chronic Low Back Pain. *IJERPH* **2022**, *19*, 1904. https://doi.org/10.3390/ijerph19031904

Academic Editors: Sergio Iavicoli, Vincenzo Denaro, Gianluca Vadalà and Fabrizio Russo

Received: 23 December 2021
Accepted: 5 February 2022
Published: 8 February 2022

1. Introduction

Sedentary behavior is characterized by an energy expenditure less than or equal to 1.5 metabolic equivalents (METs) while in a sitting or reclining position when awake [1]. Today, increasing numbers of individuals spend extended periods in a seated position at work as well as during leisure time [2,3]. Recently, sedentary workers in Thailand reported recurring low back pain, with 63% showing that their low back pain was aggravated by sitting during working hours [4]. Chronic low back pain (CLBP) has a global prevalence of 11–23% among people with low back pain [5,6]. The socioeconomic burden of CLBP stems from a prolonged loss of function, which consequently results in decreased work productivity and increased medical costs [5,7].

Deep trunk muscle fatigue may arise from continuous contraction of the trunk muscles during prolonged sitting [8–10]. This fatigue reduces muscular support to the spine and increases stress on ligaments and intervertebral discs [9,10]; consequently, it reduces intervertebral disc height [11,12]. Reductions in disc height could increase compressive stress on sensitive spinal structures [13,14] and may stimulate nociceptor activity, leading to pain [14]. Stature change measurement is a method used to reflect alterations in spinal length, and the reduction of spinal length is known as spinal shrinkage or stature loss [15]. Prolonged sitting postures could lead to stature reduction and ultimately to low back pain [9,16,17].

Trunk muscles play an essential role in contributing to spinal stability [18]. There are two types of trunk muscle systems: superficial and deep [19,20]. The internal oblique (IO), transversus abdominis (TrA), and lumbar multifidus (LM) muscles represent a deep muscle system that compensates for forces on the upper body of the spine and increases lumbar stability [18,21]. Previous studies reported changes in the muscle recruitment pattern and timing of muscle onset in people with low back pain [22,23]. Increased superficial trunk muscle activation occurs to compensate for deep trunk muscle dysfunction [24,25], in which the neural control subsystem attempts to maintain spinal stability [18,26]. Increased activation of the superficial trunk muscle can compress the spinal structure and lead to delayed stature recovery [25,27].

Previous research reported that superficial trunk muscle activity can be reduced by activating the deep trunk muscles using the abdominal drawing-in maneuver (ADIM) technique [15,28,29]. The ADIM technique is known to elicit preferential recruitment of the transversus abdominis muscle with minimal activation of the superficial trunk muscles. For this technique, participants are instructed to 'gently draw in their lower abdomen toward the spine' [13]. Saiklang et al. (2020) investigated the change in stature recovery in patients with CLBP following the immediate effect of the ADIM technique for 1 min repeated three times throughout a 41 min prolonged sitting period. The results demonstrated that the ADIM technique significantly improved stature recovery and increased TrA and IO muscle activities and TrA and IO/RA ratios compared with upright sitting without exercise [15].

To date, no investigation has reported the effect of the long-term impact of the core stabilization exercise (CSE) with the ADIM technique program focusing on deep trunk muscle on stature recovery during prolonged sitting. The CSE with ADIM technique aims to improve neuromuscular control skills, relearn normal function, and enhance endurance of the deep muscles around the lumbar spine, such as the TrA and LM muscles [28,30].

The aim of the current study was to investigate differences in stature change, pain, and trunk muscle fatigue during prolonged sitting in seated sedentary workers with CLBP between baseline and the first day after a 5-week CSE with the ADIM technique. We hypothesized that the 5-week CSE with the ADIM technique can improve deep trunk muscle endurance, reduce pain, and delay stature reduction during prolonged sitting.

2. Materials and Methods

2.1. Design and Setting

The study used a within-subject repeated-measures design. It was conducted at the research center of the Back, Neck, Other Joint Pain and Human Performance (BNOJPH) laboratory, Khon Kaen University, Thailand. Ethics approval for this research was granted before the study by the Human Research Ethics Committee (HE612220) of Khon Kaen University. The study was registered at clinicaltrials.in.th (registration number: TCTR20180823004).

2.2. Participants

Thirty participants, aged 20–39 years, were recruited via posters on bulletin boards at Khon Kaen University. Fifteen males and fifteen females were recruited to reduce the influence of gender. Inclusion criteria for the participants were established as follows: CLBP lasting more than three months, mild to moderate levels of pain on the numerical rating scale (NRS; $\leq$7 score) [31,32], low levels of activity limitation on the Roland Morris disability questionnaire (RMDQ; $\leq$12 score) [33], and reported sitting for at least two hours on any working day [9]. Participants were excluded if they had previous vertebral surgery, had been identified with a medical condition that affected spinal soft tissues, or were pregnant [32,34].

2.3. Sample Size Determination

The sample size was calculated after preliminary data collection from 12 participants (six male and six female) who performed the CSE with ADIM for 5 weeks. The mean difference of the stature changes before and after the exercise intervention was set at 3 mm. A significance level of 0.05 ($Z\alpha$ (0.05) = 1.96) and a power of 90% ($Z\beta$ (0.1) = 1.28) were used in the calculation. After an additional 15% correction for dropouts, the number of subjects was 21. Thus, the current study required at least 30 participants (15 males and 15 females for balanced gender) to achieve sufficient statistical power for the analyses.

2.4. Outcome Measurements

2.4.1. Stature Change Response

Stature change response was measured using a seated stadiometer device (certified Thai petty-patent No. 5607; Figure 1) [15,34]. The Digimatic Indicator identifies variations in stature change with a resolution of ±0.006 mm and was used to measure stature change. The device displays real-time data and repeatedly records data up to 5 Hz (ID-C 150, 1050 Digimatic Indicator, Manual No. 3061, Series No. 543, Mitutoyo, Kawasaki, Japan).

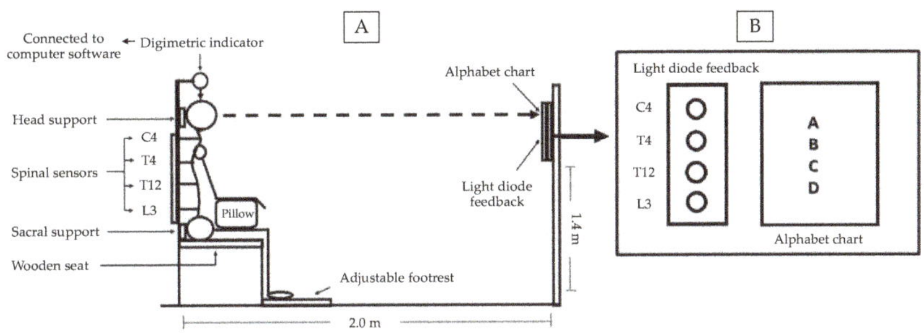

Figure 1. A seated stadiometer device: (**A**) participant position, (**B**) feedback chart. Source: [15].

The position of the participants was controlled during the measurements as follows:

(i) The Digimatic Indicator settings allowed the distal end to rest directly on the highest apex of the skull to ensure its consistent positioning throughout participant repositioning (Figure 1) [15,34].

(ii) Head positioning with the eyes kept level was maintained by coaching the participants to concentrate on a visual cue, a letter on an alphabet chart, placed at eye level on the opposite end (Figure 1B) [15,34].

(iii) The wooden seat platform and footrest were adjustable so that the participants' ankle, knee, and hip joints were positioned at 90° throughout the measurements. Heels touched the back of the footrest [35]. The head and sacral supports were adjustable to accommodate the participants' spinal posture [35]. A pillow was placed on the participants' lap to support their forearms positioned at 90° to their upper arms (Figure 1A).

(iv) The spinal alignment was controlled by sensors placed on the spinous processes of the following vertebrae: cervical spine 4, thoracic spine 4, thoracic spine 12, and lumbar spine 3. These sensors were connected to a light diode feedback located opposite the seated participant (Figure 1B) [12,15]. These sensors were used as a measure for control to ensure that the participants maintained the same posture throughout the experiment period. The process of stature change measurement was performed by Researcher P.S.

2.4.2. Trunk Muscle Fatigue

Eight pairs of Ag-AgCl disposable surface electromyography (sEMG) electrodes (EL 503) with electrical contact surface areas of 1 cm^2 and a center-to-center spacing of 2.5 cm were attached parallel to each muscle on both sides: to the rectus abdominis muscle (RA) [36], the internal oblique and transversus abdominis (IO and TrA) [37], the iliocostalis lumborum pars thoracis (ICLT) [38], and the lumbar multifidus (LM) [30] after skin abrasion and cleaning with alcohol. Electromyography (EMG) data were recorded at 2000 Hz using the Wireless Bipolar Cometa Mini Wave Plus 16-channel EMG system (Cometa, Bareggio, Italy), an online band-pass filter (10–500 Hz), and a 60 Hz notch filter (power line in Thailand). The raw EMG signal was first visually checked for electrocardiac artifacts. The raw EMG signal was processed with fast Fourier transformation to determine the median frequency (MDF) value (Hz). The decrease in the MDF of the EMG signal was taken as an indirect measure of muscle fatigue [6]. Trunk muscle fatigue was collected by Researcher P.S.

2.4.3. Pain Rating Scale

Pain intensity was assessed using an 11-point numerical rating scale (0–10 NRS). Subjective measures of pain were obtained from the NRS, employed to assess pain on a scale ranging from 0 (no pain) to 10 (worst possible pain) [31,39]. This outcome measurement was evaluated by Researcher T.C.

2.4.4. Functional Disability

The Roland Morris disability questionnaire (RMDQ) Thai version was used to assess functional disability due to low back pain [40]. This questionnaire includes 24 items [41], which were rated by Researcher T.C.

2.5. Procedure

The flowchart of the current study is presented in Figure 2. Thirty-three participants were recruited from the advertisements. After the screening process, 30 participants were included in the study. Three participants were excluded due to experiencing low back pain >7 based on the NRS. Thirty participants meeting the inclusion criteria were asked to visit the research laboratory.

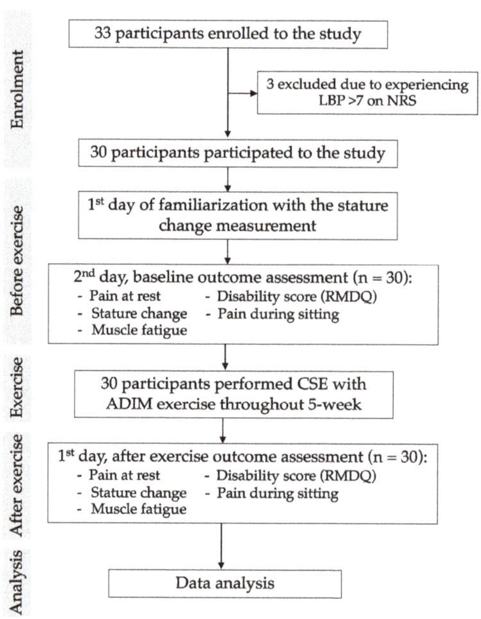

Figure 2. Flowchart of the study.

On the first day, the participants were familiarized with the stature change measurement. These involved participants practiced stepping in and out of the stadiometer until a standard deviation (SD) of <0.5 mm was achieved over ten repeated stature measurements [12,42]. Then, they were asked about their disability score (RMDQ) and pain intensity at rest (NRS).

On the second day, all participants arrived within an hour of waking, between 8 and 10 a.m. [42,43] to avoid stature loss before the test trial. They were requested to sleep for at least 8 h each night before the days of the experiment [44]. They were asked to undertake normal activities of daily living, refrain from vigorous physical activities, and refrain from alcohol consumption for 24 h before the experimental sessions [45]. After the application of surface electrodes, the participants were asked to maintain the Fowler's position (lying) for 20 min to eliminate any abnormal spinal loading that may have been present before arrival [34,45]. Then, they were asked to sit for 41 min, and the outcome measurements were collected, as shown in Figure 3.

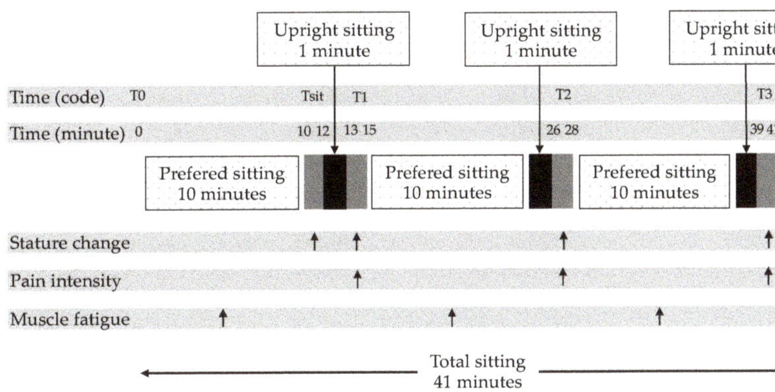

Figure 3. Overview of the prolonged sitting condition. The upright sitting periods (back zone) are presented below the axis. Arrows illustrate times of outcome measurement: stature change, pain intensity, and muscle fatigue. Participants sit in the seated stadiometer and perform the upright sitting three times (at 12–13, 25–26, and 38–39 min) throughout the prolonged sitting time of 41 min. Stature change measurements are collected at Tsit (10–12 min), T1 (13–15 min), T2 (26–28 min), T3 (39–41 min). Pain intensity measurements are collected at T1 (13–15 min), T2 (26–28 min), and T3 (39–41 min). Muscle fatigue measurements are collected at 0–10, 15–25, and 28–38 min.

Next, the participants were asked to practice CSE using the ADIM technique with the researcher R.P. When they could perform this correctly, they were required to exercise CSE with ADIM for 5 weeks. After 5 weeks of training, all participants were asked to stop their exercises completely. The outcomes were re-measured on the first day after the 5-week CSE with the ADIM technique.

2.6. Prolonged Sitting Condition

An overview of the prolonged sitting condition, with time points and their outcome measurements, is shown in Figure 3. Participants sat in the seated stadiometer, according to conditions described in 2.4.1, with the Digimatic Indicator in contact with the skull apex, marked by a waterproof pen. During the measurements, the participants remained in the same posture without speaking. To reduce errors in the spinal change measurements due to involuntary movements and slight differences in the breathing phase, all measurements were taken at the end of the expiration phase of the breathing cycle [35,46]. Each measurement set, consisting of 75 data points sampled over 15 s, was considered at time 0 and at the end of a 2 min interval, which reduced the effect of variations in the stature change assessment due to both breathing patterns and uncontrolled movements [12,34,35].

A baseline stature measurement set was recorded (T_0). During the test trials, the participants remained in a freestyle sitting position, which did not require a straight back, without a backrest for 10 min. Then, the stature change (Tsit) in the participants was measured to be used as a normalized value. Next, the participant was asked to sit upright for 1 min. The stature change and pain intensity were measured at the end of each session (T1 [at 13–15 min], T2 [at 26–28 min], and T3 [at 39–41 min]). The raw sEMG signal was processed using the triangle-Bartlett method of fast Fourier transformation to determine the median frequency (MDF) value. The sEMG data were retrieved every 10 min block of sEMG data from the 41 min sitting period (at 0–10, 15–25, and 28–38 min) for analysis, Figure 3. The total time for each test trial was 41 min. Participants were not allowed to stand during the test trials.

2.7. Core Stabilization Exercise (CSE) with ADIM Technique

The exercise program was supervised by a physical therapist with 30 years of experience (RP). This exercise program was modified from Puntumetakul et al. [28]. The details of the CSE each week are appended (Table A1). Researcher R.P. trained this exercise to all participants face-to-face in a 20 min session. The participants were re-assessed with researcher R.P. twice a week for 5 weeks at the laboratory to determine whether they could successfully perform the previous exercise. As the CSE with the ADIM technique was a milestone exercise, if the participants failed to perform the previous exercise accurately, they were retrained in the previous exercise until they succeeded. The participants were required to perform a daily set of home exercises of the same level, position, and frequency as those demonstrated during the exercise session with the physical therapist. The participants were asked to record in their logbook a daily home exercise program, including position, duration, and frequency of the exercise, as well as a record of their drug and alternative treatment throughout the study period and any adverse effects of the exercise. During the exercise, one of the researchers contacted the participants by telephone every week to motivate them to continue their daily home exercises. After 5 weeks of training, all participants were asked to stop their exercises completely.

2.8. Data Analysis

The mean and standard deviation (SD) were used to assess participants' demographics and data of stature change at each time of measurement and were calculated from the reference point of T_{sit}. Differences in stature within a condition were assessed using a one-way repeated measure ANOVA for time effect (T_1, T_2, and T_3) with the Bonferroni post-hoc analysis (significant at $p < 0.017$; 0.05/3).

The differences in trunk muscle fatigue and pain intensity within the condition for non-normally distributed data were analyzed using the Friedman test, and post hoc tests were conducted using Wilcoxon signed-rank tests. A significance level was set at $p < 0.05$ for trunk muscle fatigue and pain intensity.

Data comparisons before and after the first day of the 5-week CSE regarding pain at rest and functional disability were analyzed using the paired t-test ($p < 0.05$). During the prolonged sitting condition, the data comparison before and after the first day of 5 weeks of the exercise on stature change was analyzed using the paired t-test ($p < 0.05$). Further, the pain intensity and trunk muscle fatigue were analyzed using Wilcoxon signed-rank tests ($p < 0.05$).

All analyses were performed using SPSS version 19.0 software (SPSS Inc., Chicago, IL, USA). The Shapiro–Wilk test was performed to check the data distribution.

3. Results

3.1. Participant Characteristics

All participants achieved the preferred level of repeatability for the stature change measurements (SD $\leq$ 0.5 mm). The participants reported that they did not use any drug for reducing their low back pain and had no adverse repercussions of the exercise throughout the 5 weeks of the training. The demographic data and clinical characteristics are presented in Table 1.

Table 1. Demographic characteristics of participants.

Characteristics	Male (n = 15)	Female (n = 15)	Total (n = 30)
Age (years), mean ± SD	25.67 ± 3.35	26.07 ± 3.37	25.87 ± 3.31
Weight (kg), mean ± SD	63.93 ± 7.94	52.80 ± 4.84	58.37 ± 8.59
BMI (kg m^{-2}), mean ± SD	22.11 ± 1.90	20.95 ± 1.28	21.53 ± 1.70
Sitting height (cm), mean ± SD	87.93 ± 5.38	84.50 ± 3.39	86.22 ± 4.75
Standing height (cm), mean ± SD	169.80 ± 5.16	158.67 ± 4.70	164.23 ± 7.45
Smoking status	no	no	no
Occupation, n (%)			
-Student	12 (80)	13 (86.67)	25 (83.33)
-Office worker	3 (20)	2 (13.33)	5 (16.67)
Working time (hours/day), mean ± SD	8.60 ± 2.95	7.60 ± 2.13	8.10 ± 2.58
Period of LBP (month), mean ± SD	10.73 ± 6.18	10.53 ± 4.60	10.63 ± 5.35
Disability index score, mean ± SD	4.20 ± 1.82	4.80 ± 1.82	4.50 ± 1.81
Pain scale 24 h (score), mean ± SD	4.27 ± 1.33	4.47 ± 1.36	4.37 ± 1.33

Note: SD = Standard deviation; BMI = Body mass index.

3.2. Pain Intensity at Rest and Functional Disability

The results showed a significant reduction in pain intensity (mean difference: 2.14 ± 1.50 (95% CI: 1.57 to 2.69) at $p < 0.001$) and functional disability (mean difference: 2.33 ± 1.81 (95% CI: 1.66 to 3.01) at $p < 0.001$) between baseline and the 5-week CSE with the ADIM, as shown in Figures 4 and 5, respectively.

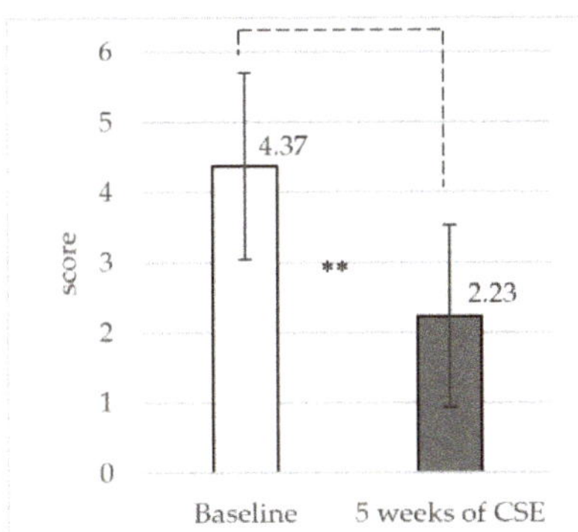

Figure 4. Mean and standard deviation (SD) of pain intensity from baseline to 5 weeks of CSE (** $p < 0.001$).

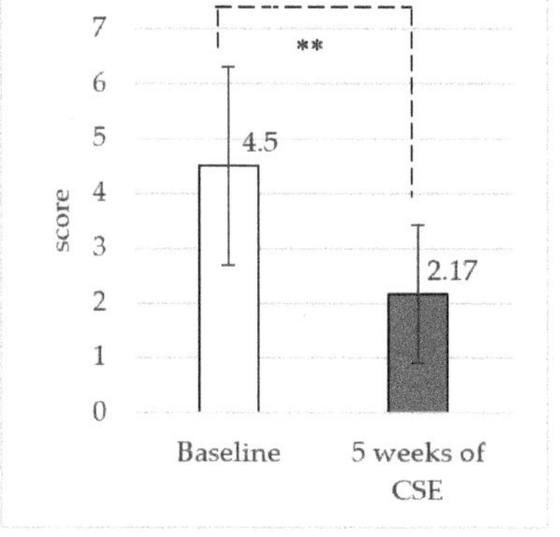

Figure 5. Mean and standard deviation (SD) of functional disability from baseline to 5 weeks of CSE (** $p < 0.001$).

3.3. *Stature Changes during Sitting*

The stature changes during sitting before and after CSE are shown in Table 2. The result of stature change after sitting for 10 min (T_{sit}) showed no significant differences between baseline and after the 5-week CSE with the ADIM technique ($p = 0.458$). This result indicates that T_{sit} between baseline and after the 5-week CSE with the ADIM technique was comparable and could be used as a reference point for the stature changes calculation at T_1, T_2, and T_3.

Table 2. The stature change during sitting before and after CSE with ADIM.

	T_{sit} (mm) Mean ± SD (95%CI)	Mean Change from T_{sit} (mm) Mean ± SD (95%CI)		
		T_1	T_2	T_3
Baseline	−4.266 ± 2.221 (−5.095 to −3.437)	−3.999 ± 1.482 (−4.553 to −3.446) ^{g*h**}	−5.782 ± 1.605 (−6.382 to −5.183) ^{f*h*}	−7.365 ± 2.180 (−8.179 to −6.552) ^{f**g*}
After 5 weeks of CSE with ADIM technique	−3.864 ± 1.986 (−4.605 to −3.122)	−2.538 ± 1.004 (−2.913 to −2.163) ^{g*h**}	−4.027 ± 1.306 (−4.515 to −3.539) ^{f**h*}	−5.367 ± 1.258 (−5.837 to −4.897) ^{f**g*}
p-value Between	0.458	**0.001**	**0.001**	**0.001**

Note: Data presented as Mean ± standard deviation (SD), T_{sit} = after sitting for 10 min, T_1 = 13th–15th min, T_2 = 26th–28th min, T_3 = 39th–41st min, f = significant difference from T_1, g = significant difference from T_2, h = significant difference from T_3 (* significant difference at *p*-value < 0.008, ** significant difference at *p*-value < 0.001).

At baseline, the results of the current study illustrated that the baseline showed a significant reduction in stature due to time (T_1, T_2, and T_3) ($p < 0.017$). In the same pattern, the results illustrated that after the 5-week CSE with the ADIM technique, there was a significant reduction in stature due to time.

Comparing baseline and the 5-week CSE with the ADIM technique, the stature changes at T_1, T_2, and T_3 were significantly improved in the 5-week CSE with the ADIM technique (Table 2).

3.4. *Pain Intensity during Sitting*

At baseline, the pain intensity of T_2 and T_3 were significantly increased compared to other time of measurements (T_{sit} and T_1). After the 5-week CSE with the ADIM technique, the pain intensity of T_2 ($p < 0.05$) and T3 ($p < 0.001$) was significantly increased from T_{sit}. At T_3, pain intensity was significantly increased from T_1 and T_2 ($p < 0.05$).

Comparing baseline and the 5-week CSE with the ADIM technique, the pain intensity at T_1, T_2, and T_3 was significantly decreased in the 5-week CSE with the ADIM technique, as shown in Table 3.

Table 3. Comparisons pain intensity during sitting before and after received CSE with ADIM.

Conditions	Times				*p*-Value within Conditions
	T_{sit}	(T_1)	(T_2)	(T_3)	
Baseline	3.00 (2.00–5.00) f*g**h**	3.00 (3.00–5.00) e*g*h*	3.50 (3.00–6.00) e**f*	3.50 (3.00–6.00) e**f*	0.001
After 5 weeks of CSE with ADIM technique	1.00 (0.00–2.00) g*h**	1.00 (0.00–2.25) g*h*	1.50 (0.00–3.00) e*h*	2.00 (0.00–3.00) e**f*g*	0.001
p-value Between	0.632	0.001	0.001	0.001	

Note: Data presented as Median (interquartile range), *p*-value from the Friedman test, ** significant difference at *p*-value < 0.001, * significant difference at *p*-value < 0.05 by the Wilcoxon signed-rank test), T_{sit} = after sitting for 10 min, T_1 = 13th–15th min, T_2 = 26th–28th min, T_3 = 39th–41st min, e = significant difference from T_{sit}, f = significant difference from T_1, g = significant difference from T_2, h = significant difference from T_3.

3.5. Trunk Muscle Fatigue during the Experiment

At baseline, the MDF in the sitting condition is shown in Table 4. The Friedman test revealed a significant difference in the MDF values in the muscles and both sides at each time of measurement during prolonged sitting. The Wilcoxon signed-rank tests showed a significant difference between the measurement times. For both sides of the TrA and IO muscles, the MDF value at the 15th–25th min was significantly decreased compared to that of the 0–10th min. A further reduction in the MDF value was observed at the 28th–38th min.

Table 4. Comparison muscle fatigue between before and after performed CSM with ADIM exercise during prolong sitting.

Muscle Fatigue (Hz.)	Right			Left		
	Baseline Median (Interquartile Range)	After 5 Weeks of CSE Median (Interquartile Range)	*p*-Value	Baseline Median (Interquartile Range)	After 5 Weeks of CSE Median (Interquartile Range)	*p*-Value
RA						
(0–10th min)	25.71 (24.99–27.25)	25.70 (25.69–25.71)	0.713	25.71 (15.72–28.95)	25.70 (25.12–25.71)	0.128
(15th–25th min)	25.70 (917.29–27.64)	25.70 (23.91–25.71)	0.992	25.71 (16.95–27.34)	25.71 (25.69–25.72)	0.926
(28th–38th min)	25.72 (25.70–28.69)	25.71 (25.69–25.72)	0.144	25.71 (25.69–28.70)	25.70 (24.72–25.71)	0.130
p-value	0.177	0.441		0.852	0.084	
TrA & IO						
(0–10th min)	42.59 (34.58–42.72) b*c*	46.71 (46.63–48.32)	0.001 **	42.70 (35.37–42.73) b*c*	46.71 (44.88–48.69)	0.001 **
(15th–25th min)	35.99 (35.69–37.99) a*	45.30 (40.80–47.96)	0.001 **	36.15 (34.19–37.32) a*	45.45 (41.49–47.23)	0.001 **
(28th–38th min)	35.71 (33.21–36.45) a*	45.70 (41.99–48.71)	0.001 **	35.95 (33.53–37.21) a*	45.99 (43.45–48.65)	0.001 **
p-value	0.001	0.058		0.001 *	0.503	
ICLT						
(0–10th min)	35.70 (33.21–36.03)	36.14 (35.69–37.94)	0.060	35.70 (34.81–35.72)	35.71 (34.39–37.76)	0.524
(15th–25th min)	35.69 (33.28–35.72)	35.72 (35.30–37.33)	0.206	35.69 (33.66–36.21)	35.71 (35.66–36.47)	0.289
(28th–38th min)	35.70 (33.85–35.71)	36.70 (33.85–35.71)	0.153	35.69 (33.58–35.71)	35.71 (35.69–37.05)	0.360
p-value	0.873	0.644		0.721	0.594	

Table 4. *Cont.*

Muscle Fatigue (Hz.)	Right			Left		
	Baseline Median (Interquartile Range)	After 5 Weeks of CSE Median (Interquartile Range)	p-Value	Baseline Median (Interquartile Range)	After 5 Weeks of CSE Median (Interquartile Range)	p-Value
LM						
(0–10th min)	49.04 (46.42–52.79)	52.65 (47.71–53.57)	0.185	49.36 (45.71–52.38)	51.90 (48.22–53.57)	0.175
(15th–25th min)	49.41 (46.96–53.57)	52.74 (48.71–53.57)	0.098	49.21 (47.71–55.70)	52.33 (48.15–55.40)	0.544
(28th–38th min)	49.21 (47.71–55.69)	52.34 (48.72–55.72)	0.082	49.84 (48.00–55.69)	52.42 (48.68–55.71)	0.237
p-value	0.695	0.341		0.273	0.125	

Note: Data presented as Median (interquartile range). Significant difference at * p-value < 0.05, ** p-value < 0.001 by Wilcoxon signed-rank test. a = significant difference from 0–10th, b = significant difference from 15th–25th, c = significant difference from 28th–38th.

After the 5-week CSE with the ADIM technique, the Friedman test did not reveal a significant difference in the MDF value in trunk muscles and both sides at all times of measurement. Compared with baseline values, the 5-week CSE with the ADIM technique showed a significant improvement in the MDF values (both sides of TrA and IO, the MDF value of 15th–25th and 28th–38th min).

4. Discussion

The aim of this study was to investigate differences in stature change, pain, and trunk muscle fatigue during prolonged sitting conditions in sedentary workers with CLBP between baseline and the 1st day after the 5-week CSE with the ADIM technique.

The results of the current study showed that the 5-week CSE with the ADIM technique provided a significant decrease in resting pain (mean difference: 2.14 ± 1.50; $p < 0.001$) and improvement in functional disability (mean difference: 2.33 ± 1.81; $p < 0.001$). The result of the current study agreed with those of previous studies that reported the potential of CSE to improve functional disability in patients with CLBP [47–49]. The results of the current study demonstrated that the CSE program might be clinically advantageous for CLBP patients with functional disability improvement by reducing pain.

During prolonged sitting, forces from bodyweight cause deformation of the elastic components of the disc and increased intra-discal pressure [50,51]. Fluid loss is known to occur when the pressure inside the disc increases and can be indicated as the major mechanism to account for the reduction in disc height and consequent stature loss [52,53]. Prolonged sitting in CLBP participants induced stature loss (mean difference −7.365 mm) at 41 min (T_3) (Table 2).

The current study revealed that bilateral TrA and IO muscle fatigue occurred earlier during sitting (approximately 15–25 min after sitting) (Table 3). Sitting for prolonged periods has been partly attributed to trunk muscle fatigue resulting from the continuous contraction of deep trunk muscles in seated postures [9,10]. During prolonged sitting, the lumbar multifidus is passively stretched, resulting in increased co-contraction of the TrA and IO muscles to balance the back muscle forces. Consequently, the TrA and IO muscles become fatigued over time [9,10].

A significant increase in low back pain in the sitting condition was found in this study, suggesting that static loading of the lumbar spine during prolonged sitting may be associated with disc compression [54,55]. Healey et al. (2005) and Rodacki et al. (2003) proposed that persistent contraction of the superficial paraspinal muscles in patients with CLBP may produce greater compressive loading, increasing disc compression and reducing stature [25,27]. Moreover, the results of the current study demonstrate that the sitting condition reduced deep trunk muscle activation. These results may explain increased low back pain [14,56]. The results of the present study align with previous studies showing that perceived body discomfort increased significantly during prolonged sitting [56].

After the 5-week CSE with the ADIM technique, participants showed significantly improved stature changes during prolonged sitting. The stature change at the T_1 (mean difference: 1.462 ± 1.752 mm; $p < 0.001$), T_2 (mean difference: 1.756 ± 1.752 mm; $p < 0.001$), and T_3 occasion (mean difference: 1.998 ± 2.653 mm; $p < 0.001$) was significantly improved before and after the 5-week CSE with the ADIM technique. Although our mean difference in stature change was only 1.99 mm, which did not reach the minimal clinically important stature change of 3 mm [12], there is evidence that changes in stature of a magnitude above 0.985 mm can be attributed to intervention effects in CLBP participants [57].

Thus, we show that the 5-week CSE with the ADIM technique can enhance recovery of disc height and reduce the loading on other spinal structures, which may facilitate a reduction in symptoms in patients with low back pain during prolonged sitting. This is consistent with the findings of Healey et al. (2005) and Lewis et al. (2014). They reported significant positive correlations between delayed stature recovery and higher levels of pain and disability [27,32].

Our results demonstrated that the 5-week CSE could improve trunk muscle endurance during prolonged sitting when compared with baseline. The results showed that both sides of TrA and IO significantly improved the MDF (at 0–10, 15–25th, and 28–38th min) when compared with baseline values. These revealed that muscle endurance increased after the 5-week CSE with the ADIM technique, which could be related to the specific effects of CSE with the ADIM technique on deep muscle activities. Macdonald et al. (2006) indicated that deep trunk muscles have a high percentage of type I muscle fibers, blood vessels, and mitochondria [58]. Therefore, CSE with the ADIM technique can improve the endurance of deep trunk muscles during prolonged sitting. Increased activity of the deep trunk muscle is thought to raise intra-abdominal pressure [59], resulting in decreased spinal loading [60,61]. This suggests that the CSE with the ADIM technique can increase deep trunk muscle activity in CLBP participants by reducing compression forces on the spine during prolonged sitting [8], leading to improved stature recovery.

The results of the five weeks of CSE with the ADIM technique significantly improved low back pain at all measurement time points (T_1, T_2, and T_3) when compared with the baseline. However, the participants reported significantly increased pain intensity throughout the increased sitting time. Static loading of the lumbar spine increases stress in spinal structures [54,55]. These results may increase low back pain [14,56]. Thus, in addition to performing CSE with the ADIM technique, participants should perform the movement during the working day to prevent lower back pain during prolonged sitting.

The current study has some limitations. First, the investigation was conducted in a laboratory; the findings of this study may have limited ecological validity, and a real-life situation may be required in future investigations. Second, the participants were young, with a small age range (aged 21–29 years). Thus, the results might not be applicable to other age groups due to the variation in degenerative stage. Third, the current study was limited to the immediate effects of the 5-week CSE with the ADIM technique. Future studies should investigate the long-term follow-up effects of this program. Fourth, although the participants in this study reported a significant increase in low back pain during prolonged sitting, a history of previous injury did not meet our exclusion criteria. Therefore, low back pain may be due to other reasons besides prolonged sitting. Adding a history of previous injury in the exclusion criteria may better clarify the cause of low back pain in future studies. Lastly, the current study included only one group that performed the pre- and post-exercise interventions. Future investigations should include a control group or a comparison of CSE with other exercises to strengthen the findings.

5. Conclusions

This study demonstrated that a 5-week CSE with the ADIM technique affects the pain at rest and functional disability in sedentary workers with CLBP. Our result showed that CSE with the ADIM technique provides a protective effect on detrimental reductions in stature change and trunk muscle fatigue during prolonged sitting in young participants under controlled conditions in a laboratory. Based on these findings, we recommend that CLBP patients (aged 21–29 years) should perform CSE with the ADIM technique at home to reduce low back pain problems due to prolonged sitting activities.

Author Contributions: P.S.: conceptualization, data curation, formal analysis, methodology, writing, and editing; R.P.: conceptualization, data curation, funding acquisition, project administration, resources, writing, and editing; T.C.: data analysis, reviewing, and editing. All authors have read and agreed to the published version of the manuscript.

Funding: This research received no external funding.

Institutional Review Board Statement: The study was conducted according to the guidelines of the Declaration of Helsinki and approved by the Ethics Committee of Center for Ethics in Human Research, Khon Kaen University (no. HE612220).

Informed Consent Statement: Informed consent was obtained from all subjects involved in the study.

Data Availability Statement: The data will be available for anyone who wishes to access them for any purpose and contract should be made via the corresponding author (rungthiprt@gmail.com).

Acknowledgments: The authors thank the participants for their willingness to participate in this study. This work was supported by the Royal Golden Jubilee (RGJ) Ph.D. Programme, the Thailand Research Fund (TRF), Thailand (PHD/0090/2558) and the Research Center in Back, Neck, Other Joint Pain and Human Performance (BNOJPH), Khon Kaen University, Thailand. Many thanks to Rose Boucaut, School of Health Sciences (Physiotherapy), University of South Australia, for English editing.

Conflicts of Interest: The authors declare no conflict of interest.

Appendix A

Table A1. Program of the CSE with ADIM technique.

Week	Exercise Protocol
1	The participants were given the instructions on how to isolate activation of the TrA muscles correctly. After that, they were asked to draw their lower abdomen gently in towards the spine (ADIM) with normal breathing control, and no movement of the spine and pelvis while in a prone lying position on a couch with a small pillow placed beneath their ankles. In addition, a pressure biofeedback device set at 70 mmHg was placed under the lower abdomen of the participants. If they were able to lower pressure 4 to 10 mmHg, it represented successful activation of the transversus abdominis muscle. The exercises progressed until the participants could perform muscle contractions for 10 s holds with 10 repetitions/set 10 sets/day.
2	The participants performed co-contraction of the TrA and LM muscles while in a crooked lying position with both hips at 45 degrees and both knees at 90 degrees. They were asked to perform the ADIM technique with floor muscles, normal breathing control, and no movement of the spine and pelvis. Furthermore, they used their index and middle fingers to palpate contraction of the TrA muscle and opposite 2 fingers palpated contraction of LM muscle. If the participants performed correctly, they could feel the tight contraction of each muscle under their fingers. The exercises progressed until the participants could perform muscle contractions for 10 s holds with 10 repetitions/set 10 sets/day.
3	The participants performed co-contraction of the TrA and LM muscles in a crooked lying position with both hips at 45 degrees and both knees at 90 degrees. Then, they abducted one leg to 45 degrees of hip abduction and held it for 10 s. After that, they adducted their leg to the starting position. After repeating this movement 5 times, they did it with the other leg/set 10 sets/day. The next exercise in this week was to train co-contraction of these muscles in a crooked lying position with both hips at 45 degrees and both knees at 90 degrees. Then, they slid a single leg down until the knee was straight, maintained it for a 10 s hold, and then slid it back up to the starting position. After repeating this movement 10 times, they did it with the other leg/set 10 sets/day.

Table A1. *Cont.*

Week	Exercise Protocol
4	The participants performed co-contraction of the TrA and LM muscles while sitting on a chair. They were asked to perform the ADIM technique with normal breathing control, and no movement of the spine and pelvis. Furthermore, they used their index and middle fingers to palpate contraction of the TrA muscle and opposite 2 fingers palpated contraction of the LM muscle. If the participants performed correctly, they could feel the tight contraction of each muscle under their fingers. The exercises progressed until the participants could perform muscle contractions for 10 s holds with 10 repetitions/set 10 sets/day. The next exercise in this week was to train co-contraction of these muscles with the trunk forward and backward while sitting on a chair and keeping their lumbar spine and pelvis in a neutral position. The second exercise in this week required 10 s holds with 10 repetitions/set 10 sets/day.
5	The participants performed co-contraction of the TrA and LM muscles during sitting on a balance board. They were asked to perform co-contraction of the muscles with the trunk forward and backward while sitting on a balance board and keeping their lumbar spine and pelvis in a neutral position. They performed each pose for 10 s holds with 10 repetitions/set 10 sets/day.

References

1. Sedentary Behaviour Research Network. Letter to the editor: Standardized use of the terms "sedentary" and "sedentary behaviours". *Appl. Physiol. Nutr. Metab.* **2012**, *37*, 540–542. [CrossRef] [PubMed]
2. Hadgraft, N.T.; Lynch, B.M.; Clark, B.K.; Healy, G.N.; Owen, N.; Dunstan, D.W. Excessive sitting at work and at home: Correlates of occupational sitting and TV viewing time in working adults. *BMC Public Health* **2015**, *15*, 899. [CrossRef]
3. Saidj, M.; Menai, M.; Charreire, H.; Weber, C.; Enaux, C.; Aadahl, M.; Kesse-Guyot, E.; Hercberg, S.; Simon, C.; Oppert, J.M. Descriptive study of sedentary behaviours in 35,444 French working adults: Cross-sectional findings from the ACTI-Cités study. *BMC Public Health* **2015**, *15*, 379. [CrossRef] [PubMed]
4. Montakarn, C.; Nuttika, N. Physical activity levels and prevalence of low back pain in Thai call-center operators. *Indian J. Occup. Environ. Med.* **2016**, *20*, 125–128. [CrossRef] [PubMed]
5. Balagué, F.; Mannion, A.F.; Pellisé, F.; Cedraschi, C. Non-specific low back pain. *Lancet* **2012**, *379*, 482–491. [CrossRef]
6. Jiménez, R.; Bennett, P.; García, O.; Vargas, C. Fatigue detection during sit-to-stand test based on surface electromyography and acceleration: A case study. *Sensors* **2019**, *19*, 4202. [CrossRef]
7. Olafsson, G.; Jonsson, E.; Fritzell, P.; Hägg, O.; Borgström, F. A health economic lifetime treatment pathway model for low back pain in Sweden. *J. Med. Econ.* **2017**, *20*, 1281–1289. [CrossRef] [PubMed]
8. Saiklang, P.; Puntumetakul, R.; Selfe, J.; Yeowell, G. An Evaluation of an Innovative Exercise to Relieve Chronic Low Back Pain in Sedentary Workers. Available online: https://pubmed.ncbi.nlm.nih.gov/33111563/ (accessed on 22 December 2021).
9. Waongenngarm, P.; Rajaratnam, B.S.; Janwantanakul, P. Internal oblique and transversus abdominis muscle fatigue induced by slumped sitting posture after 1 h of sitting in office workers. *Saf. Health Work* **2016**, *7*, 49–54. [CrossRef] [PubMed]
10. Areeudomwong, P.; Puntumetakul, R.; Kaber, D.B.; Wanpen, S.; Leelayuwat, N.; Chatchawan, U. Effects of handicraft sitting postures on lower trunk muscle fatigue. *Ergonomics* **2012**, *55*, 693–703. [CrossRef]
11. Lewis, S.E.; Fowler, N.E. Changes in intervertebral disk dimensions after a loading task and the relationship with stature change measurements. *Arch. Phys. Med. Rehabil.* **2009**, *90*, 1795–1799. [CrossRef]
12. Kanlayanaphotporn, R.; Trott, P.; Williams, M.; Fulton, I. Effects of chronic low back pain, age and gender on vertical spinal creep. *Ergonomics* **2003**, *46*, 561–573. [CrossRef]
13. Watanabe, S.; Kobara, K.; Yoshimura, Y.; Osaka, H.; Ishida, H. Influence of trunk muscle co-contraction on spinal curvature during sitting. *J. Back Musculoskelet Rehabil.* **2014**, *27*, 55–61. [CrossRef] [PubMed]
14. Fryer, J.C.J.; Quon, J.A.; Smith, F.W. Magnetic resonance imaging and stadiometric assessment of the lumbar discs after sitting and chair-care decompression exercise: A pilot study. *Spine J.* **2010**, *10*, 297–305. [CrossRef]
15. Saiklang, P.; Puntumetakul, R.; Swangnetr, M.; Boucaut, R. The immediate effect of the abdominal drawing-in maneuver technique on stature change in seated sedentary workers with chronic low back pain. *Ergonomics* **2020**, *64*, 55–68. [CrossRef]
16. Holmes, M.W.R.; De Carvalho, D.E.; Karakolis, T.; Callaghan, J.P. Evaluating abdominal and lower back muscle activity while performing core exercises on a stability ball and a dynamic office chair. *Hum. Factors* **2015**, *57*, 1149–1161. [CrossRef] [PubMed]
17. Gregory, D.E.; Dunk, M.N.; Callaghan, J.P. Stability ball versus office chair: Comparisonof muscle activation and lumbar spine posture during prolonged sitting. *Hum. Factors* **2006**, *48*, 142–153. [CrossRef]
18. Panjabi, M.M. Clinical spinal instability and low back pain. *J. Electromyogr. Kinesiol. J. Electromyogr. Kines* **2003**, *13*, 371–379. [CrossRef]
19. Bergmark, A. Stability of the lumbar spine: A study in mechanical engineering. *Acta Orthop. Scand Suppl.* **1989**, *230*, 1–54. [CrossRef]
20. Faries, M.D.; Greenwood, M. Core Training: Stabilizing the confusion. *Strength Cond J.* **2007**, *29*, 10–25. [CrossRef]
21. Kavcic, N.; Grenier, S.; McGill, S.M. Determining the stabilizing role of individual torso muscles during rehabilitation exercises. *Spine* **2004**, *29*, 1254–1265. [CrossRef] [PubMed]

22. Marshell, P.W.; Murphy, B.A. Delayed abdominal muscle onsets and self-report measures of pain and disability in chronic low back pain. *J. Electromyogr. Kinesiol.* **2010**, *20*, 833–839. [CrossRef]
23. Silfies, S.P.; Mehta, R.; Smith, S.S.; Karduna, A.R. Differences in feedforward trunk muscle activity in subgroups of patients with mechanical low back pain. *Arch. Phys. Med. Rehabil.* **2009**, *90*, 1159–1169. [CrossRef]
24. Tsao, H.; Druitt, T.R.; Schollum, T.M.; Hodges, P.W. Motor training of the lumbar paraspinal muscles induces immediate changes in motor coordination in patients with recurrent low back pain. *J. Pain* **2010**, *11*, 1120–1128. [CrossRef]
25. Rodacki, C.L.; Fowler, N.E.; Rodacki, A.L.; Birch, K. Stature loss and recovery in pregnant women with and without low back pain. *Arch. Phys. Med. Rehabil.* **2003**, *84*, 507–512. [CrossRef]
26. Singh, R.E.; Iqbal, K.; White, G.; Hutchinson, T.E. A Systematic Review on Muscle Synergies: From Building Blocks of Motor Behavior to a Neurorehabilitation Tool. *Appl. Bionics. Biomech.* **2018**, *22*, 3165368. [CrossRef]
27. Healey, E.L.; Fowler, N.E.; Burden, A.M.; McEwan, I.M. The influence of different unloading positions upon stature recovery and paraspinal muscle activity. *Clin. Biomech. Bristol. Avon.* **2005**, *20*, 365–371. [CrossRef] [PubMed]
28. Puntumetakul, R.; Areeudomwong, P.; Emasithi, A.; Yamauchi, J. Effect of 10-week core stabilization exercise training and detraining on pain-related outcomes in patients with clinical lumbar instability. *Patient Prefer. Adherence* **2013**, *7*, 1189–1199. [CrossRef] [PubMed]
29. Puntumetakul, R.; Saiklang, P.; Tapanya, W.; Chatprem, T.; Kanpittaya, J.; Arayawichanon, P.; Boucaut, R. The effects of core stabilization exercise with the abdominal drawing-in maneuver technique versus general strengthening exercise on lumbar segmental motion in patients with clinical lumbar instability: A randomized controlled trial with 12-month follow-up. *Int. J. Environ. Res. Public Health* **2021**, *18*, 7811. [CrossRef] [PubMed]
30. Areeudomwong, P.; Puntumetakul, R.; Jirarattanaphochai, K.; Wanpen, S.; Kanpittaya, J.; Chatchawan, U.; Yamauchi, J. Core stabilization exercise improves pain intensity, functional disability and trunk muscle activity of patients with clinical lumbar instability: A pilot randomized controlled study. *J. Phys. Ther. Sci.* **2012**, *24*, 1007–1012. [CrossRef]
31. Boonstra, M.A.; Stewart, R.E.; Köke, A.J.A.; Oosterwijk, R.F.A.; Swaan, J.L.; Schreurs, K.M.G.; Schiphorst, H.R. Cut-off points for mild, moderate, and severe pain on the numeric rating scale for pain in patients with chronic musculoskeletal pain: Variability and influence of sex and catastrophizing. *Front. Psychol.* **2016**, *7*, 1466. [CrossRef] [PubMed]
32. Lewis, S.; Holmes, P.; Woby, S.; Hindle, J.; Fowler, N. Changes in muscle activity and stature recovery after active rehabilitation for chronic low back pain. *Man. Ther.* **2014**, *19*, 178–183. [CrossRef] [PubMed]
33. Hall, M.A.; Maher, C.G.; Latimer, J.; Ferreira, M.L.; Costa, L.O.P. The patient-specific functional scale is more responsive than the roland morris disability questionnaire when activity limitation is low. *Eur. Spine J.* **2011**, *20*, 79–86. [CrossRef] [PubMed]
34. Phimphasak, C.; Swangnetr, M.; Puntumetakul, R.; Chatchawan, U.; Boucaut, R. Effects of seated lumbar extension postures on spinal height and lumbar range of motion during prolonged sitting. *Ergonomics* **2016**, *59*, 112–120. [CrossRef]
35. Rodacki, C.L.; Fowler, N.E.; Rodacki, A.L.; Birch, K. Technical note: Repeatability of measurement in determining stature in sitting and standing postures. *Ergonomics* **2001**, *44*, 1076–1085. [CrossRef]
36. Imai, A.; Kaneoka, K.; Okubo, Y.; Shiina, I.; Tatsumura, M.; Izumi, S. Trunk muscle activity during lumbar stabilization exercises on both a stable and unstable surface. *J. Orthop. Sports Phys. Ther.* **2010**, *40*, 369–375. [CrossRef]
37. Marshall, P.; Murphy, B. The validity and reliability of surface EMG to assess the neuromuscular response of the abdominal muscles to rapid limb movement. *J. Electromyogr. Kinesiol.* **2003**, *13*, 477–489. [CrossRef]
38. Danneels, L.A.; Cagnie, B.J.; Cools, A.M.; Vanderstraeten, G.G.; Cambier, D.C.; Witvrouw, E.E. Intra-operator and inter-operator reliability of surface electromyography in the clinical evaluation of back muscles. *Man. Ther.* **2001**, *6*, 145–153. [CrossRef]
39. Ostelo, R.W.J.G.; de Vet, J.C.W. Clinically important outcomes in low back pain. *Best Res. Clin. Rheumatol.* **2005**, *19*, 593–607. [CrossRef]
40. Roland, M.; Fairbank, J. The roland-morris disability questionnaire and the oswestry disability questionnaire. *Spine* **2000**, *25*, 3115–3124. [CrossRef]
41. Jirarattanaphochai, K.; Jung, S.; Sumananont, C.; Saengnipanthkul, S. Reliability of the Roland-Morris Disability Questionnaire (Thai version) for the evaluation of low back pain patients. *J. Med. Assoc. Thai.* **2005**, *88*, 407–411. [PubMed]
42. Healey, E.L.; Burden, A.M.; McEwan, I.M.; Fowler, N.E. Stature loss and recovery following a period of loading: Effect of time of day and presence or absence of low back pain. *Clin. Biomech. Bristol. Avon.* **2008**, *23*, 721–726. [CrossRef] [PubMed]
43. Puntumetakul, R.; Trott, P.; Williams, M.; Fulton, I. Effect of time of day on the vertical spinal creep response. *Appl. Ergon.* **2009**, *40*, 33–38. [CrossRef] [PubMed]
44. Healey, E.L.; Burden, A.M.; McEwan, I.M.; Fowler, N.E. Diurnal variation in stature: Do those with chronic low back pain differ from asymptomatic controls? *Clin. Biomech.* **2011**, *26*, 331–336. [CrossRef]
45. Fowler, N.E.; de Lourdes Rodacki, C.; Rodacki, A.L. Spinal shrinkage and recovery in women with and without low back pain. *Arch. Phys. Med. Rehabil.* **2005**, *86*, 505–511. [CrossRef] [PubMed]
46. Corlett, E.N.; Eklund, J.A.E.; Reilly, T.; Troup, J.D.G. Assessment of workload from measurements of stature. *Appl. Ergon.* **1987**, *18*, 65–71. [CrossRef]
47. Koumantakis, G.A.; Watson, P.J.; Oldham, J.A. Trunk muscle stabilization training plus general exercise versus general exercise only: Randomized controlled trial of patients with recurrent low back pain. *Phys. Ther.* **2005**, *85*, 209–225. [CrossRef]
48. Moseley, G.L.; Hodges, P.; Gandevia, S.C. Deep and superficial fibers of the lumbar multifidus muscle are differentially active during voluntary arm movements. *Spine* **2002**, *27*, 29–36. [CrossRef]

49. Goldby, L.; Moore, A.; Trew, M. A randomized controlled trial investigating the efficiency of musculoskeletal physiotherapy on chronic low back disorder. *Spine* **2006**, *31*, 1083–1093. [CrossRef]
50. Schmidt, H.; Reitmaier, S.; Graichen, F.; Shirazi-Adl, A. Review of the fluid flow within intervertebral discs-How could in vitro measurements replicate in vivo? *J. Biomech.* **2016**, *49*, 3133–3146. [CrossRef]
51. Vergroesen, P.P.A.; Van Der Veen, A.J.; Emanuel, K.S.; Van Dieën, J.H.; Smitm, T.H. The poro-elastic behaviour of the intervertebral disc: A new perspective on diurnal fluid flow. *J. Biomech.* **2016**, *49*, 857–863. [CrossRef]
52. Munster, M.M.; Brismée, J.M.; Sizer, P.S.; Browne, K.; Dewan, B.; Litke, A.; Pape, J.L.; Sobczak, S. Can 5 min of repetitive prone press-ups and sustained prone press-ups following a period of spinal loading reverse spinal shrinkage? *Physiother Theory Pract.* **2018**, *35*, 1–9. [CrossRef]
53. Pape, J.L.; Brismée, J.M.; Sizer, P.S.; Matthijs, O.C.; Browne, K.L.; Dewan, B.M.; Sobczak, S. Increased spinal height using propped slouched sitting postures: Innovative ways to rehydrate intervertebral discs. *Appl. Ergon.* **2018**, *66*, 9–17. [CrossRef] [PubMed]
54. Billy, G.G.; Lemieux, S.K.; Chow, M.X. Changes in lumbar disk morphology associated with prolonged sitting assessed by magnetic resonance imaging. *PM&R* **2014**, *6*, 790–795. [CrossRef]
55. Fryer, J.; Zhang, W. Preliminary investigation into a seated unloading movement strategy for the lumbar spine: A pilot study. *J. Bodyw. Mov. Ther.* **2010**, *14*, 119–126. [CrossRef] [PubMed]
56. Søndergaard, K.H.; Olesen, C.G.; Sondergaard, E.K.; de Zee, M.; Madeleine, P. The variability and complexity of sitting postural control are associated with discomfort. *J. Biomech.* **2010**, *43*, 1997–2001. [CrossRef] [PubMed]
57. Saiklang, P.; Puntumetakul, R.; Siritaratiwat, W.; Boucaut, R. Effect of Time of Day on the Magnitude of Stature Change Response Variability in Participants With Chronic Low Back Pain on Consecutive 2 Days. *Trends Sci.* **2021**, *18*, 678. [CrossRef]
58. MacDonald, D.A.; Moseley, G.L.; Hodges, P.W. The lumbar multifidus: Does the evidence support clinical beliefs? *Man. Ther.* **2006**, *11*, 254–263. [CrossRef]
59. Tayashiki, K.; Takai, Y.; Maeo, S.; Kanehisa, H. Intra-abdominal pressure and trunk muscular activities during abdominal bracing and hollowing. *Int. J. Sports Med.* **2015**, *37*, 134–143. [CrossRef]
60. Stokes, I.A.F.; Gardner-Morse, M.G.; Henry, S.M. Intra-abdominal pressure and abdominal wall muscular function: Spinal unloading mechanism. *Clin. Biomech.* **2010**, *25*, 859–866. [CrossRef]
61. Rodacki, C.L.; Rodacki, A.; Ugrinowitsch, C.; Zielinski, D.; Costa, R.B. Spinal unloading after abdominal exercises. *Clin. Biomech.* **2008**, *23*, 8–14. [CrossRef] [PubMed]

International Journal of
Environmental Research and Public Health

Article

Can Hip-Knee Line Angle Distinguish the Size of Pelvic Incidence?—Development of Quick Noninvasive Assessment Tool for Pelvic Incidence Classification

Shota Yamada [1,2], Takeshi Ebara [1,*], Toru Uehara [2], Taro Matsuki [1], Shingo Kimura [2], Yuya Satsukawa [2], Akira Yoshihara [2], Kazuji Aoki [3], Atsushi Inada [4] and Michihiro Kamijima [1]

1 Department of Occupational and Environmental Health, Graduate School of Medical Sciences, Nagoya City University, Nagoya 4678601, Japan; ptyamada@med.nagoya-cu.ac.jp (S.Y.); tmatsuki@med.nagoya-cu.ac.jp (T.M.); kamijima@med.nagoya-cu.ac.jp (M.K.)
2 Department of Rehabilitation, Nagoya City University West Medical Center, Nagoya 4628508, Japan; toru8toru22@gmail.com (T.U.); kimura.shingo.1234@gmail.com (S.K.); satsu.rh@gmail.com (Y.S.); football_4218@yahoo.co.jp (A.Y.)
3 Faculty of Rehabilitation Sciences, Nagoya Gakuin University, Nagoya 4568612, Japan; aokikazu@ngu.ac.jp
4 Spine Center, Department of Orthopedic Surgery, Nagoya City University West Medical Center, Nagoya 4628508, Japan; a.inada.24@west-med.jp
* Correspondence: ebara@med.nagoya-cu.ac.jp; Tel.: +81-52-853-8171

Citation: Yamada, S.; Ebara, T.; Uehara, T.; Matsuki, T.; Kimura, S.; Satsukawa, Y.; Yoshihara, A.; Aoki, K.; Inada, A.; Kamijima, M. Can Hip-Knee Line Angle Distinguish the Size of Pelvic Incidence?—Development of Quick Noninvasive Assessment Tool for Pelvic Incidence Classification. *IJERPH* **2022**, *19*, 1387. https://doi.org/10.3390/ijerph19031387

Academic Editors: Vincenzo Denaro, Sergio Iavicoli, Fabrizio Russo, Gianluca Vadalà and Paul Tchounwou

Received: 24 November 2021
Accepted: 22 January 2022
Published: 26 January 2022

Abstract: This study aimed to explore effective measurement angles for pelvic incidence (PI) classification and to develop a quick, noninvasive assessment tool for PI classification. We defined five variation types of hip–knee line (HKL) angles and tested the discrimination ability of the receiver operating characteristic (ROC) analysis using 125 photographs of upright standing posture from the right lateral side. ROC analysis revealed an applicable HKL angle defined by the line connecting the most raised part of the buttock and the central point of the knee and the midthigh line. The acceptable cut-off points for discriminating small or large PIs in terms of HKL angle were 18.5° for small PI (sensitivity, 0.91; specificity, 0.79) and 21.5° for large PI discrimination (sensitivity, 0.74; specificity, 0.72). In addition, we devised a quick noninvasive assessment tool for PI classification using the cut-offs of the HKL angle with a view to practical application. The results of intra- and inter-rater reliability ensured a substantial/moderate level of the tool (Cohen's kappa coefficient, 0.79; Fleiss's kappa coefficient, 0.50–0.54). These results revealed that the HKL angle can distinguish the size of the PI with a high/moderate discrimination ability. Furthermore, the tool indicated acceptable inter-/intra-rater reliability for practical applications.

Keywords: pelvic incidence; low back pain; hip–knee line; anthropometry; ROC curve; reliability

1. Introduction

Low back pain (LBP) is one of the most common health problems worldwide. According to a systematic review, the lifetime prevalence in the general population is 60–80% [1]. The one-year prevalence in the general population is 38%, the one-month prevalence is 23–26%, and the point prevalence is 12–14% [2–4]. According to the Global Burden of Diseases study, the average disability-adjusted life years (DALY) for all generations of LBP increased from 1.7 years to 2.5 years as of 2019 compared to 1990 [5]. Out of 310 diseases, LBP was the leading cause of the years lived with disability in 2017 [6]. Furthermore, the DALY of LBP in the working generation is at a high level of 3.9 years. Hence, LBP, the most commonly reported work-related musculoskeletal disorder, has also received attention in terms of outcomes leading to a decline in labor productivity and economic loss [7,8].

Effective measures for LBP are still controversial. Although a recent meta-analysis [9] showed some intervention effects such as exercise in combination with education, there is no solid evidence found for others such as back belts, ergonomic adjustments, or training on manual material handling [10]. One of the possible reasons for such a discrepancy in the effect of LBP can be found in nonpersonalized countermeasures without considering individual biomechanical features of the lumbar spine.

For example, pelvic incidence (PI), which is measured by lateral spine radiography, has been attracting attention as a determinant of LBP. PI is an important clinical parameter generally applied in the field of spinal surgery after Legaye et al. reported it in 1993 [11]. Since then, it has been used for PI-based spinal surgery alignment design and analysis of clinical outcomes [12,13]. PI has some features representing an angle that: (1) defines the lumbar lordosis [14,15], (2) indicates the inclination of the sacrum in the pelvis [16], (3) is fixed during puberty [17], and (4) is a constant independent of posture changes [18]. Recent research shows that PI also has no differences in any age groups and sex [19], but there are differences in ethnicity [20]. PI represents the physiological lordosis of the lumbar spine, inducing differences in mechanical stress on the intervertebral discs and facet joints depending on the degree of angle. If we could successfully grasp the angle of PI in a simple way without exposure to X-rays, PI may be available as a personalized preventive measure for LBP according to the angle of individual PIs. Thus, we devised a noninvasive indirect measurement using surrogate indicators to estimate the PI using anthropometric landmarks on the body surface. Our previous studies [21,22] revealed a surrogate angle on the body surface reflecting the PI. The surrogate angle of the PI can be defined as "the angle between the line connecting the upper edges of the greater trochanter and iliac crest, and the line connecting the upper edge of the iliac crest and the buttock at the same height as the greater trochanter". It has sufficient practical reliability to estimate the PI (R^2 = 0.63), although the measurement method requires substantial time (about 15 min/person) and physiotherapist skill for palpation. Another possible approach that can more easily determine the classification of PI might be to use a body silhouette. As shown in a commentary paper by Ramchandran et al. providing a theoretical interpretation of PI, changing PI visually reflects the waistline and buttocks due to increasing pelvic overhang with increasing sacroiliac joint angulation [23]. For further simplification, we can hypothesize that those with an anatomically large PI will have raised buttocks; in contrast, those with a small PI will have a smaller bulge in the buttocks. However, little research has been conducted to determine whether the thickness of the buttocks (silhouette) may reflect the actual size of the PI.

Therefore, we focused on the thickness of the buttocks on the silhouette to find an easy, noninvasive way to classify the stage of the PIs. The purpose of this study was to explore effective measurement angles for PI classification in terms of discrimination ability using receiver operating characteristic (ROC) curve analysis, and to identify the availability of visual buttock silhouettes (Study 1). Furthermore, the intra- and inter-rater reliability of a devised tool for simple PI classification focusing on the visual buttock silhouette was assessed (Study 2).

2. Materials and Methods (Study 1: Exploring Effective Surrogate Angles for PI Classification Focusing on the Buttocks)

2.1. Measurement Angular Definition

We discussed potential landmarks applicable for the estimation of PI based on anatomical, anthropometric, and physiotherapy aspects: (1) the most raised part of the buttock; (2) top of the head; (3) the anterior (patella), central, and posterior points of the sagittal knee; and (4) center of the thigh (the anterior–posterior diameter of the transition between the buttocks and the thigh). To verify the PI with the appearance of the body silhouette, the posterior and anterior surfaces of the knee are landmarks that can be easily confirmed, and have been used in many studies in ergonomics and clinical practice. Furthermore, the lines at the center of the knee and the center of the thigh on the lateral side (sagittal plane) are used to measure the range of motion of the hip and knee joints in clinical situations. The parietal point was adopted as a frequently used landmark for cervical and standing posture analyses.

The hip–knee line (HKL) angles were defined as the angle between the following two intersecting lines on the sagittal plane: the line connecting the most raised part of the buttock and either the anterior, central, or posterior knee; and the line connecting either knee point and either of vertical lines as the parietal, vertical, or the midthigh line. The midthigh line was defined as the femoral axis (the central part of the anterior–posterior diameter of the knee and the center of the thigh). As a combination of the lines, five variation types of HKL angles (A1, A2, B1, B2, and C) were defined as shown in Figure 1.

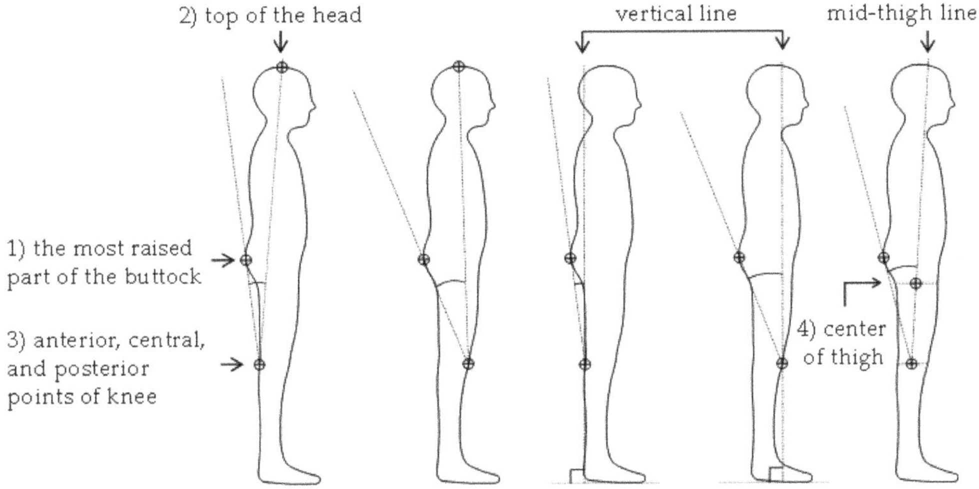

Types of HKL angles		A1	A2	B1	B2	C
knee point	anterior		✓		✓	
	posterior	✓		✓		
	central point					✓
vertical point	top of the head	✓	✓			
	vertical line from floor			✓	✓	
	central of femur					✓

Figure 1. Five variation types of hip-knee line (HKL) angles.

2.2. Procedures

The 125 voluntary participants (71 males, 54 females; average age 55.9 ± 18.9 years) were recruited in line with the following inclusion criteria: (1) normal spinal alignment with sagittal vertical axis less than 40mm, (2) standing alignment without knee flexion contracture, (3) no difference in leg length, and (4) maintained postural standing stability. We provided a sufficient explanation for this study (IRB No. 60-21-0072, Nagoya City University). They were then asked to assume a neutral posture in an upright position. Subsequently, a photograph of their upright standing posture from the right lateral side was taken using a digital camera (Power Shot S5, Canon Inc., Tokyo, Japan) fixed on a camera tripod at a height of 60 cm, with the focus distance to the subject set to 3 m.

2.3. Measurement of HKL Angles and Outcome Variable

The landmarks in each picture were obtained using the ImageJ ver. 1.48 software (NIH, Bethesda, MD, USA), and five types of HKL angles (A1, A2, B1, B2, and C) were measured using the landmarks.

The surrogate angle correlating with PI [21,22] was used to measure the outcome. The surrogate angle of the PI can be defined as "the angle between the line connecting the upper edges of the greater trochanter and iliac crest, and the line connecting the upper edge of the iliac crest and the buttock at the same height as the greater trochanter". The details of the measurement and reliability of the landmarks used to estimate the surrogate angle by palpation have been described elsewhere [22]. The surrogate angle has sufficient practical reliability for estimating PI (R^2 = 0.63).

2.4. Data Analysis and Statistical Analyses

The surrogate angles of PI obtained from the 125 photographs were classified by using quartiles: small ("S", less than 42°, indicating the first quartile), medium ("M", between 42° and 51°, the first and third quartile range), or large ("L", more than 51°, the third quartile). To explore effective measurement angles for PI classification in terms of discrimination ability using ROC analysis, the outcome variables were further converted to the two dichotomous variables determining S/ML and SM/L.

To assess the discriminating ability of the HKL angles relative to S/ML and SM/L of PI, ROC curves were created by plotting values of sensitivity for the y-axis and 1-specificity for the x-axis to calculate the area under the curve (AUC). Discriminating ability was assessed by the AUC of the ROC curve by the following criteria: 0.5–0.7 as low accuracy, 0.7–0.9 as moderate accuracy, and >0.9 as high accuracy. We considered the HKL angle with the highest AUC out of the five HKL angles as a candidate angle applicable for a practical PI classification tool using the thickness of the buttocks. Focusing on the candidate angle with the highest AUC, the Youden index of the candidate angle, indicating the maximum difference between sensitivity and specificity, was calculated to assess the optimal threshold value (cut-off point) of S/ML and SM/L for HKL angles. These analyses were conducted using SPSS version 26.0 (IBM Corp., Armonk, NY, USA).

2.5. Criteria for Selecting Cut-Off Points Applicable to Practical PI Classification Tools Using Buttock Thickness

To devise a practical PI classification tool, we empirically considered that at least the following levels were needed as an acceptable range of cut-off points: ≥0.7 sensitivity and specificity. The acceptable range should include the maximum value of the Youden index. Furthermore, to ensure the availability of visual buttock silhouettes, angular differences between thresholds of S/ML and SM/L for PI classification should be set at least more than 3°.

3. Results (Study 1)

3.1. AUCs of HKL Angles Discriminating Small or Large PIs

The average value of surrogate angles of PI was 47.0 ± 6.2° (male 45.4 ± 5.1°; female, 49.1 ± 7.0°) and BMI was 23.4 ± 3.4kg/m^2. Figure 2 shows the AUCs of the HKL angles discriminating small (<42°) or large (>51°) PIs. The HKL angle with the highest AUC value was the HKL angle C, reaching a high accuracy of 0.93 for the small and moderate PIs, and 0.82 for the large PIs. On the other hand, the HKL angle with the lowest AUC was A2, which had low accuracies of 0.61 for the small PIs and 0.58 for the large PIs.

3.2. Cut-Off Points Applicable to Practical PI Classification Tool Using the Thickness of the Buttocks

In terms of the AUC value, the HKL angle C was considered to be an appropriate candidate for a practical PI classification tool using the thickness of the buttocks. Table 1 shows the results of the sensitivity, specificity, and Youden index of the HKL angle C. Columns that are shaded in gray in the table mean out of criteria (unacceptable range of cut-off points); namely, sensitivity and specificity of >0.7. The acceptable cut-off angles relative to S/ML discrimination ranged from 18.5° to 19.5° based on the criteria for selecting the cut-off points. The acceptable range also included a maximum Youden index value of 0.74 at 19.5°. Similar trends were observed in the strata of males and females.

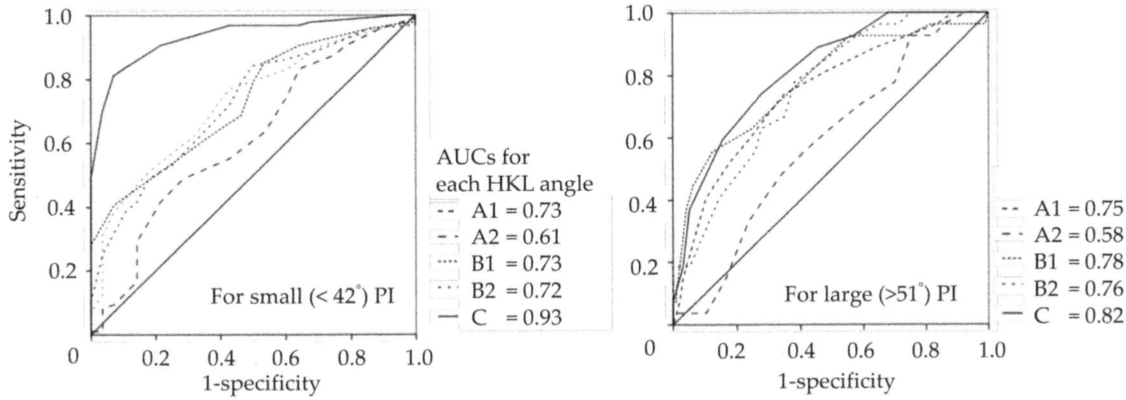

PI, pelvic incidence; AUC, area under curve; HKL angle, hip-knee line angle

Figure 2. AUCs and HKL angles discriminating small (<42°) or large (>51°) PIs.

Table 1. The results of sensitivity, specificity, and Youden index of the HKL angle C.

HKL Angle C	Cut-Off	Total (n = 125)			Male (n = 71)			Female (n = 54)		
		sen.	spec.	Y. I	sen.	spec.	Y. I	sen.	spec.	Y. I
	13.0	1.00	0.00	0.00	1.00	0.00	0.00	–	–	–
	14.5	1.00	0.04	0.04	1.00	0.05	0.05	1.00	0.00	0.00
	15.5	0.98	0.32	0.30	1.00	0.38	0.38	0.96	0.10	0.10
	16.5	0.97	0.36	0.33	1.00	0.57	0.43	–	–	–
	17.5	0.97	0.57	0.54	1.00	0.71	0.71	0.94	0.08	0.08
	18.5	0.91	0.79	0.69	0.98	0.81	0.79	0.83	0.71	0.54
	19.5	0.81	0.93	0.74	0.90	0.95	0.85	0.72	0.86	0.58
S/ML	20.5	0.70	0.96	0.66	0.76	0.95	0.71	0.63	1.00	0.63
	21.5	0.50	1.00	0.49	0.54	1.00	0.54	0.44	1.00	0.44
	22.5	0.32	1.00	0.32	0.32	1.00	0.32	0.33	1.00	0.33
	23.5	0.16	1.00	0.16	0.14	1.00	0.14	0.17	1.00	0.17
	24.5	0.09	1.00	0.08	0.08	1.00	0.08	0.09	1.00	0.09
	25.5	0.02	1.00	0.02	0.02	1.00	0.02	0.02	1.00	0.02
	26.5	0.01	1.00	0.01	0.00	1.00	0.00	–	–	–
	28.0	0.00	1.00	0.00	–	–	–	0.00	1.00	0.00
	13.0	1.00	0.00	0.00	1.00	0.00	0.00	–	–	–
	14.5	1.00	0.01	0.01	1.00	0.02	0.02	1.00	0.00	0.00
	15.5	1.00	0.11	0.11	1.00	0.13	0.13	1.00	0.08	0.08
	16.5	1.00	0.13	0.13	1.00	0.15	0.15	–	–	–
	17.5	1.00	0.20	0.20	1.00	0.25	0.25	1.00	0.11	0.11
	18.5	1.00	0.32	0.32	1.00	0.30	0.30	1.00	0.36	0.36
	19.5	0.93	0.43	0.36	1.00	0.41	0.41	0.88	0.47	0.36
SM/L	20.5	0.89	0.55	0.44	1.00	0.53	0.53	0.82	0.58	0.41
	21.5	0.74	0.72	0.46	0.90	0.70	0.61	0.65	0.75	0.40
	22.5	0.59	0.84	0.44	0.80	0.87	0.67	0.47	0.81	0.28
	23.5	0.37	0.95	0.32	0.50	0.97	0.47	0.29	0.92	0.21
	24.5	0.19	0.97	0.15	0.30	0.98	0.28	0.12	0.94	0.06
	25.5	0.07	1.00	0.07	0.10	1.00	0.10	0.06	1.00	0.06
	26.5	0.04	1.00	0.04	0.00	1.00	0.00	–	–	–
	28.0	0.00	1.00	0.00	–	–	–	0.00	1.00	0.00

Note: sen., sensitivity; spec., specificity; Y. I, Youden index, S/ML, thresholds of HKL C angles discriminating between small (<42°) and more than medium (≥42°) pelvic incidence; SM/L, thresholds of HKL C angles discriminating between less than medium (<51°) and large (≥51°) pelvic incidence. Columns shaded in gray in the table mean out of criteria (unacceptable range of cut-off points); namely, sensitivity and specificity of >0.7.

The acceptable cut-off angle for SM/L discrimination was 21.5°, with a maximum Youden index value of 0.46. For male participants, the acceptable range was from 21.5° to 22.5°, while no acceptable range was found for female participants.

3.3. Devised PI Classification Tool Using the HKL Angle

Emphasizing the ease of use and quick assessment, we developed a practical PI classification tool using the HKL angle, focusing on the thickness of the buttocks (Figure 3). The tool, which was printed on a transparent film, can be used to estimate the HKL angle sizes of S, M, and L by looking through it at the standing posture of the lateral side. Based on the acceptable range of cut-off points, the thresholds were set to 18.5° for S/ML and 21.5° for SM/L discrimination. In terms of the Youden index, the optimal cut-off angle was 19.5°; however, we applied an acceptable cut-off angle of 18.5° relative to S/ML. Angular differences between thresholds of S/ML and SM/L for PI classification should be set at least more than 3° in order to ensure the availability of visual buttock silhouettes. The tool enables quick assessment following two steps. First, align the points of (a) the central point of the anterior–posterior diameter of the knee and (b) the center of the thigh in the tool with those on the body surface of the subject. Depending on the focal length, measurement points A, B, and C were selected. Then, we determined where the most-raised point of the buttock was placed in the S, M, or L areas of the HKL angle.

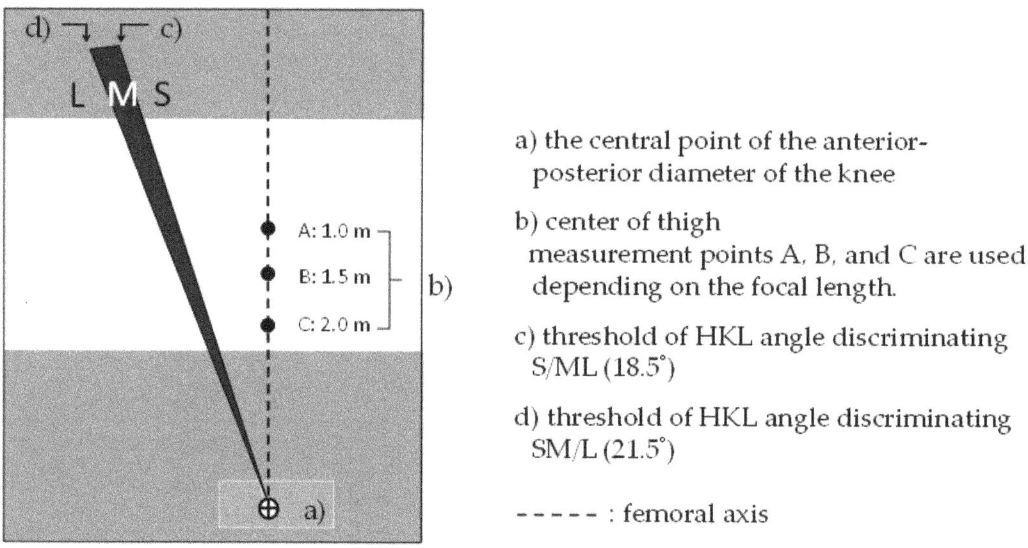

Figure 3. Practical PI classification tool using the HKL angle.

4. Materials and Methods (Study 2: Assessing Intra-/Inter-Rater Reliability of the PI Classification Tool Using the HKL Angle)

4.1. Participants

A total of 14 healthy persons (working/studying at university/hospital, mean age 37.3 ± 10.0 years, 7 males and 7 females) voluntarily participated in this study as examiners for evaluating intra- and inter-rater consistency. Informed consent was obtained from the participants after an explanation of the study.

4.2. Materials Used for the Assessment

Of the 125 photos used in Survey 1, three sets of 24 photos were randomly selected for testing. Of these test materials, one was intended for use in prior training, and the other two were used for the assessment of intra-and inter-rater reliability. The ratio of classification was 1:2:1 for S:M:L. The developed PI classification tool using the HKL angle (Figure 3) was printed on a transparent film with a length of 14.8 cm and width of 21 cm (A5 size).

4.3. Procedures

The photos of the material were randomly projected on the screen and reduced to 1/2 of the actual subject size. This means that the focal distance between the examiner and the screen needed to be converted to 1/2 size. Therefore, the examiner was requested to sit in front of the screen 75 cm away at a focal distance of 1.5 m. Subsequently, the examiners were trained in advance on how to use the tool using the photo material for prior training.

In the training session, the examiners sitting in front of the screen were instructed to hold the tool and keep their eye level horizontal with the height of the subject on the screen. In addition, they were asked to place their eyes and the tool on a vertical line to the screen. After the setting, the researchers instructed them by showing 24 prior training photos one by one on how to use the tool already represented in Section 4.3. Examples of applications of the tool for practice are shown in Figure 4. Each examiner evaluated the materials for the assessment twice, with intervals for rest after training.

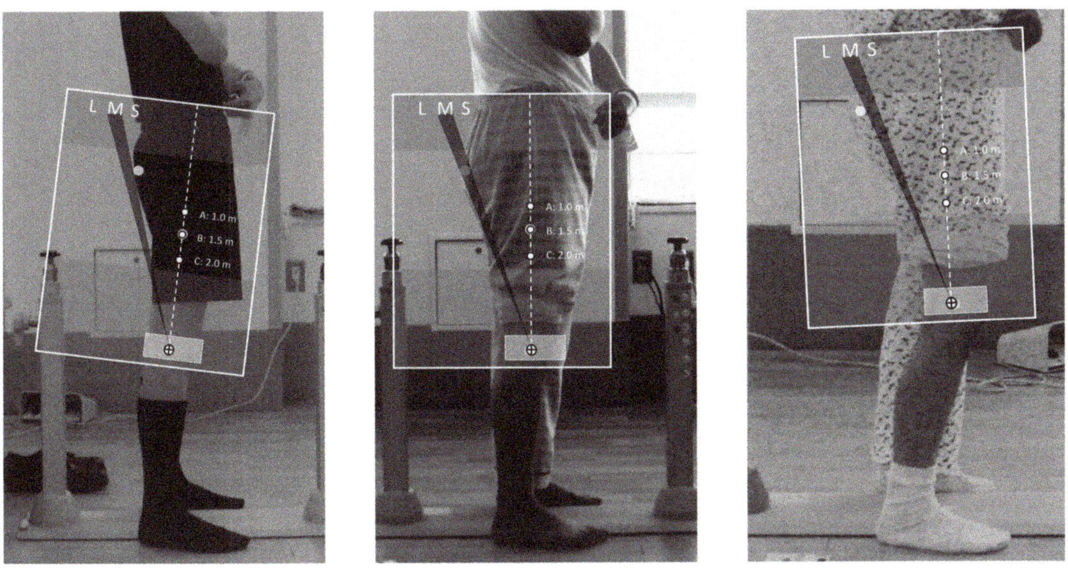

Figure 4. Application examples of the practical PI classification tool. Left photo shows in case of HKL angle classified "S", meaning small (<42°) PI. Similarly, the center photo shows in case of the HKL angle classified "M", meaning medium (42–51°) PI. The right photo shows in case of the HKL angle classified "L", meaning large (>51°) PI.

4.4. Data Analysis and Statistical Analysis

As one of the usability metrics, we measured the required time of first and second sessions to complete the classification of PI. The mean time (s/photo) of classifying 24 photos in each session was calculated. Furthermore, the correct rate was calculated as an effectiveness metric. The correct rate represented the extent to which examiners' responses agreed with an external criterion for the classification of the HKL angle. We applied the surrogate angle of the PI as an external criterion, calculated in Study 1. A paired t-test was conducted to show the significant differences in the means for each metric between sessions.

Cohen's kappa coefficient was calculated to examine the intra-rater reliability of measurements, showing the test–retest reliability of each examiner within the first and second sessions. Fleiss's kappa coefficient (quadratic) was applied as a weighted inter-rater reliability, meaning reliability between the initial test (first session) and retest (second session) measurements obtained from all examiners. Both coefficients were estimated using 95% confidence intervals (CIs). According to Landis's criteria [24], reliability levels of kappa coefficients were defined as 0 for poor, 0.01–0.20 for slight, 0.21–0.40 for fair, 0.41–0.60 for moderate, 0.61–0.80 for substantial, and 0.81–1.00 for almost perfect. These analyses were conducted using SPSS version 26.0 (IBM Corp., Armonk, NY, USA).

5. Results (Study 2)

5.1. Usability Metrics

The mean time (sec/photo) was 17.0 ± 4.7 for the first session and 13.6 ± 2.9 for the second session (Table 2); this difference was significant (mean difference, 3.4; 95% CI, 1.6–5.2; $t(13) = 4.1$, $p = 0.001$). The correct rate (%) reached 77.1 ± 6.9 for the first session and 79.8 ± 6.9 for the second session; no significant difference was found ($t(13) = 1.0$, $p = 0.32$).

Table 2. Results of usability metrics and intra-/inter-rater reliability of the PI classification tool using the HKL angle.

Sub. No.	Mean Time (s/photo)		Correct Rate (%)		Intrarater Reliability Cohen Kappa (95% CI)			Inter-Rater Fleiss's Kappa (95% CI)	
	1st	2nd	1st	2nd	Total (n = 24)	Male (n = 14)	Female (n = 10)	1st	2nd
1	20.0	13.8	88	83	0.87 (0.72–1.00)	0.93 (0.77–1.00)	0.68 (0.34–1.00)		
2	17.5	13.8	79	88	0.80 (0.61–0.99)	0.83 (0.58–1.00)	0.69 (0.38–1.00)	total 0.50 (0.47–0.53)	total 0.54 (0.51–0.57)
3	13.1	12.9	83	71	0.82 (0.67–0.97)	0.74 (0.52–0.96)	1.00 (1.00–1.00)		
4	14.6	12.0	79	83	0.87 (0.71–1.00)	0.78 (0.68–0.89)	0.78 (0.39–1.00)	male 0.43 (0.39–0.47)	male 0.50 (0.27–0.57)
5	19.6	17.1	67	83	0.75 (0.56–0.94)	0.75 (0.50–0.99)	0.69 (0.38–1.00)		
6	16.0	14.0	83	83	0.78 (0.60–0.97)	0.71 (0.48–0.95)	0.85 (0.63–1.00)	female 0.56 (0.46–0.68)	female 0.57 (0.51–0.73)
7	13.2	12.6	83	71	0.67 (0.45–0.90)	0.58 (0.27–0.88)	0.80 (0.44–1.00)		
8	11.3	10.6	79	88	0.75 (0.57–0.94)	0.71 (0.42–0.99)	0.79 (0.52–1.00)		
9	12.5	11.3	75	75	0.83 (0.67–1.00)	0.70 (0.41–0.99)	1.00 (1.00–1.00)		
10	15.0	13.1	67	79	0.67 (0.45–0.88)	0.69 (0.43–0.95)	0.59 (0.29–0.90)		
11	24.6	18.3	71	67	0.82 (0.64–0.99)	0.91 (0.75–1.00)	0.63 (0.38–0.88)		
12	21.7	10.3	67	88	0.81 (0.66–0.97)	0.89 (0.74–1.00)	0.63 (0.38–0.88)		
13	26.3	19.8	83	83	0.77 (0.58–0.95)	0.83 (0.61–1.00)	0.65 (0.43–0.87)		
14	12.8	10.4	75	75	0.82 (0.65–0.99)	0.83 (0.62–1.00)	0.76 (0.42–1.00)		
mean (sd)	17.0 (4.7)	13.6 (2.9)	77.1 (6.9)	79.8 (6.9)	0.79 (0.61–0.96)	0.78 (0.59–0.97)	0.75 (0.50–0.97)		

Note: 95% CI, 95% confidence interval; sub., subject; s, second.

5.2. Intra-/Inter-Rater Reliability

Table 2 also shows the results of intra- and inter-rater reliability of the PI classification tool using the HKL angle. The mean Cohen's kappa coefficient as intra-rater reliability was 0.79 (95% CI, 0.61–0.96), indicating "substantial" levels of test–retest consistency. No sex difference was found in intra-rater reliability, even when focusing on the differences in the stratified analysis of the male and female subjects in the photos.

The mean Fleiss's kappa coefficient as inter-rater reliability was 0.50 (95% CI, 0.47–0.53) for the first session and 0.54 (95% CI, 0.51–0.57) for the second session, both indicating "moderate" levels of consistency between examiners. We also conducted a stratified analysis, showing that raters tended to do better with female subjects than male ones (first session, 0.43 for males and 0.56 for females; second session, 0.50 for males and 0.57 for females, respectively).

6. Discussion

This was the first study to show that the HKL angle formed by lateral body landmarks focusing on the area of the femur and buttock can discriminate the classification of small or large PIs. We examined the optimal cut-off points for discriminating small or large PIs in terms of HKL angle, leading to an acceptable cut-off angle of 18.5° for S/ML and 21.5° for SM/L discrimination, considering a threshold range of more than 3° or larger angle. In addition, we devised a quick noninvasive assessment tool for PI classification using the cut-offs of the HKL angle with a view to practical application. The results of intra- and inter-rater reliability indicated certain limitations, but ensured a substantial/moderate level of the tool, meaning that it has sufficient potential for practical use. We will discuss the relationship between the HKL angle and PI, as well as the intra-and inter-rater reliability of the tool, in the following sections.

6.1. Relationship between the HKL Angle and PI Using the Visual Buttock Silhouette

HKL angle C reached a high accuracy of 0.93 for the small and moderate of 0.82 for the large PIs, whereas the HKL angle A2 showed the lowest AUC among all types of HKL angles. This difference deserves careful attention when interpreting the rationale for the HKL angles reflecting PI.

The HKL angles A1 and A2 shown in Figure 1 used the line connecting the top of the head and anterior/posterior knee points. The line was intended to reflect the alignment of the whole body, because whole-body alignment is a key clinical parameter generally applied in the field of spinal surgery to evaluate the bending characteristics of the spinal column from the standing posture of the sagittal plane. In addition, the Kendall classification [25], which is known as a representative of the classification of whole-body alignment, can be generally used to study muscle shortening and weakness in the clinical field of physiotherapy, and to correlate the relationship between the type of Kendall posture and lumbar pain [26]. Furthermore, PI is anatomically considered to have tendencies to be larger in the order of flat-back (S), sway-back (M), and ideal (L) and kyphosis–lordosis (LL) in the Kendall postures [27].

On the other hand, whole-body alignment maintains a standing posture under the influence of the movement control of many joints, such as the large angular displacement of the hip joint in a static standing position and curvature of the spinal column [28]. Moreover, the spinal column alignment compensates to maintain the whole-body balance of the head, chest, lumbar region, and lower limbs [29]. These findings provide an insight that the use of the line connecting the top of the head and anterior/posterior knee points for estimating PI, which is a part of spinal alignment indicating the inclination of the sacrum in the pelvis, might have disadvantages. This might also be explained by the differences between the HKL angles A and B. In fact, the HKL angles B1/B2 were intended to exclude the effect of whole-body alignment by using the line connecting the reference vertical line and the knee points. In addition, the HKL angle B1/B2 formed by the line connecting the raised part of the buttocks and either the anterior/posterior knee point or the reference vertical line can be considered to be an angle that simply indicates the degree of the uplift of the buttocks (sacrum). The AUCs of HKL angle B1/B2, excluding body alignment effects, were relatively high compared to HKL angle A2, regardless of the anterior/posterior knee positions. Moreover, the effect of compensation on body alignment is expected to be smaller for A1 than A2, owing to a relatively smaller angle, and leading to a relatively high AUC of A1 compared to that of A2. Therefore, it seems reasonable to suppose that HKL angle A2 showed a relatively low discriminating ability of PI due to these findings.

What needs to be emphasized here is that the HKL angle C focuses not on the whole spinal alignment, but on the limited area of the femur and buttock. It should be noted that HKL angle C defines an angle using a line connecting the central point of the anterior–posterior diameter of the transition between the buttocks and the thigh and the midpoint of the lateral joint space of the knee. This indicates the equivalent line connecting the great trochanter and the epicondyle of the femur used in the measurement of the flexion/extension angle of the hip and knee joint defined in the range of motion, or the line connecting the femoral epicondyle and the hip joint center of rotation defined in the International Society of Biomechanics' recommendation [30].

This equivalent line has the following advantages: First, humans have a strategy of maintaining standing balance, the so-called "cone of economy" with the feet as the fulcrum; in other words, standing posture can be maintained in the conical space with the foot as the apex [31]. To compensate for changes in spinal alignment, the central line of the femur plays an important role in a stable standing posture. Humans compensate for age-related changes in spinal alignment by increasing posterior pelvic tilt and changing leg alignments to maintain the gravity line over the feet [32]. This indicates the possibility that the central line of the femur sensitively represents the extent of the PI. Second, according to a recent study that investigated the correlation between the sway of the center of gravity and spinal alignment such as PI, the larger the PI, the more the center of gravity moved forward [33]. This indicated that the axis of the thigh, when viewed from the lateral side, had shifted forward, and the larger the PI, the larger the angle of the HKL angle C. Thus, the HKL angle C reflecting the central line of the femur may be linked to postural fluctuations due to anterior–posterior bending of the body, which might have resulted in the highest AUC of HKL angle C sensitively reflecting PI.

6.2. *Optimal Cut-Off Points Applicable to Practical PI Classification Tool*

The acceptable cut-off values of the HKL angle C were 18.5° (sensitivity, 0.91; specificity, 0.79) for S/ML and 21.5° (sensitivity, 0.74; specificity, 0.72) for SM/L discrimination. Although it seemed to be sufficient for discriminating abilities in terms of application to a quick, noninvasive tool in practical use, some challenges remain.

The first thing we noticed was that Youden index value at 18.5° of the HKL angle for discriminating small (<42°) PI showed 0.69, whereas the value for large (>51°) PI was as low as 0.46 at 21.5° of the HKL angle. One of the reasons for this is the PI characteristics of the Japanese population. Asians such as Japanese or Koreans generally have smaller PI values than Caucasians; for example, Kanemura reported Japanese PI as 46.7 ± 8.7° [34], and another similar study showed 46.7 ± 8.9° in 86 Japanese healthy adults aged from 23 to 59 years [35], and 47.8 ± 9.3° for the Korean population [36]. In contrast, 52.0 ± 9.0° was reported from the United States [12], 50.6 ± 10.2° [37] and 54.7 ± 10.6° [38] from France, and 50.2 ± 10.0° [11] from Belgium. Thus, the Japanese population not only has a smaller PI, but also a narrower range of standard deviation than Caucasians. Another reason might be the feature of the Japanese physique with relatively flat, smaller swelling of the buttocks compared to Westerners [35]. This means that even if the PI was large, the HKL angle tended to be underestimated in the case of focusing on the visual buttock silhouette.

Second, we should focus on the differences in the stratified analysis of male and female participants. The Youden index is relatively smaller in females than in males, indicating that the HKL angle might have disadvantages when applied to female patients. This might be due to the anatomical sex difference in the pelvis. The female pelvis has a large lateral diameter and is circular, and the anterior–posterior direction of the female is also larger than that of the male [39]. In this study, since we defined and assessed the HKL angle from the lateral side, such an anatomical feature may have biased the HKL angle in female participants. Another possibility may depend on the sample population used in this study.

In general, it is believed that there is no sex difference in PI itself, although the samples of participants included in this study had a slight tendency to have a larger PI of 49.1 ± 7.0° in females, as compared to 45.4 ± 5.1° for males. The PI obtained from the 125 photographs were classified by quartiles regardless of sex, and the large PI was tentatively defined as >51° using the threshold in the third quartile. The large PI category consequently resulted in more females than males. Such biases might have led to a relatively low discriminating ability under the condition of SM/L.

6.3. Intra-/Inter-Rater Reliability of Tool Using HKL Angle

As shown in Table 2, the mean Cohen's kappa coefficient as intra-rater reliability was 0.79 (95% CI, 0.61–0.96), indicating "substantial" levels of test–retest consistency. No sex difference was found in intra-rater reliability, even when focusing on the differences in the stratified analysis of the male and female subjects in the photos.

In contrast, some challenges seem to be in the inter-rater reliability in the results of Fleiss's kappa coefficients, which might be considered insufficient levels due to "moderate" consistency between examiners. However, it should be noted that Fleiss's kappa coefficient contains a paradox, meaning that the kappa statistic may behave inconsistently in case of strong agreement between raters, since this index assumes lower values than would have been expected [40]. Second, the stratified analysis showed tendencies to be better for female subjects than male subjects (first session, 0.43 for males and 0.56 for females; second session, 0.50 for males and 0.57 for females). This might be related to the points previously discussed regarding the anatomical sex difference of the pelvis and sampling biases resulting in the inclusion of more females in the large PI category. Third, we should focus on improving the inter-rater reliability in the second session compared to the first session. We provided sufficient training to examiners using the 24 prior training photos individually. Nevertheless, we observed a tendency to be better in the 2nd session. Furthermore, a significant improvement in the meantime was observed between sessions. These results indicated that they did not reach a plateau, so further improvements in the inter-rater reliability could be expected.

Considering the results comprehensively, the devised assessment tool for PI classification using the cut-offs of the HKL angle has sufficient potential for practical use.

6.4. Practical Implication and Limitation

This study provides useful information about how to estimate PI using the HKL angle in a validated, noninvasive way. PI is considered to be one of the possible determinants of low back pain, and has been attracting attention as a new countermeasure considering individual biomechanical features of the lumbar spine. This knowledge can contribute not only to improving the new insight of occupational health research, but also to developing new noninvasive technologies for measuring PI using image recognition with artificial intelligence.

This study had some limitations. First, the values of PI were estimated by using the surrogate angle, not measured by X-rays. The surrogate angle has sufficient reliability for estimating PI under practical use (R^2 = 0.63); however, the remaining variation of R^2 can lead to misclassification into small, medium, or large PIs. Second, the study population was limited to Japanese adults, so the proposed acceptable range of cut-off points might vary with ethnicity [41]. Third, the small sample size may lead to imbalanced distributions of the target variable so that the sample size would need to be even larger in terms of the robustness of the results [42]. Fourth, a morbidly obese patient and an underweight patient with the same PI will likely have a different HKL angle; however, this study could not provide any evidence of the effect of BMI on the HKL angles. Japanese have a relatively low BMI compared to other ethnicities; morbidly obese patients (BMI > 35) are rare in the Japanese population, and none were in the current study population. In addition, there were only five underweight participants (BMI < 18.5). Further research is warranted in various populations. Fifth, the results of the AUC and Youden index showed that the discriminating ability of HKL angles for detecting large PIs was relatively low, as was the case for female patients. Such cases may require multiple measurements with different examiners to ensure the reliability of the PI classification tool. Sixth, standard procedures for training manuals should be prepared using various photos with a wide range of ages and/or multiple features of curvature of the spinal column. Further improvement is warranted to enhance the intra- and inter-rater reliability of the tools for practical use.

7. Conclusions

To explore effective measurement angles for PI classification, we determined the applicable HKL angle defined as the angle of two intersecting lines on the sagittal plane, the line connecting the most raised part of the buttock and central point of the knee, and the midthigh line (the femoral axis). This study revealed that the HKL angle could distinguish the size of PI with high/moderate discrimination ability. Furthermore, a quick, noninvasive assessment tool devised for PI classification using the cut-offs of the HKL angle indicated acceptable inter- and intra-rater reliability, enough for practical application.

Author Contributions: Conceptualization, S.Y., T.E. and M.K.; methodology, S.Y., K.A., T.U. and T.E.; formal analysis, S.Y. and T.E.; investigation, S.Y., T.E., T.U., S.K., Y.S., A.Y. and M.K.; resources, A.I. and T.E.; data curation, S.Y., T.E., T.M. and M.K.; visualization, S.Y. and T.E.; writing—original draft, S.Y.; writing—review and editing, T.E., T.M., K.A., A.I. and M.K.; supervision, T.E., T.M., K.A. and M.K.; project administration, T.E., A.I. and M.K. All authors have read and agreed to the published version of the manuscript.

Funding: This research received no external funding.

Institutional Review Board Statement: The study was conducted according to the guidelines of the Declaration of Helsinki and approved by the Institutional Review Board of Nagoya City University Graduate School of Medical Sciences (No. 60-21-0072).

Informed Consent Statement: Informed consent was obtained from all subjects involved in the study.

Data Availability Statement: The anonymized data presented in this study are available upon request from the corresponding author. The data are not publicly available due to privacy considerations or ethical challenges.

Acknowledgments: We would like to thank the staff (physiotherapists, occupational therapists, speech therapists, pharmacists, and nurses) and patients of the Nagoya City University West Medical Center for their active consent to participate in this study. We would also like to thank the staff of Nagoya City University for their active consent to participate in this study.

Conflicts of Interest: The authors declare no conflict of interest.

References

1. Ganesan, S.; Acharya, A.S.; Chauhan, R.; Acharya, S. Prevalence and risk factors for low back pain in 1,355 young adults: A cross-sectional study. *Asian Spine J.* **2017**, *11*, 610–617. [CrossRef]
2. Andersson, G.B. Epidemiological features of chronic low-back pain. *Lancet* **1999**, *354*, 581–585. [CrossRef]
3. Manchikanti, L.; Singh, V.; Falco, F.J.; Benyamin, R.M.; Hirsch, J.A. Epidemiology of low back pain in adults. *Neuromodulation* **2014**, *17* (Suppl. S2), 3–10. [CrossRef]
4. Meucci, R.D.; Fassa, A.G.; Faria, N.M. Prevalence of chronic low back pain: Systematic review. *Rev. Saude Publica* **2015**, *49*, 73. [CrossRef]
5. Diseases, G.B.D.; Injuries, C. Global burden of 369 diseases and injuries in 204 countries and territories, 1990–2019: A systematic analysis for the Global Burden of Disease Study 2019. *Lancet* **2020**, *396*, 1204–1222.
6. Wu, A.; March, L.; Zheng, X.; Huang, J.; Wang, X.; Zhao, J.; Blyth, F.M.; Smith, E.; Buchbinder, R.; Hoy, D. Global low back pain prevalence and years lived with disability from 1990 to 2017: Estimates from the Global Burden of Disease Study 2017. *Ann. Transl Med.* **2020**, *8*, 299. [CrossRef] [PubMed]
7. Itoh, H.; Kitamura, F.; Yokoyama, K. Estimates of annual medical costs of work-related low back pain in Japan. *Ind. Health* **2013**, *51*, 524–529. [CrossRef]
8. Punnett, L.; Pruss-Utun, A.; Nelson, D.I.; Fingerhut, M.A.; Leigh, J.; Tak, S.; Phillips, S. Estimating the global burden of low back pain attributable to combined occupational exposures. *Am. J. Ind. Med.* **2005**, *48*, 459–469. [CrossRef] [PubMed]

9. Steffens, D.; Maher, C.G.; Pereira, L.S.; Stevens, M.L.; Oliveira, V.C.; Chapple, M.; Teixeira-Salmela, L.F.; Hancock, M.J. Prevention of low back pain: A systematic review and meta-analysis. *JAMA Intern. Med.* **2016**, *176*, 199–208. [CrossRef]
10. Verbeek, J.H.; Martimo, K.P.; Karppinen, J.; Kuijer, P.P.; Viikari-Juntura, E.; Takala, E.P. Manual material handling advice and assistive devices for preventing and treating back pain in workers. *Cochrane Database Syst Rev.* **2011**, 6. [CrossRef]
11. Legaye, J.; Duval-Beaupere, G.; Hecquet, J.; Marty, C. Pelvic incidence: A fundamental pelvic parameter for three-dimensional regulation of spinal sagittal curves. *Eur. Spine J.* **1998**, *7*, 99–103. [CrossRef] [PubMed]
12. Schwab, F.; Patel, A.; Ungar, B.; Farcy, J.P.; Lafage, V. Adult spinal deformity-postoperative standing imbalance: How much can you tolerate? An overview of key parameters in assessing alignment and planning corrective surgery. *Spine* **2010**, *35*, 2224–2231. [CrossRef]
13. Schwab, F.J.; Blondel, B.; Bess, S.; Hostin, R.; Shaffrey, C.I.; Smith, J.S.; Boachie-Adjei, O.; Burton, D.C.; Akbarnia, B.A.; Mundis, G.M.; et al. Radiographical spinopelvic parameters and disability in the setting of adult spinal deformity: A prospective multicenter analysis. *Spine* **2013**, *38*, e803–e812. [CrossRef]
14. Chen, R.Q.; Hosogane, N.; Watanabe, K.; Funao, H.; Okada, E.; Fujita, N.; Hikata, T.; Iwanami, A.; Tsuji, T.; Ishii, K.; et al. Reliability analysis of spino-pelvic parameters in adult spinal deformity: A comparison of whole spine and pelvic radiographs. *Spine* **2016**, *41*, 320–327. [CrossRef]
15. Tyrakowski, M.; Yu, H.; Siemionow, K. Pelvic incidence and pelvic tilt measurements using femoral heads or acetabular domes to identify centers of the hips: Comparison of two methods. *Eur. Spine J.* **2015**, *24*, 1259–1264. [CrossRef] [PubMed]
16. Lowe, T.; Berven, S.H.; Schwab, F.J.; Bridwell, K.H. The SRS classification for adult spinal deformity: Building on the King/Moe and Lenke classification systems. *Spine* **2006**, *31*, s119–s125. [CrossRef]
17. Bao, H.; Liabaud, B.; Varghese, J.; Lafage, R.; Diebo, B.G.; Jalai, C.; Ramchandran, S.; Poorman, G.; Errico, T.; Zhu, F.; et al. Lumbosacral stress and age may contribute to increased pelvic incidence: An analysis of 1625 adults. *Eur. Spine J.* **2018**, *27*, 482–488. [CrossRef] [PubMed]
18. Place, H.M.; Hayes, A.M.; Huebner, S.B.; Hayden, A.M.; Israel, H.; Brechbuhler, J.L. Pelvic incidence: A fixed value or can you change it? *Spine J.* **2017**, *17*, 1565–1569. [CrossRef] [PubMed]
19. Mac-Thiong, J.M.; Roussouly, P.; Berthonnaud, E.; Guigui, P. Age- and sex-related variations in sagittal sacropelvic morphology and balance in asymptomatic adults. *Eur. Spine J.* **2011**, *20* (Suppl. S5), 572–577. [CrossRef]
20. Chen, H.F.; Zhao, C.Q. Pelvic incidence variation among individuals: Functional influence versus genetic determinism. *J. Orthop. Surg. Res.* **2018**, *13*, 59. [CrossRef]
21. Yamada, S.; Ebara, T.; Uehara, T.; Kimura, S.; Aoki, K.; Inada, A.; Kamijima, M. Reliability of anthropometric landmarks on body surface for estimating pelvic incidence without lateral X-ray. *Occup Environ Med.* **2021**, 3. [CrossRef]
22. Yamada, S.; Ebara, T.; Uehara, T.; Kimura, S.; Satsukawa, Y.; Aoki, K.; Inada, A.; Kamijima, M. Exploring the anthropometric parameters for improving reliability of estimation model of pelvic incidence. In Proceedings of the 46th International Society for the Study of the Lumbar Spine Annual Meeting, Kyoto, Japan, 3–7 June 2019; p. 461.
23. Ramchandran, S. Pelvic Incidence (PI) is more easily understood as the pelvic base angle (PBA). *Spine Res.* **2017**, 3. [CrossRef]
24. Landis, J.R.; Koch, G.G. The measurement of observer agreement for categorical data. *Biometrics* **1977**, *33*, 159–174. [CrossRef]
25. Kendall, F.P.; McCreary, E.K.; Provance, P.G.; Rodgers, M.; Romani, W. *Muscles: Testing and Function with Posture and Pain*, 5th ed.; Williams & Wilkins: Baltimore, MD, USA, 2005.
26. Guimond, S.; Massrieh, W. Intricate correlation between body posture, personality trait and incidence of body pain: A cross-referential study report. *PLoS ONE* **2012**, *7*, e37450. [CrossRef]
27. Yamada, S.; Ebara, T.; Uehara, T.; Inada, A.; Kamijima, M. Can postural pattern assessment be used to estimate pelvic incidence?: Reliability and validity of simple classification tools for postural pattern assessment. *Jpn. J. Ergon.* **2021**, *57*, 288–293.
28. Day, B.L.; Steiger, M.J.; Thompson, P.D.; Marsden, C.D. Effect of vision and stance width on human body motion when standing: Implications for afferent control of lateral sway. *J. Physiol.* **1993**, *469*, 479–499. [CrossRef] [PubMed]
29. Czaprowski, D.; Stolinski, L.; Tyrakowski, M.; Kozinoga, M.; Kotwicki, T. Non-structural misalignments of body posture in the sagittal plane. *Scoliosis Spinal Disord.* **2018**, *13*, 6. [CrossRef] [PubMed]
30. Wu, G.; Siegler, S.; Allard, P.; Kirtley, C.; Leardini, A.; Rosenbaum, D.; Whittle, M.; D'Lima, D.D.; Cristofolini, L.; Witte, H.; et al. Standardization; Terminology Committee of the International Society of, B. ISB recommendation on definitions of joint coordinate system of various joints for the reporting of human joint motion—Part I: Ankle, hip, and spine. *Int. Soc. Biomech. J. Biomech* **2002**, *35*, 543–548. [CrossRef]
31. Savage, J.W.; Patel, A.A. Fixed sagittal plane imbalance. *Global. Spine J.* **2014**, *4*, 287–296. [CrossRef] [PubMed]
32. Schwab, F.; Lafage, V.; Boyce, R.; Skalli, W.; Farcy, J.P. Gravity line analysis in adult volunteers: Age-related correlation with spinal parameters, pelvic parameters, and foot position. *Spine* **2006**, *31*, e959–e967. [CrossRef]
33. Segi, N.; Nakashima, H.; Ando, K.; Kobayashi, K.; Seki, T.; Ishizuka, S.; Takegami, Y.; Machino, M.; Ito, S.; Koshimizu, H.; et al. Spinopelvic imbalance is associated with increased sway in the center of gravity: Validation of the "cone of economy" concept in healthy subjects. *Global. Spine J.* **2021**, 21925682211038897. [CrossRef] [PubMed]
34. Kanemura, T. Sagittal spino-pelvic alignment in an asymptomatic japanese population. *J. Spine Res.* **2011**, *2*, 52–58.
35. Endo, K.; Suzuki, H.; Nishimura, H.; Tanaka, H.; Shishido, T.; Yamamoto, K. Characteristics of sagittal spino-pelvic alignment in Japanese young adults. *Asian Spine J.* **2014**, *8*, 599–604. [CrossRef]

36. Lee, C.S.; Chung, S.S.; Kang, K.C.; Park, S.J.; Shin, S.K. Normal patterns of sagittal alignment of the spine in young adults radiological analysis in a Korean population. *Spine* **2011**, *36*, e1648–e1654. [CrossRef] [PubMed]
37. Roussouly, P.; Gollogly, S.; Berthonnaud, E.; Dimnet, J. Classification of the normal variation in the sagittal alignment of the human lumbar spine and pelvis in the standing position. *Spine* **2005**, *30*, 346–353. [CrossRef] [PubMed]
38. Vialle, R.; Levassor, N.; Rillardon, L.; Templier, A.; Skalli, W.; Guigui, P. Radiographic analysis of the sagittal alignment and balance of the spine in asymptomatic subjects. *J. Bone Joint. Surg. Am.* **2005**, *87*, 260–267. [CrossRef]
39. Jānis Šavlovskis, K.R. The Bony Pelvis & Gender Differences in Pelvic Anatomy. Available online: https://www.anatomystandard.com/Pelvis/Pelvis.html (accessed on 18 January 2022).
40. Rosa, F.; Piero, Q. Fleiss' kappa stastic wuthout paradoxes. *Qual. Quant.* **2014**, *49*, 463–470.
41. Handa, V.L.; Lockhart, M.E.; Fielding, J.R.; Bradley, C.S.; Brubaker, L.; Cundiff, G.W.; Ye, W.; Richter, H.E. Pelvic Floor Disorders, Network, Racial differences in pelvic anatomy by magnetic resonance imaging. *Obstet. Gynecol.* **2008**, *111*, 914–920. [CrossRef]
42. Branco, P.; Torgo, L.; Ribeiro, R. A survey of predictive modelling under imbalanced distributions. *ACM Comput. Surv.* **2017**, *49*, 1–50. [CrossRef]

International Journal of
Environmental Research and Public Health

MDPI

Article

The Prevalence and Severity of Sick Leave due to Low Back Disorders among Workers in Slovenia: Analysis of National Data across Gender, Age and Classification of Economic Activities

Dorjana Zerbo Šporin [1], Žiga Kozinc [1,2], Ticijana Prijon [3] and Nejc Šarabon [1,4,5,*]

1 Faculty of Health Sciences, University of Primorska, Polje 42, SI-6310 Izola, Slovenia; dorjana.zerbosporin@fvz.upr.si (D.Z.Š.); ziga.kozinc@fvz.upr.si (Ž.K.)
2 Andrej Marušič Institute, University of Primorska, Muzejski trg 2, SI-6000 Koper, Slovenia
3 National Institute of Public Health, Trubarjeva 2, SI-1000 Ljubljana, Slovenia; Ticijana.Prijon@nijz.si
4 Human Health Department, InnoRenew CoE, Livade 6, SI-6310 Izola, Slovenia
5 Laboratory for Motor Control and Motor Behavior, S2P, Science to Practice, Ltd., Tehnološki Park 19, SI-1000 Ljubljana, Slovenia
* Correspondence: nejc.sarabon@fvz.upr.si

Abstract: Musculoskeletal disorders are the most common work-related health problems. As low back disorders (LBD) are the most problematic, the aim of this study was to provide an in-depth analysis of the nationwide data on sick leaves due to work-related LBDs among workers in Slovenia in 2015–2019 by gender, age and various economic activities (NACE Rev 2 classification). We retrospectively analyzed the Slovene national data for sick leave (SL) rates due to the LBDs between 2015 and 2019. The analyzed SL outcomes were (i) index of temporary disability as a diagnosis-specific loss of calendar days (all calendar days except Sundays) per employee, (ii) frequency of spells as the number of SL cases per 100 employees in one year and (iii) severity as the average duration of one absence from work due to a health condition. A high prevalence of sick leaves due to LBDs in Slovenia was present among young male workers in "mining and quarrying". In the next age group (20.0–44.9 years), LBD is most frequent in "water supply; sewerage, waste management and remediation activities". Particular attention should be paid to ''agriculture, forestry and fishing" which shows a large average sick leave duration and probably a more demanding course of LBDs.

Keywords: musculoskeletal pain; absenteeism; epidemiology; workplace; back pain

Citation: Zerbo Šporin, D.; Kozinc, Ž.; Prijon, T.; Šarabon, N. The Prevalence and Severity of Sick Leave due to Low Back Disorders among Workers in Slovenia: Analysis of National Data across Gender, Age and Classification of Economic Activities. *IJERPH* **2022**, *19*, 131. https://doi.org/10.3390/ijerph19010131

Academic Editors: Vincenzo Denaro, Sergio Iavicoli, Fabrizio Russo and Gianluca Vadalà

Received: 13 November 2021
Accepted: 22 December 2021
Published: 23 December 2021

1. Introduction

Musculoskeletal disorders (MSDs) are the most common work-related health problem in the European Union (EU) [1]. According to the Centre for Disease Control and Prevention, work-related MSDs are injuries or disorders of the muscles, nerves, tendons, joints, cartilage and spinal discs in which (i) the work environment and performance of work contribute significantly to the condition and/or (ii) the condition is made worse or persists longer due to work conditions [2]. Various factors may contribute to MSDs, such as physical, organizational, psychosocial, sociodemographic and individual factors [1]. Psychosocial risks, especially in combination with physical risks, may cause or aggravate MSDs [3].

In 2015, three out of five workers in the EU-28 reported MSDs. Sixty-three percent of workers in Slovenia reported that they suffered from one or more MSDs in the past 12 months [1]. While the prevalence of musculoskeletal conditions increases with age, younger people are also affected [4]. In addition to the normal degenerative aging process, poor work environment and performance of work contribute significantly to the condition [2]. MSDs concern workers in all sectors and occupations, but a high prevalence of self-reported MSDs is most often reported by workers employed in the sector of construction, water supply and agriculture, forestry and fishing, and is above average among human health and social work activities [1]. MSDs are one of the most significant diseases contributing to work disability [5] and they lead to high costs to enterprises and society [1].

With an aim to reduce the incidence of MSDs and thus the economic and social consequence of MSDs and according to the objectives of the EU Strategic Framework of Health and Safety at Work 2014–2020 [6], in 2020, the project "Promotion of activities to prevent musculoskeletal disorders and psychosocial risk in the workplace (2020–2021)" (OP20.05955) in Slovenia was established. This national project that is jointly run by the National Institute of Public Health (lead partner) and the University of Primorska, Faculty of Health Science (project partner) implements a holistic approach to preventing and managing the incidence of work-related MSDs and psychosocial risk using a range of systematic activities: workplace ergonomics, physical activity and psychological interventions, all included in a friendly e-tool (www.pkmo.si, accessed on 11 November 2021).

In the project's first phase, a retrospective analysis of sick leave (SL) incidence due to the most common work-related MSDs was carried out based on health statistics. In the period 2015–2019, the average loss of calendar days due to work-related MSDs was 2.3 per employee, with eight annual cases per 100 employees. The average duration of one MSDs related SL was 33.5 days. Within MSDs, the highest percentage of SL was detected for work-related low back disorders (LBDs) (56.6% of all lost working days). In the observed period, the average loss of calendar days due to LBDs was 1.3 per employee with four annual cases per 100 employees. The average duration of one LBD-related sick leave (SL) was 30.7 days. The loss of calendar days per employee was higher in females than in males (1.53 and 1.38, respectively) and reached its peak (2.7) in the 45–64 age group [7,8].

LBDs are common in all age groups [9], with the lifetime prevalence as high as 75 to 84% [10] and point prevalence ranging from 1.4 to 20.0% across populations and geographical locations [11]. Risk factors for work-related musculoskeletal disorders are psychosocial, individual and biomechanical [12]. Although LBDs may occur as a result of specific pathologies, such as damage to the intervertebral disc, damage to the intervertebral joints or nerve impingements [13], in most cases, the underlying cause remains unknown [14,15]. Therefore, the prevention and treatment of LBDs remain a huge challenge [16,17]. Chronic pain syndromes have a severe negative impact on the patients and people around them, affecting their social life and overall life quality [18,19]. Medical conditions also represent a large burden for healthcare systems. For instance, reports from Sweden estimated the annual costs due to LBDs at 740 million euros [20]. Studies from around the world have reported that the majority of the costs related to LBDs are indirect (i.e., due to sick leaves and reduced work ability) [21,22]. In the working-age population, between 20 and 40% of persons suffer from LBDs annually [23]. Effective interventions for the prevention and rehabilitation of LBDs would likely unload the public healthcare systems across the world, in addition to helping the patients directly. Several studies have already shown a great potential of physical activity and exercise in the treatment and prevention of LBDs [24–26]. In addition to exercise, educating workers has also been suggested as an effective approach to preventing LBDs [27]. However, given the abovementioned impacts of LBDs on the individual, their family and society, further work needs to be done to prevent and treat LBDs more effectively.

As LBDs are evidently the most common work-related MSDs, the aim of this study was to provide an in-depth analysis of the nationwide data on sick leaves due to work-related LBDs among workers in Slovenia in 2015–2019 by gender, age and various economic activities. The scope was to identify and latterly present the standing out data that could be useful to decision-making in developing LBDs control strategies in Slovenia and globally.

2. Materials and Methods

2.1. Study Population and Data Collection

We retrospectively analyzed the Slovene national data for sick leave (SL) rates due to the most common work-related low back disorders (LBDs) in relation to gender, four age groups (15–19.9 years, 20–44.9 years, 45–64.9 years and 65+ years) and following the incidence of LBDs across different economic activities according to the NACE Rev 2 classification [28]. Table 1 shows the study sample (i.e., the number of employed personnel across calendar years for both genders). In Table 2, the data is further broken down by age groups, gender and economic activities. Persons from the group over 65 years in our study are still employed. The right to an old age pension in Slovenia is regulated by the Pension and Disability Insurance Act. The law lays down the conditions under which a person who satisfies them joins compulsory pension. In order to qualify for an old age pension, one must meet the following general conditions: (i) 60 years of age and 40 years of pensionable service without purchase or (ii) 65 years of age and a 15-year insurance period [29]. It is not mandatory that the employee must also terminate the employment relationship under these conditions. An employment contract may be concluded by persons who have reached the age of 15 [30]. Within young workers (15.0–19.9 years and 20.0–44.9 years), persons performing student work are compulsorily insured [29].

Table 1. Active workers included in the study by gender, age and calendar year.

	Male	Female	All
2015	443,641	360,996	804,637
2016	446,863	370,346	817,209
2017	463,451	382,003	845,454
2018	478,148	394,624	872,772
2019	492,475	401,754	894,229

Data regarding SL represent an important source of information on the health status of the working population. The purpose of data collection is to monitor and analyze the pattern of sick leave or temporary absence from work as a result of disease, injury or other health conditions. The National Institute for Public Health (NIJZ) is the institution responsible for collecting official data on SL in the working population in which employed and self-employed persons working in Slovenia are included (Table 1). Data collection has a legal basis in the Health Care Databases Act (ZZPPZ—Ur. l. RS 65/00, database NIJZ3) and Personal Data Protection act (ZVOP-1—Ur. l. RS 94/07), which gives the basis for further processing of data for scientific research or historical or statistical purposes. The source of data is the Certificate of justified abstinence from work due to health conditions (eBOL). The data is obtained from health care providers regarding the decisions of personal general practitioners who ensure the need of absence from work. After the 30th day of SL, the decision must be confirmed by an appointed doctor from the Health Insurance Institute of Slovenia [31]. All data were anonymous at all stages of the study.

2.2. Classification of Economic Activities

The study population was segregated according to the NACE Rev 2—Statistical classification of economic activities in the European Community [28]. According to this classification, the activities are separated into 21 categories: (A) Agriculture, forestry and fishing, (B) Mining and quarrying, (C) Manufacturing, (D) Electricity, gas, steam and air conditioning supply, (E) Water supply; sewerage, waste management and remediation activities, (F) Construction, (G) Wholesale and retail trade; repair of motor vehicles and motorcycles, (H) Transportation and storage, (I) Accommodation and food service activities, (J) Information and communication, (K) Financial and insurance activities, (L) Real estate activities, (M) Professional, scientific and technical activities, (N) Administrative and support service activities, (O) Public administration and defense; compulsory social security, (P) Education, (Q) Human health and social work activities, (R) Arts, entertainment and recreation, (S) Other service activities, (T) Activities of households as employers; undifferentiated goods- and services-producing activities of households for own use and (U) Activities of extraterritorial organizations and bodies.

Table 2. The average size of the study population by age, gender and economic activities from 2015 to 2019.

Age gr.	15–19.9			20–44.9			45–64.9			65+		
Ec. Act.	**M**	**F**	**TOTAL**	**M**	**F**	**TOTAL**	**M**	**F**	**TOTAL**	**M**	**F**	**TOTAL**
A—agriculture, forestry and fishing	43	2	45	4629	2054	6683	5759	3474	9233	35	8	43
B—mining and quarrying	12	0	12	1238	178	1416	800	164	965	6	1	7
C—manufacturing	755	90	845	79,124	32,639	111,763	55,898	30,735	86,633	399	58	457
D—electricity, gas, steam, air cond. supply	5	0	5	3017	814	3831	3128	791	3920	49	6	55
E—water suppl; sewer., wst. manag., remed. act	17	1	18	3562	1014	4576	3816	1025	4841	21	2	23
F—construction	371	12	383	33,094	3211	36,305	22,196	2085	24,281	191	6	197
G—wholesale, retail; repair of mot. vehicles	237	110	347	33,306	36,835	70,140	19,618	20,971	40,589	252	87	339
H—transportation and storage	73	6	79	24,050	4898	28,948	18,268	4401	2269	236	9	245
I—accommodation and food ser. activities	122	114	236	9819	13,389	23,157	4401	7881	12,283	73	56	129
J—information and communication	8	2	10	12,572	5810	18,383	5477	2796	8272	76	21	97
K—financial and insurance activities	10	0	1	4483	7140	11,622	3385	6430	9816	27	12	39
L—real estate activities	5	1	6	1558	1063	2620	2367	1087	3454	40	9	49
M—professional, scientific, technical act	31	10	41	17,762	16,527	34,289	11,552	9365	20,917	449	144	593
N—administrative and support service act;	192	67	259	12,806	10,487	23,293	7418	6702	14,120	58	22	80
O—public admin, defense	11	2	13	11,872	13,080	24,952	11,286	13,528	24,815	169	69	239
P—education	15	18	33	8604	30,541	39,145	8151	27,914	36,065	305	125	430
Q—human health and social work; activities;	65	140	204	7304	27,087	34,391	4543	23,215	27,757	333	233	566
R—arts, entertainment and recreation	67	5	72	4429	4653	9082	3376	3169	6544	92	28	120
S—other service activities	26	93	119	2715	6938	9625	2033	3735	5766	63	59	121

M—male; F—female; A—agriculture, forestry and fishing; B—mining and quarrying; C—manufacturing; D—electricity, gas, steam, air cond. supply; E—water suppl; sewer., wst. manag., remed. act; F—construction; G—wholesale, retail; repair of mot. vehicles; H—transportation and storage; I—accommodation and food ser. activities; J—information and communication; K—financial and insurance activities; L—real estate activities; M—professional, scientific, technical act.; N—administrative and support service act; O—public admin, defense; compulsory soc. sec.; P—education; Q—human health and social work; activities; R—arts, entertainment and recreation; S—other service activities.

2.3. Outcome Measures

For the analysis, we received anonymous data as numbers representing SL rates by gender, age and NACE Rev 2 classification of economic activities. The analyzed SL rates were: (i) index of temporary disability as a diagnosis-specific loss of calendar days (all calendar days except Sundays) per employee, (ii) frequency of spells as the number of SL cases per 100 employees in one year and (iii) severity as the average duration of one absence from work due to a health condition. The number of SL cases is considered to be the number of completed SL duo to LBDs in a calendar year (1 January–31 December) regardless of when SL started [31,32]. The average values for SL rates from 2015 till 2019 were used for analysis.

According to the MKB-10-AM, diagnoses for the most common work-related LBDs are classified as follows: (i) lumbar spine intervertebral disc defects (M51.0–M51.9), (ii) ischialgia (M54.3), (iii) lumboischialgia (M54.4) and (iv) lumbalgia (M54.5) [8].

3. Results

The overall results for days of absence, number of cases and average case duration because of sick leave due to low back disorders are presented separated by age, gender and NACE Rev 2 classification of economic activities, in Table S1. For clarity, we separated the graphical results by age groups.

Figure 1 presents the data for the 15–19.9 age group. Regarding the loss of calendar days, males working in "mining and quarrying" (B) and "public administration and defence; compulsory social security" (O) activities presented by far the largest numbers (2.70 and 1.81 days per individual, respectively). For the males in other activities, as well as females in all activities, much lower values were documented (all < 0.5). The males in the abovementioned activities (B and O) also had the highest SL frequency (6.5 and 5.5 per 100 persons, respectively). Additionally, both males (4.4) and females (2.4) in "administrative and support service activities (N)" also presented with high numbers of annual cases of sick leaves. In the remaining activities, the number of annual cases was ≤2.0. Finally, the average severity of the sick leave was also the highest in the previously mentioned activities (B and O) in males (42.0 and 33.3, respectively). For the remaining groups that had documented cases, the severity of sick leaves was 4.5–11.3 days. Note that some activities did not have any documented cases in this age group (refer to the middle section of Figure 1, depicting the number of cases, and Table S1 for details).

Figure 2 presents the data for the 20–44.9 age group. Regarding the sick leave days, "the male "mining and quarrying" (B) group stood out, with 2.57 days per person. The results for the remaining activities ranged from 0.25 to 1.35 in males and from 0.35 to 1.30 in females. It seems that the groups that had no cases in the 15.0–19.9 group tended to be less affected in this group as well (i.e., activities "agriculture, forestry and fishing" (A), "electricity, gas, steam and air conditioning supply" (D), "information and communication" (J), "financial and insurance activities" and (K) and "real estate activities" (L)). Males in "mining and quarrying" (B) again had the highest number of cases (8.2 cases per 100 persons). Moreover, both genders in "administrative and support service activities" (N) and "public administration and defence; compulsory social security" (O) also had high numbers of cases (5.5 and 7.1 for males, and 4.3 and 5.6 for females). Another group with a high number of cases were males in "water supply; sewerage, waste management and remediation activities" (E), with 5.5 cases per 100 persons. Interestingly, the average duration of the sick leave was the highest in "agriculture, forestry and fishing" (A) (43.8 days for males and 49.9 days for females). Females tended to be on sick leave for a slightly longer time in most groups. No other group stood out with a particularly low or high average duration of sick leave compared to others, although the range of data was fairly wide (13.8–31.2 days for males and 15.7–37.2 days for females).

Figure 3 presents the data for the 45.0–64.9 age group. As in the two previous age groups, males in "mining and quarrying" (B) had the most sick leave days (5.0 days per person). However, contrary to the previous age groups, the next highest scores were present in 4 groups for females: "manufacturing" (C), "accommodation and food service activities" (I), "administrative and support service activities" (N) and "human health and social work activities" (Q) (3.3, 3.1, 3.5 and 2.9 days per person, respectively). Several groups presented with a high number of cases, often exceeding 5.0 per 100 persons. Three groups presented with a notably lower number of cases in both genders: "agriculture, forestry and fishing" (A) (2.4 and 3.1 cases for males and females, respectively), "professional, scientific and technical activities" (M) (1.8 and 2.1 cases for males and females, respectively) and "other service activities" (S) (1.6 and 2.4 cases for males and females, respectively). As in the previous age group, the average duration of sick leave was the highest in "agriculture, forestry and fishing" (A) (86.3 days for males and 82.4 days for females). While the average duration of sick leave was fairly similar across other groups, somewhat higher numbers were documented for both genders in "construction" (F) (50.6 days for males, 61.2 days for females), "accommodation and food service activities" (I) (55.2 days for males, 53.6 days for females), "other service activities" (S) (56.5 days for males, 65.2 days for females) as well as in "mining and quarrying" (B) for males (58.1 days).

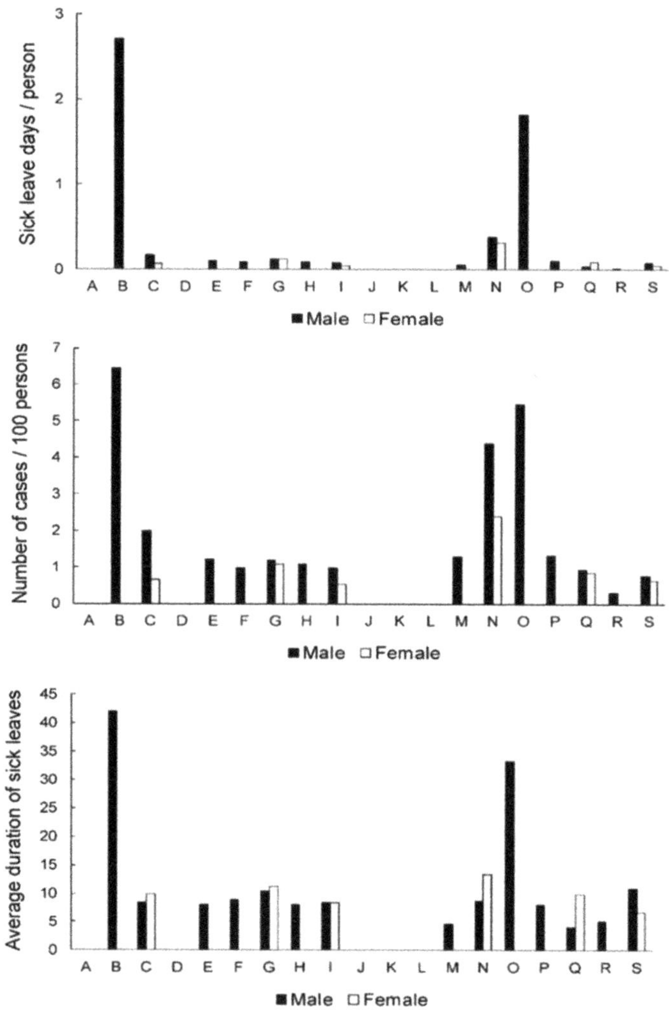

Figure 1. Sick leave days per person (index of temporary disability,) frequency of cases and severity (average case duration) of sick leave across genders and activities in 15.0–19.9-year-old group. A—agriculture, forestry and fishing; B—mining and quarrying; C—manufacturing; D —electricity, gas, steam, air cond. supply; E—water suppl; sewer., wst. manag., remed. act; F—construction; G—wholesale, retail; repair of mot. vehicles; H—transportation and storage; I—accommodation and food ser. activities; J—information and communication; K—financial and insurance activities; L—real estate activities; M—professional, scientific, technical act.; N—administrative and support service act; O—public admin, defense; compulsory soc. sec.; P—education; Q—human health and social work; activities; R—arts, entertainment and recreation; S—other service activities.

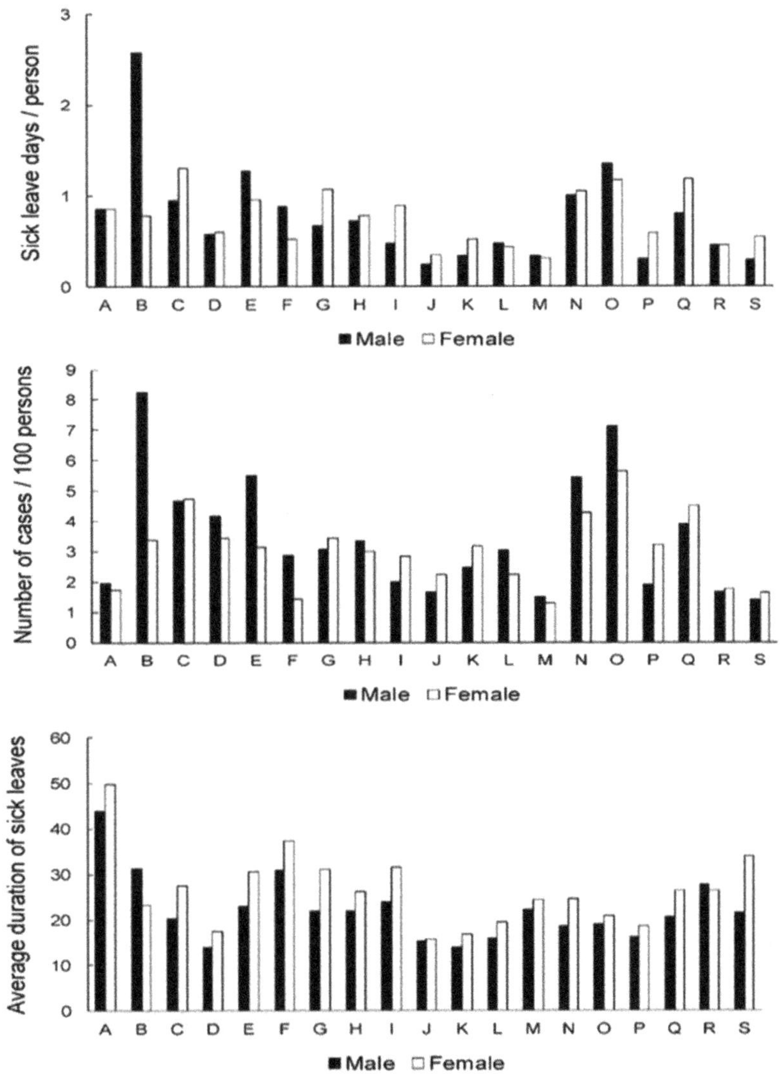

Figure 2. Sick leave days per person (index of temporary disability,) frequency of cases and severity (average case duration) of sick leave across genders and activities in 20.0–44.9-year-old group. A—agriculture, forestry and fishing; B—mining and quarrying; C—manufacturing; D —electricity, gas, steam, air cond. supply; E—water suppl; sewer., wst. manag., remed. act; F—construction; G—wholesale, retail; repair of mot. vehicles; H—transportation and storage; I—accommodation and food ser. activities; J—information and communication; K—financial and insurance activities; L—real estate activities; M—professional, scientific, technical act.; N—administrative and support service act; O—public admin, defense; compulsory soc. sec.; P—education; Q—human health and social work; activities; R—arts, entertainment and recreation; S—other service activities.

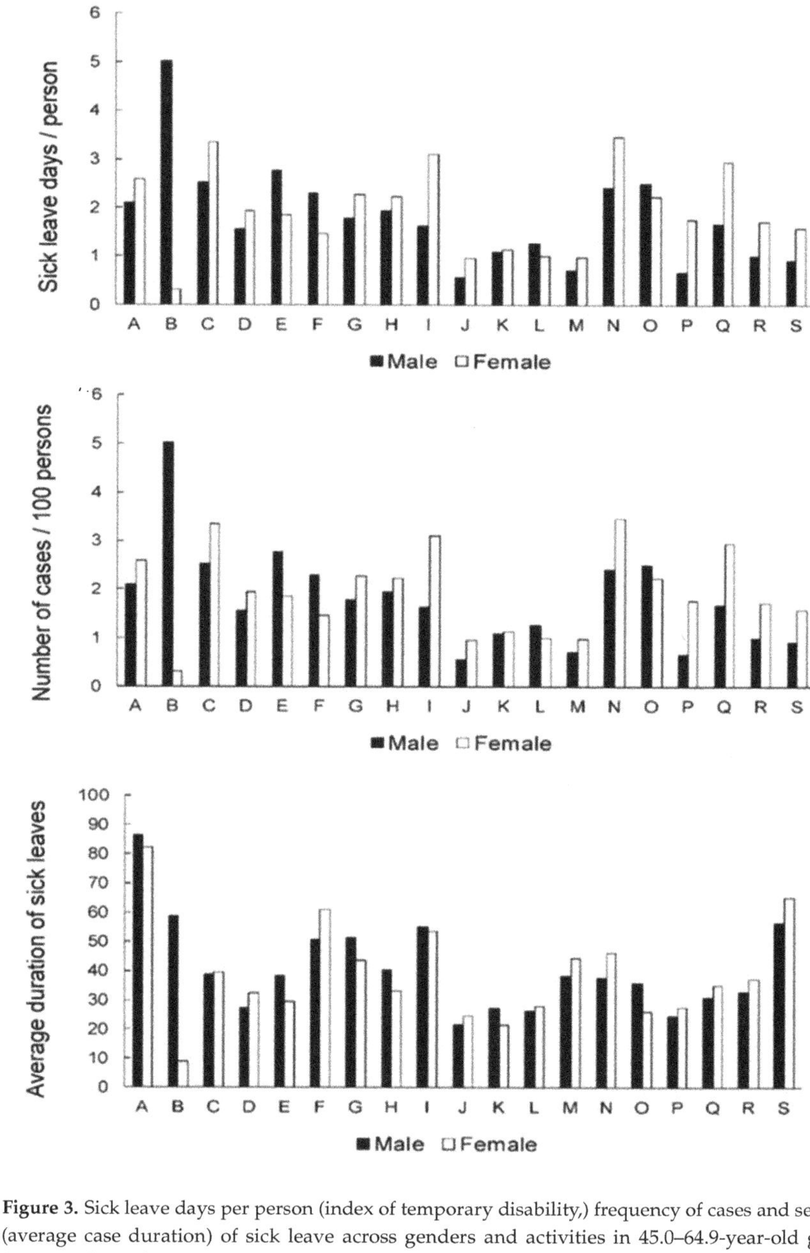

Figure 3. Sick leave days per person (index of temporary disability,) frequency of cases and severity (average case duration) of sick leave across genders and activities in 45.0–64.9-year-old group. A—agriculture, forestry and fishing; B—mining and quarrying; C—manufacturing; D —electricity, gas, steam, air cond. supply; E—water suppl; sewer., wst. manag., remed. act; F—construction; G—wholesale, retail; repair of mot. vehicles; H—transportation and storage; I—accommodation and food ser. activities; J—information and communication; K—financial and insurance activities; L—real estate activities; M—professional, scientific, technical act.; N—administrative and support service act; O—public admin, defense; compulsory soc. sec.; P—education; Q—human health and social work; activities; R—arts, entertainment and recreation; S—other service activities.

Figure 4 presents the data for the >65 age group. In this group, the females in "administrative and support service activities" (N) presented with particularly high numbers of sick leave days (15.2 days per person). In addition, high numbers were observed for females in "construction" (F) (4.2 days per person) and the males in "administrative and support service activities" (N) and "agriculture, forestry and fishing" (A) (5.1 and 4.6 days per person, respectively). The frequency of cases was more comparable across activities, with the highest number of cases for both genders reported for "public administration and defence; compulsory social security" (O) (5.8 and 5.2 cases per 100 persons for males and females, respectively). An extremely high average of severity of sick leave was observed for males in "agriculture, forestry and fishing" (A) and females in "administrative and support service activities" (N) (401.0 and 411.5 days, respectively).

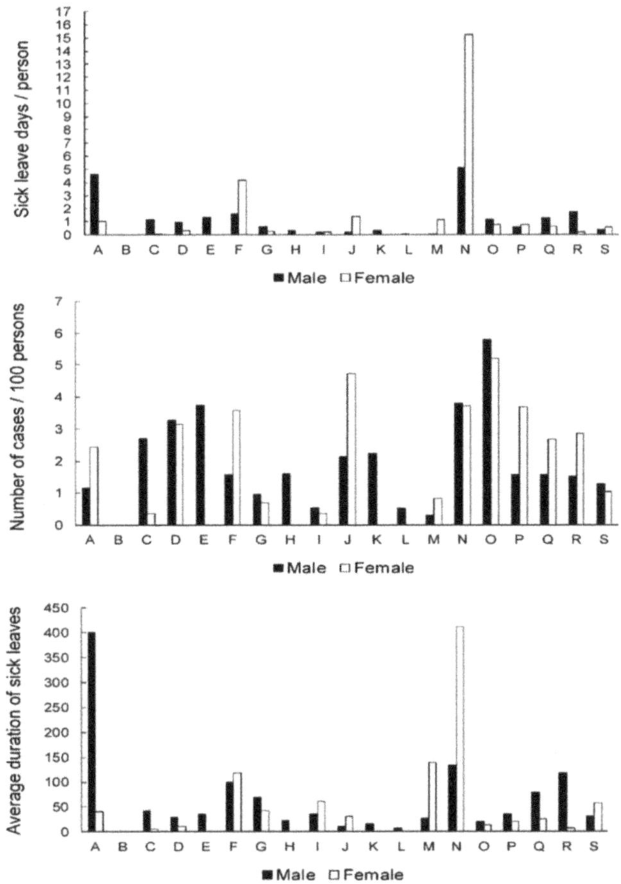

Figure 4. Sick leave days per person (index of temporary disability,) frequency of cases and severity (average case duration) of sick leave across genders and activities in >65-year-old group. A—agriculture, forestry and fishing; B—mining and quarrying; C—manufacturing; D —electricity, gas, steam, air cond. supply; E—water suppl; sewer., wst. manag., remed. act; F—construction; G—wholesale, retail; repair of mot. vehicles; H—transportation and storage; I—accommodation and food ser. activities; J—information and communication; K—financial and insurance activities; L—real estate activities; M—professional, scientific, technical act.; N—administrative and support service act; O—public admin, defense; compulsory soc. sec.; P—education; Q—human health and social work; activities; R—arts, entertainment and recreation; S—other service activities.

Figure 5 presents the cumulative data across all economic activities. Sick leave days and frequency of cases seem to show very similar patterns, with the 45.0–64.9 years old group being affected the most and the 15.0–19.0 group being affected the least. However, in terms of the duration of sick leave, there is a clear increasing trend with age. There seem to be very few differences between sexes. Males in the older two groups tend to have slightly longer average sick leave, while the opposite is the case in the younger two groups.

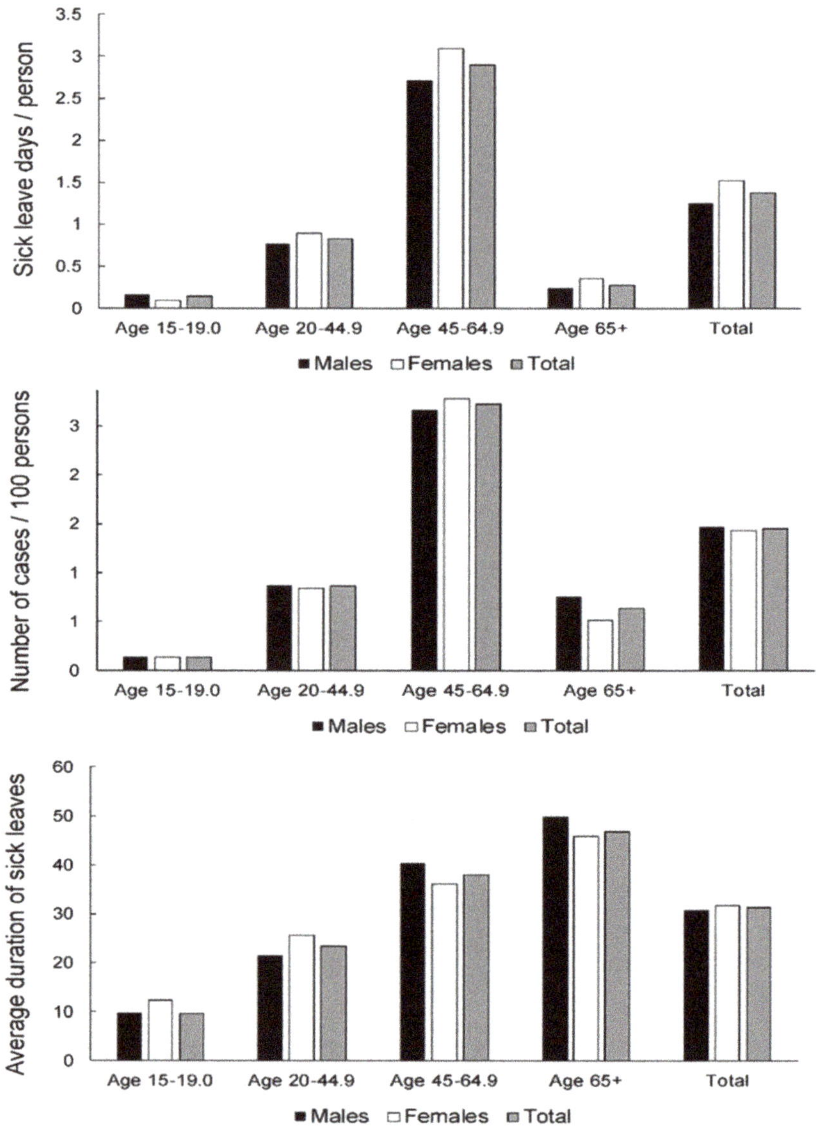

Figure 5. Sick leave days per person (index of temporary disability,) frequency of cases and severity (average case duration) of sick leave across genders and age groups across all economic activities.

4. Discussion

Musculoskeletal disorders are the most common work-related health problems in the EU [1] and are one of the most significant diseases contributing to work disability [5]. In addition to aging [4], poor working environments are also responsible for the occurrence of MSDs [2]. Low back disorders are the most predominant MSDs over the world [23,33] and in Slovenia as well [7,8]. The aim of this study was to provide an in-depth analysis of the nationwide data on sick leave rates due to work-related LBDs in 2015–2019 by gender, age and various economic activities.

Among young employees (15.0–19.9 years), males working in "mining and quarrying" and "public administration and defence; compulsory social security" activities presented by far the largest numbers for LBDs specific loss of calendar days per employee and the number of cases per 100 employees (Figure 1). These activities also show the longest average duration of LBD-related sick leave. This is understandable, given that in other activities of this age group, there are little or no cases of sick leave. However, it should be pointed out that the number of young male workers in "mining and quarrying" and "public administration and defence; compulsory social security" is very low (Table 2); therefore, the sick leave duration in these activities is largely determined by a small number of individual cases. "Public administration and defence; compulsory social security" includes activities of a governmental nature, normally carried out by the public administration. It covers legislative activities, taxation, national defense, public order and safety, immigration services, foreign affairs and the administration of government programs [28]. In this activity, several physical and psychosocial risk factors for MSDs have been reported [34].

According to the larger number of young employees in "administrative and support service activities" compared to, for example, "mining and quarrying" (Table 2), the presented higher frequency of cases (Figure 1 is somewhat contradictory regarding LBDs cases among young workers in this activity. "Administrative and support services" are covering activities that support general business [28]. According to LBDs in sub-activities, we can find those that involve physically demanding work, e.g., cleaning activities, those linked to sedentary working environment e.g., office administrative support and those which could be connected to an elevated psychosocial risk e.g., travel agency and private security activities. Interestingly, "administrative and support service activities" remain endangered for LBDs in all age groups (Figures 2–4), but the prevalence and severity of sick leave seem to be increasing in females during aging. We advise finding out what are the risk factors for LBDs in "administrative and support service activities" starting from the very beginning in young employees.

Returning to "mining and quarrying," accompanied by unfavorable sick leave rates due to LBDs in males for almost all age groups, we might assume that demanding physical work increases the risk for LBDs in this activity. The high incidence of low back pain among miners is believed to be due to frequent awkward postures, manual handling and other heavy work and exposure to vibrations in the working environment [35,36]. Because of the demanding physical work, miners in Slovenia are included in occupational insurance. Under the terms of Pension and Disability Insurance Act ZPIZ-2 legislation [29], occupational insurance includes insured persons who perform particularly difficult and unhealthy work and persons who perform work that cannot be successfully performed professionally after a certain age. On the basis of collected funds, those insured acquire the right to count the insurance period with an added period, meaning to acquire the right of earlier retirement. Therefore, at age 65+ we can find no sick leave cases (Figure 4) in "mining and quarrying," which means that sick leave cases are not documented but LBDs in elderly miners probably persist.

Compared to the younger (15.0–19.0 age) group, sick leave rates for LBDs in workers from the 20.0–44.9 age group are expressed in all NACE Rev2 economic activities (Figure 2). In addition to the previously mentioned activities (that have evidently high SL rates for LBDs in almost all age groups), an increased frequency of cases for males in "water supply; sewerage, waste management and remediation activities" occurs in the 20.0–44.9 age group. Studies related to waste management reported forced postures in waste collection movements, mainly in lifting and unloading tasks [37]. In this age group, as in 45.0–64.9 years old (Figure 3), we found interesting the large increase in the average duration of sick leave due to LBDs in both genders for ''agriculture, forestry and fishing" in contrast to the relatively low number of recorded LBDs cases in this activity (Figure 2). From the Republic of Slovenia statistical office (SiStat) data for December 2019 [38], it is evident that a large proportion (82.3%) of the 24.364 employees in ''agriculture, forestry and fishing" are self-employed. It could be assumed that due to the socio-economic circumstances arising from their form of employment, fewer workers utilize sick leave for LBDs or only when the disease is already advanced, which is reflected in a low frequency but a high duration of sick leave (Figure 2). A similar deviation (low frequency, but long duration of sick leaves) (Figure 2) was found in "construction," a primarily physically demanding working environment; however, a relatively lower percentage (37%) of people were self-employed in this activity [38]. Workers in "agriculture, forestry, and fishing" are exposed to MSDs risk factors such as repetitive movements, lifts, vibrations and awkward postures [39].

Female workers in the 20.0–44.9 age group with LBDs mostly remain on sick leave longer than males (Figure 2). This observation could reflect a more demanding course of the disease in female workers at this age. Female gender was identified as a predictor for longer sick leave in acute low back pain [40]. This was not the case in our study at age 45.0–64.9 years in "manufacturing" and in "human health and social work activities," where an increase in the frequency of sick leaves due to LBDs occurs mainly among females, but the average duration of sick leaves remains comparable between genders (Figure 3). It seems that 45.0–64.9-year-old females in these activities are more susceptible to LBDs than age-matched males but do not show a more severe course of the disease. In "human health and social work activities," risk factors for LBDs could be linked to physically demanding work, e.g., handling patients and psychosocial risk factors, which are both reported frequently in the healthcare sector [41]. In "manufacturing," forceful physical activity and material handling tasks represent the main risk [42].

The average sick leave episode duration due to LBDs increases for both genders in this group as well as in "accommodation and food service activities and "other service activities". "Accommodation and food service activities" are particularly demanding due to physical factors, e.g., manual handling, repetitions, awkward and static postures and psychosocial risk factors [43].

In the transition between different age groups, we found that activities with no sick leave cases—''information and communication," "financial and insurance activities" and "real estate activities" (Figure 1)—in younger workers tended to be less affected in all age groups (Figures 2–4), with the exception of the number of sick leave cases for ''information and communication" and "financial and insurance activities" in the oldest group (Figure 4); however, these data are to be interpreted with caution due to the small number of employees (Table 2). Although there is an expected increase of musculoskeletal condition with age [4], this does not appear to be reflected in sick leave due to LBDs within the aforementioned activities. It would make sense to follow these activities to identify good practices that inhibit the growth of LBDs during workers aging and translate them into more critical activities.

In the oldest group (65+ age), we found some outstanding sick leave rates for LBDs in economic activities such as "administrative and support service activities," "construction," "agriculture, forestry and fishing," and "public administration and defence; compulsory social security" (Figure 4), but we focused on SL rates related to LBDs only for economic activities in which more than 300 workers remain employed. Among such activities are manufacturing; wholesale and retail trade and repair of motor vehicles and motorcycles; professional, scientific, technical activities; education; and human health and social work activities (Table 2). The highest frequency due to LBDs was found for females in education and human health and social work activities. The reason is most likely a higher number of females compared to males who remain employed in these activities even after the age of 65 (Table 2) and in connection with the expected increase in musculoskeletal condition with age [4] and the duration of exposure to the working environment [2].

Comparisons to previous studies are difficult because this is the first nationwide data analysis on LBDs that used NACE Rev 2 classification. A large 2010 survey in the Netherlands identified the following groups to have an increase of incurring LBDs: male healthcare practitioners, female and younger healthcare support workers and female farming, fishing and forestry workers [44]. While some previous studies have indicated possible sex differences [23], our study shows that (a) the differences in LBD prevalence between sexes are negligible in the overall population and (b) large differences between the sexes may be seen in certain economic activities, which deserves further attention. Other research has been focusing on specific risk factors rather than individual economic activities. These include psychosocial, individual and biomechanical factors. For instance, load lifting, trunk bending and awkward postures have been associated with LBDs [45,46]. It can be speculated that performing such tasks heavily contributes to LBD incidence and severity in certain activities, as observed in this study (e.g., agriculture, forestry and fishing). Moreover, studies have linked LBDs to several psychosocial variables, such as job satisfaction, support in the workplace, job freedom and overtime work [47,48], which can be important in any economic activity. While risk factors for LBDs seem to be well documented, this study reveals which economic activities seem to be the most problematic and deserve further attention in terms of extensive research and practical implementation.

We must consider that individual NACE Rev2 economic activities cover different sub-activities that are not necessarily similar in type of work performed. Nevertheless, the results of our study can help decision-makers target the proper economic activity when developing strategies for reducing LBDs. The main aim of our study was not to provide a list of risk factors but to identify the most problematic economic activates that deserve immediate attention. Within the individual activities, it could be useful to identify the most exposed sub-activities in the future. However, we did not have available data to determine this in our study.

It is known from the existing literature that a. higher prevalence of LBDs is associated with a physically demanding or sedentary type of work, especially if psychosocial risk factors are also present. The added value of our study is the magnitude of the data regarding the sick leave due to LBDs as we used nationwide data to analyze LBDs in various economic activities. The results of our study are useful in developing LBD control strategies in Slovenia and globally focusing on individual economic activities.

Limitations of the Study

The main limitation of the study is that we do not know the exact number of workers working at particular types of workplaces within an individual NACE Rev2 economic activity. For example, in "mining and quarrying," we do not know the exact number of miners as this activity includes all the workers who work in mining and quarrying, including, for example, office workers in mining companies.

Another limitation is that we do not know when the retirement of workers occurs. Worse sick leave rates in a particular activity for older workers could indicate that they retire later, thus having more LBDs than those already retired from other activities.

5. Conclusions

The purpose of the present study was to provide national data on sick leaves related to LBDs in order to identify economic activities that deserve the most attention in the future. A high prevalence of sick leave due to LBDs in Slovenia was present among young male workers in "mining and quarrying". They are joint in the next age group (20.0–44.9 years) in the frequent occurrence of LBDs by men from "water supply; sewerage, waste management and remediation activities". In the activities "administrative and support service activities," a higher frequency in LBD-related sick leave is already manifested among younger employees and later followed by a longer average absence from work, especially in females. To better understand this phenomenon, further research is needed to identify biopsychological influences on the incidence of LBDs in "administrative and support service activities".

Particular attention should be paid to "agriculture, forestry and fishing," which showed a large average sick leave duration and thus probably a more demanding course of LBDs already in the case of young workers (20.0–44.9 years) of both genders. A similar phenomenon is seen in "construction" activities.

We also advise the identification of good practices in activities in Slovenia that are less affected by LBDs: "information and communication," "financial and insurance activities" and "real estate activities" and transferring them into more affected activities. According to gender, we have to pay attention to the longer average sick leave duration due to LBDs in younger females (20.0–44.9 years) and to the higher LBDs prevalence in females from "manufacturing" and "human health and social work activities".

Supplementary Materials: The following supporting information can be downloaded at: https://www.mdpi.com/article/10.3390/ijerph19010131/s1; Table S1. The overall results for days of absence, number of cases and average case duration because of sick leave due to low back disorders are presented separated by age, gender and NACE Rev 2 classification of economic activities.

Author Contributions: Conceptualization, all authors; methodology, T.P. software, T.P.; validation, all authors.; formal analysis, all authors.; investigation, all authors, resources, D.Z.Š. and T.P., data curation, D.Z.Š. and Ž.K.; writing—original draft preparation, D.Z.Š. and Ž.K.; writing—review and editing, T.P. and N.Š.; visualization, all authors.; supervision, T.P. and N.Š.; project administration, D.Z.Š. and T.P.; funding acquisition, D.Z.Š. and T.P. All authors have read and agreed to the published version of the manuscript.

Funding: The project "Promotion of activities for the prevention of musculoskeletal disorders and psychosocial risks at work" (OP20.05955) (hereinafter: the project) contribute to reduce the incidence of musculoskeletal disorders and psychosocial risks at work in Slovenia. The project is funded by the European Union from the European Social Fund and the Republic of Slovenia, the Ministry of Labor, Family, Social Affairs and Equal Opportunities. The project is implemented under the Operational Program for the Implementation of European Cohesion Policy 2014–2020, under Priority Axis 8: "Promoting employment and transnational labor mobility", 8.3 Priority investments: "Active and healthy aging", 8.3.1 the specific objective: "Extending and improving the work activity of older people involved in the measures".

Institutional Review Board Statement: The distribution of the data in this paper was approved by the Ethical Committee of the National Institute of Public Health (Approval number: 6310-1/2021-35 (241)).

Informed Consent Statement: Patient consent was waived, as National Institute of Public Health of Slovenia collects, analyses and disseminates data on sick leave of employed and self-employed workers which are enlisted in compulsory health insurance in Slovenia. In our study the data was anonymous in all stages, we operated only with mean/frequency numbers. No patient information was used at any stage.

Data Availability Statement: All data is available in the paper and within the Supplementary Materials.

Conflicts of Interest: The authors declare no conflict of interest.

References

1. De Kok, J.; VroonHhof, P.; Snijders, J.; Roullis, G.; Clark, M.; Peerebom, K.; van Dorst, P.; Isusi, I. European Agency for Safety and Health at Work: Work-related musculoskeletal disorders: Prevalence, costs and demographics in the EU European Risk Observatory Report. Available online: https://osha.europa.eu/en/publications/summary-msds-facts-and-figures-overview-prevalence-costs-and-demographics-msds-europe/view (accessed on 10 November 2021).
2. CDC Work-Related Musculoskeletal Disorders & Ergonomics. Available online: https://www.cdc.gov/workplacehealthpromotion/health-strategies/musculoskeletal-disorders/index.html (accessed on 8 November 2021).
3. EU OSHA Discussion Paper: Return to Work after MSD-Related Sick Leave in the Context of Psychosocial Risk at Work. Available online: https://osha.europa.eu/en/publications/return-work-after-msd-related-sick-leave-context-psychosocial-risks-work/view (accessed on 2 November 2021).
4. WHO Musculoskeletal Conditions. Available online: https://www.who.int/news-room/fact-sheets/detail/musculoskeletal-conditions (accessed on 2 November 2021).
5. Shannan, E.M. Work disabilities and musculoskeletal disease. *Int. J. Rheum. Dis.* **2019**, *22*, 965–966. [CrossRef] [PubMed]

6. European Commission. Communication from the Commission to the European Parliament, the Council, the European Economic and Social Committee and the Committee of the Regions on a EU Strategic Framework on Health and Safety at Work 2014–2020. Available online: https://eur-lex.europa.eu/legal-content/EN/ALL/?uri=CELEX:52014DC0332 (accessed on 2 November 2021).
7. Prijon, T. Najpogostejša z Delom Povezana Kostno-Mišična Obolenja po Anatomskih Regijah. Primerjalna Analiza Zdravstvenega Absentizma v Letih 2015 in 2019. Available online: https://nijz.si/sites/www.nijz.si/files/uploaded/pkmo_najpogostejsa-kmo-po-anatomskih-regijah.pdf (accessed on 2 November 2021).
8. Prijon, T. Zdravstveni Absentizem Zaradi z Delom Povezanih Kostno-Mišičnih Obolenj in Duševnih Stresnih Motenj v Sloveniji. Primerjalna Analiza Začasne Nezmožnosti za Delo v Letih 2015 in 2019 v Okviru Projekta "Promocija Aktivnosti za Preprečevanje Kostno-Mišičnih. Available online: https://nijz.si/sites/www.nijz.si/files/uploaded/pkmo_analiza_bs_zaradi_z_delom_povezanih_kmo_in_dusevnih-stresnih_motenj.pdf (accessed on 10 November 2021).
9. Hartvigsen, J.; Hancock, M.; Kongsted, A.; Louw, Q.; Ferreira, M.L.; Genevay, S.; Hoy, D.; Karppinen, J.; Pransky, G.; Sieper, J.; et al. What low back pain is and why we need to pay attention. *Lancet* **2018**, *391*, 2356–2367. [CrossRef]
10. Thiese, M.S.; Hegmann, K.T.; Wood, E.M.; Garg, A.; Moore, J.S.; Kapellusch, J.; Foster, J.; Ott, U. Prevalence of low back pain by anatomic location and intensity in an occupational population. *BMC Musculoskelet. Disord.* **2014**, *15*, 1–11. [CrossRef]
11. Fatoye, F.; Gebrye, T.; Odeyemi, I. Real-world incidence and prevalence of low back pain using routinely collected data. *Rheumatol. Int.* **2019**, *39*, 619–626. [CrossRef]
12. Da Costa, B.R.; Vieira, E.R. Risk factors for work-related musculoskeletal disorders: A systematic review of recent longitudinal studies. *Am. J. Ind. Med.* **2009**, *53*, 285–323. [CrossRef] [PubMed]
13. Petersen, T.; Laslett, M.; Juhl, C.B. Clinical classification in low back pain: Best-evidence diagnostic rules based on systematic reviews. *BMC Musculoskelet. Disord.* **2017**, *18*, 1–23. [CrossRef]
14. Borczuk, P. An evidence-based approach to the evaluation and treatment of low back pain in the emergency department. *Emerg. Med. Pr.* **2013**, *15*, 1–23.
15. Buchbinder, R.; van Tulder, M.; Öberg, B.; Costa, L.M.; Woolf, A.; Schoene, M.; Croft, P.; Lancet Low Back Pain Series Working Group. Low back pain: A call for action. *Lancet* **2018**, *391*, 2384–2388. [CrossRef]
16. Clark, S.; Horton, R. Low back pain: A major global challenge. *Lancet* **2018**, *391*, 2302. [CrossRef]
17. Foster, N.E.; Anema, J.R.; Cherkin, D.; Chou, R.; Cohen, S.P.; Gross, D.P.; Ferreira, P.H.; Fritz, J.M.; Koes, B.W.; Peul, W.; et al. Prevention and treatment of low back pain: Evidence, challenges, and promising directions. *Lancet* **2018**, *391*, 2368–2383. [CrossRef]
18. Dueñas, M.; Ojeda, B.; Salazar, A.; Mico, J.A.; Failde, I. A review of chronic pain impact on patients, their social environment and the health care system. *J. Pain Res.* **2016**, *9*, 457–467. [CrossRef]
19. Froud, R.; Patterson, S.; Eldridge, S.; Seale, C.; Pincus, T.; Rajendran, D.; Fossum, C.; Underwood, M. A systematic review and meta-synthesis of the impact of low back pain on people's lives. *BMC Musculoskelet. Disord.* **2014**, *15*, 50. [CrossRef] [PubMed]
20. Olafsson, G.; Jonsson, E.; Fritzell, P.; Hägg, O.; Borgström, F. Cost of low back pain: Results from a national register study in Sweden. *Eur. Spine J.* **2018**, *27*, 2875–2881. [CrossRef]
21. Geurts, J.W.; Willems, P.C.; Kallewaard, J.W.; Van Kleef, M.; Dirksen, C. The Impact of Chronic Discogenic Low Back Pain: Costs and Patients' Burden. *Pain Res. Manag.* **2018**, *2018*, 1–8. [CrossRef] [PubMed]
22. Walker, B.F.; Muller, R.; Grant, W.D. Low back pain in Australian adults: The economic burden. *Asia-Pac. J. Public Health* **2003**, *15*, 79–87. [CrossRef]
23. Hoy, D.; Bain, C.; Williams, G.; March, L.; Brooks, P.; Blyth, F.; Woolf, A.; Vos, T.; Buchbinder, R. A systematic review of the global prevalence of low back pain. *Arthritis Rheum.* **2012**, *64*, 2028–2037. [CrossRef]
24. Coulombe, B.J.; Games, K.E.; Neil, E.R.; Eberman, L.E. Core stability exercise versus general exercise for chronic low back pain. *J. Athl. Train.* **2017**, *52*, 71–72. [CrossRef]
25. Gomes-Neto, M.; Lopes, J.M.; Conceição, C.S.; Araujo, A.; Brasileiro, A.; Sousa, C.; Carvalho, V.O.; Arcanjo, F.L. Stabilization exercise compared to general exercises or manual therapy for the management of low back pain: A systematic review and meta-analysis. *Phys. Ther. Sport* **2017**, *23*, 136–142. [CrossRef] [PubMed]
26. Owen, P.J.; Miller, C.T.; Mundell, N.L.; Verswijveren, S.J.J.M.; Tagliaferri, S.D.; Brisby, H.; Bowe, S.J.; Belavy, D.L. Which specific modes of exercise training are most effective for treating low back pain? Network meta-analysis. *Br. J. Sports Med.* **2020**, *54*, 1279–1287. [CrossRef]
27. Sowah, D.; Boyko, R.; Antle, D.; Miller, L.; Zakhary, M.; Straube, S. Occupational interventions for the prevention of back pain: Overview of systematic reviews. *J. Saf. Res.* **2018**, *66*, 39–59. [CrossRef]
28. European Communities NACE Rev. 2. Statistical Classification of Economic Activities in the European Community. EUROSTAT Methodologies and Working Papers. 2008. Available online: https://ec.europa.eu/eurostat/documents/3859598/5902521/KS-RA-07-015-EN.PDF (accessed on 3 December 2021).
29. Pension and Disability Insurance Act (Official Gazette of the Republic of Slovenia; 2013; p. No. 96/12, 39/13, 99/13-ZSVarPre-C, 101/13-Z. Available online: https://www.google.com/url?sa=t&rct=j&q=&esrc=s&source=web&cd=&ved=2ahUKEwjJusH8-vb0AhWvrlYBHR-ICZ8QFnoECAIQAQ&url=http%3A%2F%2Fpisrs.si%2FPis.web%2FnpbDocPdf%3FidPredpisa%3DZAKO8017%26idPredpisaChng%3DZAKO6280%26type%3Ddoc%26lang%3Den&usg=AOvVaw2lITH0Ub88LNqD552ChYmO (accessed on 3 December 2021).

30. Employment Relationships; Official Gazette of the Republic of Slovenia. 2008. Available online: https://www.eui.eu/Projects/InternationalArtHeritageLaw/Documents/NationalLegislation/Slovenia/culturalheritageprotectionact.pdf (accessed on 3 December 2021).
31. National Institute of Public Health Bolniški Stalež. Posredovanje Podatkov o Začasni Odsotnosti z Dela Zaradi Bolezenskih Razlogov. Available online: https://www.nijz.si/sites/www.nijz.si/files/uploaded/podatki/podatkovne_zbirke_raziskave/bs/mn_bs_2019.pdf (accessed on 3 December 2021).
32. National Institute of Public Health Zdravstveni Statistični Letopis. Spremljanje Zdravstvenega Stanja Delavcev. Available online: https://www.nijz.si/sl/publikacije/zdravstveni-statisticni-letopis-2012 (accessed on 4 December 2021).
33. Meucci, R.D.; Fassa, A.G.; Faria, N.M.X. Prevalence of chronic low back pain: Systematic review. *Rev. Saude Publica* **2015**, *49*, 1. [CrossRef] [PubMed]
34. Public Administration and Defence; Compulsory Social Security statistics in Great Britain 2021. Available online: https://www.hse.gov.uk/statistics/industry/public-admin.pdf (accessed on 7 December 2021).
35. Brinckmann, P.; Frobin, W.; Biggemann, M.; Tillotson, M.; Burton, K.; Burke, C.; Dickinson, C.; Krause, H.; Pangert, R.; Pfeifer, U.; et al. Quantification of overload injuries to thoracolumbar vertebrae and discs in persons exposed to heavy physical exertions or vibration at the workplace Part II Occurrence and magnitude of overload injury in exposed cohorts. *Clin. Biomech.* **1998**, *13*, S1–S36. [CrossRef]
36. McPhee, B. Ergonomics in minning. *Occup. Med.* **2004**, *54*, 297–303. [CrossRef] [PubMed]
37. Hita-Gutiérrez, M.; Gómez-Galán, M.; Díaz-Pérez, M.; Callejón-Ferre, Á.-J. An overview of reba method applications in the world. *Int. J. Environ. Res. Public Health* **2020**, *17*, 2635. [CrossRef]
38. Republic of Slovenia Statistical Office Persons in Employment By Activities (NACE Rev. 2), Slovenia, Monthly. Available online: https://pxweb.stat.si/SiStatData/pxweb/en/Data/-/0700921S.px/ (accessed on 5 December 2021).
39. Reduce Occumational Musculoskeletal Disorders: Agriculture, Forestry and Fishing. Available online: https://www.cdc.gov/niosh/about/strategicplan/musag.html (accessed on 6 December 2021).
40. Steenstra, I.; Verbeek, J.; Heymans, M.; Bongers, P. Prognostic factors for duration of sick leave in patients sick listed with acute low back pain: A systematic review of the literature. *Occup. Environ. Med.* **2005**, *62*, 851–860. [CrossRef]
41. Discussion Paper: Musculoskeletal Disorders in the Healthcare Sector. Available online: https://www.google.com/url?sa=t&rct=j&q=&esrc=s&source=web&cd=&ved=2ahUKEwiByLmm_fb0AhWksFYBHaZ5Cp8QFnoECAsQAQ&url=https%3A%2F%2Fhealthy-workplaces.eu%2Fsites%2Fdefault%2Ffiles%2Fpublications%2Fdocuments%2FDiscussion_paper_MSDs_in_health_care_sector.pdf&usg=AOvVaw3qvM3kQEj8NL9a8UKGhP3U (accessed on 8 December 2021).
42. Reduce Occupational Musculoskeletal Disorders: Manufacturing. Available online: https://www.cdc.gov/niosh/about/strategicplan/muscmanuf.html (accessed on 9 December 2021).
43. EU OSHA Work Related Musculoskeletal Disorders in the Service and Retail Sectors. Available online: https://osha.europa.eu/en/publications/e-fact-12-work-related-musculoskeletal-disorders-service-and-retail-sectors. (accessed on 7 December 2021).
44. Yang, H.; Haldeman, S.; Lu, M.L.; Baker, D. Low Back Pain Prevalence and Related Workplace Psychosocial Risk Factors: A Study Using Data From the 2010 National Health Interview Survey. *J. Manip. Physiol. Ther.* **2016**, *39*, 459–472. [CrossRef]
45. Wai, E.K.; Roffey, D.M.; Bishop, P.; Kwon, B.K.; Dagenais, S. Causal assessment of occupational bending or twisting and low back pain: Results of a systematic review. *Spine J.* **2010**, *10*, 76–88. [CrossRef]
46. Wai, E.K.; Roffey, D.M.; Bishop, P.; Kwon, B.K.; Dagenais, S. Causal assessment of occupational lifting and low back pain: Results of a systematic review. *Spine J.* **2010**, *10*, 554–566. [CrossRef]
47. Waters, T.R.; Dick, R.B.; Davis-Barkley, J.; Krieg, E.F. A cross-sectional study of risk factors for musculoskeletal symptoms in the workplace using data from the General Social Survey (GSS). *J. Occup. Environ. Med.* **2007**, *49*, 172–184. [CrossRef]
48. Waters, T.R.; Dick, R.B.; Krieg, E.F. Trends in work-related musculoskeletal disorders: A comparison of risk factors for symptoms using quality of work life data from the 2002 and 2006 general social survey. *J. Occup. Environ. Med.* **2011**, *53*, 1013–1024. [CrossRef]

International Journal of
Environmental Research and Public Health

MDPI

Article

Validity of a Screening Tool for Patients with a Sub-Threshold Level of Lumbar Instability: A Cross-Sectional Study

Arisa Leungbootnak [1], Rungthip Puntumetakul [1,*], Jaturat Kanpittaya [2], Thiwaphon Chatprem [1] and Rose Boucaut [3]

1 Research Center of Back, Neck, Other Joint Pain and Human Performance (BNOJPH), Department of Physical Therapy, Faculty of Associated Medical Sciences, Khon Kaen University, Khon Kaen 40002, Thailand; rung.smd28@gmail.com (A.L.); thiwaphon.ao@gmail.com (T.C.)
2 Department of Radiology, Faculty of Medicine, Khon Kaen University, Khon Kaen 40002, Thailand; jatkan@gmail.com
3 International Centre for Allied Health Evidence (iCAHE), School of Health Sciences (Physiotherapy), University of South Australia, Adelaide, SA 5001, Australia; rose.boucaut@unisa.edu.au
* Correspondence: rungthiprt@gmail.com; Tel.: +66-834196186

Abstract: Lumbar instability (LI) comprises one subgroup of those with chronic low back pain (CLBP); it indicates the impairment of at least one of the spinal stabilizing systems, and radiographic criteria of translation and rotation are used for its diagnosis. Previous studies have developed and tested a screening tool for LI where patients with sub-threshold lumbar instability (STLI) were detected in the initial stage of lumbar pathology using radiographs as a gold standard for diagnosis. The radiographic measurement in STLI lies between the range of translation and rotation of the LI and asymptomatic lumbar motion. However, there are no studies indicating the validity and cut-off points of the screening tool for STLI. The current study aimed to determine the validity of an LI screening tool to support the diagnostic process in patients with STLI. This study design was cross-sectional in nature. A total of 135 participants with CLBP, aged between 20 and 60 years, who had undergone flexion and extension radiographs, answered a screening tool with 14 questions. The cut-off score for identifying STLI using the screening tool was at least 6/14 positive responses to the LI questions. The findings suggested that the LI screening tool we tested is effective for the detection of STLI. The tool can be used in outpatient settings.

Keywords: sub-threshold lumbar instability; non-radiological lumbar instability; lumbar instability; radiography; lumbar translation; lumbar rotation; screening tool; X-ray; sensitivity; specificity

Citation: Leungbootnak, A.; Puntumetakul, R.; Kanpittaya, J.; Chatprem, T.; Boucaut, R. Validity of a Screening Tool for Patients with a Sub-Threshold Level of Lumbar Instability: A Cross-Sectional Study. *IJERPH* **2021**, *18*, 12151. https://doi.org/10.3390/ijerph182212151

Academic Editors: Vincenzo Denaro, Sergio Iavicoli, Fabrizio Russo and Gianluca Vadalà

Received: 1 October 2021
Accepted: 15 November 2021
Published: 19 November 2021

Publisher's Note: MDPI stays neutral with regard to jurisdictional claims in published maps and institutional affiliations.

1. Introduction

Lumbar instability (LI) is defined as excessive translation and rotation motion values of each lumbar segment compared with normal values [1], and dysfunction of one of the spinal stabilizing systems, which consist of three subsystems: passive, active, and neural [2,3]. LI is considered to be a subgroup of mechanical low back pain (MLBP) [4]. Previous articles have demonstrated that 12–57% [1,5–9] of patients with CLBP have lumbar instability. Patients with LI report a number of symptoms, including: pain [10], muscle spasms [11,12], abnormal movement quality [13], functional disability [5,10,12], and diminished quality of life [12]. The severity of LI can evolve and may require surgical stabilization [14].

LI can be measured using a flexion–extension radiograph as the gold standard for diagnosis. Patients with LI were defined by White and Panjabi (1990) using flexion–extension radiographic characteristics. Their criteria were: a sagittal plane translation greater than 4.5 mm, or sagittal plane rotation of greater than 15° at L1/L2, L2/L3, or L3/L4, greater than 20° at L4/L5, or greater than 25° at L5/S1 [15].

Staub and colleagues reported that, for asymptomatic volunteers, with an age range of 18 to 82 years, the value of sagittal translation at each lumbar segment was: 1.9 mm at L1/L2, 2.4 mm at L2/L3, 2.7 mm at L3/4, 2.8 mm at L4/L5, and 0.5 mm at L5/S1. The sagittal plane rotation ranges were 11.0° at L1/L2, 12.6° at L2/L3, 13.3° at L3/4, 14.7° at L4/L5, and 12.8° at L5/S1 [16]. However, there remains a gap in the current clinical knowledge about the range of segmental motion in asymptomatic participants and those with a diagnosis of LI which has not yet been studied. This was the reason for developing the new concept of the "sub-threshold level of lumbar instability (STLI)", defined as the sagittal translation and rotation of each lumbar segment that lies between the asymptomatic or normal range and the range leading to an X-ray diagnosis of LI. The STLI prevalence was reported as 78% amongst participants with CLBP in a previous study that was reported at an international conference on integrative medicine in 2019 at Mae Fah Luang University (Arisa Leungbootnak) [17]. This group of patients with STLI has not been reviewed, although a review is necessary as they are at risk of developing LI, as LI is rarely addressed by conservative treatment. The early detection of patients with STLI may facilitate early appropriate treatment and thereby slow the patient's progression to LI.

The gold-standard radiograph measurement has some limitations in terms of assessment, including the cost of radiograms, time required for examination and interpretation, and radiation exposure [18,19]. The detection of patients with STLI in the community remains difficult due to the limitations of X-ray assessment. Thus, a screening tool (questionnaire) for LI patients was developed. The signs and symptoms of LI were studied originally in 1982 by Kirkaldy-Willis and Farfan [20]. More recently, a specific lumbar instability screening tool was first reported by Cook et al. (2006), who performed a Delphi study with 168 physical therapists [21] and developed a screening tool based on subjective signs and symptoms. The screening was also associated with clinical lumbar instability, and was used to inform the diagnosis in many studies where the criteria related to LI were of interest [9,21–25]. A Thai version was subsequently created by a group of Thai physical therapists [26], who evaluated its construct validity; in 2020, the same group used the questionnaire as a tool for screening patients with LI among CLBP patients [23]. They found that a score of at least 7/14 correlated with having LI; this score was used as an inclusion criterion for participants with LI among patients with CLBP. The tool was also translated (from English into Brazilian-Portuguese) [22] and tested for reliability [9,21,22,24–26].

Questionnaires to evaluate the signs and symptoms of patients with lumbar instability are easy to use in clinical practice [21]; however, to date, no studies have used a specific LI screening tool for patients with STLI in the early STLI stages. The results of this study could be beneficial for the early detection of STLI in patients with CLBP using minimally invasive assessment methods; these patients may be at the pre-stage of LI.

2. Materials and Methods

2.1. Design and Setting

The study design was cross-sectional in nature. Participant testing was conducted at the Department of Associated Medical Science, Khon Kaen University, Thailand. The study was approved by the Ethics Committee for Human Research (HE582228) approved on 12 September 2018 at Khon Kaen University, Thailand. This study has been registered in Thai Clinical Trials at http://www.thaiclinicaltrials.org/show/TCTR20180920003 (accessed on 20 September 2019).

2.2. Participants

Participants with CLBP were recruited via poster and social media announcements. The inclusion criteria were: (1) having intermittent low back pain for at least 12 weeks, (2) participant age range 20 to 60 years, and (3) a pain numeric rating scale (NRS) score between 4 and 7 [27–29]. Participants were excluded if they had: (1) a contraindication to radiographic assessment, such as pregnancy, (2) acute fracture, tumor, or infection, (3) a history of serious neurological or psychiatric disease, (4) previous lumbar fusion surgery, or (5) a diagnosis of radiological lumbar instability (LI), spondylolisthesis, or lumbar disc herniation [1,5,30]. The investigators gained informed written consent from the participants prior to assessment.

2.3. Sample Size Determination

This is the first study to specifically investigate STLI, so the rate of disease came from the results of our previous study, reported at an international conference on integrative medicine (Arisa Leungbootnak, 7 October 2019) in 2019 at Mae Fah Luang University, Thailand (prevalence of STLI = 78%) [17]. The sample size calculation was conducted based on a significance level (α) of 0.01 and a precision of estimation value (e) of 0.10. According to acceptable sensitivity in the physical therapy field, (P) was set as 80% [31], meaning 135 participants were required.

2.4. Instruments

2.4.1. Radiographic Assessment and Measurement

Flexion–extension radiography is the most common form of examination used in studies to provide an imaging diagnosis of lumbar intervertebral instability [32]. The method of measuring the sagittal translation and rotation in the radiographs was previously described by Iguchi [30]. The sagittal translation and rotation were measured during full flexion and extension, and calculated using the formula (sagittal rotation (degrees) = A − (−a) and translation (mm) = B − (−b)), as shown in Figures 1 and 2.

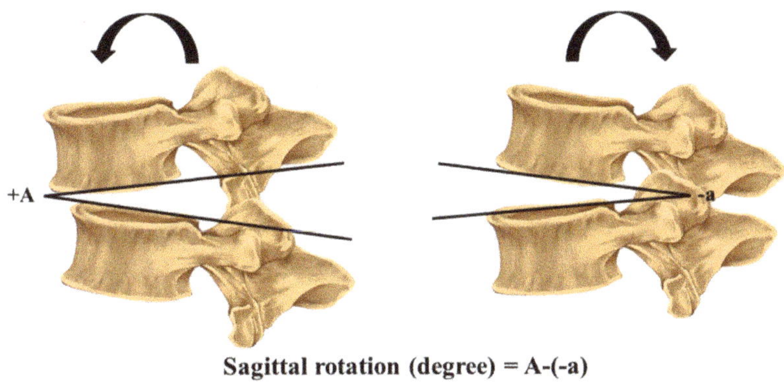

Figure 1. Sagittal rotation: a baseline is drawn passing the anterior and posterior endplate of the upper and lower vertebrae where the difference between the flexion and extension radiograph is measured and calculated.

The measurement of sagittal translation and rotation motion of each lumbar segment level from L1 to S1 was compared with ranges set for diagnosing STLI. Values for STLI were defined as: translation range from 1.9 to 4.5 mm at L1/L2, 2.4 to 4.5 mm at L2/L3, 2.7 to 4.5 mm at L3/4, 2.8 to 4.5 mm at L4/L5, and 0.5 to 4.5 mm at L5/S1, or a sagittal plane rotation range from 11.0° to 15° at L1/L2, 12.6° to 15° at L2/L3, 13.3° to 15° at L3/4, 14.7° to 20° at L4/L5, or 12.8° to 25° at L5/S1 [15,16]. The diagnosis of LI required showing either rotation or translation in two segments or rotation and translation in one segment to be classified as STLI [1]. This definition was used to reduce the false-positive rates from the X-ray measurement process [23].

A trained observer (researcher A.L.), who was blinded to other information, measured the translation and rotation of each lumbar segment in all participants using digitalized radiographs from a picture archive and communication (PACS) on a computer at Srinagarind Hospital, Khon Kaen, Thailand. The researcher (A.L.) was trained qualified to assess the radiographs by a radiologist with 30 years of experience. The evaluation practice of A.L.'s X-ray measurement consisted of using the program, pinpointing margins and spinal borders, and drawing the measurement line. The measurement process was practiced until

A.L.'s accuracy was rated as satisfactory by the radiologist. Next, A.L. measured each level of the lumbar spine three times. The mean of each of the three measurements was compared with the STLI diagnostic criteria. The intra-rater reliability of X-ray measures was assessed based on the radiographs of 10 participants selected at random. The within researcher ICC (A.L.), measured three times, was 0.998 (95%CI: 0.994–0.999) in translation and 0.994 (95%CI: 0.982–0.998) in rotation. The standard error of measurement (SEM) was 0.025 for translation and 0.058 for rotation. These were a lesser variation than the X-ray measurement [33].

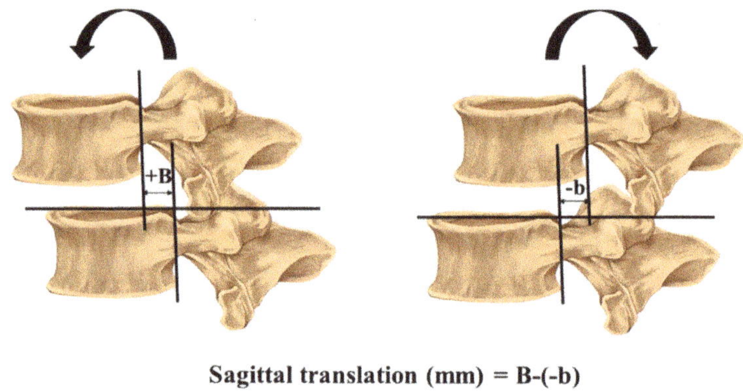

Figure 2. Sagittal translation: a horizontal line is drawn passing the superior endplate of the lower vertebra. Two vertical lines are drawn passing the posterior edge of the upper vertebra and lower vertebra. The distance between the two vertical lines is considered during flexion and extension, and the difference between the two distances from flexion and extension is used in the formula to calculate the segmental translation.

2.4.2. Screening Tool

The specific lumbar instability screening tool had 14 questions and was written in English (Table 1) [23]. This tool was translated and tested for content validity, criteria-related validity, and rater reliability in relation to lumbar instability as reported in previous studies [23,26]. The possible total questionnaire scores ranged from 0 (not correlated with instability) to 14 (strongly correlated with instability). The questions relate to participants' pain characteristics, positional alteration, muscle spasms, and injury history.

2.5. Measurement Procedure

The 135 participants, all diagnosed with CLBP, were split into two groups: those with and without STLI based on X-ray findings. Data were collected from March 2019 through July 2019. The participants were assessed at one visit. On that day, participants signed an informed consent form before starting the study and were evaluated for LBP symptoms. Researcher R.P. asked participants for their demographic information: age, gender, BMI, duration of low back pain, pain scale rating, and smoking history, and then the 14 screening tool questions. Then, participants were evaluated by an orthopedic doctor who ordered an X-ray assessment. X-rays were taken in the full lateral flexion and extension positions, and were performed by a radiologist. The X-ray images were measured by the trained observer (A.L.) after data collection was completed. A.L. was blinded to all other information about the patients.

2.6. Statistical Analyses

The data were analyzed and grouped based on participants with and without STLI. Descriptive statistics were used, involving calculations of the mean, standard deviation, and percentage. Significance was set at a p-value of less than 0.05 when comparing between groups. Intra-observer reliability was measured using the intraclass correlation coefficient (ICC) for the radiographic measurements.

The receiver operating characteristic (ROC) curve was used to determine the possible cut-off score of lumbar instability screening. Sensitivity, specificity, positive likelihood ratio (LR+), and negative likelihood ratio (LR-) for each cut-off score were calculated. The cut-off score that reached the maximum of sensitivity was taken as the maximum score. The area under the ROC curve (AUG) was used to evaluate the discriminatory ability of the 14 items of the screening tool. Statistical analysis was conducted using STATA ver. 10.0 (StataCorp, College Station, TX, USA).

Table 1. Lumbar instability screening tool.

Question 14 items	Yes (1)	No (0)
1. Patient reports his/her back has collapsed.		
2. Patient frequently self-manipulates to decrease their symptoms.		
3. Patient's back pain symptoms alternate periodically.		
4. Patient has a history of complaints of stiffness and sudden back pain when twisting or bending their back.		
5. Patient's back pain has been provoked by changing posture, for example standing up from sitting, etc.		
6. Patient has increased back pain when returning to upright after forward bending.		
7. Sudden or minor movements increase patient's back pain.		
8. Patient gets worse when sitting on a chair without a backrest and gets better when sitting on a chair with backrest.		
9. Patient reports being in a static posture for a long time has an effect on their back problem.		
10. Patient's back pain is worsening.		
11. Patient wears a brace or corset to temporarily alleviate back pain.		
12. Patient with back problems regularly experiences muscle spasms.		
13. Patient avoids or hesitates to move when they have back symptoms.		
14. Patient has a past history of back injury.		
Total score		

3. Results

3.1. Participant Characteristics

Of the 135 participants recruited, 113 participants (83.70%) had STLI. The mean age of all the participants was 35.58 ± 12.02 years. The participants' gender was 60.74% female. The average age of the participant groups with STLI and without STLI was 35.60 ± 12.46 and 35.45 ± 9.70 years old, respectively. All the continuous data were similar between the groups ($p > 0.05$), as shown in Table 2.

Table 2. Demographic characteristics of participants.

Variable	Total (n = 135)	With STLI (n = 113)	Without STLI (n = 22)	p-Value
Age (years)	35.58 ± 12.02	35.60 ± 12.46	35.45 ± 9.70	0.95
Gender (%)				
Male	53 (39.26)	45 (39.82)	8 (36.36)	0.82
Female	82 (60.74)	68 (60.18)	14 (63.64)	
BMI (kg/m^2)	22.16 ± 2.10	22.29 ± 2.10	21.48 ± 2.04	0.10
Duration of current symptoms (months)	27.33 ± 27.07	26.40 ± 27.54	32.14 ± 24.57	0.37
NRS (pain)	4.63 ± 0.94	4.57 ± 0.88	4.95 ± 1.21	0.17
Smoking history (%)				
Yes	12 (8.89)	10 (8.85)	2 (9.09)	1.00
No	123 (91.11)	103 (91.15)	20 (90.91)	

Note: STLI: sub-threshold level of lumbar instability; BMI: body mass index; NRS: numeric rating scale.

3.2. The Screening Tool-Specific STLI Cut-Off Scores

The current study showed that the cut-off score for identifying STLI requires a total of at least six positive responses of a possible score of 14 from the screening tool-specific questions. The sensitivity, specificity, positive likelihood ratio, and negative likelihood ratio were 99.12%, 18.18%, 1.21, and 0.05, respectively, as shown in Table 3. The high sensitivity indicates the potential screening value because the measure rarely misses subjects who have the condition [34]. The area under the ROC curve was 0.73, reflecting sufficient diagnostic accuracy [35].

Table 3. The screening tool-specific STLI cut-off scores.

Cut-Off Value	Sensitivity (%)	Specificity (%)	LR+	LR−	AUG (95%CI)
≥5	100.00	0.00	1.00		
≥6	99.12	18.18	1.21	0.05	
≥7	90.27	31.82	1.32	0.31	
≥8	69.91	68.18	2.20	0.44	
≥9	46.02	72.73	1.69	0.74	
≥10	31.86	95.45	7.01	0.71	0.73 (0.61–0.84)
≥11	17.70	100.00		0.82	
≥12	5.31	100.00		0.95	
≥13	2.65	100.00		0.97	
≥14	0.88	100.00		0.99	
14	0.00	100.00		1.00	

4. Discussion

The impairment or dysfunction of the active, passive, or neural control subsystems in CLBP patients causes LI. Delayed detection of patients with LI can lead to more structural degeneration, which is challenging to improve with conservative treatment. The early-stage detection of LI by physical therapists may help the patient by preventing further structural damage early on; for example, exercise stabilization, which focuses on deep trunk muscle training, can improve or delay the development of lumbar instability [36–38]. In the present study, a subjective questionnaire cut-off score for identifying STLI was found to be 6/14.

The Delphi survey results from Cook et al. (2006) listed 14 screening questions to determine the clinical lumbar instability in the CLBP participants [21]. The current study results showed that at least six positive answers to the fourteen listed lumbar instability questions among the CLBP participants led to a diagnosis of STLI. This result was similar to Kumar (2018) and Puntumetakul et al. (2014), who selected 7/13 positive responses to similar screening questions that could act as the criteria for diagnosing lumbar segmental instability [9,24]. However, when compared with the study by Chatprem et al. in 2020, our result was lower than their cutoff score (7/14) [23]. This may be because the participants in this study had less translation and rotation (STLI) than those in their (LI) study.

In the current study, three key symptoms were frequently reported by the participants with CLBP, that led to a positive diagnosis of STLI: (1) back pain alternating occasionally (97.35%); (2) worsening pain when sitting on a chair without a backrest and less pain when sitting on a chair with a backrest (96.46%); and (3) increasing back pain when maintaining prolonged static postures (95.58%) (Figure 3). The symptom of back pain alternating periodically may arise from the spinal unit moving together while the passive subsystem is in fixed alignment. During spinal movement, the passive subsystem is unable to restrain shear forces that arise from muscle contraction and gravity. A possible reason underlying this symptom is that one or more of the spinal subsystems that cause instability may be dysfunctional due to injury or degeneration [6]. In the sitting position, many people experience musculoskeletal discomfort, principally in the lower back and buttock areas, where the discomfort increases significantly during prolonged sitting [39]. The responses in the current study differed from some of those reported in a farming population by Puntumetakul et al. (2014) [9]. The responses in that study consisted of: (1) frequent episodes of muscle spasm (93%); (2) worsening symptoms with sustained postures (90%); and (3) frequent episodes of symptoms (88%). The differences between these two studies included: the participant occupation, number and age of participants, and use of different criteria to determine the presence of lumbar instability (clinical symptoms in the former study but X-ray evaluation in the current study). Nevertheless, the characteristic of worsening symptoms with sustained postures was the same in both studies.

An optimal cut-off point for LI patients of at least 7 of 14 scores was reported by Chatprem et al. in 2020 [23]. The 7/14 score was based on the maximum summation of sensitivity and specificity, which differed by one point from the current study in choosing the highest sensitivity for screening [34]. However, they reported three items as receiving 100% positive responses in LI, and two out of their three fit with the top responses in our study. These symptoms were prolonged sitting and sustained posture. The possible reasons for the lower score in the STLI participants may be twofold. First, the number of lumbar instability participants (STLI, 113 or 83.70%) was higher in the current study than the number of patients with LI in Chatprem's study. Second, the STLI criteria in the current study required less translation and rotation compared with LI. Less severe LI may lead to a lower score when comparing LI and STLI. However, considering their study in terms of the highest sensitivity for screening, a score of at least 7 of 14 remains the cut-off point.

Our participant demographic characteristics, particularly gender, have no difference between those with chronic low back pain with STLI and without STLI. Therefore, this minimizes other factors that may influence our results. The current study demonstrated that a screening tool specific for LI is beneficial in accurately identifying STLI among CLBP participants. In a clinical setting, choosing the cut-off scores of at least 6/14 positive points can be used to screen patients with CLBP for STLI responses. Interestingly, patients with positive STLI also had positive results on items 3, 8, and 9 of the screening tool questionnaire. Moreover, the effect size was large [40].

This study has useful implications for clinicians; however, there are some limitations. First, the participants experiencing pain while the flexion–extension radiographs were taken may have shown a lower range of instability than they actually had. Second, the sample size was small. Third, our participants had a wide age range, meaning they may have had a varied state of disc degeneration. Fourth, there may have been an overlap with the upper range of normal asymptomatic participants with those in the STLI group. Future studies could consider using a magnetic resonance image assessment to determine the stage of disc degeneration, to help improve the strength of the results.

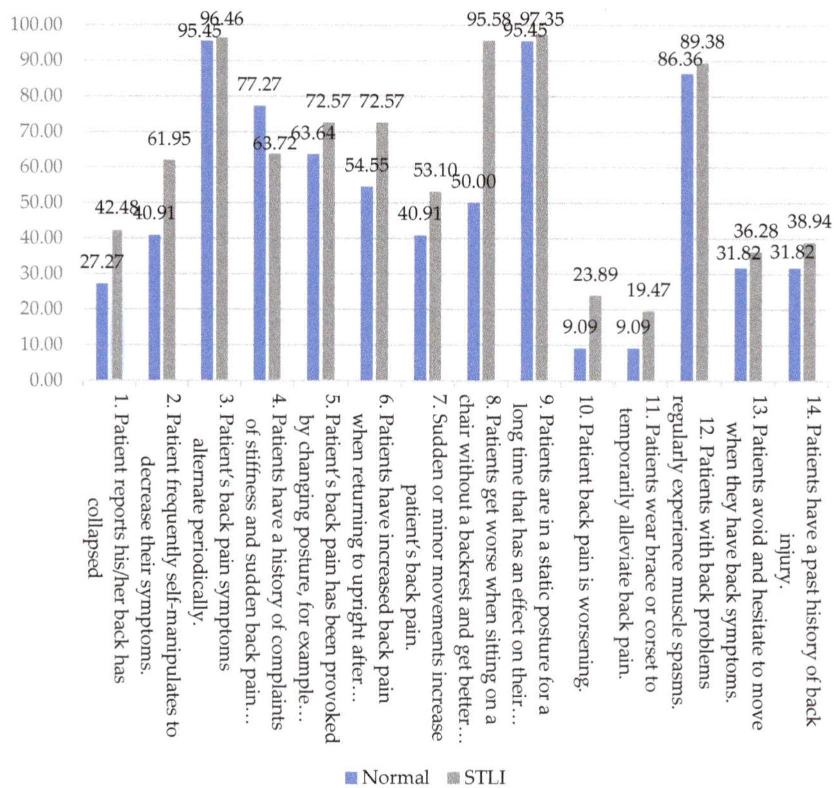

Figure 3. Comparison of the percentage of positive answers to the lumbar instability screening tool questions in chronic low back pain patients with and without the sub-threshold level of lumbar instability.

5. Conclusions

The findings of the current study revealed useful information for identifying STLI using a specific screening tool for patients with LI. Although this screening tool is effective for assessing the lumbar spinal instability in an outpatient clinic without the need for sophisticated and expensive equipment, high specificity tests, such as the passive lumbar extension test (PLE), are still recommended to confirm the diagnosis of STLI.

Author Contributions: A.L.: conceptualization, data curation, formal analysis, methodology, writing, and editing; R.P.: conceptualization, data curation, funding acquisition, project administration, resources, writing, and editing; J.K.: resources; T.C.: resources; R.B.: reviewing and editing. All authors have read and agreed to the published version of the manuscript.

Funding: This research received no external funding.

Institutional Review Board Statement: The study was conducted according to the guidelines of the Declaration of Helsinki, and approved by the Ethics Committee of Center for Ethics in Human Research, Khon Kaen University (no. HE582228) approved on 12 September 2018.

Informed Consent Statement: Informed consent was obtained from all subjects involved in the study.

Data Availability Statement: The data will be available for anyone who wishes to access them for any purpose and contact should be made via the Corresponding author rungthiprt@gmail.com.

Acknowledgments: The authors thank the participants who participated in the study, and the doctor and radiologist who provided the facilities.

Conflicts of Interest: The authors declare no conflict of interest.

References

1. Fritz, J.M.; Piva, S.R.; Childs, J.D. Accuracy of the clinical examination to predict radiographic instability of the lumbar spine. *Eur. Spine J.* **2005**, *14*, 743–750. [CrossRef]
2. Panjabi, M.M. The stabilizing system of the spine. Part I. Function, dysfunction, adaptation, and enhancement. *J. Spinal Disord.* **1992**, *5*, 383–389. [CrossRef]
3. Panjabi, M.M. Clinical spinal instability and low back pain. *J. Electromyogr. Kinesiol.* **2003**, *13*, 371–379. [CrossRef]
4. Last, A.R.; Hulbert, K. Chronic low back pain: Evaluation and management. *S. Afr. Fam. Pract.* **2010**, *52*, 184–192. [CrossRef]
5. Abbott, J.H.; McCane, B.; Herbison, P.; Moginie, G.; Chapple, C.; Hogarty, T. Lumbar segmental instability: A criterion-related validity study of manual therapy assessment. *BMC Musculoskelet. Disord.* **2005**, *6*, 56. [CrossRef]
6. Ahn, K.; Jhun, H.J. New physical examination tests for lumbar spondylolisthesis and instability: Low midline sill sign and interspinous gap change during lumbar flexion-extension motion Orthopedics and biomechanics. *BMC Musculoskelet. Disord.* **2015**, *16*, 97. [CrossRef]
7. Hicks, G.E.; Fritz, J.M.; Delitto, A.; McGill, S.M. Preliminary development of a clinical prediction rule for determining which patients with low back pain will respond to a stabilization exercise program. *Arch. Phys. Med. Rehabil.* **2005**, *86*, 1753–1762. [CrossRef]
8. Kasai, Y.; Morishita, K.; Kawakita, E.; Kondo, T.; Uchida, A. A new evaluation method for lumbar spinal instability: Passive lumbar extension test. *Phys. Ther.* **2006**, *86*, 1661–1667. [CrossRef]
9. Puntumetakul, R.; Yodchaisarn, W.; Emasithi, A.; Keawduangdee, P.; Chatchawan, U.; Yamauchi, J. Prevalence and individual risk factors associated with clinical lumbar instability in rice farmers with low back pain. *Patient Prefer. Adherence* **2014**, *9*, 1–7. [CrossRef]
10. Vanti, C.; Conti, C.; Faresin, F.; Ferrari, S.; Piccarreta, R. The Relationship Between Clinical Instability and Endurance Tests, Pain, and Disability in Nonspecific Low Back Pain. *J. Manip. Physiol. Ther.* **2016**, *39*, 359–368. [CrossRef]
11. Kotilainen, E.; Valtonen, S. Acta Neurochirurgica Clinical Instability of the Lumbar Spine After Mierodiseectomy. *Acta Neurochir.* **1993**, *125*, 120–126. [CrossRef]
12. Tang, S.; Qian, X.; Zhang, Y.; Liu, Y. Treating low back pain resulted from lumbar degenerative instability using Chinese Tuina combined with core stability exercises: A randomized controlled trial. *Complementary Ther. Med.* **2016**, *25*, 45–50. [CrossRef]
13. Panjabi, M. The stabilizing system of the spine. Part II. Neutral zone and instability hypothesis. *J. Spinal Disord.* **1992**, *5*, 390–396. [CrossRef] [PubMed]
14. Dang, L.; Zhu, J.; Liu, Z.; Liu, X.; Jiang, L.; Wei, F.; Song, C. A new approach to the treatment of spinal instability: Fusion or structural reinforcement without surgery? *Med. Hypotheses* **2020**, *144*, 109900. [CrossRef] [PubMed]
15. White, M.M.; Panjabi, A. *Clinical Biomechanics of the Spine*, 2nd ed.; Lippincott: Philadelphia, PA, USA, 1990; Volumes 23–45, pp. 342–360.
16. Staub, B.N.; Holman, P.J.; Reitman, C.A.; Hipp, J. Sagittal plane lumbar intervertebral motion during seated flexion-extension radiographs of 658 asymptomatic nondegenerated levels. *J. Neurosurg. Spine* **2015**, *23*, 731–738. [CrossRef] [PubMed]
17. Leungbootnak, A. The range of lumbar spine motion in Thai patients with non-radiological lumber instability (pilot study). In Proceedings of the 1st International Conference on Integrative Medicine, Chiang Rai, Thailand, 7 October 2019; pp. 55–62.
18. Alqarni, A.M.; Schneiders, A.G.; Hendrick, P.A. Clinical tests to diagnose lumbar segmental instability:A systematic review. *J. Orthop. Sports Phys. Ther.* **2011**, *41*, 130–140. [CrossRef]
19. Ozcete, E.; Boydak, B.; Ersel, M.; Kiyan, S.; Uz, I.; Cevrim, O. Comparison of conventional radiography and digital computerized radiography in patients presenting to emergency department. *Turk. J. Emerg. Med.* **2015**, *15*, 8–12. [CrossRef] [PubMed]
20. Kirkaldy-Willis, W.H.; Farfan, H. Instability of the lumbar spine. *Clin. Orthop. Relat. Res.* **1994**, *165*, 110–123. [CrossRef]
21. Cook, C.; Brismée, J.M.; Sizer, P.S. Subjective and objective descriptors of clinical lumbar spine instability: A Delphi study. *Man. Ther.* **2006**, *11*, 11–21. [CrossRef]
22. Araujo, A.C.; Da Cunha Menezes Costa, L.; De Oliveira, C.B.S.; Morelhão, P.K.; De Faria Negrão Filho, R.; Pinto, R.Z.; Costa, L.O.P. Measurement Properties of the Brazilian-Portuguese Version of the Lumbar Spine Instability Questionnaire. *Spine* **2017**, *42*, E810–E814. [CrossRef]
23. Chatprem, T.; Puntumetakul, R.; Boucaut, R.; Wanpen, S.; Chatchawan, U. A Screening Tool for Patients With Lumbar Instability: A Criteria-related Validity of Thai Version. *Spine* **2020**, *45*, E1431–E1438. [CrossRef]
24. Kumar, S.P. Efficacy of segmental stabilization exercise for lumbar segmental instability in patients with mechanical low back pain: A randomized placebo controlled crossover study. *N. Am. J. Med. Sci.* **2011**, *3*, 456–461. [CrossRef]
25. Saragiotto, B.T.; Maher, C.G.; New, C.H.; Catley, M.; Hancock, M.J.; Cook, C.E.; Hodges, P.W. Clinimetric testing of the Lumbar Spine instability questionnaire. *J. Orthop. Sports Phys. Ther.* **2018**, *48*, 915–922. [CrossRef]

26. Chatprem, T.; Puntumetakul, R.; Yodchaisarn, W.; Siritaratiwat, W.; Boucaut, R.; Sae-jung, S. A Screening Tool for Patients With Lumbar Instability: A Content Validity and Rater Reliability of Thai Version. *J. Manip. Physiol. Ther.* **2020**, *43*, 515–520. [CrossRef]
27. Wennberg, P.; Möller, M.; Sarenmalm, E.K.; Herlitz, J. Evaluation of the intensity and management of pain before arrival in hospital among patients with suspected hip fractures. *Int. Emerg. Nurs.* **2020**, *49*, 100825. [CrossRef]
28. Fejer, R.; Jordan, A.; Hartvigsen, J. Categorising the severity of neck pain: Establishment of cut-points for use in clinical and epidemiological research. *Pain* **2005**, *119*, 176–182. [CrossRef]
29. Hanley, M.A.; Masedo, A.; Jensen, M.P.; Cardenas, D.; Turner, J.A. Pain interference in persons with spinal cord injury: Slassification of mild, moderate, and severe pain. *J. Pain* **2006**, *7*, 129–133. [CrossRef]
30. Maigne, J.Y.; Lapeyre, E.; Morvan, G.; Chatellier, G. Pain immediately upon sitting down and relieved by standing up is often associated with radiologic lumbar instability or marked anterior loss of disc space. *Spine* **2003**, *28*, 1327–1334. [CrossRef]
31. Van der Wulf, P.; Meyne, W.; Hagmeijer, R.H.M. Clinical tests of the sacroiliac joint. A systematic methodological review. Part 2: Validity. *Man. Ther.* **2000**, *5*, 89–96. [CrossRef]
32. Leone, A.; Guglielmi, G.; Cassar-Pullicino, V.N.; Bonomo, L. Lumbar intervertebral instability: A review. *Radiology* **2007**, *245*, 62–77. [CrossRef]
33. Saiklang, P.; Puntumetakul, R.; Swangnetr Neubert, M.; Boucaut, R. Effect of time of day on height loss response variability in asymptomatic participants on two consecutive days. *Ergonomics* **2019**, *62*, 1542–1550. [CrossRef]
34. Goetzinger, K.R.; Tuuli, M.G.; Odibo, A.O. Statistical analysis and interpretation of prenatal diagnostic imaging studies, part 3: Approach to study design. *J. Ultrasound Med.* **2011**, *30*, 1415–1423. [CrossRef]
35. Simundic, A.M. Measures of Diagnostic Accuracy: Basic Definitions. *Ejifcc* **2009**, *19*, 203–211.
36. Puntumetakul, R.; Areeudomwong, P.; Emasithi, A.; Yamauchi, J. Effect of 10-week core stabilization exercise training and detraining on pain-related outcomes in patients with clinical lumbar instability. *Patient Prefer. Adherence* **2013**, *7*, 1189–1199. [CrossRef]
37. Puntumetakul, R.; Saiklang, P.; Tapanya, W.; Chatprem, T.; Kanpittaya, J.; Arayawichanon, P.; Boucaut, R. The effects of core stabilization exercise with the abdominal drawing-in maneuver technique versus general strengthening exercise on lumbar segmental motion in patients with clinical lumbar instability: A randomized controlled trial with 12-month follow-up. *Int. J. Environ. Res. Public Health* **2021**, *18*, 7811. [CrossRef]
38. Javadian, Y.; Akbari, M.; Talebi, G.; Taghipour-Darzi, M.; Janmohammadi, N. Influence of core stability exercise on lumbar vertebral instability in patients presented with chronic low back pain: A randomized clinical trial. *Casp. J. Intern. Med.* **2015**, *6*, 98–102.
39. Søndergaard, K.H.E.; Olesen, C.G.; Søndergaard, E.K.; de Zee, M.; Madeleine, P. The variability and complexity of sitting postural control are associated with discomfort. *J. Biomech.* **2010**, *43*, 1997–2001. [CrossRef]
40. Salgado, J.F. Transforming the area under the normal curve (AUC) into cohen's *d*, pearson's r_{pb}, odds-ratio, and natural log odds-ratio: Two conversion tables. *Eur. J. Psychol. Appl. Leg. Context* **2018**, *10*, 35–47. [CrossRef]

International Journal of
Environmental Research and Public Health

MDPI

Article

Fraction and Number of Unemployed Associated with Self-Reported Low Back Pain: A Nation-Wide Cross-Sectional Study in Japan

Kimiko Tomioka [1,*], Teruyo Kitahara [2], Midori Shima [1] and Keigo Saeki [1]

[1] Nara Prefectural Health Research Center, Nara Medical University, Kashihara 634-8521, Nara, Japan; mshima@naramed-u.ac.jp (M.S.); saekik@naramed-u.ac.jp (K.S.)
[2] Division of Occupational and Environmental Health, Department of Social Medicine, Shiga University of Medical Science, Otsu 520-2192, Shiga, Japan; teruyo@belle.shiga-med.ac.jp
* Correspondence: tkimiko@naramed-u.ac.jp

Abstract: This study examined a cross-sectional association between self-reported low back pain (LBP) and unemployment among working-age people, and estimated the impact of self-reported LBP on unemployment. We used anonymized data from a nationally representative survey (24,854 men and 26,549 women aged 20–64 years). The generalized estimating equations of the multivariable Poisson regression models stratified by gender were used to estimate the adjusted prevalence ratio (PR) and 95% confidence interval (CI) for unemployment. The population attributable fraction (PAF) was calculated using Levin's method, with the substitution method for 95% CI estimation. The prevalence of self-reported LBP was 9.0% in men and 11.1% in women. The prevalence of unemployment was 9.3% in men and 31.7% in women. After adjusting for age, socio-economic status, lifestyle habits, and comorbidities, the PR (95% CI) for the unemployment of the LBP group was 1.32 (1.19–1.47) in men and 1.01 (0.96–1.07) in women, compared with the respective non-LBP group. The PAF (95% CI) of unemployment associated with self-reported LBP was 2.8% (1.6%, 4.2%) in men. Because the total population of Japanese men aged 20–64 in 2013 was 36,851 thousand, it was estimated that unemployment in 1037 thousand of the Japanese male working population was LBP-related.

Keywords: low back pain; unemployment; gender difference; population attributable fraction; cross-sectional studies

Citation: Tomioka, K.; Kitahara, T.; Shima, M.; Saeki, K. Fraction and Number of Unemployed Associated with Self-Reported Low Back Pain: A Nation-Wide Cross-Sectional Study in Japan. *IJERPH* **2021**, *18*, 10760. https://doi.org/10.3390/ijerph182010760

Academic Editors: Sergio Iavicoli, Vincenzo Denaro, Gianluca Vadalà and Fabrizio Russo

Received: 21 July 2021
Accepted: 10 October 2021
Published: 13 October 2021

1. Introduction

The Global Burden of Disease Study 2016 reported that low back pain (LBP) has become the leading cause of years lived with disability [1]. Additionally, LBP has a high economic burden. In Japan, musculoskeletal disorders, including LBP, account for 7.7% of the total medical expenses (about JPY 2.3 trillion) in 2016, ranking third after cardiovascular diseases and neoplasms [2]. According to a recent report in the United States, the annual cost of LBP in 2016 was an estimated USD 134.5 billion of the total healthcare spending [3]. According to a systematic review on the overall costs associated with LBP, indirect costs (sick leave, early retirement, lost household productivity, and presenteeism, etc.) are estimated to be approximately six times higher than direct medical costs [4]. Therefore, LBP is considered to be a major global health problem.

Occupation is a worldwide cause of LBP; globally, 37% of LBP is attributed to occupational risk factors [5]. A large number of studies report that work-related LBP is more prevalent in occupations with heavy lifting, whole-body vibration, forceful movements, and awkward postures [5,6], such as nursing [7], caregiving [8], transport [9], construction [10], and manufacturing [11]. However, it is difficult to determine whether LBP is associated with unemployment through a workplace survey of workers only.

Previous studies reveal that the prevalence of self-reported LBP is significantly higher in the unemployed than in working people [12], and that self-reported LBP is a predictor of health-related job resignations [13] or sickness absences [14]. However, the findings of prior research did not quantify the importance of unemployment associated with LBP as a public health issue. Additionally, previous studies [12–14] did not perform stratified analyses by gender. Gender-specific analyses are essential in evaluating the impact of self-reported LBP on unemployment because of the differences in the percentage of employment [14,15] and the prevalence of LBP [5,6] by gender. Therefore, we made an attempt to estimate the gender-specific burden of unemployment attributable to LBP. Our study will be able to demonstrate how important it is to improve the workplace for people with LBP and to enhance LBP care in terms of employment assistance.

In this study, we used anonymized data from the Comprehensive Survey of Living Conditions (CSLC); the CSLC is a representative sample of Japanese people including both the employed and the unemployed, and the CSLC 2013 is the latest data available, as of the end of August 2021 [16]. The aims of this study were twofold. The first was to investigate a cross-sectional association between self-reported LBP and unemployment among working-age people. The second was to estimate the fraction and number of unemployed people associated with self-reported LBP in Japanese working-age people. To clarify gender difference in the association between self-reported LBP and unemployment and its impact, we performed stratified analyses by gender.

2. Materials and Methods

2.1. Data and Study Participants

We used anonymized data from the CSLC 2013 conducted by the Ministry of Health, Labour and Welfare of Japan [16]. The CSLC is a cross-sectional nationwide survey which collects data on national lifestyle, health, and welfare. The details of the CSLC 2013 are explained elsewhere [17]. Briefly, the CSLC 2013 targeted all households (approximately 300,000 households) and household members (approximately 740,000 persons) in the 5530 districts stratified and randomly selected from the 2010 census ward. The proportion of respondents was 79.6%. Anonymized data had confidentiality measures such as resampling and top coding. Resampling referred to a survey technique that re-extracts approximately one-sixth of the data from the original CSLC using the same procedure as the original survey. The purpose of resampling was to eliminate concerns about identification of individuals, through shrinking the data and deleting dates of birth and addresses. In the CSLC 2013, questions about lifestyle habits were limited to people aged over 20. Additionally, people in hospital or with a long-term need of care were exempt from answering questions about health status and lifestyles. Therefore, among 97,345 anonymized data, we excluded 45,942 persons from our analyses based on the following criteria: under 20 years old, over 65 years old, in hospital, in school, certification of long-term care need, as well as missing data on age, hospital admission, working status, and/or self-reported LBP. Therefore, this study included 51,403 persons (24,854 men and 26,549 women) (Figure 1). The CSLC was conducted in June 2013, but the employment status was asked in May 2013.

2.2. Measurements

2.2.1. Exposure: Self-Reported Low Back Pain (LBP)

The first question was "Did you have any sort of subjective symptoms of a disorder or disease during the past few days?". Next, persons with subjective symptoms were asked about their symptoms. The second question allowed for multiple answers and presented 42 symptoms. Those who chose "low back pain" in the second question were defined as persons with self-reported LBP. The definition of self-reported LBP in this study was an experience of subjective LBP during the several days prior to being surveyed.

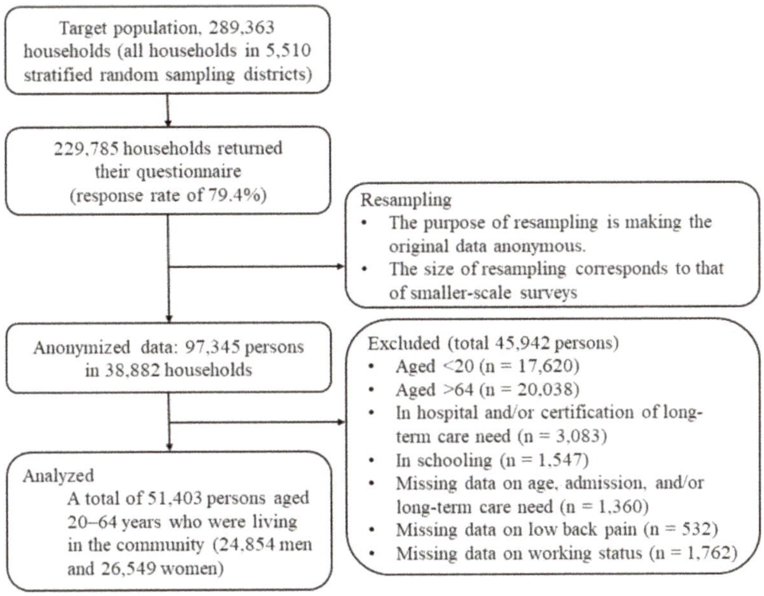

Figure 1. Selection of study participants.

2.2.2. Outcome: Employment Status

Employment status in this study was evaluated using the answer to the question "Did you have any paid work during May of 2013?". A respondent who answered "Yes" was considered to be employed. On the other hand, a respondent who answered "No" was considered to be unemployed. Only for unemployed persons, did the CSLC ask the question "Are you currently looking for work?". Using the answers to these questions, unemployed persons were classified into two groups: unemployed looking for work and unemployed not looking for work.

2.2.3. Covariates

Epidemiological evidence suggested that LBP was affected by many factors, reporting that self-reported LBP was common among older age groups [12,18–20], widowed or separated individuals [18], people with lower educational level [12,18], people from poorer economic backgrounds [12], drinkers [20], current smokers [6,18–21], individuals with sleep problems [12,20,22], and those with medical comorbidities [12,18]. Additionally, unemployment was associated with low socio-economic status (SES), cigarette smoking, heavy drinking, and chronic physical illnesses [23]. Therefore, the following variables were included as covariates that could be potential confounders of the associations between LBP and unemployment: age, SES, lifestyle habits, and health status. SES included marital status, family size, housing tenure, equivalent household expenditures, and education. Lifestyle habits included alcohol intake, smoking status, and sleep duration. Health status included comorbidities.

Marital status was categorized into married, never married, and widowed/divorced. Family size was categorized into 1, 2, 3–4, and >4. Housing tenure was dichotomized as owners–occupiers versus renters. Regarding equivalent household expenditures, because there was no gender difference, equivalent household expenditures were divided into three groups by the tertiles as a whole and not by gender: low (lower tertile, JPY 10.5 thousand or less), middle (middle tertile, JPY 10.6–15.6 thousand), and high (upper tertile, JPY 15.7 thousand or more). Regarding education, in Japan, mandatory education was from grade one to nine, and high school education was from grade ten to twelve. Post-high school education was defined as higher education. According to the Japanese school education system [24], education (years of schooling) was categorized into <10 (i.e., junior high graduates), 10–12 (i.e., high school graduates/dropouts), 13–15 (i.e., graduates/dropouts of junior/technical colleges), and >15 (i.e., college graduates and graduate school graduates). For alcohol consumption, in the CSLC, the questionnaire asked about frequency (daily, 5–6 days a week, 3–4 days a week, 1–2 days a week, 1–3 days a month, seldom drink, quit, do not drink). Participants were divided into the following four drinking groups: non-drinkers (quit or do not drink), social drinkers (1–3 days a month or seldom drink), occasional drinkers (1–4 days a week), and almost daily drinkers (more than 5 days a week). Smoking status was categorized into never smokers, ex-smokers, and current smokers. For sleep duration, the CSLC instructed participants to choose from six options for the average amount of sleep per day for the past month: less than 5 h, 5 h or more and less than 6 h, 6 h or more and less than 7 h, 7 h or more and less than 8 h, 8 h or more and less than 9 h, and 9 h or more. A Japanese cohort study reported that more than 7 h of sleep significantly increased the risk of all-cause mortality [25], and in a US population-based study, insufficient sleep duration was defined as less than 6 h per day [26]. In this study, sleep duration (sleeping hours per day) wasd categorized into <6, 6–7, and >7. Comorbidities were defined as participants having at least one disease under treatment for hypertension, diabetes mellitus, cerebrovascular disease, heart disease, and cancer. We confirmed that there were no multicollinearity problems in all covariates (i.e., none of the covariate variables had a variance inflation factor greater than 5.0) and entered all covariates into the final model.

To deal with missing data on the covariates, a category entitled "missing" was used for values that were missing in responses to questions on the covariates [27]. Further details about the covariates are shown in Table S1.

2.3. Statistical Analysis

Statistical analyses were performed by using the IBM SPSS Statistics Ver. 24 for Windows (Armonk, New York, NY, USA). The level of significance was 0.05 (two-tailed test). We performed analyses stratified by gender.

Data comparisons between the two groups were tested using the chi-squared test for categorical variables and the *t*-test for continuous variables.

We used the generalized estimating equations of the multivariable Poisson regression model to calculate a prevalence ratio (PR) and a 95% confidence interval (CI) for unemployment. We examined the cross-sectional association between self-reported LBP and unemployment. The independent variable was the presence or absence of self-reported LBP. The dependent variable was unemployment. Additionally, we investigated the relationship of self-reported LBP with unemployment with employment hope. The independent variable was the presence or absence of self-reported LBP, and the dependent variable was unemployment with employment hope.

We used four models to examine the cross-sectional association between self-reported LBP and unemployment. Model 1 was a crude model. Model 2 was adjusted for age. Model 3 was further adjusted for SES including marital status, family size, housing tenure, equivalent household expenditures, and education. The final Model 4 was adjusted as for Model 3 plus lifestyle habits (i.e., alcohol intake, smoking status, and sleep duration) and health status (i.e., comorbidities). For the estimation of PR for those who were unemployed but hoped for employment according to their self-reported LBP status, we used one model; all covariates (i.e., age, SES, lifestyle habits, and health status) were added simultaneously.

To evaluate the potential impact of self-reported LBP on unemployment, we calculated the population attributable fraction (PAF) of unemployment in the Japanese working-age population that could be attributed to the presence of self-reported LBP. We used the Levin formula [28] for the PAF for dichotomous exposures (i.e., the presence or absence of self-reported LBP), and used the substitution method [29] to calculate 95% CI for PAF. The PAF for unemployment was calculated as PAF = [Pe (PR−1)]/[Pe (PR−1) + 1]; where Pe was the prevalence of self-reported LBP in the study participants and PR was the adjusted PR for unemployment of the LBP group controlling for all covariates.

From the population estimation [30], we obtained data on the total population of Japanese men aged 20 to 64 as of June 2013, which was the survey month of the CSLC 2013. Population estimation was the basic statistical data on the country prepared by the Ministry of Internal Affairs and Communications to capture monthly and yearly population data in the middle year of the census. Using the total population of Japanese men aged 20–64 years and the PAF obtained in this study, we estimated how many of the total number of unemployed men aged 20 to 64 would be considered unemployed due to self-reported LBP.

2.4. Additional Analyses

Self-reported LBP particular to women included menstrual pain [7] and musculoskeletal pain due to excess strain on the lower back as a result of pregnancy [19]. Therefore, among women, a sensitivity analysis was conducted on the final model, excluding those who answered that they had menstrual pain (n = 806) and those who answered that they were going to hospital due to pregnancy (n = 154). Moreover, women were exposed to a heavy burden of care responsibilities which included the care of young children and a family member in need of nursing care, and both childcare and long-term care were considered risk factors for LBP [19]. Therefore, among women, a supplementary analysis was performed by adding the presence or absence of preschoolers and the presence or absence of living with persons requiring long-term care to the covariates.

2.5. Ethics

Based on Article 36 of the Statistics Act, we received approval of use for academic purposes from the Japanese Ministry of Health, Labour and Welfare (approval number 20009), and were provided data without any information that would identify individuals.

3. Results

3.1. Characteristics of the Study Participants

A total of 10,730 of the study participants were unemployed, of whom 2323 were men and 8407 women. The prevalence of unemployed women (31.7%) was more than three times higher than that of men (9.3%, $p < 0.001$). The prevalence of self-reported LBP during the past few days was 9.0% in men and 11.1% in women, showing a significant gender difference ($p < 0.001$).

Among both genders, compared to individuals in employment, unemployed persons were significantly older, more likely to spend less, to have a low education, to have comorbidities, and to sleep longer, while they were less likely to be daily drinkers and current smokers (Table 1).

3.2. Association between Self-Reported LBP and Unemployment by Gender

Among men, in the crude model (Model 1), people with self-reported LBP tended to have a higher prevalence of unemployment, compared to those without self-reported LBP (crude PR = 1.45, 95% CI 1.29 to 1.63). After the adjustment for age (Model 2), this association was attenuated but remained significant (age-adjusted PR = 1.33, 95% CI 1.19 to 1.49). The additional adjustment for SES brought a consistent result with a significant association (Model 3). In the final model (Model 4), where the data were adjusted for all covariates including lifestyle habits and health status, the association remained significant, showing that men with self-reported LBP had a significantly higher PR for unemployment than men without self-reported LBP (adjusted PR = 1.32, 95% CI 1.19 to 1.47).

Among women, in the crude model (Model 1), unemployment did not exhibit a significant difference between participants with and without self-reported LBP (crude PR = 1.02, 95% CI 0.96 to 1.08). A non-significant association remained by additional adjustment for age, SES, lifestyle habits, and health status (Model 4), showing that the presence or absence of self-reported LBP was not associated with unemployment in women (adjusted PR = 1.01, 95% CI 0.96 to 1.07) (Table 2).

Table 1. Participant characteristics according to employment status (employed or unemployed), by gender.

	Men (n = 24,854)			Women (n = 26,549)		
	Employed (n = 22,531) n (%)	Unemployed (n = 2323) n (%)	p-Value *	Employed (n = 18,142) n (%)	Unemployed (n = 8407) n (%)	p-Value *
Age: 45 years or older	11,016 (48.9)	1474 (63.5)	<0.001	8635 (47.6)	4690 (55.8)	<0.001
Marital status: not married	7101 (31.5)	1420 (61.1)	<0.001	6975 (38.4)	1526 (18.2)	<0.001
Family size: one (i.e., living alone)	2605 (11.6)	389 (16.7)	<0.001	1558 (8.6)	372 (4.4)	<0.001
Housing tenure: renters	7084 (31.4)	682 (29.4)	0.041	5267 (29.0)	2599 (30.9)	0.002
Household expenditures: low (<10.6)	7045 (31.3)	902 (38.8)	<0.001	5684 (31.3)	2758 (32.8)	0.017
Education: <10 years of schooling	1324 (5.9)	373 (16.1)	<0.001	780 (4.3)	624 (7.4)	<0.001
Alcohol intake: ≥5 days a week	8211 (36.4)	588 (25.3)	<0.001	2524 (13.9)	855 (10.2)	<0.001
Smoking status: current smokers	9077 (40.3)	832 (35.8)	<0.001	2683 (14.8)	920 (10.9)	<0.001
Sleep duration: <6 h a day	8950 (39.7)	627 (27.0)	<0.001	7904 (43.6)	3134 (37.3)	<0.001
Comorbidities: present	2605 (11.6)	471 (20.3)	<0.001	1309 (7.2)	960 (11.4)	<0.001
Self-reported LBP: present	1950 (8.7)	292 (12.6)	<0.001	1996 (11.0)	948 (11.3)	0.515

Note: LBP, low back pain. * p-values from chi-squared test. Not married included never married, widowed, and divorced. Household expenditures meant monthly equivalent household expenditures (unit: JPY one-thousand). Comorbidities included hypertension, diabetes mellitus, cerebrovascular disease, heart disease, and cancer.

Table 2. Prevalence ratio and population attributable fraction for unemployment associated with self-reported LBP, by gender.

LBP Status	n	% of the Unemployed	Model 1 PR (95% CI)	Model 2 PR (95% CI)	Model 3 PR (95% CI)	Model 4 PR (95% CI)	PAF (95% CI)
Men (n = 24,854)							
No LBP	22,612	9.0%	1.00	1.00	1.00	1.00	
LBP	2242	13.0%	1.45 (1.29 to 1.63)	1.33 (1.19 to 1.49)	1.29 (1.16 to 1.44)	1.32 (1.19 to 1.47)	2.8% (1.6% to 4.2%)
Women (n = 26,549)							
No LBP	23,605	31.6%	1.00	1.00	1.00	1.00	
LBP	2944	32.2%	1.02 (0.96 to 1.08)	0.98 (0.92 to 1.03)	0.98 (0.93 to 1.04)	1.01 (0.96 to 1.07)	0.13% (−0.48% to 0.77%)

Note: CI, confidence interval; LBP, low back pain; PAF, population attributable fraction; PR, prevalence ratio. LBP status means presence or absence of self-reported LBP. Model 1 is a crude model. Model 2 is adjusted for age (per 5-year increase). Model 3 is adjusted for age and socio-economic status (i.e., marital status, family size, housing tenure, equivalent household expenditures, and education). Model 4 is adjusted for the variables in Model 3 plus lifestyle habits (i.e., alcohol intake, smoking status, and sleep duration) and health status (i.e., comorbidities).

3.3. PAFs for Unemployment Associated with Self-Reported LBP by Gender

The PAF of unemployment due to self-reported LBP was 2.8% (95% CI 1.6% to 4.2%) for men and 0.13% (95% CI −0.48% to 0.77%) for women (Table 2). According to the population estimate released by the Ministry of Internal Affairs and Communications, the total population of Japanese men aged 20–64 as of June 2013 was 36,851 thousand. Therefore, it was estimated that unemployment in 1037 thousand Japanese men aged between 20 and 64 years were related to self-reported LBP.

3.4. Relationship of Self-Reported LBP with Unemployed Looking for Work by Gender

After adjustment for all covariates, among women as well as men, people with self-reported LBP had a significantly higher PR among the unemployed hoping for employment than people without self-reported LBP (adjusted PR = 1.49, 95% CI 1.29 to 1.73 in men; adjusted PR = 1.22, 95% CI 1.12 to 1.33 in women) (Table 3).

Table 3. Adjusted prevalence ratios for unemployed looking for work according to presence or absence of self-reported low back pain, by gender.

LBP Status		*n*	% of Unemployed Looking for Work	Adjusted PR * (95% CI)
Men (*n* = 24,854)				
	No LBP	22,612	5.5%	1.00
	LBP	2242	8.0%	1.49 (1.29 to 1.73)
Women (*n* = 26,549)				
	No LBP	23,605	14.3%	1.00
	LBP	2944	16.4%	1.22 (1.12 to 1.33)

Note: CI, confidence interval; LBP, low back pain; PR, prevalence ratio. LBP status means presence or absence of self-reported LBP.
* Adjusted for age (per 5-year increase), marital status, family size, housing tenure, equivalent household expenditures, education, alcohol intake, smoking status, sleep duration, and comorbidities.

3.5. Additional Analyses for Women

To consider female-specific self-reported LBP, two additional analyses were conducted. First, from a sensitivity analysis limited to women without menstrual pain and pregnancy, similar results were observed. Second, a supplementary analysis was performed with the covariates plus the presence or absence of preschoolers and for living with a care recipient, and similar results were obtained (see Table S2).

4. Discussion

Our study had two main findings. First, a significant association between self-reported LBP and unemployment was observed only in men, but not in women, independently of potential confounders such as age, SES, lifestyle habits, and health status; the prevalence of unemployment was significantly higher in men with self-reported LBP than in those without self-reported LBP. Second, we quantified the importance of unemployment associated with self-reported LBP as a public health issue by estimating the PAF and found that self-reported LBP accounted for 2.8% of unemployed men aged 20 through 64 years. In this study, unemployment in 1037 thousand of the Japanese male working population was estimated to be LBP-related.

Our results are consistent with previous studies, showing that self-reported LBP is associated with unemployment [12,13,31]. This study also reveals the gender differences in this association. However, the mechanisms are unclear. First, previous studies report that women are prone to LBP, but we cannot find reports that women are at an increased risk of leaving their jobs due to LBP. Numerous studies show that women have a higher prevalence of LBP than men [6,11–13,32]. For this mechanism, women have a lower level of muscle strength and lung function [33], a lower pain threshold [34], and a higher level of engagement in domestic work than men [19]. In contrast, a significant association between musculoskeletal pain and unemployment was observed only in men [35], and among employees with LBP and/or neck and shoulder pain; being female is significantly associated with a higher risk of work disability but not associated with a risk of unemployment [36]. Next, men who develop LBP may have difficulty finding a new job, because men often work in occupations at high risk of LBP. Hooftman et al. indicated that men are more likely than women to be engaged in back-straining work such as handling heavy objects weighing 20 kg or more and long hours of driving [37]. According to a report on employment services in Japan [38], construction, manufacturing, and transport are among the top five industries with the highest number of mid-career hires. These industries are reported to have a high risk of LBP [9–11] and to be male-dominated; the percentage of male workers by industry is 85.6% for construction, 70.5% for manufacturing, and 81.8% for transport [39]. To summarize the above, although women are more vulnerable to LBP than men [19,33,34], unemployed men with LBP may have more difficulty in finding new employment because it is more difficult for men than for women to find work that does not put a strain on the lower back [37–39].

Since, as shown in this study, there are a considerable number of men unemployed due to LBP, we should take some measures to address this issue. Our results based on the PAF suggest that if this association is causal, the elimination of self-reported LBP could reduce the risk of unemployment related to self-reported LBP by an estimated 2.8% (95% CI 1.6% to 4.2%) in men. The PAF of 2.8% is equivalent to the estimate that unemployment in 1037 thousand Japanese men aged 20 to 64 years would be LBP-related. Based on the results in Table 3, among both genders, people with self-reported LBP had a significantly higher PR for the unemployed looking for work than people without self-reported LBP. This result suggested that if the LBP problem was resolved, these people may be able to work, regardless of gender. For prevention strategies of LBP, exercise, education, back belts, and ergonomic interventions in the workplace became widespread [40]. With regard to LBP prevention measures in the workplace, it was necessary and effective to take a systematic approach after conducting a risk assessment [41]. Moreover, in Western countries, mass-media campaigns aimed at changing the public's beliefs about LBP (i.e., changing from the maladaptive belief that rest is important for LBP management because movement creates physical damage to the spine to the sound belief that one should stay active during LBP episodes) were studied with some success [40,42]. Such public health interventions were a possible solution to the LBP problems of unemployed people.

Our study has the following strengths. First, the study's results are based on a nationally representative sample of Japan, which has a high proportion of respondents: about 80%. This ensures the generalizability of the results of this study. Second, because the CSLC gathers enough information about SES, lifestyle habits, and health status, we are able to control for potential confounding. However, the study does have some limitations. First, because this study is a cross-sectional design, we cannot identify a causal relationship. A longitudinal study based on a nationally representative survey is needed to clarify the causal relationship between self-reported LBP and unemployment. Second, the CSLC asked individuals a question about the presence or absence of subjective symptoms of LBP during the preceding several days. Because many studies evaluate chronic LBP such as the one-year prevalence of LBP [6,10,14,19,22], our study's results are not comparable with those of previous studies. Furthermore, self-reported LBP may include non-spinal causes of LBP, such as visceral causes (i.e., urinary, gynecological, and digestive disorders), osteoarthritis of the hip, fibromyalgia, obesity, and mood disorders [43]. Because this study has clinical and diagnostic implications, future studies are needed to confirm the association between clinical LBP and unemployment. Third, the definition of having a job (employed) in this study was that the employee had a job with income during May 2013; in the case of a person who did not have a job with income during the same month, his/her status was defined as having no job (unemployed). According to this definition, respondents were classified as unemployed if they worked until April 2013, and as employed if they started working in May 2013. It is unlikely that less than a month of unemployment or work experience would affect self-reported LBP. Although this study aims to investigate the association between self-reported LBP and unemployment, it should be noted that the definition of employment status in this study has a limitation. Fourth, this study uses the data from the 2013 survey, and while this is old data (8 years), it is the latest available data as of August 2021. As the summary tables of the 2016 and 2019 surveys are open to the public [44], we compared the data of the 2013, 2016, and 2019 surveys based on the published spreadsheet in order to verify the representativeness of the results of this study. Because the published tabulation data only lists the number of people who chose LBP as their symptom of most concern, the percentage of those who chose LBP cannot be compared to recent surveys. The percentage of people aged 20–64 with self-reported LBP as the symptom of most concern in 2013, 2016, and 2019 was 4.1%, 4.0%, and 3.8% for men, and 3.6%, 3.4%, and 3.4% for women; the proportion of people with self-reported LBP remains almost unchanged. In contrast, the percentage of unemployed people aged 20 to 64 in 2013, 2016, and 2019 was 11.8%, 10.2%, and 8.8% for men and 33.6%, 30.1%, and 26.4% for women (we should note that the percentages obtained do not match because the criteria for unemployed people to be included or excluded were different between the anonymized data and the aggregated results based on raw data). The proportion of unemployed people declined gradually from 2013 to 2019, particularly for women, where there was a substantial decline. However, it was reported that the number of unemployed people increased for both men and women; there was an especially large increase in the number of females unemployed due to the spread of the COVID-19 since 2020 [45]. Therefore, because the association between unemployment and LBP may be altered by the female workforce participation and a COVID-19-related unemployment, further research is needed to validate the gender-specific association between self-reported LBP and unemployment using the latest data. Fifth, it is difficult to compare the results obtained in this study with the results of previous studies. The reasons for this are that, as far as we know, there is no previous study examining the fraction and number of unemployed people associated with self-reported LBP. Moreover, the definition of self-reported LBP adopted in this study is different from the definition of self-reported LBP in previous studies [6,10,14,19,22]. However, it is consistent with the results of previous studies reporting the association between self-reported LBP and unemployment and suggests that LBP is the most important factor in disability and social costs [1,3,4,12–14,31]. Therefore, it is considered that the results of this study are supported by the results of previous studies.

Our findings have important policy implications. In Japan, securing a labor force is an urgent issue due to the rapidly declining birthrate and aging population. The government has implemented measures to encourage women and older people to participate in the labor force and provide employment support for those who could not get a job during the period when it was difficult to find employment after the burst of the bubble economy (currently those in their late 30s to late 40s) [46]. The results of this study show that by taking new measures to solve the LBP problem of unemployed men aged 20 to 64, it is possible to encourage them to return to work and secure the male labor force of a working-age population of about 1 million.

5. Conclusions

Independent of relevant confounding factors, a significant association between self-reported LBP and unemployment was observed only in men; men with self-reported LBP were more likely to be unemployed compared to men without self-reported LBP. Additionally, the estimated PAF for unemployment associated with self-reported LBP was about 3.0% in men. Our findings suggested that LBP measures for unemployed men between the age of 20 and 64 could enable some of the LBP-related non-workers to return to the workforce.

Supplementary Materials: The following are available online at https://www.mdpi.com/article/10.3390/ijerph182010760/s1, Table S1: Distribution of covariates by gender, Table S2: Additional analyses among women.

Author Contributions: Conceptualization, K.T., M.S., and K.S.; Methodology, K.T., and T.K.; Formal Analysis, K.T., and K.S.; Data Curation, K.T.; Writing—Original Draft Preparation, K.T.; Writing—Review and Editing, T.K., and M.S.; Supervision, K.S.; Funding Acquisition, K.T. All authors have read and agreed to the published version of the manuscript.

Funding: This research was funded by a Grant-in-Aid for Scientific Research (No. 17K09209) from the Japan Society for the Promotion of Science.

Institutional Review Board Statement: In this study, we received approval of use for academic purposes from the Japanese Ministry of Health, Labour and Welfare (approval number 20009), based on Article 36 of the Statistics Law of Japan. The data provided had personally identifiable information removed.

Informed Consent Statement: Not applicable for studies using existing data.

Data Availability Statement: Data are available from the Ministry of Health, Labour and Welfare, Japan (https://www.mhlw.go.jp/toukei/itaku/tokumei.html, accessed on 15 September 2021) for researchers who obtain approval to use the anonymous data in accordance with Article 36 of the Statistics Law of Japan.

Acknowledgments: We wish to thank the participants of this study and support staff who make the study possible. We also thank Heather Hill for her English language editing.

Conflicts of Interest: K.S. received research grants from YKK AP Inc.; Ushio Inc.; Tokyo Electric Power Company; EnviroLife Research Institute Co., Ltd.; Sekisui Chemical Co., Ltd.; LIXIL Corp.; and KYOCERA Corp. The funders had no role in the design of the study; in the collection, analyses, or interpretation of data; in the writing of the manuscript, or in the decision to publish the results. The other authors declare no conflict of interest.

References

1. GBD 2016 Disease and Injury Incidence and Prevalence Collaborators. Global, regional, and national incidence, prevalence, and years lived with disability for 328 diseases and injuries for 195 countries, 1990–2016: A systematic analysis for the Global Burden of Disease Study 2016. *Lancet* **2017**, *390*, 1211–1259. [CrossRef]
2. Ministry of Health, Labour and Welfare, Japan. Estimates of National Medical Care Expenditure 2016. Available online: https://www.mhlw.go.jp/toukei/list/37-21.html (accessed on 15 September 2021).
3. Dieleman, J.L.; Cao, J.; Chapin, A.; Chen, C.; Li, Z.; Liu, A.; Horst, C.; Kaldjian, A.; Matyasz, T.; Scott, K.W.; et al. US Health Care Spending by Payer and Health Condition, 1996–2016. *JAMA* **2020**, *323*, 863–884. [CrossRef]
4. Dagenais, S.; Caro, J.; Haldeman, S. A systematic review of low back pain cost of illness studies in the United States and internationally. *Spine J.* **2008**, *8*, 8–20. [CrossRef]
5. Punnett, L.; Prüss-Utün, A.; Nelson, D.I.; Fingerhut, M.A.; Leigh, J.; Tak, S.; Phillips, S. Estimating the global burden of low back pain attributable to combined occupational exposures. *Am. J. Ind. Med.* **2005**, *48*, 459–469. [CrossRef] [PubMed]
6. Shiri, R.; Falah-Hassani, K.; Heliövaara, M.; Solovieva, S.; Amiri, S.; Lallukka, T.; Burdorf, A.; Husgafvel-Pursiainen, K.; Viikari-Juntura, E. Risk Factors for Low Back Pain: A Population-Based Longitudinal Study. *Arthritis. Care. Res.* **2019**, *71*, 290–299. [CrossRef] [PubMed]
7. Ghilan, K.; Al-Taiar, A.; Yousfi, N.A.; Zubaidi, R.A.; Awadh, I.; Al-Obeyed, Z. Low back pain among female nurses in Yemen. *Int. J. Occup. Med. Environ. Health* **2013**, *26*, 605–614. [CrossRef] [PubMed]
8. Yoshimoto, T.; Oka, H.; Fujii, T.; Kawamata, K.; Kokaze, A.; Koyama, Y.; Matsudaira, K. Survey on chronic disabling low back pain among care workers at nursing care facilities: A multicenter collaborative cross-sectional study. *J. Pain Res.* **2019**, *12*, 1025–1032. [CrossRef] [PubMed]
9. Joseph, L.; Standen, M.; Paungmali, A.; Kuisma, R.; Sitilertpisan, P.; Pirunsan, U. Prevalence of musculoskeletal pain among professional drivers: A systematic review. *J. Occup. Health* **2020**, *62*, e12150. [CrossRef]
10. Umer, W.; Antwi-Afari, M.F.; Li, H.; Szeto, G.P.Y.; Wong, A.Y.L. The prevalence of musculoskeletal symptoms in the construction industry: A systematic review and meta-analysis. *Int. Arch. Occup. Environ. Health* **2018**, *91*, 125–144. [CrossRef]
11. Inoue, G.; Uchida, K.; Miyagi, M.; Saito, W.; Nakazawa, T.; Imura, T.; Shirasawa, E.; Akazawa, T.; Orita, S.; Inage, K.; et al. Occupational Characteristics of Low Back Pain Among Standing Workers in a Japanese Manufacturing Company. *Workplace Health Saf.* **2020**, *68*, 13–23. [CrossRef]
12. Shmagel, A.; Foley, R.; Ibrahim, H. Epidemiology of Chronic Low Back Pain in US Adults: Data from the 2009-2010 National Health and Nutrition Examination Survey. *Arthritis. Care. Res.* **2016**, *68*, 1688–1694. [CrossRef] [PubMed]
13. Lallukka, T.; Mänty, M.; Cooper, C.; Fleischmann, M.; Kouvonen, A.; Walker-Bone, K.E.; Head, J.A.; Halonen, J.I. Recurrent back pain during working life and exit from paid employment: A 28-year follow-up of the Whitehall II Study. *Occup. Environ. Med.* **2018**, *75*, 786–791. [CrossRef] [PubMed]

14. Morken, T.; Riise, T.; Moen, B.; Hauge, S.H.; Holien, S.; Langedrag, A.; Pedersen, S.; Saue, I.L.; Seljebø, G.M.; Thoppil, V. Low back pain and widespread pain predict sickness absence among industrial workers. *BMC Musculoskelet. Disord.* **2003**, *4*, 21. [CrossRef] [PubMed]
15. Di Gessa, G.; Corna, L.M.; Platts, L.G.; Worts, D.; McDonough, P.; Sacker, A.; Price, D.; Glaser, K. Is being in paid work beyond state pension age beneficial for health? Evidence from England using a life-course approach. *J. Epidemiol. Community Health* **2017**, *71*, 431–438. [CrossRef] [PubMed]
16. Ministry of Health, Labour and Welfare, Japan. Overview of the 2013 Comprehensive Survey of Living Conditions. Available online: https://www.mhlw.go.jp/toukei/saikin/hw/k-tyosa/k-tyosa13/index.html (accessed on 15 September 2021).
17. Tomioka, K.; Shima, M.; Saeki, K. Association between heaviness of cigarette smoking and serious psychological distress is stronger in women than in men: A nationally representative cross-sectional survey in Japan. *Harm. Reduct. J.* **2021**, *18*, 27. [CrossRef]
18. Bento, T.P.F.; Genebra, C.V.D.S.; Maciel, N.M.; Cornelio, G.P.; Simeão, S.F.A.P.; Vitta, A. Low back pain and some associated factors: Is there any difference between genders? *Braz. J. Phys. Ther.* **2020**, *24*, 79–87. [CrossRef]
19. Garcia, J.B.; Hernandez-Castro, J.J.; Nunez, R.G.; Pazos, M.A.; Aguirre, J.O.; Jreige, A.; Delgado, W.; Serpentegui, M.; Berenguel, M.; Cantemir, C. Prevalence of low back pain in Latin America: A systematic literature review. *Pain Physician* **2014**, *17*, 379–391. [CrossRef]
20. Tani, N.; Ohta, M.; Higuchi, Y.; Akatsu, J.; Kumashiro, M. Lifestyle and subjective musculoskeletal symptoms in young male Japanese workers: A 16-year retrospective cohort study. *Prev. Med. Rep.* **2020**, *20*, 101171. [CrossRef]
21. Power, C.; Frank, J.; Hertzman, C.; Schierhout, G.; Li, L. Predictors of low back pain onset in a prospective British study. *Am. J. Public Health* **2001**, *91*, 1671–1678. [CrossRef]
22. Mork, P.J.; Vik, K.L.; Moe, B.; Lier, R.; Bardal, E.M.; Nilsen, T.I.L. Sleep problems, exercise and obesity and risk of chronic musculoskeletal pain: The Norwegian HUNT study. *Eur. J. Public Health* **2014**, *24*, 924–929. [CrossRef] [PubMed]
23. Cook, D.G.; Cummins, R.O.; Bartley, M.J.; Shaper, A.G. Health of unemployed middle-aged men in Great Britain. *Lancet* **1982**, *1*, 1290–1294. [CrossRef]
24. Ministry of Education, Culture, Sports, Science and Technology, Japan. School Education Act. Available online: https://elaws.e-gov.go.jp/document?lawid=418AC0000000120 (accessed on 15 September 2021).
25. Tamakoshi, A.; Ohno, Y.; JACC Study Group. Self-reported sleep duration as a predictor of all-cause mortality: Results from the JACC study, Japan. *Sleep* **2004**, *27*, 51–54.
26. Jean-Louis, G.; Turner, A.D.; Seixas, A.; Jin, P.; Rosenthal, D.M.; Liu, M.; Avirappattu, G. Epidemiologic Methods to Estimate Insufficient Sleep in the US Population. *Int. J. Environ. Res. Public Health* **2020**, *17*, 9337. [CrossRef]
27. White, I.R.; Thompson, S.G. Adjusting for partially missing baseline measurements in randomized trials. *Stat. Med.* **2005**, *24*, 993–1007. [CrossRef]
28. Levin, M.L. The occurrence of lung cancer in man. *Acta. Unio. Int. Contra. Cancrum.* **1953**, *9*, 531–541. [PubMed]
29. Daly, L.E. Confidence limits made easy: Interval estimation using a substitution method. *Am. J. Epidemiol.* **1998**, *147*, 783–790. [CrossRef]
30. Statistics Bureau, Ministry of Internal Affairs and Communications, Japan. Population Estimation. Available online: http://www.stat.go.jp/data/jinsui/ (accessed on 15 September 2021).
31. Nwaru, C.A.; Nygård, C.H.; Virtanen, P. Musculoskeletal pain and re-employment among unemployed job seekers: A three-year follow-up study. *BMC Public Health* **2016**, *16*, 531. [CrossRef] [PubMed]
32. Yang, H.; Haldeman, S.; Lu, M.L.; Baker, D. Low Back Pain Prevalence and Related Workplace Psychosocial Risk Factors: A Study Using Data From the 2010 National Health Interview Survey. *J. Manipulative Physiol. Ther.* **2016**, *39*, 459–472. [CrossRef]
33. Das, B. Gender differences in prevalence of musculoskeletal disorders and physiological stress among the brick field workers of West Bengal, India. *Work* **2019**, *63*, 389–403. [CrossRef] [PubMed]
34. Ellermeier, W.; Westphal, W. Gender differences in pain ratings and pupil reactions to painful pressure stimuli. *Pain* **1995**, *61*, 435–439. [CrossRef]
35. Bingefors, K.; Isacson, D. Epidemiology, co-morbidity, and impact on health-related quality of life of self-reported headache and musculoskeletal pain—A gender perspective. *Eur. J. Pain* **2004**, *8*, 435–450. [CrossRef] [PubMed]
36. Mather, L.; Ropponen, A.; Mittendorfer-Rutz, E.; Narusyte, J.; Svedberg, P. Health, work and demographic factors associated with a lower risk of work disability and unemployment in employees with lower back, neck and shoulder pain. *BMC Musculoskelet. Disord.* **2019**, *20*, 622. [CrossRef] [PubMed]
37. Hooftman, W.E.; van der Beek, A.J.; Bongers, P.M.; van Mechelen, W. Is there a gender difference in the effect of work-related physical and psychosocial risk factors on musculoskeletal symptoms and related sickness absence? *Scand. J. Work Environ. Health* **2009**, *35*, 85–95. [CrossRef] [PubMed]
38. Ministry of Health, Labour and Welfare, Japan. Report on Employment Service in 2013. Available online: https://www.mhlw.go.jp/toukei/youran/roudou-nenpou2013/02.html (accessed on 15 September 2021).
39. Ministry of Internal Affairs and Communications, Japan. Labor Force Survey 2013. Available online: https://www.stat.go.jp/data/roudou/index2.html#kekka (accessed on 15 September 2021).

40. Foster, N.E.; Anema, J.R.; Cherkin, D.; Chou, R.; Cohen, S.P.; Gross, D.P.; Ferreira, P.H.; Fritz, J.M.; Koes, B.W.; Peul, W.; et al. Prevention and treatment of low back pain: Evidence, challenges, and promising directions. *Lancet* **2018**, *391*, 2368–2383. [CrossRef]
41. International Organization for Standardization (ISO). ISO/TR 12296:2012. Ergonomics—Manual Handling of People in the Healthcare Sector. 2012. Available online: https://www.iso.org/obp/ui/#iso:std:iso:tr:12296:ed-1:v1:en (accessed on 15 September 2021).
42. Gross, D.P.; Russell, A.S.; Ferrari, R.; Battié, M.C.; Schopflocher, D.; Hu, R.; Waddell, G.; Buchbinder, R. Evaluation of a Canadian back pain mass media campaign. *Spine* **2010**, *35*, 906–913. [CrossRef] [PubMed]
43. Goel, S. Non-spinal Causes of Back Pain: An "Undiagnosed" Diagnosis. *J. Med. Res. Innov.* **2019**, *3*, e000172. [CrossRef]
44. Ministry of Health, Labour and Welfare, Japan. Summary of Results of the Comprehensive Survey of Living Conditions. Available online: https://www.mhlw.go.jp/toukei/list/20-21kekka.html (accessed on 15 September 2021).
45. Gender Equality Bureau Cabinet Office. Study Group Report on Impacts and Challenges for Women due to COVID-19 Pandemic. Available online: https://www.gender.go.jp/kaigi/kento/covid-19/index.html (accessed on 9 October 2021).
46. Ministry of Health, Labour and Welfare, Japan. Information on Support for the Employment Ice Age Generation. Available online: https://www.mhlw.go.jp/shushoku_hyogaki_shien/ (accessed on 15 September 2021).

International Journal of
Environmental Research and Public Health

MDPI

Article

Influence of Work-Related Safety and Health Guidelines on Knowledge and Prevalence of Occupational Back Pain among Rehabilitation Nurses in Saudi Arabia: A 6-Month Follow-Up Study

Ahmad H. Alghadir [1], Hani Al-Abbad [2], Syamala Buragadda [1] and Amir Iqbal [1,*]

1 Rehabilitation Research Chair, College of Applied Medical Sciences, King Saud University, Riyadh 11433, Saudi Arabia; alghadir@ksu.edu.sa (A.H.A.); syamala3110@yahoo.co.in (S.B.)
2 Physical Therapist, King Fahad Medical City, Riyadh 12231, Saudi Arabia; haniala733@gmail.com
* Correspondence: physioamir@gmail.com or ajamaluddin@ksu.edu.sa; Tel.: +966-1469-6010; Fax: +966-1469-3589

Abstract: Background: Nurses are frequently involved in different types of patient handling activities in different departments of the hospitals. Mishandling the patients causes accumulative stress on their spine that results in occupational back pain (OBP), substantial morbidity, and incurred cost. Objectives: This study aimed to observe the influence of work-related safety and health guidelines on knowledge and prevalence of occupational back pain among rehabilitation nurses in Saudi Arabia. Methodology: This cohort study was conducted with the inclusion of a total of 116-registered rehabilitation nurses (97-female, 19-male, mean age = 39.6-years) from different regions of Saudi Arabia. After the invitation, these nurses attended an ergonomic workshop focusing on work-related safety and patient handling guidelines, risk assessment, and control of OBP. A self-administered questionnaire was used to assess the knowledge, risk, and prevalence of OBP at baseline and 6-months follow-up. Results: The perceived knowledge score significantly improved (95% CI; t = 4.691; $p < 0.001$; Cohen's $d = 0.72$) at 6-month follow-up (mean ± SD = 81.6 ± 18.2) from its baseline score (mean ± SD = 68.2 ± 19.2). Likewise, the prevalence score of OBP markedly reduced from 71.5% (baseline) to 65.0% (6-month follow-up). Conclusion: The level of knowledge highly improved and the prevalence of OBP markedly reduced within a span of 6-month among rehabilitation nurses in Saudi Arabia after attending an ergonomic workshop. Importantly, the nurses learned and geared up themselves for practicing the safe patient handling guidelines to avoid occupational back pain in the future. Therefore, rehabilitation nurses should update their knowledge and awareness about occupational safety and health guidelines, risk assessments, and control of OBP at a regular interval for increasing the knowledge and reducing the prevalence of OBP among them.

Keywords: occupational safety and health; risk assessment; occupational disorder; knowledge; low back pain; rehabilitation nurses; patient care

Citation: Alghadir, A.H.; Al-Abbad, H.; Buragadda, S.; Iqbal, A. Influence of Work-Related Safety and Health Guidelines on Knowledge and Prevalence of Occupational Back Pain among Rehabilitation Nurses in Saudi Arabia: A 6-Month Follow-Up Study. *IJERPH* **2021**, *18*, 8711. https://doi.org/10.3390/ijerph18168711

Academic Editor: Paul B. Tchounwou

Received: 28 June 2021
Accepted: 13 August 2021
Published: 18 August 2021

1. Introduction

Occupational back problems among nurses is an area of interest for many authors in different countries as they account for substantial morbidity and cost [1]. Nurses are frequently involved in different patients handling activities that require either prolonged sustained postures or repetitive movement that have an accumulative stress on the spine [2,3]. Many risk factors have been identified as contributing to this problem including increased physical work demands, nurses' skills in patient handling, poor ergonomics in patient care, unavailability of assistive devices, psychosocial factors, and work organizational factors [4,5]. Interventions to help prevent or reduce this problem have shown controversial results. Among these interventions were patients handling education and training which is the most common method, ergonomics intervention, lifting teams, stress management, exercises, and provision of assistive devices [6,7].

The literature is abundant with studies that used single-factor interventions that showed limited impact on outcomes [8–11]. However, studies utilizing multidimensional strategies specifically based on risk assessment and control are more likely to be effective [12,13]. The aim of the risk assessment process is to identify the potential risks involved in patient handling and subsequently control them. The risk control process requires taking all available steps to eliminate hazards, if it is not possible to isolate them, they must be minimized and closely monitored for their effectiveness [14,15]. Four key risk factors need to be assessed [15]. First, the load refers to "patient characteristics" that can affect the handling risk. Second, the individual refers to the capabilities of the caregiver that can influence their capacity to carry out the job safely. Third, the task refers to the nature of the task as different tasks with different requirements, each needing proper assessment and a unique approach. Fourth, the working environment that impacts how the task is performed. The process takes place at the levels of the workplace and in relation to the handling of each patient [15].

The influence of nurses' awareness and knowledge about safe patient handling practice and compliance to standard guidelines is scarce in the literature and advised by many authors to design such an educational program scheduled with their work as to avoid the work-related injuries among them [16,17]. To our knowledge, there are not enough studies that have approached multidirectional strategies, including organized courses with hands-on-workshop to improve the knowledge and awareness about risk assessments and their control in reducing the prevalence of work-related low back pain among nurses. This study fulfills that scarcity by estimating the effectiveness of a well-organized ergonomics workshop focused on safe patient handling guidelines and measures to reduce the prevalence of occupational back pain among rehabilitation nurses. Furthermore, this study points out the importance of adhering to a particular safe patient's handling guidelines and measures aiming to be free from occupational back pain among rehabilitation nurses.

Objectives

The objective of this study was to educate and increase the level of knowledge and awareness about the risk assessments and control of OBP among rehabilitation nurses in Saudi Arabia through following the safe patient handling guidelines. In addition, it also aimed to evaluate the impact of the ergonomics workshop on the level of knowledge and the prevalence of OBP among them. Two research questions/hypotheses directed this study as follows:

Did the level of acquired and perceived knowledge about the risk assessments and control of OBP among rehabilitation nurses in Saudi Arabia increase after attending the ergonomics workshop?

Did the improved acquired and perceived knowledge (about the risk assessment and control of OBP) after attending the ergonomics workshop play an important role in reducing the prevalence of OBP among rehabilitation nurses in Saudi Arabia?

2. Materials and Methods

2.1. Study Design

A cohort study design with six-months follow-up was used in this study.

2.2. Ethical Consideration

This study was fully complied with the ethical standard for human research and approved by the Ethics Sub-Committee at King Saud University (file ID: RRC-2017-003; dated: 23-02-2017) and also complied with the Helsinki Declaration as revised in 2013.

2.3. Setting

An invitation to attend the ergonomics workshop titled "workshop on patient's handling and occupational back pain among rehabilitation nurses" was sent via email to different hospitals in Riyadh city, Saudi Arabia.

2.4. Participants

Practicing nurses handling patients and head and in-charge nurses were the targeted participants in this study. Inclusion criteria were included nurses working in rehabilitation, having the age more than thirty years, and continuous practicing experience of more than two years. Nurses who had morbid obesity and health-related problems prohibiting the provision of patient care were excluded from the study. Out of the 156 nurses who registered for the workshop, 116 were screened and recruited for the study based on the inclusion and exclusion criteria.

2.5. Procedure and Measurements

A one-day ergonomics workshop was conducted at a specified center of our university. It was organized and accredited for a 7 h continuing medical education (CME) by our university. The course delivered in the ergonomic workshop was divided into theoretical and practical sessions. The theoretical topics included anatomy and risk of injury, biomechanics, and the concept of patient handling based on the "New Zealand Patient Handling Guidelines-the LITEN UP Approach 2003", (Wellington, New Zealand, 2003) evidence-based patient handling, controversial techniques, and hazardous tasks, walking aids and patient handling assistive devices, back care, and exercises [15,17]. The practical session covered the use of walking aids and assistive devices, techniques of safe patient handling including moving the patient in bed, bed positioning, sitting to the edge of the bed, standing, and sitting, bed to wheelchair transfer, transferring a patient on lying surface, assisting a fallen patient. The concept of risk assessment and control was the theme of instruction during the entire workshop. The workshop was delivered by four instructors experienced in back care and patient handling. The participants were provided with a 30 pages' manual covering all aspects of the workshop.

2.6. Outcome Measures

A "self-administered questionnaire" of two pages was given to the participants on the day of the workshop as a baseline measure of their knowledge, risk, and exposure to occupational back pain. The initial version of the questionnaire was developed based on a literature review. The questionnaire was then reviewed by a panel of experts with more than 10 years of experience. Modifications were made based on the recommendation of the panel. A convenience sample consisting of 20 non-participating nurses was asked to fill out the final version of the questionnaire on two occasions, two days apart. Test-retest reliability was assessed using intra-class correlation coefficient, which ranged between 0.85–0.9, indicating high reliability of the questionnaire.

A cover letter stated the objective of the study and assured the participants that the data obtained are confidential. Each questionnaire was coded with a number that corresponded to a master list of names. The questionnaire was composed of four main sections with primarily "close ended questions". The first section was designed to obtain demographic information such as gender, age, height, weight, educational level, years of professional experience, and type of working area. The second section aimed to evaluate the perceived level of knowledge and awareness of the participants about different parameters derived from standard guidelines of safe patient handling and previous training on patient handling. The third section inquired about the participant's physical exposure including the number of patients handled on a daily basis, the percentage of time devoted for patients moving and transfer, and the handling tasks practiced during work. The Standardized Nordic Questionnaire was used in the fourth section to assess the amount of back injury through information on the number of days with back problems during the past year [18].

A 6-month follow-up questionnaire was presented to the participants through their e-mail contacts. The baseline questionnaire items were preserved with two additional items in the second section asking about the perceived implementation of knowledge gained from the workshop and reasons for not implementing the knowledge gained in the workshop into practice. The third section was replaced with a quiz of 10 true/false questions to examine the knowledge gained at the workshop. The Nordic questionnaire was modified to ask about back injuries during the past 6-months.

2.7. Statistical Analysis

Data analysis for all the variables was done using the statistical software SPSS (IBM SPSS Statistics for Windows, v. 21, Armonk, NY, USA, IBM Corp). Mean differences from baseline to 6-months follow-up and descriptive statistics were calculated by applying a paired t-test. Further, Cohen's *d* [19] test was used to see the effect size of the intervention (the ergonomic workshop) on the perceived knowledge among the participants. Percentage change and composite means of the test scores were used to evaluate the prevalence of OBP, implementation of perceived knowledge, and acquired knowledge for all the participants. The level of significance (α) was set at 0.05 for all the statistical analyses.

3. Results

3.1. Demographic Characteristics

The analysis was conducted on the subset sample of participants that responded to the second questionnaire. A total of 116-participants attended an ergonomics workshop to receive an educational intervention (the ergonomic workshop) but only 84-participants returned the questionnaire with a response rate of 72.4% at 6-months follow-up via e-mail. The mean age of the participant nurses was 39.6 $\pm$ 8.60 years, with a BMI of 26 $\pm$ 4.70 kg/m^2. Their average clinical work experience was 5 $\pm$ 1.50 years. In addition, data for professional characteristics including educational level, area/department of practice, and year of professional experiences are presented below (see Table 1).

Table 1. Professional characteristic of the respondents.

Professional Characteristics	Respondents (%)
Educational level	
Diploma	25
Bachelor	52
Post graduate diploma	8
Clinical specialty	5
Other	1
Area of practice	
Medical units	1
Surgical units	31
Neurology/neurosurgery	2
Out-patient clinics	6
Intensive care units	3
Neuro rehabilitation	25
Other units (OBG, emergency etc.)	29
Rotations	3
Professional experience (years)	
2–4	7
4–6	7
6–8	13
8–10	16
>10 years	57

The outcome measures of this study were perceived and acquired knowledge and the prevalence of OBP, and the independent variables were educational intervention in the ergonomic workshop and physical exposure parameters.

3.2. Perceived Knowledge

As shown in Table 2, the score of perceived knowledge among nurses was significantly improved (13.4 ± 9.5; $p < 0.05$) from its baseline score (68.2 ± 19.2) to 6-months of the follow-up score (81.6 ± 18.2). Furthermore, an item-wise and overall comparison of perceived knowledge can be seen between baseline and 6-month follow-up (see Figures 1 and 2).

Table 2. Item-wise differences in baseline vs. follow-up data of perceived knowledge.

Items of Perceived Knowledge (11-Items)	Baseline (%)	Follow Up (%) (6-Months)	Mean Difference (%) (95% CI)
Handling policy and plan	64	79	15
Reporting incidence	79	83	4
Reviewing handling steps	68	78	10
Asking for help	96	96	0
Patients with special needs	82	95	13
Using assistive devices	77	94	17
Using body weight	88	99	11
Using back exercises	56	80	24
Using relaxation breaks	36	60	24
Management of mild LBP	66	94	28
Involvement in fitness program	38	39	1
Overall total scores (mean ± SD) of perceived knowledge (%)	68.2 ± 19.2	81.6 ± 18.2	13.4 ± 9.5

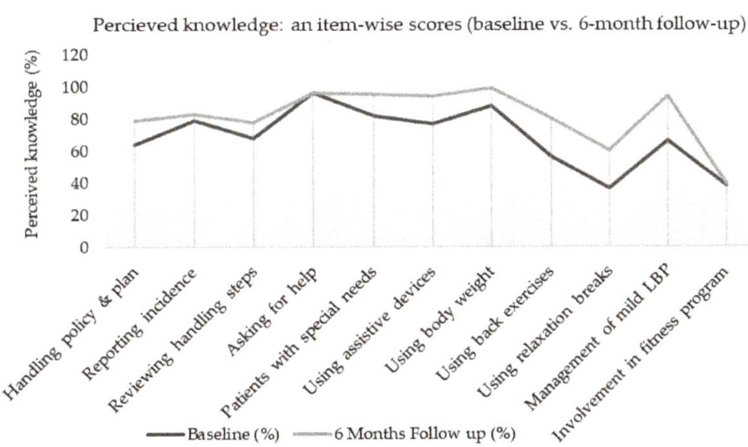

Figure 1. Item-wise comparison of perceived knowledge scores between baseline and 6-month follow-up.

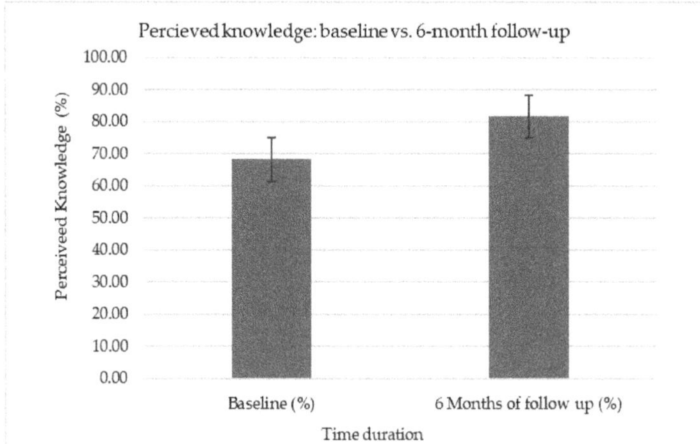

Figure 2. Overall comparison of perceived knowledge scores between baseline and 6-month follow-up.

In addition, a Cohen's *d* test applied indicating a large effect-size (Cohen's $d = 0.72$) of an ergonomics workshop on the perceived knowledge among rehabilitation nurses (see Table 3).

Table 3. Comparison of the effects between pre- and post-ergonomic workshop on the perceived knowledge (Cohen's *d* and paired *t*-test).

Variables	Mean ± SD	Paired *t*-Test		Cohen's *d*
		t-Value	*p*-Value	
Pre-workshop	68.2 ± 19.2	4.691	0.001 **	95% CI [3.18, −4.21] $d = 0.72$
Post-workshop	81.6 ± 18.2			
(M2-M1)/SD pooled	13.4/18.71			

** extremely significant if $p < 0.001$; small effect if $d = 0.20$; medium effect if $d = 0.40$; large effect if $d = 0.60$ [19].

3.3. Acquired Knowledge

Test yourself. Ten questions were asked to assess the extent of knowledge gained from the course delivered in the workshop and the composite score mean obtained was 7.67 ± 1.10. Their test score indicated that the level of knowledge acquired from the course delivered in the workshop was above the average. Prior to distributing the questionnaire to the participants, we instructed them to fill the answers at their own acquired knowledge from the course without referring back to the manual or other nursing professionals.

3.4. Perceived Implementation of Knowledge

The majority of participants (98.0%) declared "my manual handling knowledge improved after taking the ergonomics workshop". Around 90.0% of participants agreed that "I applied gained knowledge into my daily work" and that probably affected the percentage reduction in the prevalence of their occupational back pain among them.

3.5. Prevalence of OBP

The prevalence of OBP symptoms reduced with a difference of 6.5% (65% at 6-months follow-up) from its baseline scores (71.5%). This confirms that the nurses applied the gained (13.4 ± 9.5; $p < 0.05$; Cohen's $d = 0.72$) perceived knowledge (after the ergonomic workshop) into their clinical practices while handling the patients. Moreover, this also confirms the impact/role of the ergonomics workshop (an educational intervention) on reducing the prevalence of OBP among rehabilitation nurses in Saudi Arabia.

4. Discussion

The present study was designed to investigate the impact of a workshop-based educational program on the knowledge and awareness about potential risk factors on the prevalence of occupational back pains. The prevalence of low back pain in Saudi Arabia ranges from 53.2% to 79.2% and has multifactorial risk factors such as vitamin deficiency, obesity, sprains, stretching, and bending activities [20]. In addition, the number of patients, number of working hours in patient handling, patients care with poor ergonomics, and lack of adoption of "no lift policies" by health organizations were marked as major risk factors for back pain related to occupation [21]. Similarly, Mitchell et al. (2008) reported that noticeable health problems were identified "higher" among the nurses working longer than "20 h per week" in the hospitals, and about 60% of them were taken treatment, medicine, or reduction in activities [22]. It is widely known that physical exposure is the strongest risk factor of back pain among nurses and other health care workers [16]. However, means of prevention remain controversial. Nurses' injuries develop with the use of improper techniques in lifting and handling, which cause unnecessary stress and load on the spine and results in LBP; once an injury takes place, re-injury is inevitable [23,24]. Likewise, many studies supported this view that direct patient handling during the provision of care such as shifting (bed sheath changing, side rotation, up and down), lifting (heavy patients), mobility (assisted walking), transfer (bed to wheelchair, wheelchair to vehicle, etc.) and toilet care activities constitute a major risk for occupational low back pain among nurses in the hospitals [7,12,25].

The literature is abundant with evidence on back injuries among nurses [26–28]. Such injuries are associated with sick leaves [29]. LBP has been linked to worker's compensation claims and disability insurance in Western countries [30]. There is strong evidence from four high-quality studies and eight moderate studies that training intervention has no impact on working practice and injury rate. However, other studies showed that training interventions have mixed (positive and negative) short-term results [31]. An intervention based on a risk assessment program is most likely to be beneficial in reducing risk factors during patient handling [32]. We believe that there is a strong need to address new interventional preventive strategies emphasizing on risk assessment and control principle that looks at this problem from a multidimensional view.

Our findings confirm the roles of knowledge and awareness and the importance of education in reducing the risk of back pain among nurses [33,34]. Moreover, this is in line with previous studies [12,25]. The decline in self-reported back pain from 71.5% to 65% observed in this study 6-months after attending a workshop on safe handling techniques reflects the need for ongoing training. The value of introducing a patient handling policy and compliance with guidelines at work remains to be explored. We support the current moderate evidence that supports utilizing multidimensional strategies specifically based on risk assessment and control strategies that could potentially produce favorable results [6,7,35].

The risk of musculoskeletal injuries is mostly associated with dependent patient care and is usually secondary to manual patient handling. Knowledge about how and when to use assistive devices is necessary to avoid back injuries as high forces are required to transfer patients [36]. In contrast, ergonomic training proved ineffective in the prevention of back injuries with manual patient lifting [37]. On the other hand, it has been shown that awareness of transfer techniques along with physical fitness training may reduce disability due to low back pain [4,38]. The majority of the participants (90%) in our study agreed that knowledge and awareness about the handling techniques improved after taking the course and were being applied in their daily tasks. However, the measurement is only subjective and an objective method such as video recording the handling techniques would be more reliable.

Although the content and format of the educational intervention delivered in the ergonomics workshop utilized the best available guidelines for patient handling and back care among nurses, ensuring the implementation of the acquired knowledge and skills into clinical practice was not feasible. This is mainly due to the inability to impact policymakers within the different organizations to supervise and ensure adherence to the workshop recommendations of safe practice. The risk assessment and control model that the workshop adopted requires commitment at all levels of the organization. This commitment needs to be visible where staff need to be involved in decisions. Thus, it is suggested that training on its own is not enough for bringing about change and must be supported with effective health and safety systems to ensure compliance with safe practice.

This study showed a connection between gained knowledge through an educational intervention received from an ergonomic workshop and decreased prevalence of occupational back pain among rehabilitation nurses in Saudi Arabia. Therefore, conducting the ergonomics workshop (educational intervention program) dedicated to increasing the level of knowledge about risk assessment and control of OBP among nurses should be encouraged. Thus, the results of this study can be generalized among rehabilitation nurses to prevent/avoid work-related physical injuries while conducting patient handling/care activities such as shifting, lifting, mobility, toilet care, transfer activities, and so on. In addition, the report of this study can be generalized among hospital/organizational policies makers, so that, the organizational policies might be more focused on clear, constructive directives "including international standardized guidelines" and vigilance in injury prevention for rehabilitation nurses involved in patients handling activities. Moreover, there is a need for "preventive plans including ergonomic advice" and a mechanism of their implementation in patient care units for avoiding occupational back pain and injuries among rehabilitation nurses.

Limitation

Besides the value and importance of this study, there are few limitations also which require to be addressed in future studies. The participants tested their acquired knowledge by solving a questionnaire of 10 items of 10 marks provided to them. Their composite mean value was above average (7.67 ± 1.10 or 76.7%) which indicated a gain of knowledge because of attending the ergonomics workshop. Although, before filling the questionnaire, a written instruction was given to them to choose only the appropriate responses without using the workshop course manual. Similarly, a post-workshop evaluation at 6 months for the perceived knowledge was also conducted under strict supervision rather than believing in the honesty of the participants. However, this study did not try to ensure conducting the evaluation procedure at immediate and 6-months post-workshop under strict supervision using a professional standby video recording camera or recording through a CCTV or mobile camera. In addition, the study was limited to not including more nurses from more cities/provinces of the country. Therefore, a future study is required to address the shortcomings of this study to become part of the perfect solution for following safe patient handling and controlling the incidences of OBP among rehabilitation nurses. In addition, there is a need of conducting future studies focusing on the role of education and training in combating the barriers of safe patient handling and practices including working with overweight patients, more numbers of patients, being rushed or short-staffed, exhaustible working hours, certain physical requests from the patients that compromise nurse's safety, and so on, among rehabilitation nurses.

5. Conclusions

The report of the study concludes by answering both questions/hypotheses. First, the level of knowledge about the risk assessments and control of OBP among rehabilitation nurses in Saudi Arabia increased after attending the ergonomic workshop focusing on work-related safety, risk assessments and control, and patient handling guidelines. Second, the increased level of knowledge after attending the ergonomics workshop since 6-months before markedly reduced the prevalence of OBP among the nurses. Importantly, the nurses learned and geared up themselves to practice the safe patient handling guidelines to avoid occupational back pain in the future. Additionally, the implication may further reduce the prevalence of disability (due to back pain), morbidity, and incurred costs, resulting in an overall improvement in activities of daily living (ADLs). Therefore, the rehabilitation nurses should update their knowledge and awareness about safety, risk assessments and control, and patient handling guidelines at regular intervals for increasing the knowledge and reducing the prevalence of OBP among them.

Author Contributions: Conceptualization, A.H.A. and A.I.; data curation, H.A.-A. and S.B.; formal analysis, A.I.; funding acquisition, A.H.A.; methodology, A.H.A., H.A.-A., S.B. and A.I.; project administration, A.H.A.; resources, A.H.A., H.A.-A. and S.B.; supervision, A.H.A. and A.I.; writing—original draft, S.B. and A.I.; writing—review and editing, A.H.A., H.A.-A., S.B. and A.I. All authors have read and agreed to the published version of the manuscript.

Funding: This study was funded by the Deanship of Scientific Research (Vice Deanship of Scientific Research Chairs), King Saud University.

Institutional Review Board Statement: The study was conducted according to the guidelines of the Declaration of Helsinki, and approved by the Ethics Sub-Committee at King Saud University (file ID: RRC-2017-003; dated: 23-02-2017).

Informed Consent Statement: Informed consent was obtained from all subjects involved in the study.

Data Availability Statement: The data presented in this study are available on request from the corresponding author. The data are not publicly available due to reason of privacy.

Acknowledgments: The authors are grateful to the Deanship of Scientific Research, King Saud University for funding through Vice Deanship of Scientific Research Chairs.

Conflicts of Interest: The authors declare no conflict of interest.

References

1. Davis, K.G.; Kotowski, S.E. Prevalence of musculoskeletal disorders for nurses in hospitals, long-term care facilities, and home health care: A comprehensive review. *Hum. Factors* **2015**, *57*, 754–792. [CrossRef]
2. Adhikari, S.; Dhakal, G. Prevalent Causes of Low Back Pain and its Impact among Nurses Working in Sahid Gangalal National Heart Centre. *J. Nepal. Health Res. Counc.* **2014**, *12*, 167–171.
3. Fronteira, I.; Ferrinho, P. Do nurses have a different physical health profile? A systematic review of experimental and observational studies on nurses' physical health. *J. Clin. Nurs.* **2011**, *20*, 2404–2424. [CrossRef]
4. Boughattas, W.; El Maalel, O.; Maoua, M.; Bougmiza, I.; Kalboussi, H.; Brahem, A.; Chatti, S.; Mahjoub, F.; Mrizak, N. Low Back Pain among Nurses: Prevalence, and Occupational Risk Factors. *Occup. Dis. Environ. Med.* **2017**, *5*, 26. [CrossRef]
5. Al-samawi, M.A.G.; Awad, H.M.A.A. Incidences of Low Back Pain among Nurses Working in Elmak Nimer University Hospital—Shendi-Sudan 2015. *Nurs. Health* **2015**, *3*, 129–138. [CrossRef]
6. Mousavi Ouri, A.; Sedghi Goyaghaj, N.; Monjamed, Z.; Alamdarloo, A. The Study of Nursing Personnel, s Workload Caused by Patient Transferring. *J. Clin. Nurs. Midwifery* **2016**, *5*, 34–43.
7. Dawson, A.P.; McLennan, S.N.; Schiller, S.D.; Jull, G.A.; Hodges, P.W.; Stewart, S. Interventions to prevent back pain and back injury in nurses: A systematic review. *Occup. Environ. Med.* **2007**, *64*, 642–650. [CrossRef]
8. Dollard, M.F.; LaMontagne, A.D.; Caulfield, N.; Blewett, V.; Shaw, A. Job stress in the Australian and international health and community services sector: A review of the literature. *Int. J. Stress Manag.* **2007**, *14*, 417. [CrossRef]
9. Hoskins, A.B. Occupational injuries, illnesses, and fatalities among nursing, psychiatric, and home health aides, 1995–2004. *Retrieved* **2006**, *9*, 2009.
10. Hignett, S.; Crumpton, E. Competency-based training for patient handling. *Appl. Ergon.* **2007**, *38*, 7–17. [CrossRef]
11. Teeple, E.; Collins, J.E.; Shrestha, S.; Dennerlein, J.T.; Losina, E.; Katz, J.N. Outcomes of safe patient handling and mobilization programs: A meta-analysis. *Work* **2017**, *58*, 173–184. [CrossRef]
12. Szeto, G.; Wong, T.; Law, R.; Lee, E.; Lau, T.; So, B.; Law, S. The impact of a multifaceted ergonomic intervention program on promoting occupational health in community nurses. *Appl. Ergon.* **2013**, *44*, 414–422. [CrossRef] [PubMed]
13. Nelson, A.; Matz, M.; Chen, F.; Siddharthan, K.; Lloyd, J.; Fragala, G. Development and evaluation of a multifaceted ergonomics program to prevent injuries associated with patient handling tasks. *Int. J. Nurs. Stud.* **2006**, *43*, 717–733. [CrossRef] [PubMed]
14. Lin, C.H.; Lin, Y.-W.; Yang, C.M.; Hung, L.C.; Wang, J.Y.; Lin, M.H. Occupational health and safety hazards faced by healthcare professionals in Taiwan: A systematic review of risk factors and control strategies. *SAGE Open Med.* **2020**, *8*. [CrossRef]

15. Worksafe, A. The New Zealand Patient Handling Guidelines. The Liten up Approach. Available online: www.acc.co.nz/injury-prevention/saferindustries/health/phg:2003 (accessed on 23 February 2021).
16. Bin Homaid, M.; Abdelmoety, D.; Alshareef, W.; Alghamdi, A.; Alhozali, F.; Alfahmi, N.; Hafiz, W.; Alzahrani, A.; Elmorsy, S. Prevalence and risk factors of low back pain among operation room staff at a Tertiary Care Center, Makkah, Saudi Arabia: A cross-sectional study. *Ann. Occup. Environ. Med.* **2016**, *28*, 1. [CrossRef] [PubMed]
17. Abolfotouh, S.M.; Mahmoud, K.; Faraj, K.; Moammer, G.; ElSayed, A.; Abolfotouh, M.A. Prevalence, consequences and predictors of low back pain among nurses in a tertiary care setting. *Int. Orthop.* **2015**, *39*, 2439–2449. [CrossRef] [PubMed]
18. Kuorinka, I.; Jonsson, B.; Kilbom, A.; Vinterberg, H.; Biering-Sørensen, F.; Andersson, G.; Jørgensen, K. Standardised Nordic questionnaires for the analysis of musculoskeletal symptoms. *Appl. Ergon.* **1987**, *18*, 233–237. [CrossRef]
19. Wolf, F.M. *Meta-Analysis: Quantitative Methods for Research Synthesis*; Sage: New York, NY, USA, 1986; Volume 59.
20. Awaji, M. Epidemiology of Low Back Pain in Saudi Arabia. *J. Adv. Med. Pharm. Sci.* **2016**, *6*, 1–9. [CrossRef]
21. Shieh, S.-H.; Sung, F.-C.; Su, C.-H.; Tsai, Y.; Hsieh, V.C.-R. Increased low back pain risk in nurses with high workload for patient care: A questionnaire survey. *Taiwan. J. Obstet. Gynecol.* **2016**, *55*, 525–529. [CrossRef]
22. Mitchell, T.; O'Sullivan, P.B.; Burnett, A.F.; Straker, L.; Rudd, C. Low back pain characteristics from undergraduate student to working nurse in Australia: A cross-sectional survey. *Int. J. Nurs. Stud.* **2008**, *45*, 1636–1644. [CrossRef] [PubMed]
23. Sikiru, L.; Hanifa, S. Prevalence and risk factors of low back pain among nurses in a typical Nigerian hospital. *Afr. Health Sci.* **2010**, *10*, 26.
24. Karahan, A.; Kav, S.; Abbasoglu, A.; Dogan, N. Low back pain: Prevalence and associated risk factors among hospital staff. *J. Adv. Nurs.* **2009**, *65*, 516–524. [CrossRef] [PubMed]
25. Szeto, G.P.Y.; Law, K.Y.; Lee, E.; Lau, T.; Chan, S.Y.; Law, S.W. Multifaceted ergonomic intervention programme for community nurses: Pilot study. *J. Adv. Nurs.* **2010**, *66*, 1022–1034. [CrossRef]
26. Tosunoz, I.K.; Oztunc, G. Low back pain in nurses. *Int. J. Caring Sci.* **2017**, *10*, 1728–1732.
27. Mekonnen, T.H. Work-related factors associated with low back pain among nurse professionals in east and west Wollega zones, Western Ethiopia, 2017: A cross-sectional study. *Pain Ther.* **2019**, *8*, 239–247. [CrossRef] [PubMed]
28. O'Brien, W.H.; Horan, K.A.; Singh, S.R.; Moeller, M.M.; Wasson, R.S.; Jex, S.M.; Matthews, R.A.; Barratt, C.L. Relationships among training, mindfulness, and workplace injuries among nurse aides working in long-term care settings. *Occup. Health Sci.* **2019**, *3*, 45–58. Available online: https://doi.org/10.1007/s41542-018-0031-7 (accessed on 25 February 2021). [CrossRef]
29. Feng, C.-K.; Chen, M.-L.; Mao, I.-F. Prevalence of and risk factors for different measures of low back pain among female nursing aides in Taiwanese nursing homes. *BMC Musculoskelet. Disord.* **2007**, *8*, 52. [CrossRef] [PubMed]
30. Qin, J.; Kurowski, A.; Gore, R.; Punnett, L. The impact of workplace factors on filing of workers' compensation claims among nursing home workers. *BMC Musculoskelet. Disord.* **2014**, *15*, 29. [CrossRef]
31. Hignett, S. Intervention strategies to reduce musculoskeletal injuries associated with handling patients: A systematic review. *Occup. Environ. Med.* **2003**, *60*, e6. [CrossRef]
32. Kindblom-Rising, K.; Wahlström, R.; Nilsson-Wikmar, L.; Buer, N. Nursing staff's movement awareness, attitudes and reported behaviour in patient transfer before and after an educational intervention. *Appl. Ergon.* **2011**, *42*, 455–463. [CrossRef]
33. Jaromi, M.; Nemeth, A.; Kranicz, J.; Laczko, T.; Betlehem, J. Treatment and ergonomics training of work-related lower back pain and body posture problems for nurses. *J. Clin. Nurs.* **2012**, *21*, 1776–1784. [CrossRef] [PubMed]
34. Raeisi, S.; Hosseini, M.; Attarchi, M.S.; Golabadi, M.; Rezaei, M.S.; Namvar, M. The association between job type and ward of service of nursing personnel and prevalence of musculoskeletal disorders. *Razi J. Med. Sci.* **2013**, *20*, 1–10.
35. Shawashi, T.O.; Subih, M.M.; Al Hadid, L.A.R.; Abu Adas, M. Occupational-related back pain among Jordanian nurses: A descriptive study. *Int. J. Nurs. Pract.* **2015**, *21*, 108–114. [CrossRef] [PubMed]
36. Al-Eisa, E.; Al-Abbad, H. Occupational back pain among rehabilitation nurses in Saudi Arabia: The influence of knowledge and awareness. *Workplace Health Saf.* **2013**, *61*, 401–407. [CrossRef] [PubMed]
37. Edlich, R.; Winters, K.L.; Hudson, M.A.; Britt, L.; Long, W.B., III. Prevention of disabling back injuries in nurses by the use of mechanical patient lift systems. *J. Long Term Eff. Med. Implant.* **2004**, *14*, 521–533. [CrossRef] [PubMed]
38. Warming, S.; Ebbeh⊘j, N.; Wiese, N.; Larsen, L.; Duckert, J.; T⊘nnesen, H. Little effect of transfer technique instruction and physical fitness training in reducing low back pain among nurses: A cluster randomised intervention study. *Ergonomics* **2008**, *51*, 1530–1548. [CrossRef] [PubMed]

International Journal of
Environmental Research and Public Health

MDPI

Article

Effectiveness of Soft versus Rigid Back-Support Exoskeletons during a Lifting Task

Mathilde Schwartz [1,2], Jean Theurel [1] and Kévin Desbrosses [1,*]

[1] Working Life Department, French National Research and Safety Institute for the Prevention of Occupational Accidents and Diseases (INRS), 54500 Vandœuvre-les-Nancy, France; mathilde.schwartz@inrs.fr (M.S.); jean.theurel@inrs.fr (J.T.)

[2] Développement, Adaptation & Handicap (DevAH), University of Lorraine, 54000 Nancy, France

* Correspondence: kevin.desbrosses@inrs.fr; Tel.: +33-38-350-9893

Abstract: This study investigated the influence of passive back-support exoskeletons (EXO_{BK}) design, trunk sagittal inclination (TSI), and gender on the effectiveness of an exoskeleton to limit erector spinae muscle (ES) activation during a sagittal lifting/lowering task. Twenty-nine volunteers performed an experimental dynamic task with two exoskeletons (two different designs: soft (SUIT) and rigid (SKEL)), and without equipment (FREE). The ES activity was analyzed for eight parts of TSI, each corresponding to 25% of the range of motion (lifting: P1 to P4; lowering: P5 to P8). The impact of EXO_{BK} on ES activity depended on the interaction between exoskeleton design and TSI. With SKEL, ES muscle activity significantly increased for P8 (+36.8%) and tended to decrease for P3 (−7.2%, p = 0.06), compared to FREE. SUIT resulted in lower ES muscle activity for P2 (−9.6%), P3 (−8.7%, p = 0.06), and P7 (−11.1%), in comparison with FREE. Gender did not influence the effect of either back-support exoskeletons on ES muscle activity. These results point to the need for particular attention with regard to (1) exoskeleton design (rigid versus soft) and to (2) the range of trunk motion, when selecting an EXO_{BK}. In practice, the choice of a passive back-support exoskeleton, between rigid and soft design, requires an evaluation of human-exoskeleton interaction in real task conditions. The characterization of trunk kinematics and ranges of motion appears essential to identify the benefits and the negative effects to take into account with each exoskeleton design.

Keywords: musculoskeletal disorders; workload; wearable assistive device; occupational back-support exoskeleton; EMG; handling task; low back pain

Citation: Schwartz, M.; Theurel, J.; Desbrosses, K. Effectiveness of Soft versus Rigid Back-Support Exoskeletons during a Lifting Task. *IJERPH* **2021**, *18*, 8062. https://doi.org/10.3390/ijerph18158062

Academic Editors: Vincenzo Denaro, Sergio Iavicoli, Fabrizio Russo and Gianluca Vadalà

Received: 24 June 2021
Accepted: 27 July 2021
Published: 29 July 2021

1. Introduction

Occupational exoskeletons are wearable devices generally aimed at supporting users in performing their work tasks, by generating appropriate force/torque on one or multiple human joints. There has been increasing interest in employing exoskeletons for workplace ergonomics, particularly with the aim of reducing physical workload [1] and risk of musculoskeletal disorders (MSD) for workers [2]. Among these technologies, back-support exoskeletons (EXO_{BK}) have been specifically designed in view of preventing the occurrence of low back pain (LBP). Much evidence has been reported in the scientific literature about the efficiency of passive EXO_{BK} in relation to limiting lumbar muscular stress during handling operations involving the trunk flexion/extension in the sagittal plane [1–3]. For examples, during laboratory studies, reductions in back muscle activity ranged from −10 to −44% for repetitive lifting [4] and from −10 [5,6] to −57% [7] in static postures. Research carried out in real work conditions has also revealed significant reductions (−20 to −25%) in spinal muscle activity when using a passive EXO_{BK} [8]. The recent systematic review on industrial back-support exoskeletons [3] evaluated a mean reduction in erector spinae muscles activity of −18% during lifting and −36% during static bending for passive EXO_{BK}.

However, considerable disparities in the protocols of previous studies—including exoskeleton designs (e.g., flexible (soft) versus stiff (rigid) components), postures adopted (i.e., trunk inclination, and hip and knee flexion), loads lifted, task modalities (static versus dynamic), and populations—likely account for the substantial differences in the consequences on back muscle activity when using these systems [3]. As a result, these differences have created confusion regarding the benefits to be expected when using an EXO_{BK}, in terms of back muscle loading. Three sources of variability seem to be of particular importance.

First, the differences in the magnitude of reductions of muscular activity can vary depending on the passive EXO_{BK} designs [6,9]. More particularly, the assistive torque delivered by passive EXO_{BK} can be ensured either by elastic garments (exosuits, or soft exoskeletons) arranged in parallel with the spinal muscles, or by rigid spring elements (rigid exoskeletons) that act as hip extensions [10]. These components are tensioned when the trunk bends forward in the sagittal plane, and they then prompt the body to straighten up and adopt its initial position. However, to our knowledge, there is no information available on the influence of soft versus rigid exoskeleton designs on the efficiency of assistive devices in reducing spinal muscle activity during a similar trunk sagittal bending task [3]. Contrary to exosuits, the tensioning mechanisms of rigid EXO_{BK} are generally located at the hip level [6,9,11], thus partially neglecting independent hip and lumbar flexion in humans. In addition, the location of the contact point between the device and the user could also have impacts on posture and hence on the activity of muscles involved in posture regulation. For example, it appears that using a soft exoskeleton can significantly affect spine kinematics during lifting, by reducing lumbar and thoracic flexion [12]. Abdoli et al. [13] found a significant increase in plantar flexion during lifting tasks. Using a rigid EXO_{BK} (Laevo V1), Bosch, van Eck, Knitel, and de Looze [11] showed an increase in knee extension during forward bending.

Second, the benefits obtained from EXO_{BK} regarding back muscle activity appear to be dependent on the posture adopted (i.e., trunk sagittal inclination) during the experimental tasks performed. In the study by Lamers et al. [14], the reductions in back muscle activity observed with a passive custom-made EXO_{BK} varied from −23 to −43% for leaning tasks, performed at 30 and 90° respectively, in flexion (with a 4 kg load). In the study by Koopman, Kingma, Faber, de Looze, and van Dieen [7], the reductions in back muscle activity observed during static bending tasks with a passive EXO_{BK} (Laevo V2) varied from −11 to −57%, depending on the experimental postures imposed (five different hand heights). Moreover, it appeared that the relationships between the amplitude of muscle activity reduction and trunk posture were highly variable across subjects, probably as a function of individual kinematics and/or anthropometric characteristics [7,15].

Third, gender may also affect the relative benefits of using a passive exoskeleton. Indeed, some studies have shown different changes on lumbar muscles activity between men and women [6,9,16]. For example, So, Cheung, Liu, Tang, Tsoi, and Wu [16] noted a reduction in EMG activity of the erector spinae (ES) muscles with an EXO_{BK} for men but no change for women. Alemi, Madinei, Kim, Srinivasan, and Nussbaum [6] also reported an interaction between EXO_{BK} models and gender. In their study, the trunk extensors muscles activity was significantly decreased for women when using the Laevo V2.5 and BackX exoskeletons during a repetitive lifting/lowering task, while men noted a decrease only with the BackX.

Considering these sources of variability in back muscles activity, the objective of this study was to assess the influence of (1) the exoskeleton design (soft versus rigid exoskeleton), (2) the posture adopted (different trunk sagittal inclination), (3) the gender, and (4) the interactions between these parameters on the effectiveness of using a passive EXO_{BK} in terms of lumbar muscle activation reductions during a dynamic lifting/lowering task.

2. Materials and Methods

2.1. Participants

Twenty-nine volunteers (15 men: 23 ± 3 years, 179 ± 6 cm, 77 ± 7 kg; and 14 women: 22 ± 2 years, 167 ± 4 cm, 58 ± 9 kg) without back pathologies volunteered to participate in this study. All the participants followed a standardized training protocol including functional and handling tasks with and without an exoskeleton. The training protocol consisted of three 20 min sessions for each assistive condition (wearing soft and rigid exoskeletons and without wearing it). Participants gave their written consent after receiving detailed information on the objectives, protocol, and possible risks of the study. The experimental protocol received approval from the ethical committee (no. IDRCB 2017-1702538-45). Each volunteer participated in the present study following a medical examination.

2.2. Experimental Design

The participants had to perform a load lifting/lowering task (LLT) with a soft back-exosuit (SUIT), a rigid back-exoskeleton (SKEL), and without assistance (FREE). The LLT was a standardized task consisting in lifting a load (8 kg) from a low platform to a high one, and vice versa for 30 s at an imposed rate, using a rhythmic beep (20 beeps/minute, in order to perform 5 full cycles). One full cycle included both actions of lifting and lowering the load. Both platforms were facing the participant so as to limit the movement to the sagittal plane. These platforms were installed according to the anthropometric characteristics of the participants, at ankle height and shoulder height minus 14 cm (this value corresponded to the height of the load handles). The high platform was positioned behind the low one to obtain complete elbow extension in the sagittal plane (Figure 1). The platform's position was strictly identical for the three experimental exoskeleton conditions (FREE, SKEL, and SUIT). The participants were instructed to hold their knee almost straight (i.e., from 5 to 10° flexion) but without locking the joint. Position of the feet was fixed, at shoulder width, for each participant by using tape on the floor to ensure a similar placement between conditions. The participants repeated the task twice for the three exoskeleton conditions (FREE, SKEL, and SUIT). The order of these exoskeleton conditions was randomized over subjects. A minimal recovery period of 30 s was imposed after each trial, and 5 min after each exoskeleton condition.

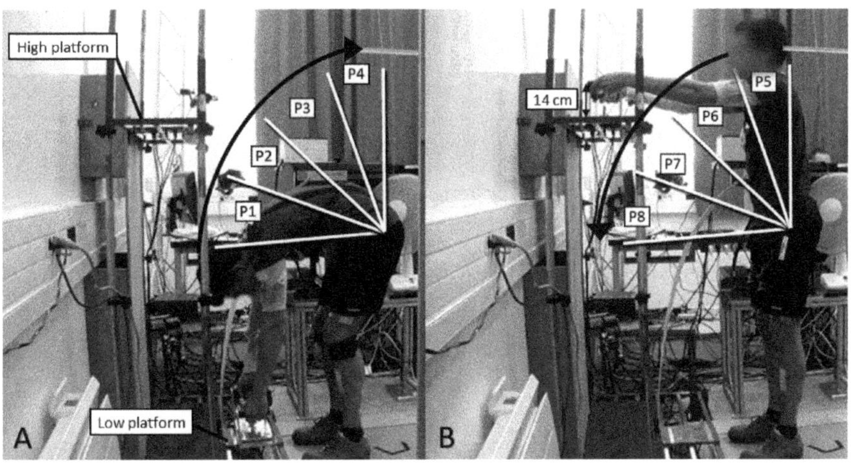

Figure 1. Experimental task performed. The range of motion of trunk sagittal inclination (TSI) was divided into four equal parts for the lifting action (**A**: low (P1) to high platform (P4)) and for the lowering action (**B**: high (P5) to low platform (P8)).

2.3. Exoskeletons

Soft back-support exoskeleton or exosuit (SUIT): a passive textile-designed assistive device (Corfor®-V2, Villemus, France) was used. This exosuit has been designed to assist the ES muscles during manual material handling and static bending posture, using elastic energy stored during trunk flexion and expended during trunk extension. Two elastic elements are attached to shoulder straps at the upper ends, and under the knees at the lower ends (Figure 2A).

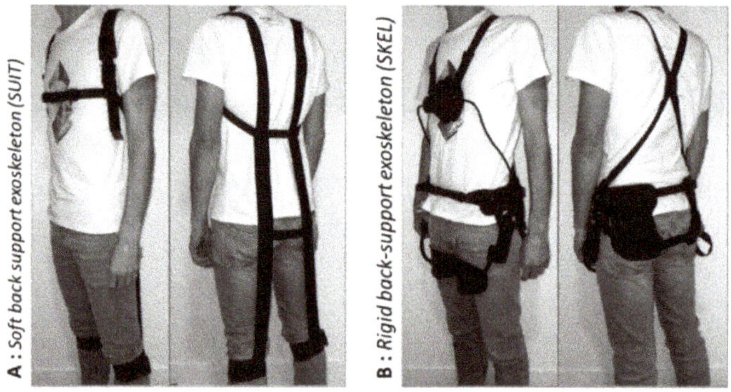

Figure 2. The two exoskeletons used; (**A**) soft back-support exoskeleton or exosuit (SUIT) (Corfor®-V2, Villemus, France), and (**B**) rigid back-support exoskeleton (SKEL) (Laevo®-V1, Deft, The Netherlands).

Rigid back-support exoskeleton (SKEL): a passive back-support exoskeleton (Laevo®-V1, Deft, The Netherlands) was used. This exoskeleton consists of 2 types of pad: one chest pad and two upper leg pads. On both sides of the body, the pads are connected through a circular tube with spring like characteristics. This exoskeleton is intended to assist the ES muscles, using energy stored in springs during trunk flexion and expended during trunk extension (Figure 2B).

2.4. Data Acquisition and Analyses

2.4.1. Trunk Sagittal Inclination (TSI)

TSI was recorded using one wireless magneto-inertial measurement unit (MIMU, firmware version 2.0.8, Xsens, Enschede, The Netherlands). The sensor was placed on the trunk on the flat portion of the sternum. Data were recorded at 50 Hz, filtered using a 5 Hz low pass filter, and synchronized with all the other recorded data. The range of motion was calculated for the lifting action, and divided into four equal TSI parts (=25% of range of motion) (low (P1) to high platform (P4)) (Figure 1A). In the same manner, the range of motion was calculated for the lowering action, and divided into four equal TSI parts (high (P5) to low platform (P8)) (Figure 1B). In terms of TSI, P1 (lifting) corresponded to P8 (lowering), P2 to P7, P3 to P6, and P4 to P5. Throughout the movement, TSI covered a range from approximately 5 to 95° (0° corresponding to the gravity axis).

2.4.2. Electromyography (EMG)

The EMG of the erector spinae (ES) muscles was continuously recorded on both sides (Cometa, Wave Plus™, Bareggio, Italy). Two single-use surface electrodes (BlueSensor N-00-S, Ambu) were placed on the skin in accordance with SENIAM recommendations [17]. The inter-electrode distance was 20 mm. The skin was prepared to maintain impedance lower than 5 kΩ. EMG signals were recorded at 2000 Hz, amplified (×1000), and filtered with a 10–500 Hz bandpass. A 30 Hz high pass filter was applied to remove the heart rate artefacts [18].

Before the first experimental task, two isometric submaximal contractions of ES muscles were performed. The participants were lying on a table, where only the lower body (hip and leg) was supported. They had to maintain a horizontal static posture for 5 s with the trunk facing the ground. All contractions were separated by a 1 min recovery. The highest 500 ms RMS (Root Mean Square) value was used as the reference value (RMS_{REF}).

During each experimental condition, RMS was calculated over successive periods of 40 ms sliding windows in 0.5 ms steps. An averaged RMS value was calculated for each TSI part (from P1 to P8). Then, RMS values were expressed in percentage of the RMS_{REF}, averaged over both sides of the body. In order to avoid any disturbance of the movement linked to the start or to the end of the task, cycles 1 and 5 were not selected for data analysis.

2.5. Statistical Analysis

The results are presented as means ± standard deviations (SD). EMG data were log-transformed for statistical analysis in order to achieve a normal distribution and were back-transformed to original units for presentation in the text and figures. For TSI and ES muscle activity, a three-way repeated ANOVA was used to assess the effect of exoskeleton (FREE, SKEL and SUIT), TSI parts (P1 to P8), gender (men and women), and their interactions. Significant effects were analyzed using post-hoc Tukey HSD pairwise comparisons. A 5% significance level was adopted ($p < 0.05$). Commercial software was used for theses analyses (Statgraphics Centurion XVI).

3. Results

Statistical analyses revealed significant ($p < 0.05$) main effects: (1) of the exoskeleton on both ES muscle activity and TSI, (2) of the TSI parts on the ES muscles activity and naturally on TSI, and (3) of the gender on the ES muscle activity (Table 1). Statistical analyses also evidenced interaction effect between (1) exoskeletons and TSI parts for ES muscle activity and TSI, (2) exoskeletons and gender for TSI, and (3) gender and TSI parts for ES muscles activity.

Table 1. Statistical fixed effects on TSI and ES muscle activity from the ANOVA.

		TSI		ES Muscles Activity	
Main Effects	**DoF**	**F**	***p***	**F**	***p***
Exoskeleton	2	18.88	**<0.001**	16.60	**<0.001**
Gender	1	2.71	0.1	121.24	**<0.001**
TSI part	7	6528.25	**<0.001**	519	**<0.001**
Interaction effects	**DoF**	**F**	***p***	**F**	***p***
Exoskeleton x TSI part	14	3.16	**<0.001**	5.39	**<0.001**
Exoskeleton x Gender	2	16.63	**<0.001**	1.96	0.14
Gender x TSI part	7	1.17	0.31	6.72	**<0.001**
Exoskeleton x Gender x TSI part	14	0.21	1	1.41	0.14

DoF: degrees of freedom. Significant effects ($p < 0.05$) are presented in bold type.

3.1. Main Effect

3.1.1. Exoskeleton Effect

Regarding overall movement (P1 to P8), the averaged ES muscle activity was significantly ($p < 0.05$) lower with the use of SUIT (52.5 ± 31.8% RMS_{REF}) than without equipment (FREE: 56.5 ± 33.0% RMS_{REF}) and SKEL (56.7 ± 31.9% RMS_{REF}) (Figure 3). Concerning TSI, the averaged value was significantly greater for SKEL (50.9 ± 26.3°) than for FREE (49.0 ± 28.0°) and SUIT (49.1 ± 27.8°).

3.1.2. Gender Effect

A significantly greater ES muscle activity was reported in women (60.7 ± 36.2% RMS_{REF}) as compared to men (49.8 ± 26.8% RMS_{REF}) on overall movement.

3.1.3. TSI Part Effect

As required by our protocol, a main TSI part effect was obtained on TSI values (see Figure 4). Concerning averaged ES muscle activity, values were dependent on TSI parts: 34.3 ± 25.1% RMS_{REF} for P1, 63.8 ± 25.5 for P2, 83.6 ± 31.5 for P3, 88.1 ± 30.5 for P4, 57.8 ± 21.2 for P5, 44.9 ± 19.4 for P6, 45.8 ± 21.2 for P7, and 24.1 ± 17.3 for P8.

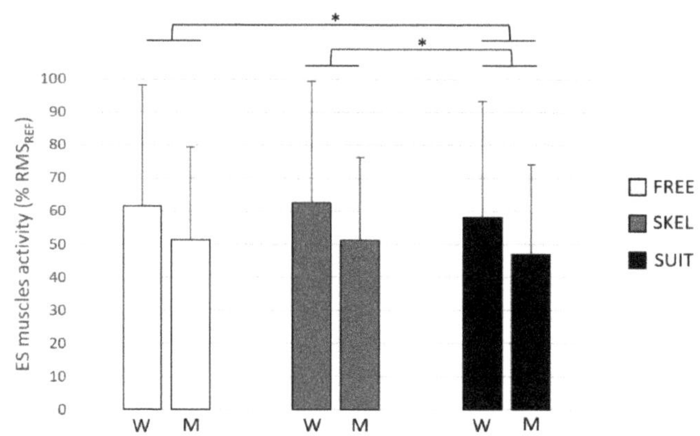

Figure 3. Averaged ES muscles activity (% RMS_{REF}) and standard deviation in women (W) and men (M), with the use of the rigid exoskeleton (SKEL, gray), the exosuit (SUIT, black), and without equipment (FREE, white). *: significant differences between exoskeletons ($p < 0.05$). No interaction between exoskeleton and gender was reported.

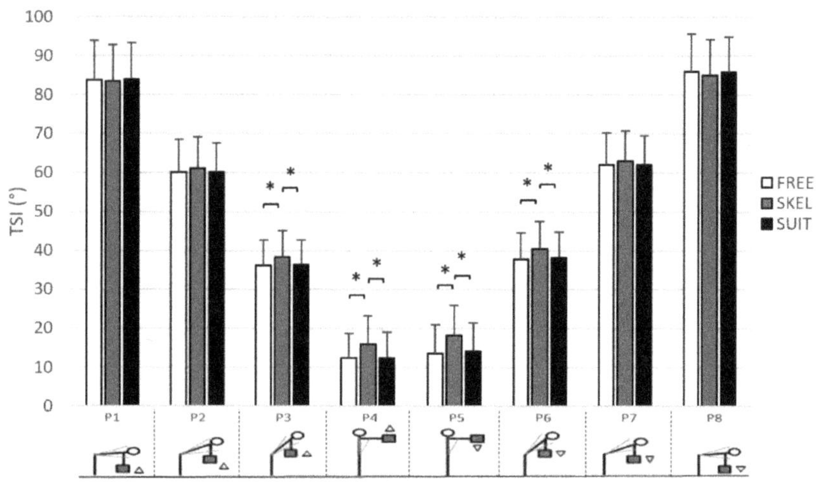

Figure 4. Averaged trunk sagittal inclination (TSI, in °) and standard deviation for each TSI part (from P1 to P8) of the lifting/lowering task (LLT) with the rigid exoskeleton (SKEL, gray), the exosuit (SUIT, black), and without equipment (FREE, white). 0° = straight posture. *: significant differences between exoskeletons ($p < 0.05$).

3.2. Interaction Effect

3.2.1. Exoskeleton × TSI Part Interaction

With the use of SKEL, the averaged TSI was significantly greater than without equipment (FREE) and SUIT over half the range of motion (from P3 to P6) (Figure 4). There is not significant TSI difference between FREE and SUIT for all TSI parts.

Significantly higher ES muscle activity was observed with the use of SKEL (30.1 ± 19.0% RMS_{REF}) during P8 compared to FREE (22.0 ± 16.9% RMS_{REF}) (Figure 5). The use of SUIT resulted in significantly lower ES muscle activity compared to FREE for P2 (59.5 ± 25.1% RMS_{REF} versus 66.9 ± 23.1% RMS_{REF}) and P7 (43.2 ± 21.0% RMS_{REF} vs. 47.8 ± 19.3% RMS_{REF}) (Figure 5). During P3, the use of the two exoskeletons induced a slight but no significant ($p = 0.06$) lower ES muscle activity compared to FREE (SKEL: 82.0 ± 32.9% RMS_{REF},

SUIT: 80.7 ± 28.6% RMS_{REF}, FREE: 88.4 ± 32.7% RMS_{REF}). For P1 and P8 TSI part, ES muscle activity was significantly lower for SUIT (P1: 31.5 ± 24.1% RMS_{REF} and P8: 20.2 ± 14.2% RMS_{REF}) than for SKEL (P1: 37.8 ± 24.2% RMS_{REF} and P8: 30.1 ± 19.0% RMS_{REF}).

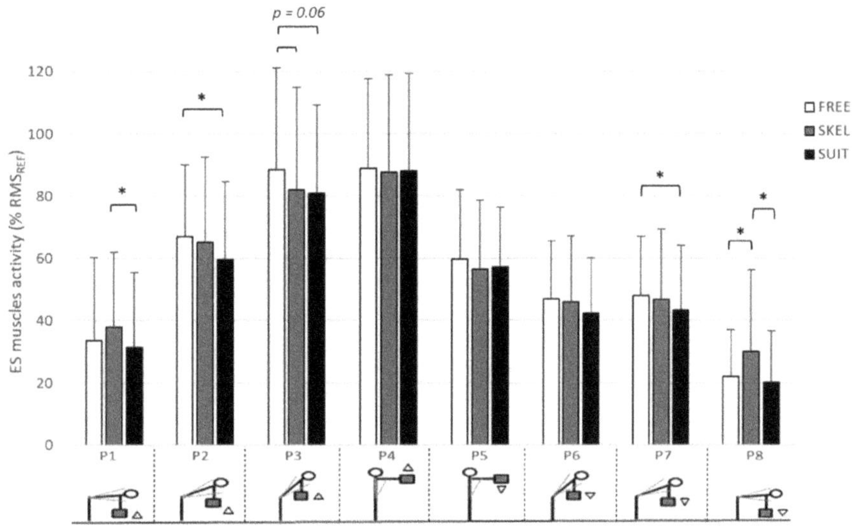

Figure 5. Averaged ES muscle activity (% RMS_{REF}) and standard deviation for each TSI part (from P1 to P8) with the rigid exoskeleton (SKEL, gray), the exosuit (SUIT, black), and without equipment (FREE, white). *: significant differences between exoskeletons ($p < 0.05$).

3.2.2. Exoskeleton × Gender

Without equipment (FREE), the averaged TSI on the overall movement was significantly greater in women (49.7 ± 28.8°) than in men (48.3 ± 27.2°). With SKEL, the averaged TSI was significantly lower in women (49.5 ± 27.0°) than in men (52.1 ± 25.6°). No significant difference was observed for the SUIT (49.0 ± 28.2° and 49.3 ± 27.4°, respectively, in women and men). For women, no significant difference was reported between the three-exoskeleton conditions, whereas for men, significant differences were observed between each condition.

3.2.3. Gender × TSI Part

ES muscles activity was significantly greater in women than in men over most of the range of motion (from P3 to P8), with the exception of P1 and P2.

4. Discussion

The present study investigated the influence of exoskeleton design, trunk sagittal inclination (TSI), and gender on the impact of using wearable assistive devices on lumbar muscle activity during a dynamic forward lifting and lowering task (LLT). During LLT, the use of both EXO_{BK} resulted in significant changes in the averaged ES muscle activity compared to the control condition (FREE). These modifications depended not only on the exoskeleton design (SUIT versus SKEL), but also on the interaction between the exoskeleton design and parts of TSI. Gender did not modify these results.

4.1. Exoskeletons Design and TSI Effects

The use of SUIT resulted in a relative decrease of −7% of ES muscle activity during the overall LLT. These results appear to agree with those in the literature which also mostly reports significant reductions in back muscle activity during bending tasks with the use of a soft EXO_{BK} [2]. Nevertheless, the amplitude of the reductions of ES muscle activity here seems slightly lower in comparison with previous observations. For example, the reductions in back muscle activity when using exosuits during lifting tasks generally ranged from −10 to −40% [4,13,19–22]. However, most of the previous studies using exosuits did not specifically focus on the sagittal inclination of the trunk during the lifting task. When analyzing the present results for each TSI part, it appears that the decreases in ES muscle activity involved in the use of SUIT varied with phases of the movement (from −0.9 to −11.1%). More specifically, the relative reductions of ES muscle activity were not significant where the participants stood almost straight (P4 and P5, TSI ≈ 5–25°), and when they were strongly leaning (P1 and P8, TSI ≈ 75–95°). The beneficial effects of SUIT on ES muscle activity even appeared to be directly dependent on the parts of the TSI: approximately −8.7% for P3 (TSI ≈ 25–50°) (p = 0.06), and −11.1 and −9.6% for P2 and P7, respectively (TSI ≈ 50–75°). For the most extreme parts of the movement, the elastic garment may not be in its optimal operating range: not stretched enough on one side (P4 and P5) and close to its maximum stretch on the other side (P1 and P8), thereby reducing its effectiveness. Additionally, although not evaluated, a hysteresis could be present for P1 as demonstrated previously by Koopman, Kingma, Faber, de Looze, and van Dieen [7] for a rigid EXO_{BK}, which can also reduce the support of SUIT. Finally, the contribution of ES muscle is reduced for large hip flexions. It is therefore likely that the effect of the system is relatively less visible in this position for this muscle. A more complete analysis of the back muscle chain could make it possible to assess whether other muscle activities were modified by SUIT at these trunk inclinations.

Contrary to most of the previous studies using exosuits, where a squat technique was advised for lifting tasks [14,21,22], the participants of the present study performed LLT with a stooped posture. Thus, it can be assumed that the tension (assistive force) in the elastic garment depends on the spine curvature and thus on the range of motion during stooped bending. Lamers, Yang, and Zelik [14] reported similar observations with the use of another soft EXO_{BK}. The latter authors showed that the amplitude of back muscle activity reductions, compared to control conditions, increased with the TSI (−23% for a TSI close to 30°, −27% close to 60°, and −43% close to 90°). Comparing the amplitude of the decrease in ES muscle between the latter study [14] and the present one nevertheless remains difficult. Lamers, Yang, and Zelik [14] indeed recorded muscle activity only during static holding, and not during dynamic lifting. The modalities of both tasks can require the activation of different muscles. In addition, exoskeleton support can change between dynamic versus static action modalities [7].

Contrary to SUIT, the use of SKEL did not significantly change the result in ES muscle activity compared to the control condition, except for P8 where EMG increased. This result seems to run counter to the most common observations reported in the literature, on the impact of EXO_{BK} on ES muscle activity [2]. However, to our knowledge, only a few studies have specifically investigated the consequences of a rigid EXO_{BK} on back muscle activity during a dynamic bending task over a wide range of motions [6,7,9,23]. Nevertheless, other studies have also reported a lack of significant reductions of ES muscle activity when using rigid EXO_{BK} devices in stooped postures (TSI close to 70°) [5], and even slight increases of ES muscle activity when using the same exoskeleton (Laevo) in standing positions [7].

As the support delivered by such an exoskeleton is a function of angle [7], we expected that the effects of the SKEL during LLT would be related to the TSI. However, the benefits involved by SKEL in terms of back muscle activity reduction did not appear to be related to the increase of TSI. The slight and not significant (p = 0.06) reduction in ES muscle activity would seem to occur only during lifting, over P3 (−7.2%). The range of postures (≈25–50°) for which this exoskeleton involved this reduction in back muscle activity appears similar to that of previous studies that carried out experiments with the same exoskeleton. For example, Koopman, Kingma, Faber, de Looze, and van Dieen [7] recorded significant reductions of ES muscle activity (approximately −15%) during a stooped bending task with an exoskeleton at only 50% of the range of motion (0% corresponding to the floor, and 100% to standing upright). The latter results suggest that using such an exoskeleton involves reductions of ES muscle activity only over restricted ranges of trunk inclination, and not over a wide range of dynamic bending.

In terms of design, SKEL consists of two chest pads and two upper leg pads connected through a circular tube with spring-like characteristics. The flexion axis of the system is located at the level of the transversal axis of the hip. As a result, the support should be activated by hip flexion, and not directly by trunk inclination. However, in the present study, the participants adopted a stooped posture during LLT. Trunk inclination was not only associated with hip flexion, but also with spine flexion (i.e., curvature). Contrary to SUIT, SKEL cannot be directly tensioned by the spine curvature. Moreover, the use of SKEL may even tend to limit spine flexion [5,7]. In the present study, the lifting/lowering technique used by the subjects could thus partly explain the differences in the effectiveness of SKEL versus SUIT to reduce back muscle activation throughout the full range of motion.

Furthermore, counter intuitively, the use of SKEL resulted in a significant increase of ES muscle activity compared to FREE, when the TSI ranged from $\approx$75 to 95°, during the lowering phases (i.e., for P8). Several hypotheses can be made to explain this observation. First, this result could be linked to a change of spine kinematics and/or the coordination of spine and hip extensor muscles, when using SKEL during LLT, compared to FREE. Previous studies evidenced several changes in postural kinematics when using rigid EXO_{BK} [5], notably the Laevo [7,11], during dynamic lifting and static holding tasks. For example, Ulrey and Fathallah [5] reported a significant reduction of thoracic and lumbar flexion when using a similar rigid exoskeleton during static forward bending, in comparison with a control condition. Koopman, Kingma, Faber, de Looze, and van Dieen [7] also reported a significant reduction of hip flexion when using the Laevo exoskeleton during static holding tasks whereas the TSI remained similar both with and without an exoskeleton. In the present study, only TSI was recorded using an inertial unit located on the thoracic plexus. This hypothesis could not be verified. On the other hand, it can be assumed that the use of SKEL involved changes in hip and spine flexion, thereby affecting both the coordination and activity of the hip (e.g., gluteus maximus and biceps femoris) and spine (e.g., lumbar erector spinae and thoracic erector spinae) extensor muscles. In this case, the increase of ES muscle activity with SKEL could be due to an increase of the relative contribution of these muscles to spine erection. However further research is needed to confirm this hypothesis. Finally, considering the large hysteresis present in this device [7], we can assume that the support delivered by SKEL was particularly limited during the change of motion direction.

Ultimately, the interaction effect between the exoskeleton and TSI on ES muscle activity could be related to the technique used during the lifting/lowering task, and more particularly to the action needed to tension the passive EXO_{BK}. It is probable that the effectiveness of SKEL to reduce back muscle efforts was related to hip flexion, while the effectiveness of SUIT was dependent on spine flexion. The present experimental protocol should be duplicated during a lifting task performed with a squat technique, involving a greater mobilization of the hip. The respective efficiency of both exoskeletons (soft versus rigid) would probably be different.

4.2. Gender Effects

In this study, we found a gender effect on the muscle activity of ES with a higher mean relative value for women. The weight of the load to be handled was the same for men and women (8 kg), and is likely to explain this difference, due to the lower average maximum voluntary force in women. Furthermore, no interaction between gender and EXO_{BK} was observed on back muscle activity. This result contrasts with previous studies showing differences in back extensor muscle activity between men and women when using an exoskeleton [6,9,16]. For example, Alemi, Madinei, Kim, Srinivasan, and Nussbaum [6] observed a reduction in muscle activity with the use of the two rigid EXO_{BK} tested (Laevo V2.5 and BackX) for women but only a reduction for one of the two EXO_{BK} (BackX) for men. Moreover, in this last study, these reductions were greater for women than for men.

Another study [16], assessing a gender effect, showed a decrease in back extensor EMG activity with a rigid EXO_{BK} (BackX) only for men.

Firstly, we assume that these disparities in results between studies could be related to the task performed and the lifting technique used: lifting tasks with a free technique for Alemi, Madinei, Kim, Srinivasan, and Nussbaum [6], cardiopulmonary resuscitation chest compressions for So, Cheung, Liu, Tang, Tsoi, and Wu [16], and a LLT with a stoop technique for the present study. As gender can influence motor coordination during lifting tasks [24], particularly with greater hip flexion in women, it is possible that these protocols had a different impact on the biomechanical responses between men and women.

Secondly, it is probable that the design of exoskeletons could also contribute to this gender effect. Our results showed an interaction on trunk kinematics between gender and EXO_{BK}. Women did not show any differences on the average TSI between the different exoskeleton conditions (FREE, SKEL, and SUIT), while men did. This gender effect may be due to variation in motor coordination [24] but also to different strategies for using exoskeletons. Men seemed to lean forward more with the exoskeletons, and this was more pronounced with the rigid one, as if they were seeking assistance through greater trunk support. The absence of this observation in women may be related to a greater perception of discomfort on their body when using exoskeletons. Although we did not evaluate this parameter, previous studies have reported differences in local discomfort between men and women when using EXO_{BK} [6,9,25]. Kozinc et al. [26] also pointed out that the discomfort threshold for pressure on the chest, thighs, or pelvis was lower in women. This parameter, related to the design of the exoskeleton, could therefore participate in the gender effect observed for the TSI.

In the present study, as there is no interaction between gender and EXO_{BK} on back muscles activity, the results on the effectiveness in terms of muscle activity reductions of the two EXO_{BK} tested appear to be valid for both men and women. However, to better understand these gender effects, further studies need to be conducted on the complete kinematics of the spine and hip during the use of different EXO_{BK}.

4.3. Limitations

Several parameters of the experimental protocol had to be standardized in order to limit bias in the measurement of dependent variables. However, these same parameters can also be considered as limitations in the interpretation of the results. For example, both exoskeletons were examined during a dynamic lifting task. Since the dynamics of the torso (i.e., velocity/acceleration) could have an influence on ES muscle activity, this parameter was controlled by the frequency of the lift, which was identical for each experimental condition. As a result, the interpretation of the present results should be limited to this particular lifting speed. Nevertheless, this lifting speed was determined by the average of the free cadence chosen by the participants during the preliminary experiments. It can be assumed that this lifting speed was relatively realistic with regard to real handling tasks. In order to limit the impact of the interindividual heterogeneity in the present results, only young people participated in this study performed in the laboratory, using a highly controlled experimental task. Therefore, these conditions are not representative of the actual work context. The main objective was to test the interaction between exoskeleton design and TSI part on back muscle activity. It was thus necessary to perfectly control the experimental conditions and to study the exoskeletons while using them strictly for the purpose they have been designed. Finally, the use of normalized ranges of motion also appeared essential here to reduce interindividual differences and facilitate intercondition comparisons. Consequently, this methodological choice substantially reduced the practical applications of these results.

Furthermore, some of the methodological choices limited the scope of the analysis performed in this study. For example, only the ES muscle was examined during this study, while several other muscle groups contribute to the flexion/extension of the spine and the hip during stooped bending tasks. In practical terms, and for comparison with previous studies, the measure of ES is probably the most relevant choice. The ES muscle is systematically studied for back-support exoskeleton evaluations during laboratory and field research [3]. Moreover, the excessive effort of ES muscle is related to the occurrence of LBP. However, fundamentally, the study of the other muscles involved in the hip and spine mobilization would have provided useful information on changes in muscle coordination with the exoskeleton over the full range of motion. In addition, the use of only a single MIMU did not allow analyzing the coordination of pelvic and trunk movements. These measures would have allowed studying the influence of the exoskeleton design on the kinematics of the spine and pelvis. The lower limb kinematics were not studied either. Only knee flexion was controlled visually by the researcher during the experimentation, thus it cannot be excluded that slight changes occurred.

5. Conclusions

The present study showed that the impact of using an EXO_{BK} during a dynamic lifting/lowering task with an 8 kg load on ES muscle activity depended on both the exoskeleton design and TSI. The impact of the soft EXO_{BK} on the back muscle workload appeared beneficial between 25 to 75° of TSI during this task. On the contrary, the use of the rigid EXO_{BK} did not involve significant reductions of ES muscle activity during this dynamic task, but a significant increase when the TSI exceeded 75°. Gender did not affect these results on back muscle activity. It seems that the technique used during the lifting/lowering task could influence the performance of the EXO_{BK} in reducing back muscle activity. This technique could depend on the exoskeleton design, and more particularly on the action needed to tension it. It is probable that the effectiveness of the rigid exoskeleton to reduce back muscle efforts is related to hip flexion, while the effectiveness of the soft exoskeleton is dependent on spine flexion.

In practice, the choice of a passive back-support exoskeleton requires prior characterization of the tasks to which workers are exposed. Trunk kinematics and ranges of motion have to be considered in priority. Moreover, evaluations of human-exoskeleton interaction in task conditions remain essential to identify the benefits and potential negative effects of the exoskeleton on back muscle effort. These evaluations should also consider the influence of gender on the consequences induced by the use of exoskeletons.

Author Contributions: Conceptualization, J.T. and K.D.; methodology, M.S., J.T. and K.D.; software, M.S.; validation, M.S., J.T. and K.D.; formal analysis, M.S. and K.D.; investigation, M.S. and K.D.; resources, M.S. and K.D.; data curation, M.S.; writing—original draft preparation, M.S. and J.T.; writing—review and editing, M.S., J.T. and K.D.; visualization, M.S.; supervision, J.T. and K.D.; project administration, M.S., J.T. and K.D. All authors have read and agreed to the published version of the manuscript.

Funding: This research received no external funding.

Institutional Review Board Statement: The study was conducted according to the guidelines of the Declaration of Helsinki, and approved by the Ethics Committee of Strasbourg (ID-RCB 2017-1702538-45; 12 January 2018).

Informed Consent Statement: Informed consent was obtained from all participants involved in the study.

Data Availability Statement: The data presented in this study are available on request from the corresponding author. The data are not publicly available due to ethical regulations.

Acknowledgments: We thank Morel O., Remy O. and Reno G. for their technical support and Gaudez C. for providing medical assistance.

Conflicts of Interest: The authors declare no conflict of interest.

References

1. De Looze, M.P.; Bosch, T.; Krause, F.; Stadler, K.S.; O'Sullivan, L.W. Exoskeletons for industrial application and their potential effects on physical work load. *Ergonomics* **2016**, *59*, 671–681. [CrossRef]
2. Theurel, J.; Desbrosses, K. Occupational Exoskeletons: Overview of Their Benefits and Limitations in Preventing Work-Related Musculoskeletal Disorders. *IISE Trans. Occup. Ergon. Hum. Factors* **2019**, 1–17. [CrossRef]
3. Kermavnar, T.; de Vries, A.W.; de Looze, M.P.; O'Sullivan, L.W. Effects of industrial back-support exoskeletons on body loading and user experience: An updated systematic review. *Ergonomics* **2021**, *64*, 685–711. [CrossRef] [PubMed]
4. Frost, D.M.; Abdoli, E.M.; Stevenson, J.M. PLAD (personal lift assistive device) stiffness affects the lumbar flexion/extension moment and the posterior chain EMG during symmetrical lifting tasks. *J. Electromyogr. Kinesiol. Off. J. Int. Soc. Electrophysiol. Kinesiol.* **2009**, *19*, e403–e412. [CrossRef] [PubMed]
5. Ulrey, B.L.; Fathallah, F.A. Effect of a personal weight transfer device on muscle activities and joint flexions in the stooped posture. *J. Electromyogr. Kinesiol. Off. J. Int. Soc. Electrophysiol. Kinesiol.* **2013**, *23*, 195–205. [CrossRef] [PubMed]
6. Alemi, M.M.; Madinei, S.; Kim, S.; Srinivasan, D.; Nussbaum, M.A. Effects of Two Passive Back-Support Exoskeletons on Muscle Activity, Energy Expenditure, and Subjective Assessments During Repetitive Lifting. *Hum. Factors* **2020**, 18720819897669. [CrossRef] [PubMed]
7. Koopman, A.S.; Kingma, I.; Faber, G.S.; de Looze, M.P.; van Dieen, J.H. Effects of a passive exoskeleton on the mechanical loading of the low back in static holding tasks. *J. Biomech.* **2019**, *83*, 97–103. [CrossRef]
8. Graham, R.B.; Agnew, M.J.; Stevenson, J.M. Effectiveness of an on-body lifting aid at reducing low back physical demands during an automotive assembly task: Assessment of EMG response and user acceptability. *Appl. Ergon.* **2009**, *40*, 936–942. [CrossRef]
9. Madinei, S.; Alemi, M.M.; Kim, S.; Srinivasan, D.; Nussbaum, M.A. Biomechanical assessment of two back-support exoskeletons in symmetric and asymmetric repetitive lifting with moderate postural demands. *Appl. Ergon.* **2020**, *88*, 103156. [CrossRef]
10. Toxiri, S.; Näf, M.B.; Lazzaroni, M.; Fernández, J.; Sposito, M.; Poliero, T.; Monica, L.; Anastasi, S.; Caldwell, D.G.; Ortiz, J. Back-Support Exoskeletons for Occupational Use: An Overview of Technological Advances and Trends. *IISE Trans. Occup. Ergon. Hum. Factors* **2019**, 1–13. [CrossRef]
11. Bosch, T.; van Eck, J.; Knitel, K.; de Looze, M. The effects of a passive exoskeleton on muscle activity, discomfort and endurance time in forward bending work. *Appl. Ergon.* **2016**, *54*, 212–217. [CrossRef]
12. Sadler, E.M.; Graham, R.B.; Stevenson, J.M. The personal lift-assist device and lifting technique: A principal component analysis. *Ergonomics* **2011**, *54*, 392–402. [CrossRef] [PubMed]
13. Abdoli, E.M.; Agnew, M.J.; Stevenson, J.M. An on-body personal lift augmentation device (PLAD) reduces EMG amplitude of erector spinae during lifting tasks. *Clin. Biomech.* **2006**, *21*, 456–465. [CrossRef] [PubMed]
14. Lamers, E.P.; Yang, A.J.; Zelik, K.E. Feasibility of a Biomechanically-Assistive Garment to Reduce Low Back Loading During Leaning and Lifting. *IEEE Trans. Biomed. Eng.* **2018**, *65*, 1674–1680. [CrossRef]
15. Zhang, H.H.; Kadrolkar, A.; Sup, F.C. Design and Preliminary Evaluation of a Passive Spine Exoskeleton. *J. Med. Devices* **2016**, *10*. [CrossRef]
16. So, B.C.L.; Cheung, H.H.; Liu, S.L.; Tang, C.I.; Tsoi, T.Y.; Wu, C.H. The effects of a passive exoskeleton on trunk muscle activity and perceived exertion for experienced auxiliary medical service providers in cardiopulmonary resuscitation chest compression. *Int. J. Ind. Ergon.* **2020**, *76*. [CrossRef]
17. Hermens, H.J.; Freriks, B. *SENIAM 5—The State of the Art on Sensors and Sensor Placement Procedures for Surface Electromyography: A Proposal for Sensor Placement Procedures*; Roessingh Research and Development b.v: Enschede, The Netherlands, 1997.
18. Drake, J.D.M.; Callaghan, J.P. Elimination of electrocardiogram contamination from electromyogram signals: An evaluation of currently used removal techniques. *J. Electromyogr. Kines* **2006**, *16*, 175–187. [CrossRef]
19. Abdoli, E.M.; Stevenson, J.M. The effect of on-body lift assistive device on the lumbar 3D dynamic moments and EMG during asymmetric freestyle lifting. *Clin. Biomech.* **2008**, *23*, 372–380. [CrossRef]
20. Godwin, A.A.; Stevenson, J.M.; Agnew, M.J.; Twiddy, A.L.; Abdoli-Eramaki, M.; Lotz, C.A. Testing the efficacy of an ergonomic lifting aid at diminishing muscular fatigue in women over a prolonged period of lifting. *Int. J. Ind. Ergon.* **2009**, *39*, 121–126. [CrossRef]
21. Lotz, C.A.; Agnew, M.J.; Godwin, A.A.; Stevenson, J.M. The effect of an on-body personal lift assist device (PLAD) on fatigue during a repetitive lifting task. *J. Electromyogr. Kinesiol. Off. J. Int. Soc. Electrophysiol. Kinesiol.* **2009**, *19*, 331–340. [CrossRef]
22. Whitfield, B.H.; Costigan, P.A.; Stevenson, J.M.; Smallman, C.L. Effect of an on-body ergonomic aid on oxygen consumption during a repetitive lifting task. *Int. J. Ind. Ergon.* **2014**, *44*, 39–44. [CrossRef]
23. Baltrusch, S.J.; van Dieen, J.H.; van Bennekom, C.A.M.; Houdijk, H. The effect of a passive trunk exoskeleton on functional performance in healthy individuals. *Appl. Ergon.* **2018**, *72*, 94–106. [CrossRef] [PubMed]
24. Pan, F.; Firouzabadi, A.; Zander, T.; Schmidt, H. Sex-dependent differences in lumbo-pelvic coordination for different lifting tasks: A study on asymptomatic adults. *J. Biomech* **2020**, *102*, 109505. [CrossRef] [PubMed]
25. Kim, S.; Madinei, S.; Alemi, M.M.; Srinivasan, D.; Nussbaum, M.A. Assessing the potential for "undesired" effects of passive back-support exoskeleton use during a simulated manual assembly task: Muscle activity, posture, balance, discomfort, and usability. *Appl. Ergon.* **2020**, *89*, 103194. [CrossRef] [PubMed]
26. Kozinc, Z.; Babic, J.; Sarabon, N. Human pressure tolerance and effects of different padding materials with implications for development of exoskeletons and similar devices. *Appl. Ergon.* **2021**, *93*, 103379. [CrossRef] [PubMed]

International Journal of
Environmental Research and Public Health

MDPI

Article

Multi-Dimensional Impact of Chronic Low Back Pain among Underserved African American and Latino Older Adults

Mohsen Bazargan [1,2,3,4], Margarita Loeza [1,4], Tavonia Ekwegh [5], Edward K. Adinkrah [2], Lucy W. Kibe [3], Sharon Cobb [5], Shervin Assari [1,2] and Shahrzad Bazargan-Hejazi [6,7,*]

1 Department of Family Medicine, Charles R Drew University of Medicine and Science (CDU), Los Angeles, CA 90059, USA; mohsenbazargan@cdrewu.edu (M.B.); MLoeza@mednet.ucla.edu (M.L.); shervinassari@cdrewu.edu (S.A.)
2 Department of Public Health, CDU, Los Angeles, CA 90059, USA; edwardadinkrah1@cdrewu.edu
3 Physician Assistant Program, CDU, Los Angeles, CA 90059, USA; lucykibe@cdrewu.edu
4 Department of Family Medicine, UCLA, Los Angeles, CA 90095, USA
5 School of Nursing, CDU, Los Angeles, CA 90059, USA; tavoniaekwegh@cdrewu.edu (T.E.); sharoncobb@cdrewu.edu (S.C.)
6 Department of Psychiatry, UCLA, Los Angeles, CA 90095, USA
7 Department of Psychiatry, CDU, Los Angeles, CA 90059, USA
* Correspondence: shahrzadbazargan@cdrewu.edu

Abstract: Chronic low back pain is one of the most common, poorly understood, and potentially disabling chronic pain conditions from which older adults suffer. The existing low back pain research has relied almost exclusively on White/Caucasian participant samples. This study examines the correlates of chronic low back pain among a sample of underserved urban African American and Latino older adults. Controlling for age, gender, race/ethnicity, education, living arrangement, and number of major chronic conditions, associations between low back pain and the following outcome variables are examined: (1) healthcare utilization, (2) health-related quality of life (HR-QoL) and self-rated quality of health; and (3) physical and mental health outcomes. Methods: We recruited nine hundred and five (905) African American and Latino older adults from the South Los Angeles community using convenience and snowball sampling. In addition to standard items that measure demographic variables, our survey included validated instruments to document HR-QoL health status, the Short-Form McGill Pain Questionnaire-2, Geriatric Depression Scale, sleep disorder, and healthcare access. Data analysis includes bivariate and 17 independent multivariate models. Results: Almost 55% and 48% of the Latino and African American older adults who participated in our study reported chronic low back pain. Our data revealed that having low back pain was associated with three categories of outcomes including: (1) a higher level of healthcare utilization measured by (i) physician visits, (ii) emergency department visits, (iii) number of Rx used, (iv) a higher level of medication complexity, (v) a lower level of adherence to medication regimens, and (vi) a lower level of satisfaction with medical care; (2) a lower level of HR-QoL and self-assessment of health measured by (i) physical health QoL, (ii) mental health QoL, and (iii) a lower level of self-rated health; and (3) worse physical and mental health outcomes measured by (i) a higher number of depressive symptoms, (ii) a higher level of pain, (iii) falls, (iv) sleep disorders, (v) and being overweight/obese. Discussion: Low back pain remains a public health concern and significantly impacts the quality of life, health care utilization, and health outcomes of underserved minority older adults. Multi-faceted and culturally sensitive interventional studies are needed to ensure the timely diagnosis and treatment of low back pain among underserved minority older adults. Many barriers and challenges that affect underserved African American and Latino older adults with low back pain simply cannot be addressed in over-crowded EDs. Our study contributes to and raises the awareness of healthcare providers and health policymakers on the necessity for prevention, early diagnosis, proper medical management, and rehabilitation policies to minimize the burdens associated with chronic low back pain among underserved older African American and Latino patients in an under-resourced community such as South Los Angeles.

Citation: Bazargan, M.; Loeza, M.; Ekwegh, T.; Adinkrah, E.K.; Kibe, L.W.; Cobb, S.; Assari, S.; Bazargan-Hejazi, S. Multi-Dimensional Impact of Chronic Low Back Pain among Underserved African American and Latino Older Adults. *IJERPH* **2021**, *18*, 7246. https://doi.org/10.3390/ijerph18147246

Academic Editors: Vincenzo Denaro, Sergio Iavicoli, Fabrizio Russo and Gianluca Vadalà

Received: 11 April 2021
Accepted: 2 July 2021
Published: 6 July 2021

Keywords: back pain; chronic; underserved; African American; Latino; older adults

1. Background

A recent study examining the most common global conditions found that low back pain ranked as the fourth most common condition in primary care visits in developed countries [1]. Indeed, chronic low back pain is one of the most common, poorly understood, and potentially disabling chronic pain conditions from which older adults suffer [2]. A review of the literature clearly shows that both the incidence and prevalence of severe and chronic low back pain increase with older age [3]. There is evidence that older adults with chronic low back pain may be at a higher risk of cardiovascular disease [4], serious falls [5], low HR-QoL [6–8], depression [9], poorer postural control and interference with activity [10,11], low physical activity levels [12], functional and mobility impairments [13], diminished energy capacity and slower gait speed [14], kinesiophobia [15], a low sense of self-efficacy [16], disability [17], appetite impairment [18], substance use [19], and sleep disorders [20]. Older adults living with low back pain often describe their golden years as "not so golden after all" [21].

The health disparity among African Americans and Latinos has been recognized and documented in the literature. People of color often fare worse than their White counterparts across many measures of health status such as life expectancy and the burden of chronic disease. However, existing low back pain research has relied almost exclusively on White/Caucasian participant samples [22–25]. Knowledge on the impact of chronic low back pain among underserved African American and Latino older adults is limited. Examination of baseline data from the Back pain Outcomes using Longitudinal Data (BOLD) registry (the largest inception cohort to date of seniors presenting to a primary care provider for back pain) shows that being of the African American race and older age were associated with worse functional disability as measured by the Roland–Morris Disability Questionnaire [22]. Additional empirical data suggest that there are racial disparities in the use of opioids for back pain among older adults [26]. Furthermore, one nationally representative sample of Medicare claims showed that the provision of care for uncomplicated low back pain varied according to nonclinical factors: mainly demographic characteristics, socioeconomic status, and area of residence. These findings highlight the need to ensure that the provision of care is based on clinically relevant characteristics and is not biased by sociodemographic factors [27].

This study examines the correlates of the chronic low back pain among a sample of underserved urban African American and Latino older adults. Controlling for age, gender, race/ethnicity, education, living arrangement, and number of major chronic conditions, associations between low back pain and the following outcome variables are examined: (1) physical HR-QoL; (2) Mental HR-QoL; (3) depression symptoms; (4) insomnia/sleep difficulty; (5) self-rated health status; (6) number of Rx medication used; (7) adherence to drug regimens; (8) complexity of medication; (9) falls; (10) ED utilization; (11) hospital admissions; (12) office-based provider visits; and (13) level of pain.

Since previous studies have clearly demonstrated that an array of independent and interdependent relationships exist between social determinants of health and chronic low back pain [28] and given our limited understanding of chronic low back pain among underserved minority older adults, there is a clear need to systematically explore the prevalence, correlates and conditions that may contribute to worse outcomes among this segment of our population. Given the nature of diseases in minority groups, findings from our study will add to the evidence used in making public health decisions. Information on the risk factors of chronic low back pain is essential for policymakers. Understanding the impact of chronic low back pain is crucial for healthcare resource planning among these populations.

2. Methods

2.1. Recruitment

We recruited nine hundred and five (905) participants from senior centers, senior housing centers, faith-based organizations, and apartment complexes in the South Los Angeles community using convenience and snowball sampling. We visited 52 sites throughout the study. We included participants who self-identified as African American or Latino aged 55 years or older and who committed to being in the area for at least one year. Recruitment was completed during regular office hours, usually in the day room or dining room areas. Face-to-face structured interviews were terminated in early February 2020 due to the COVID-19 pandemic. Home to over one million residents, South Service Planning Area 6 (SPA6) of Los Angeles County is disproportionately harmed by health disparities compared to the rest of Los Angeles County [29].The study was approved by the Charles R. Drew University of Medicine and Science (CDU) Institutional Review Board.

2.2. Measurements

Survey instruments: In addition to the standard items that measure demographic variables, our survey included validated instruments to document HR-QoL [30], health status, the Short-Form McGill Pain Questionnaire-2 (SF-MPQ-2) [31,32], depression (GDS) [33,34], sleep disorders, [35,36] and healthcare access.

Demographics characteristics: We used age, gender, educational attainment, marital status, and race/ethnicity as the covariates in this study. Educational attainment was obtained as a continuous variable (number of years). Higher scores indicated more years of education. We asked our participants whether they were married or lived with a partner, which was analyzed categorically as either married/lived with a partner or not married/do not live with a partner. We also asked our participants whether they lived alone or if there was any other member of the family such as a partner or a spouse who lived with them, which was analyzed categorically as either living alone or living with at least 1 other individual.

Financial strain: This variable was measured using five items: "In the past 12 months, how frequently were you unable to: (1) buy the amount of food your family should have, (2) buy the clothes you feel your family should have, (3) pay your rent or mortgage, (4) pay your monthly bills (5) make ends meet?" Items were on a 5-level response scale ranging from 1 (never) to 5 (always). A total "financial strain" score was calculated, with an average score of five items, ranging from 1 to 5. A high score was indicative of greater financial difficulty. These items are consistent with Pearlin's list of chronic financial difficulties experienced by low SES individuals (Cronbach alpha = 0.92) [37].

Accessibility of medical care: This variable was measured using five items: (1) In the last 12 months, have you needed to see a doctor but could not see one? (2) Overall, how difficult is it for you to get medical care? (3) How difficult would it be for you to get a routine physical exam? (4) How difficult is it for you to visit a doctor when you need medical care? (5) Overall, how satisfied are you with how available medical care is for you? Principal component analysis was used to identify potential factors underlying this 5-item instrument. Only one factor was produced, which was able to explain almost 50% of the variance. All items had primary loadings over 0.4. A lower score on this index reflects a higher level of perceived difficulty accessing medical care.

ED Utilization, Office-Based Provider Visits, and Hospital Admissions: Participants were asked how many times they had utilized ED, stayed overnight at hospital as a patient, and visited a provider at their office in the last 12 months.

Fall incidents: Fall incidents were assessed with one single item asking participants to report a single fall or multiple falls in the year before the interview (yes = 0; no = 1).

Major chronic conditions: Participants reported whether they have been diagnosed with any of the following conditions: high blood pressure, diabetes mellitus, heart-related conditions, cancer, stroke, and/or chronic obstructive pulmonary disease (COPD).

Self-rated Health Status (SRH): Participants reported SRH using a single question: "In general, would you say your health is (1) Excellent; (2) Very good; (3) Good; (4) Fair (5) Poor?" Knowing that self-rated health is differently shaped by social determinants across ethnic groups, it is important to include perceived-self status as one of the indices that measures the health condition of participants [38]. In addition, this item is repeatedly used in large-scale national surveys and predicts mortality risk among Latino and other ethnic groups [39,40]. However, National Health Interview Surveys indicate that self-rated health predicts mortality risk less well for older Hispanics than their non-Hispanic White counterparts [41].

Pain Severity: This study employed the definition as stated by the American Academy of Family Physicians, which stipulates low back pain (LBP) as, "pain, muscle tension, or stiffness localized below the costal margin and above the inferior gluteal folds, with or without sciatica, and is defined as chronic when it persists for 12 weeks or more" [42]. Low back pain was measured using the four subscales outlined in the Short-Form McGill Pain Questionnaire-2 (SF-MPQ-2) [31,32]. Participants self-reported the level to which they experienced each of the four subscales, which are composed of 22 pain items experienced in the past week using an 11-point numeric rating scale. Examination of the psychometric properties of the SF-MPQ-2 among Hispanic and non-Hispanic White patients with pain shows that this measure seems to be used equivalently across these 2 ethnic groups [43]. This pain subscale used in our sample had high internal reliability (Cronbach's alpha = 0.950 in the African American population; Cronbach's alpha = 0.961 in the Latino population; Cronbach's Alpha = 0.952 in both the Latino and African American populations) and excellent validity ($p < 0.05$, 95% CI [−2.39294, −1.85743]).

Health-Related Quality of Life Survey (SF-12): The SF-12 Health Survey measures include two subscales that measure mental and physical functioning and overall health-related quality of life [30]. The SF-12 is a multipurpose short-form survey with 12 questions, all selected from the longer SF-36 Health Survey [44]. The SF-12 is a validated measure for assessing the health status of low-income minority populations [45]. We computed the physical and mental health composite scores (PCS & MCS) using twelve questions that have scores ranging from 0 to 100, where a 0 score indicates the lowest level of health and 100 indicates the highest level of health. The PCS and MCS scores tend to vary by age: PCS tends to decrease while MCS tends to increase with age. The age-specific mean difference score (difference score) is the amount by which a person's score differs from their age group's mean score. For individual scores, someone who scores higher than the mean indicates a person with better health status than most others their age. Conversely, scores lower than the mean indicate a person with poorer health than most others their age [46].

Depressive Symptoms: This study used the 15-item Short Geriatric Depression Scale (GDS) to evaluate depression [33,34]. Responses were assessed by "yes" or "no" responses. A summary score was calculated with a potential range between 0 and 15, in which a higher score indicated more depressive symptoms. The GDS-Short Form has excellent reliability and validity, and it has been used extensively to measure depression among older adults in both community and clinical settings [33,34].

Sleep Quality: Sleep quality was measured using the 7-item Insomnia Severity Index (ISI) [35,36]. Each item was on a Likert scale with possible answers ranging from 0 to 4. We calculated a total score ranging from 0 to 28, with a higher score reflecting worse sleep quality. Research shows that the ISI has adequate validity and internal consistency (reliability) to evaluate sleep difficulties.

Medication Use: Medication use was assessed by recording an inventory of all the medications each participant was taking within two weeks prior to the interviews. Participants were asked to bring all over-the-counter and prescribed medications to the interview. From the container label, he interviewer transcribed the name of the medication, strength of the drug, expiration date, instructions, special warnings, providers' information, etc.

This medication methodology, established by Sorensen and colleagues [47–49] has been adopted by our research team previously [24,50–54].

Medication regimen complexity index (MRCI): This study employed the MRCI, a tool for quantifying multiple features of drug regimen complexity. The tool was developed and validated by George and colleagues [55]. Medication regimen complexity is a theoretical concept independent of pharmacologic, clinical, and demographic factors. The MRCI quantifies the complexity of medication regimens according to dosage forms, dosing frequencies, and additional directions. The MRCI is an open index without an upper limit for the number of drugs that could be prescribed to a patient or the number of additional instructions possible in a particular regimen. It is a reliable and valid tool with potential applications in clinical practice and research [55]. All components of the MRCI have been independently confirmed to influence patient adherence [56].

Blood Pressure: An OMRON HEM 907XL Intellisense Professional blood pressure monitor was used to measure participants' blood pressure. Our team physician and education specialist measured the blood pressure of all of the participants. Clinically validated for accuracy, the OMRON HEM 907XL Intellisense Professional blood pressure monitor has been designed for clinical use in a professional setting. All blood pressure measures were taken twice with one minute between measures. An average of the two measures was recorded. The monitor was calibrated twice during the study according to the manufacturer's instructions.

Body-Mass Index (BMI): The survey included a self-report of height and weight. Participants were classified according to their BMI as: normal weight (BMI 20.0–24.9 kg/m^2); overweight (BMI 25–29.9 kg/m^2); or obese (BMI 6 30 kg/m^2).

2.3. Data Analysis

Our analysis had three parts. The first section was a descriptive analysis. This descriptive work reported the means and standard deviation for the continuous measures and the frequency and percentages for the categorical variables. Second, we conducted Pearson correlation coefficients, an independent t-test, and ANOVA to examine the bivariate association between socio-demographic variables, other relevant variables, and low back-pain. Finally, we used (1) multivariate generalized linear models (GLM) with Poisson distribution and log link; (2) multiple linear regression; and (3) multiple binary and multinomial logistic regression to examine the independent association of low back pain on various outcomes. Controlling for age, gender, race/ethnicity, education, living arrangement, and number of major chronic conditions, the association between low back pain and several outcome variables was examined. For GLM and logistic regression, odds ratio (OR) and 95% confidence intervals are reported, and for liner regression, standardized beta coefficients are reported. For multivariate analysis, p values of less than 0.05 were considered significant. The Bonferroni correction method was used to counteract the problem of multiple comparisons between the bivariate association.

3. Results

Table 1 (1st column) reports the characteristics of the study sample. This study included 740 (81.8%) African American and 165 (18.2%) Latino individuals who were between the ages of 55 and 96 years of age (mean = 71.50 ± 8.36). Approximately 34% of the participants were 75 years of age or older, with 44% self-reporting living alone. One-third of the sample never completed high school. Regarding health status, we noted the following health conditions/illnesses: diabetes mellitus (38%), hypertension (89%), heart-related conditions (28%), and COPD (27%). The majority of the participants (79%) were overweight or obese, with 44% having a BMI of 30 and over.

Table 1. Characteristic of sample a bivariate associations between chronic low back pain and other related outcomes (n = 905).

Sample Characteristics	Total Sample	Chronic Low-Back Pain		Sig.
		No	Yes	
	N (%) Mean ± SD	N (%) Mean ± SD	N (%) Mean ± SD	
Gender				
Male	317 (35.0)	178 (57)	135 (43)	0.009
Female	588 (65.0)	277 (48)	303 (52)	
Race/Ethnicity				
Hispanics	165 (18.2)	70 (45)	85 (55)	0.113
Non-Hispanic Blacks	740 (81.8)	385 (52)	353 (48)	
Living Arrangement				
Living Alone	397 (44)	215 (54)	182 (46)	0.087
Live with other(s)	496 (56)	240 (48)	256 (52)	
Falls in last 12 months				
No	357 (71)	155 (44)	195 (56)	0.006
Yes	145 (29)	45 (31)	100 (69)	
Sleep Disorder				
No	646 (72)	382 (59)	264 (41)	0.000
Yes	247 (28)	73 (30)	174 (70)	
BMI				
<25.0	103 (20)	42 (42)	59 (58)	0.023
25–29.9	178 (35)	83 (48)	91 (52)	
≥30	224 (44)	75 (34)	145 (66)	
ER Admissions				
No	548 (61)	307 (57)	234 (43)	0.000
Yes	354 (39)	148 (42)	204 (58)	
Hospital Admissions				
No	663 (73)	354 (54)	301 (46)	0.002
Yes	241 (27)	101 (42)	137 (58)	
	Mean ± SD	**Mean ± SD**	**Mean ± SD**	**Sig.**
Age (Years: 55–96)	71.5 ± 8.36	72.6 ± 7.88	70.2 ± 8.64	0.000
Education (Years: 1–16)	11.9 ± 3.32	12.12 ± 3.20	11.7 ± 3.35	0.046
Financial Strains (Alway:1–Rarely: 5)	4.05 ± 1.22	4.36 ± 1.02	3.74 ± 1.31	0.000
Major Chronic Conditions (0–6)	2.08 ± 1.11	1.91 ± 1.04	2.27 ± 1.14	0.000
Systolic blood pressure	133 ± 20.0	135 ± 20.5	132 ± 20.0	0.095
Satisfaction with Medical Care	2.72 ± 0.95	2.65 ± 1.00	2.84 ± 0.95	0.003
Physical Health Quality of Life	40.1 ± 12.2	45.1 ± 10.71	34.9 ± 11.40	0.000
Mental Health Quality of Life	51.9 ± 11.6	54.4 ± 9.41	49.4 ± 13.05	0.000
Depressive Symptoms	2.79 ± 2.90	1.98 ± 2.31	3.63 ± 3.20	0.000
Severity of Pain	2.06 ± 2.30	1.05 ± 1.52	3.17 ± 2.46	0.000
Number of Rx Used	5.75 ± 3.25	5.31 ± 3.16	6.20 ± 3.28	0.000
Number of OTC Used	1.11 ± 1.84	1.13 ± 1.77	1.08 ± 1.91	0.659

Table 1. *Cont.*

Sample Characteristics	Total Sample	Chronic Low-Back Pain		Sig.
		No	Yes	
	N (%) Mean ± SD	N (%) Mean ± SD	N (%) Mean ± SD	
Sleep Disorder index	4.99 ± 6.09	3.50 ± 4.84	7.11 ± 7.02	0.000
Self-rated Health	3.17 ± 1.01	2.90 ± 1.01	3.45 ± 0.95	0.000
Physician Visits	5.32 ± 3.21	4.88 ± 3.18	5.80 ± 3.19	0.000
ED Admissions	0.75 ± 1.50	0.54 ± 1.13	0.97 ± 1.79	0.000
Hospital Admissions	0.71 ± 1.27	0.42 ± 1.23	0.61 ± 1.32	0.026
Adherence with Medication	5.88 ± 2.13	6.32 ± 1.85	5.61 ± 2.25	0.001
Medication Complexity	11.67 ± 7.23	10.55 ± 6.77	12.83 ± 7.49	0.000

3.1. *Bivariate Analysis:*

Table 1 (column 2–4) shows bivariate correlations between low back pain and other related variables. These correlations are based on the chi-square, *t*-test, and ANOVA F-test. Female, younger, less educated, and obese participants reported more low back pain than their counterparts. In unadjusted models, associations between reporting low back pain and the following variables were detected: falls, sleep disorders, dissatisfaction with healthcare, a higher number of office-based physician visits, ED and hospital admissions, a lower level of HR-QoL, a higher number of depressive symptoms, a lower level of adherence with medication, a higher number of Rx medication use, and a lower level of self-reported health status. No associations between, low back pain and systolic blood pressure, over the counter (OTC) medication use, living arrangement, and ethnicity were detected.

3.2. *Multivariate Analysis*

Table 2 contains the condensed results of 17 independent multivariate analyses. Each analysis controlled for age, gender, race/ethnicity, living arrangement, financial strain, and number of major chronic conditions while examining the independent impact of low back pain on 17 outcome variables. This table reveals that having low back pain was associated with three categories of outcomes: (1) a higher level of healthcare use measured by (i) physician visits, (ii) ED admissions, (iii) a higher level of Rx use, (iv) a higher level of medication complexity, (v) a lower level of adherence to medication regimens, and (vi) a lower level of satisfaction with medical care; (2) a lower level of QoL and self-assessment of health measured by (i) a lower level of physical health QoL, (ii) a lower level of mental health QoL, and (iii) a lower level of self-rated health; and (3) worse physical and mental health outcomes measured by (i) a higher number of depressive symptoms, (ii) a higher level of pain, (iii) falls, (iv) sleep disorders, (v) and being obese.

Table 2. Multivariate analysis of associations between chronic low back pain and outcome variables.

Dependent/Outcome Variable	Model	Independent Variable: Chronic Low Back Pain (No vs. Yes) Adjusting for Age, Gender, Education, Living Arraignment, Financial Strains and Number of Chronic Conditions					
		B	Beta	95% CI Unstandardized Coefficients	Exp. (B) OR	95% CI Exp. (B) OR	Sig.
Health Care Utilization							
Physician Visits	MLR	0.717	0.112	0.282–1.151			0.001
ED Admissions	GLM-P				1.490	1.152–1.928	0.002
Hospital Admissions	GLM-P				1.269	0.872–1.848	0.214
Number of Rx Used	MLR	0.508	0.078	0.103–0.914			0.014
Number of OTC Used	GLM-P				1.016	0.807–1.280	0.892
Medication Complexity	MLR	1.569	0.109	0.664–2.475	N/A	N/A	0.001
Adherence with Medication	MLR	−0.410	−0.094	−0.808–−0.012	N/A	N/A	0.044
Satisfaction with Medical Care	MLR	0.133	0.071	0.009–0.257			0.035
QoL Self-rated Health							
Physical Health Quality of Life	MLR	−7.593	−0.312	−9.033–−6.153	N/A	N/A	0.000
Mental Health Quality of Life	MLR	−3.102	−0.134	−4.616–−1.588	N/A	N/A	0.000
Self-rated Health	MLR	0.364	0.179	0.233–0.494	N/A	N/A	0.000
Physical and Mental Health Outcomes							
Systolic Blood Pressure	MLR	−2.325	−0.057	−6.168–1.517	N/A	N/A	0.235
Depressive Symptoms	MLR	0.881	0.152	0.538–1.224	N/A	N/A	0.000
Severity of Pain	LT-LR	0.563	0.409	0.485–0.640	N/A	N/A	0.000
Falls in last 12 months							
No	BLR	N/A	N/A	N/A	1.000	Ref	0.004
Yes					1.619	1.056–2.483	
Sleep Disorder							
No	BLR	N/A	N/A	N/A	1.000	Ref	0.000
Yes					2.66	1.721–4.117	
BMI							
<25	MNLR	N/A	N/A	N/A	0.863	0.518–1.438	0.573
25–30					0.858	0.407–0.968	0.035
>30					1.00	Ref	

MLR: Multiple linear regression; LT-LR: log transferred linear regression; BLR: binary logistic regression; MNLR: multinomial logistic regression; GLM-P: generalized linear regression with Poisson distribution.

4. Discussion

Almost 55% and 48% of Latino and African American older adults in our study reported chronic low back pain, a much higher prevalence than the national average for White, Black, and Hispanic older adults. Data from the latest (2018) National Health Interview Survey (NHIS) revealed that the prevalence of low back-pain among White, Black, and Hispanic men aged 55 years and older during the preceding 12 months was 35.2%, 34.7%, and 32.1% respectively. The prevalence for White, Black, and Latino women in a similar age group was 37.3%, 40.0%, and 40.3%, respectively. Our findings show a much higher prevalence of low back pain for both African American and Latino underserved men and women compared to NHIS data for the same age group. In our study, the prevalence for African American and Latino men was 43.6% and 40.4% and for women, it was 50.2% and 61.1%, respectively. It is very important to note that our data was collected from one of the most underserved and under-resourced areas in the State of California known as South Service Planning Area 6 (SPA6) of the County of Los Angeles. The much higher prevalence of low back pain among our sample of Latinos and African Americans points to urgent interventions among this and other underserved communities of minority older adults.

Low Back Pain and Healthcare Utilization: Our findings indicate that African American and Latino older adults who suffer from low back pain used ED services more frequently and had more office-based provider visits than their counterparts with no low back pain. More than 47% of African Americans and 31% of Latinos aged 55 years and older who suffered from low back pain used ED services and were admitted to the hospital at least once within the 12 months prior to the interview. Multivariate analysis showed that after adjusting for age, gender, education, living arraignment, financial strain, and number of chronic conditions, participants with low back pain had 49% increased odds of using ED services, whereas admission to the hospital was no longer significant. Further, multivariate analysis revealed that after adjusting for other variables, participants with low back pain had 51% odds of visiting office-based providers at least every two months compared to their counterparts with no back pain.

A systematic review of the literature indicates that low back pain is consistently a top presenting complaint in the emergency setting [57]. The National Electronic Injury Surveillance System shows that among all individuals in the US presenting with low back pain in emergency departments, older patients were found to be at a greater risk of hospital admission for the treatment of low back pain [58]. However, several studies documented racial disparities in the treatment of low back pain in both the ED and in the primary care setting [26,59–64]. National data show that controlling for potential confounders, non-White patients who presented at the ED for back pain were less likely than their White counterparts to receive analgesia and more likely to wait longer for opiate medication [60]. One recent study conducted by Kohen and colleagues (2018) among 600 adult patients with low back pain from three ED departments in Michigan revealed that racial disparities and psychosocial factors had concerning relationships with clinical decision-making in the ED among patients with low back pain, indicating that Caucasian race was one factor associated with advanced imaging [59].

Low Back Pain and Medication Use and Adherence to Drug Regimens: Our study documented a strong association between low back pain and the use of Rx but not OTC. On average, African American and Latino older adults with low back pain used almost one (0.89) additional prescription medication than their counterparts without low back pain. In addition, we documented that, after adjusting for other relevant factors, non-adherence to drug regimens is more prevalent among our sample of underserved African American and Latino older adults with low back pain. Medication-related challenges, including polypharmacy (excessive and unnecessary use of medication), inappropriate medication use, and non-adherence to drug regimens among minority adults, particularly among African Americans, have recently received attention among researchers [53,65–72]. Limited available data showed increased medication-related challenges among minority older adults compared to their White counterparts [53,65–72]. Polypharmacy remains an important issue among underserved older minority adults [54]. African Americans with polypharmacy, particularly those with hyper-polypharmacy, are experiencing higher levels of psychological distress, which itself is a known risk factor for poor adherence to medications [73]. Polypharmacy also has been linked to depressive symptoms in U.S.-born older Mexican Americans [74].

Low Back Pain and Accessibility of Medical Care: Another interesting result of this study was that at both the bivariate and multivariate levels, participants with low back pain were more likely to express their dissatisfaction with the availability of and satisfaction with medical care than their counterparts with no back-pain. Minority older adults forgoing needed medical care due to barriers associated with cost, type or lack of health care coverage, mistrust, racism, and discrimination are at risk of missing the needed care that may be necessary for the management of pain and other chronic conditions [75]. The sources of disparities in pain management among ethnic and racial minorities are complex, involving patient, health care provider, and health care system factors [76]. Recent findings (2020) noted that pain is often poorly recognized, inadequately assessed, and unsuccessfully managed among people in mainstream society, but this is particularly the case for people who have been historically, economically, and socially marginalized [77]. A qualitative study by Wallace and colleagues (2021) suggested that pain is entangled with and shaped by experiences of inadequate and ineffective health care as well as experiences of discrimination, stigma and dismissal, and the impacts of these and other factors intersecting experiences [78]. Equity-oriented responses to chronic pain would recognize pain not only as a biomedical issue but as a social justice issue [78] and therefore improving access and satisfaction with medical care among minority older adults with pain is the first step toward mitigating some of the racial/ethnic disparities observed in pain among this segment of our population.

Low Back Pain and Falls: Multivariate analysis of our data revealed that, after adjusting for other variables, participants with low back pain had 62% odds of having experienced a fall at least once within last 12 months compared to their counterparts with no back pain. The Centers for Disease Control and Prevention (CDC) reports that falls are currently the leading cause of both fatal and non-fatal injury and trauma among older adults [79]. Previous studies show that older adults with a higher number of chronic diseases were more likely to have fallen in the past year compared to those with fewer chronic diseases [80]. In addition, multiple studies have documented that older adults with chronic low back pain are at risk for falling [81–84].

While extensive research has been conducted on falls and fall-related risk factors, the majority is primarily among non-Hispanic White populations. There are mixed findings in the literature in regard to racial differences and falls. Although some studies have found that older non-Hispanic Whites are more likely to fall than older African Americans [85–87], a few revealed no racial differences in fall rates among these two groups [88–90]. However, examining ethnic differences in fall rates and fall circumstances in older community-dwelling White and African-American women, Faulkner and colleagues documented that although the circumstances of falling differed for older White and African American women, there were no differences in the frequencies of falling [88]. Similarly, using California Health Interview Survey data, another recent study reported that African American older adults were most likely to experience ≥ 2 falls in the past year (14.2%), followed by Hispanics (13.8%), non-Hispanic Whites (12.8%), and Asian Americans (7.6%) [80]. This study argued that further research is needed to explore factors associated with fall risks across racial/ethnic groups in order to design culturally appropriate interventions to reduce falls among this segment of our population [80]. Knowing that African American and Latino older adults suffer from the poor management of multiple conditions [91] and have a greater number of chronic conditions, including untreated or under-treated low back pain, higher and more severe falls among this segment of our population are inevitable if no appropriate interventions are implemented.

Low Back Pain and Sleep Disorder: Our study showed that, after adjusting for other relevant variables, including the number of chronic conditions, the association between low back pain and sleep disorders remained significant. The association between low back pain and sleep complaints (such as disruptive sleep and problems initiating and maintaining sleep) among older adults has been well-stablished [2,92]. However, the examination of sleep disorders among minority older adults are very limited [93–97]. Furthermore, there are few studies evaluating sleep quality among minority older adults and its relationship with pain [96]. A recent study conducted among 1664 older adults revealed significant racial/ethnic differences in overall sleep quality and in individual sleep components among older adults [93]. Interventional studies show that different types of therapies are effective in reducing the pain related to disability, quality of sleep, and depression in older adults with chronic pain and low back pain [98,99]. However, future studies are needed to understand the role of different social, psychological, physical as well as pharmacological therapy in relation to sleep disturbance and low back pain among African American older adults. Sleep evaluation should be routinely included in studies of under-resourced minority older adults with chronic pain, particularly if they have low back pain.

Low Back Pain and Depression: Our data revealed a strong association between the Geriatric Depression Scale (GDS) and low back pain, even after adjusting for other relevant variables, among underserved African American and Latino older adults. A review of the literature shows that pain is a consistent risk factor for the development of depressive symptoms [100], and the presence of chronic pain is an independent contributor to depressive symptoms, even when controlling for each individual chronic illness, and the overall number of illnesses [101]. A review of epidemiologic studies conducted by Zis and colleagues showed that among older adults, chronic pain increases the risk of depression between 2.5 and 4.1 times. Moreover, patients suffering from a major depressive disorder are six times more likely to suffer from neuropathic pain and three times more likely to suffer from non-neuropathic pain [102]. Zis and colleagues hypothesized that the common pathogenic factor between chronic pain and depression could be represented by chronic, subclinical neuroinflammation [102]. The risk of disabling back pain increases in older age [9,103,104].

There are very limited studies that have examined the association between pain or low back pain and depression among minority older adults [105,106]. Data from the Health and Retirement Study (HRS) showed that severe pain was more strongly associated with a greater risk of depressive symptoms among Black and Hispanic older adults than among White older adults. The increased risk of mental disorders that racial and ethnic minorities face due to severe chronic pain is particularly important in the context the aging process [107]. Data from Community Aging in Place: Advancing Better Living for Elders (CAPABLE) showed that depression fully mediated the relationship between pain intensity and functional limitation among low-income older adults [105]. Another study showed that beyond the obvious physical manifestations and limitations of the pain experience, there was a robust relationship between depression and pain among a sample of African American women [106].

Low Back Pain and Blood Pressure: Our study documented no association between low back pain and blood pressure among our sample of underserved African American and Latino older adults. The association between hypertension and low back pain has not been clearly documented. There are a few of studies that have demonstrated an inverse relationship between high blood pressure and the prevalence of low back pain and osteoarthritis [108]. Indeed, it is still unclear how hypertension and hypertensive medicine influence sensitivity to pain. However, it is stablished that pain sensitivity is frequently diminished in hypertensive individuals. A European study showed that in a large unselected population sample with a very wide age range, the presence of chronic pain was associated with decreased heart rate variability and baroreflex sensitivity relative to the absence of chronic pain, supporting a model in which the risk of hypertension in the chronic pain population is derived, in part, from the diminished heart rate variability and baroreflex sensitivity associated with chronic pain [109]. More definitive studies on the relationship between hypertension, low back pain, and osteoarthritis, particularly among older minority adults, are needed.

Low Back Pain and Quality of Life: While several studies have documented an association between pain and quality of life [6,8,110–115], the current study used a community-based sample of under-resourced African American and Latino older adults to document the significant association between chronic low back pain and both the physical and mental health quality of life among this segment of our population. Cedraschi and colleagues (2016) investigated the self-reporting of low back pain and associated health-related quality of life in a population sample of 3042 community-dwelling older adults aged ≥65. They suggested that low back pain may be a risk factor for frailty, defined by higher levels of functional limitations, psychological difficulties, and social restrictions, hence, globally impaired HRQoL [116]. Several cross-sectional and longitudinal studies conducted in European countries have also provided very similar results, indicating a strong association between low back pain and HR-QoL [7,117,118]. Finally, a recent systematic reviews of 35 research paper published from 1985 to 2018 with a total sample of 135,059 older adults aged between 60 and 102 years confirmed a high prevalence of low back pain in older adults and functional disability that affects quality of life and factors important for independence [119]. It is important to note that this systematic review found a significant absence of studies among older Latino and African Americans. More studies on the effect of low back pain on quality of life are needed in this population.

Low Back Pain and Body Mass Index: Accumulative evidence from longitudinal and cross-sectional studies link the occurrence of musculoskeletal diseases, including chronic low back pain, to the presence of obesity and being overweight [120–122]. Our study shows that 66% of African American and Latino older participants with a BMI $\geq$ 30 reported low back pain versus 34% with a BMI < 25. The association between low back pain and BMI remains significant even after adjusting for other related variables. The epidemic of obesity and being overweight remains a heavy burden among minority populations. Crespo and colleagues argued that although obesity and its consequences affect all age groups and ethnic subgroups, minority older adults are disproportionately affected [123]. A cross-sectional analysis of adults over 65 in the 2013 National Health Interview Study (NHIS) (N = 7714) showed that Non-Hispanic Whites (25%) had the lowest percentage of obesity compared to Latinos (27%) and African Americans (36%) [124]. Specifically, high obesity rates persist across the lifespan of African American women, from young to older adulthood, with class 3 (severe) obesity present at nearly twice the rate in women and men of other races in older adulthood [125]. A third National Health and Nutrition Examination Survey that involved 5724 adults over 60 shows that the prevalence of significant back pain increases markedly with increased BMI, particularly among African American and Mexican American older adults [126]. The relationship between BMI and back pain was stronger among Mexican Americans (44% increase from normal weight 12% to obesity class III 31%) and African Americans (43% from normal weight 29% to obesity class III 23%) than Whites (12% increase from normal weight 21% to obesity class III 24%) [126]. Our study along with limited available data for older minority adults regarding the association between low back pain and BMI point to the urgent need for interventional studies to address the relationship between being overweight and pain (knee, back, and hip) among older minority adults.

5. Study Limitations

Several limitations were noted in this study. First, the cross-sectional research design as employed in this study attributes no causal relationships between chronic low back pain and the variables analyzed. While chronic low back pain bears a significant relationship with fall risk, a vice versa association may also exist where the risk of falls impacts chronic low back pain. Another limitation of this study is that we did not collect any data regarding occupational histories prior to the onset of low back pain. Participants who have experienced years of heavy labor may be more likely to suffer from low back pain, and this may influence the outcomes and conclusions of this research. Additionally, participants were required to recall their history of chronic low back pain without providing and/or reviewing their medical records. This could introduce recall bias. Additionally, data was obtained using a convenience sampling method; results may either under- or over-represent the Latino and African American sampling frame. Despite these limitations, the findings contribute to the body of knowledge on chronic low back pain in under-resourced Latino and African American populations.

6. Implications

From this study, underserved African American and Latino older adults suffering from chronic low back pain are more likely to use emergency department services, prescription medication(s), express dissatisfaction with medical care, and be obese. When further compared with their White counterparts, they are also likely to have depressive symptoms, poor physical and mental quality of life, and experience sleep disorders as well as to fall over the course of 12 months. However, an association with blood pressure readings within this population cannot established through this study.

Our study contributes to and raises awareness of clinicians, healthcare providers, and health policymakers on the necessity for prevention, early diagnosis, proper medical management, and rehabilitation policies to minimize the burdens associated with chronic low back pain among underserved older African American and Latino patients in under-resourced communities. To treat this complex musculoskeletal disorder, there needs to be a multidisciplinary approach of resource allocation and training with a focus on health equity, mental health, and healthcare access. Objectively measuring the quality of life of minority older adults suffering from chronic low back pain patients is necessary to establish objectives and treatment plans to prevent the progression of recurring and worsening chronic lower back pain. Primary care providers can make a difference in the overall outcomes and improvements of health equity and back health. This can be accomplished in a variety of ways through (1) providing recommendations for back health, (2) health promotion initiatives that focus on proper body mechanics, (3) expansion of chiropractic services, and (4) expansion of data analytics to translate data into actionable information to guide and inform standardized public policy on musculoskeletal health.

At the population level, it is important to monitor the evolution of chronic low back pain using data such as socio-demographic characteristics, health care utilization, health care satisfaction metrics, and physical and mental health outcomes to assess the impact on older Black and Latino adults. At the individual level, chronic low back pain is more than a musculoskeletal disease; it affects all aspects of the lives of minority adults, including mental health, quality of sleep, health-related quality of life and BMI as well as health care utilization and adherence to drug regimens. Primary care physicians are often the first to identify minority older patients with chronic low back pain and can have a significant impact on their care. A multi-disciplinary and inter-professional approach to chronic low back pain should be an integral part of care to prevent disability and improve quality of life among minority older adults. These interventions may mitigate inappropriate ED visits and reduce health care expenditures. These health care resources can then be redirected to the prevention of chronic low back pain and the improvement of the quality of life of older adults. More research is needed to study the impact of chronic low back pain on the quality of life, health care utilization, and health outcomes among minority older adults in underserved communities.

Author Contributions: M.B. and S.B.-H. were involved in the conception and design of the study. M.B. and S.A. performed the data analysis and interpretation. M.B., S.B.-H., L.W.K., M.L., T.E., E.K.A., S.A. and S.C. drafted the initial manuscript together. L.W.K. and S.B.-H. also completed the modifications necessary to make the manuscript acceptable for publication. All authors revised the manuscript critically for intellectual content and approved the submitted version. All authors have read and agreed to the published version of the manuscript.

Funding: This study was supported by the Centers for Medicare and Medicaid Services (CMS) grant 1H0CMS331621 to the Charles R. Drew University of Medicine and Science (PI: M. Bazargan). Additionally, Drs. Cobb, Loeza, Adinkrah, and Ekwegh were supported by the National Institutes of Health (NIH) under award R25 MD007610 (PI: M. Bazargan). The statements contained do not necessarily reflect those of the CMS and NIMHD.

Institutional Review Board Statement: This study was conducted according to the guidelines of the Declaration of Helsinki, and approved by the Institutional Review Board of the Charles R. Drew University of Medicine and Sciences, protocol # 1489890 & 1563927.

Informed Consent Statement: Prior to the interviews, all potential participants were given information about the study including the potential risks and benefits. The information described the nature and the purpose of the study. All participants signed a consent form. Ample opportunity was allowed for all participants and their family members to ask questions about the consent form and to discuss its meaning. Illiterate participants provided consent in the presence of a reliable or legally authorized family member.

Data Availability Statement: Personal identification details of the participants were separated from the completed questionnaires. The data were stored in a locked room of the corresponding author at the Charles R. Drew University of Medicine and Science (CDU). No information relating to identifiable individuals was disseminated. The data sets used and analyzed in the current study are available from the corresponding author for collaborative studies.

Conflicts of Interest: The authors declare no conflict of interest.

Ethics Approval: Ethical approval for the study was obtained from the CDU Institutional Review Board. Participants provided informed consent. The authors assert that all procedures contributing to this work comply with the ethical standards of the relevant national and institutional committees.

Consent for Publication: The consent form explained that those who participated also gave their consent to utilize the information in a non-identified format for scientific and popular publications.

References

1. Finley, C.R.; Chan, D.S.; Garrison, S.; Korownyk, C.; Kolber, M.R.; Campbell, S.; Eurich, D.T.; Lindblad, A.J.; Vandermeer, B.; Allan, G.M. What are the most common conditions in primary care? Systematic review. *Can. Fam. Physician* **2018**, *64*, 832–840.
2. Rudy, T.E.; Weiner, D.K.; Lieber, S.J.; Slaboda, J.; Boston, J.R. The impact of chronic low back pain on older adults: A comparative study of patients and controls. *Pain* **2007**, *131*, 293–301. [CrossRef] [PubMed]
3. Wong, A.Y.; Karppinen, J.; Samartzis, D. Low back pain in older adults: Risk factors, management options and future directions. *Scoliosis Spinal Disord.* **2017**, *12*, 14. [CrossRef]
4. Coyle, P.C.; O'Brien, V.A.; Edwards, D.G.; Pohlig, R.T.; Hicks, G.E. Markers of Cardiovascular Health in Older Adults with and Without Chronic Low Back and Radicular Leg Pain: A Comparative Analysis. *Pain Med.* **2021**, *22*, 1353–1359. [CrossRef] [PubMed]
5. Kimachi, K.; Kimachi, M.; Takegami, M.; Ono, R.; Yamazaki, S.; Goto, Y.; Onishi, Y.; Sekiguchi, M.; Otani, K.; Konno, S.-I. Level of Low Back Pain–Related Disability Is Associated with Risk of Subsequent Falls in an Older Population: Locomotive Syndrome and Health Outcomes in Aizu Cohort Study (LOHAS). *Pain Med.* **2019**, *20*, 2377–2384. [CrossRef]
6. Hicks, G.E.; Sions, J.M.; Velasco, T.O. Hip symptoms, physical performance, and health status in older adults with chronic low back pain: A preliminary investigation. *Arch. Phys. Med. Rehabil.* **2018**, *99*, 1273–1278. [CrossRef]
7. Leopoldino, A.A.O.; Megale, R.Z.; Diz, J.B.M.; Moreira, B.S.; Felicio, D.C.; Queiroz, B.Z.; Bastone, A.C.; Ferreira, M.L.; Pereira, L.S.M. Influence of Frailty Status on Pain, Disability, and Quality of Life in Older Adults with Acute Low Back Pain: Results from the Back Complaints in the Elders (BACE-Brazil) Study. *Can. J. Aging* **2020**, 1–9. [CrossRef]
8. Rundell, S.D.; Resnik, L.; Heagerty, P.J.; Kumar, A.; Jarvik, J.G. Comparing the performance of comorbidity indices in predicting functional status, health-related quality of life, and total health care use in older adults with back pain. *J. Orthop. Sports Phys. Ther.* **2020**, *50*, 143–148. [CrossRef] [PubMed]
9. Lobo, C.C.; Vilar-Fernández, J.M.; Losa-Iglesias, M.E.; López-López, D.; Rodríguez-Sanz, D.; Palomo-López, P.; Becerro-de Bengoa-Vallejo, R. Depression symptoms among older adults with and without subacute low back pain. *Rehabil. Nurs. J.* **2019**, *44*, 47–51. [CrossRef]
10. Da Silva, R.A.; Vieira, E.R.; Fernandes, K.B.; Andraus, R.A.; Oliveira, M.R.; Sturion, L.A.; Calderon, M.G. People with chronic low back pain have poorer balance than controls in challenging tasks. *Disabil. Rehabil.* **2018**, *40*, 1294–1300. [CrossRef]
11. Rundell, S.D.; Sherman, K.J.; Heagerty, P.J.; Mock, C.N.; Jarvik, J.G. The clinical course of pain and function in older adults with a new primary care visit for back pain. *J. Am. Geriatr. Soc.* **2015**, *63*, 524–530. [CrossRef] [PubMed]
12. Tagliaferri, S.D.; Armbrecht, G.; Miller, C.T.; Owen, P.J.; Mundell, N.L.; Felsenberg, D.; Thomasius, F.; Belavy, D.L. Testing the deconditioning hypothesis of low back pain: A study in 1182 older women. *Eur. J. Sport Sci.* **2020**, *20*, 17–23. [CrossRef] [PubMed]
13. Coyle, P.C.; Velasco, T.; Sions, J.M.; Hicks, G.E. Lumbar mobility and performance-based function: An investigation in older adults with and without chronic low back pain. *Pain Med.* **2017**, *18*, 161–168. [CrossRef] [PubMed]
14. Coyle, P.C.; Pugliese, J.M.; Sions, J.M.; Eskander, M.S.; Schrack, J.A.; Hicks, G.E. Energy impairments in older adults with low back pain and radiculopathy: A matched case-control study. *Arch. Phys. Med. Rehabil.* **2018**, *99*, 2251–2256. [CrossRef]
15. Ishak, N.A.; Zahari, Z.; Justine, M. Kinesiophobia, pain, muscle functions, and functional performances among older persons with low back pain. *Pain Res. Treat.* **2017**, *2017*, 3489617. [CrossRef]
16. Stensland, M.L.; Sanders, S. "It has changed my whole life": The systemic implications of chronic low back pain among older adults. *J. Gerontol. Soc. Work* **2018**, *61*, 129–150. [CrossRef]
17. Jarvik, J.G.; Gold, L.S.; Tan, K.; Friedly, J.L.; Nedeljkovic, S.S.; Comstock, B.A.; Deyo, R.A.; Turner, J.A.; Bresnahan, B.W.; Rundell, S.D. Long-term outcomes of a large, prospective observational cohort of older adults with back pain. *Spine J.* **2018**, *18*, 1540–1551. [CrossRef]
18. Bosley, B.N.; Weiner, D.K.; Rudy, T.E.; Granieri, E. Is chronic nonmalignant pain associated with decreased appetite in older adults? Preliminary evidence. *J. Am. Geriatr. Soc.* **2004**, *52*, 247–251. [CrossRef]
19. Hogans, B.B.; Siaton, B.C.; Taylor, M.N.; Katzel, L.I.; Sorkin, J.D. Low Back Pain and Substance Use: Diagnostic and Administrative Coding for Opioid Use and Dependence Increased in US Older Adults with Low Back Pain. *Pain Med.* **2021**, *22*, 836–847. [CrossRef]
20. Morelhao, P.K.; Pinto, R.Z.; Tufik, S.; Andersen, M.L. Sleep Disturbance and Low Back Pain in Older Adults: A Bidirectional Relationship? *Pain Med.* **2020**, *21*, 1303–1304. [CrossRef]

21. Stensland, M.L.; Sanders, S. Not so golden after all: The complexities of chronic low back pain in older adulthood. *Gerontologist* **2018**, *58*, 923–931. [CrossRef] [PubMed]
22. Jarvik, J.G.; Comstock, B.A.; Heagerty, P.J.; Turner, J.A.; Sullivan, S.D.; Shi, X.; Nerenz, D.R.; Nedeljkovic, S.S.; Kessler, L.; James, K. Back pain in seniors: The Back pain Outcomes using Longitudinal Data (BOLD) cohort baseline data. *BMC Musculoskelet. Disord.* **2014**, *15*, 134. [CrossRef] [PubMed]
23. Carey, T.S.; Freburger, J.K.; Holmes, G.M.; Jackman, A.; Knauer, S.; Wallace, A.; Darter, J. Race, care seeking, and utilization for chronic back and neck pain: Population perspectives. *J. Pain* **2010**, *11*, 343–350. [CrossRef]
24. Bazargan, M.; Yazdanshenas, H.; Gordon, D.; Orum, G. Pain in community-dwelling elderly African Americans. *J. Aging Health* **2016**, *28*, 403–425. [CrossRef]
25. Cobb, S.; Bazargan, M.; Assari, S. Correlates of Pain, Opioid Use, and Psychotropic Medications Among Older African Americans. *Innov. Aging* **2020**, *4*, 785. [CrossRef]
26. King, C.; Liu, X. Racial and ethnic disparities in opioid use among US adults with back pain. *Spine* **2020**, *45*, 1062–1066. [CrossRef] [PubMed]
27. Lind, K.E.; Flug, J.A. Sociodemographic Variation in the Use of Conservative Therapy Before MRI of the Lumbar Spine for Low Back Pain in the Era of Public Reporting. *J. Am. Coll. Radiol.* **2019**, *16*, 560–569. [CrossRef]
28. Karran, E.L.; Grant, A.R.; Moseley, G.L. Low back pain and the social determinants of health: A systematic review and narrative synthesis. *Pain* **2020**, *161*, 2476–2493. [CrossRef]
29. Los Angeles County Department of Public Health; Office of Health Assessment and Epidemiology. *Key Indicators of Health by service Planning Area*; Los Angeles County Department of Public Health: Los Angeles County, CA, USA, 2013.
30. Hays, R.D.; Eastwood, J.A.; Kotlerman, J.; Spritzer, K.L.; Ettner, S.L.; Cowan, M. Health-related quality of life and patient reports about care outcomes in a multidisciplinary hospital intervention. *Ann. Behav. Med. A Publ. Soc. Behav. Med.* **2006**, *31*, 173–178. [CrossRef]
31. Melzack, R. The short-form McGill Pain Questionnaire. *Pain* **1987**, *30*, 191–197. [CrossRef]
32. Dworkin, R.H.; Turk, D.C.; Revicki, D.A.; Harding, G.; Coyne, K.S.; Peirce-Sandner, S.; Bhagwat, D.; Everton, D.; Burke, L.B.; Cowan, P.; et al. Development and initial validation of an expanded and revised version of the Short-form McGill Pain Questionnaire (SF-MPQ-2). *Pain* **2009**, *144*, 35–42. [CrossRef]
33. Burke, W.J.; Roccaforte, W.H.; Wengel, S.P. The Short Form of the Geriatric Depression Scale: A Comparison with the 30-Item Form. *Top. Geriatr.* **1991**, *4*, 173–178. [CrossRef]
34. Greenberg, S.A. How to try this: The Geriatric Depression Scale: Short Form. *Am. J. Nurs.* **2007**, *107*, 60–69. [CrossRef]
35. Yu, D.S. Insomnia Severity Index: Psychometric properties with Chinese community-dwelling older people. *J. Adv. Nurs.* **2010**, *66*, 2350–2359. [CrossRef]
36. Sierra, J.C.; Guillen-Serrano, V.; Santos-Iglesias, P. Insomnia Severity Index: Some indicators about its reliability and validity on an older adults sample. *Rev. Neurol.* **2008**, *47*, 566–570. [PubMed]
37. Laaksonen, E.; Lallukka, T.; Lahelma, E.; Ferrie, J.E.; Rahkonen, O.; Head, J.; Marmot, M.G.; Martikainen, P. Economic difficulties and physical functioning in Finnish and British employees: Contribution of social and behavioural factors. *Eur. J. Public Health* **2011**, *21*, 456–462. [CrossRef] [PubMed]
38. Assari, S.; Lankarani, M.M. Demographic and socioeconomic determinants of physical and mental self-rated health across 10 ethnic groups in the United States. *Int. J. Epidemiol. Res.* **2017**, *3*, 185. [PubMed]
39. Finch, B.K.; Hummer, R.A.; Reindl, M.; Vega, W.A. Validity of self-rated health among Latino (a) s. *Am. J. Epidemiol.* **2002**, *155*, 755–759. [CrossRef]
40. Fayers, P.M.; Hays, R.D. *Assessing Quality of Life in Clinical Trials: Methods and Practice*; Oxford University Press: New York, NY, USA, 2005.
41. Woo, H.; Zajacova, A. Predictive strength of self-rated health for mortality risk among older adults in the United States: Does it differ by race and ethnicity? *Res. Aging* **2017**, *39*, 879–905. [CrossRef]
42. Chou, R. Low back pain (chronic). *BMJ Clin. Evid.* **2010**, *2010*, 1116.
43. Zinke, J.L.; Lam, C.S.; Harden, R.N.; Fogg, L. Examining the cross-cultural validity of the english short-form McGill Pain Questionnaire using the matched moderated regression methodology. *Clin. J. Pain* **2010**, *26*, 153–162. [CrossRef] [PubMed]
44. Kosinski, M.; Keller, S.D.; Ware Jr, J.E.; Hatoum, H.T.; Kong, S.X. The SF-36 Health Survey as a generic outcome measure in clinical trials of patients with osteoarthritis and rheumatoid arthritis: Relative validity of scales in relation to clinical measures of arthritis severity. *Med. Care* **1999**, *37*, MS23–MS39. [CrossRef]
45. Larson, C.O.; Schlundt, D.; Patel, K.; Hargreaves, M.; Beard, K. Validity of the SF-12 for use in a low-income African American community-based research initiative (REACH 2010). *Prev. Chronic Dis.* **2008**, *5*, A44. [PubMed]
46. Interpreting the SF-12. Available online: http://www.health.utah.gov/opha/publications/1996hss/sf12/sf12hi1.pdf (accessed on 30 November 2017).
47. Sorensen, L.; Stokes, J.A.; Purdie, D.M.; Woodward, M.; Roberts, M.S. Medication management at home: Medication risk factor prevalence and inter-relationships. *J. Clin. Pharm.* **2006**, *31*, 485–491. [CrossRef] [PubMed]
48. Sorensen, L.; Stokes, J.A.; Purdie, D.M.; Woodward, M.; Roberts, M.S. Medication management at home: Medication-related risk factors associated with poor health outcomes. *Age Ageing* **2005**, *34*, 626–632. [CrossRef]

49. Sorensen, L.; Stokes, J.A.; Purdie, D.M.; Woodward, M.; Elliott, R.; Roberts, M.S. Medication reviews in the community: Results of a randomized, controlled effectiveness trial. *Br. J. Clin. Pharm.* **2004**, *58*, 648–664, Erratum in **2005**, *59*, 376. [CrossRef]
50. Yazdanshenas, H.; Bazargan, M.; Orum, G.; Loni, L.; Mahabadi, N.; Husaini, B. Prescribing patterns in the treatment of hypertension among underserved African American elderly. *Ethn. Dis.* **2014**, *24*, 431–437.
51. Yazdanshenas, H.; Bazargan, M.; Smith, J.; Martins, D.; Motahari, H.; Orum, G. Pain Treatment of Underserved Older African Americans. *J. Am. Geriatr. Soc.* **2016**, *64*, 2116–2121. [CrossRef] [PubMed]
52. Bazargan, M.; Yazdanshenas, H.; Han, S.; Orum, G. Inappropriate Medication Use Among Underserved Elderly African Americans. *J Aging Health* **2016**, *28*, 118–138. [CrossRef]
53. Bazargan, M.; Smith, J.; Yazdanshenas, H.; Movassaghi, M.; Martins, D.; Orum, G. Non-adherence to medication regimens among older African-American adults. *BMC Geriatr.* **2017**, *17*, 163. [CrossRef]
54. Bazargan, M.; Smith, J.; Movassaghi, M.; Martins, D.; Yazdanshenas, H.; Salehe Mortazavi, S.; Orum, G. Polypharmacy among underserved older African American adults. *J. Aging Res.* **2017**, *2017*. [CrossRef] [PubMed]
55. George, J.; Phun, Y.-T.; Bailey, M.J.; Kong, D.C.; Stewart, K. Development and validation of the medication regimen complexity index. *Ann. Pharmacother.* **2004**, *38*, 1369–1376. [CrossRef] [PubMed]
56. McDonald, M.V.; Peng, T.R.; Sridharan, S.; Foust, J.B.; Kogan, P.; Pezzin, L.E.; Feldman, P.H. Automating the medication regimen complexity index. *J. Am. Med Inform. Assoc.* **2013**, *20*, 499–505. [CrossRef]
57. Edwards, J.; Hayden, J.; Asbridge, M.; Gregoire, B.; Magee, K. Prevalence of low back pain in emergency settings: A systematic review and meta-analysis. *BMC Musculoskelet. Disord.* **2017**, *18*, 143. [CrossRef]
58. Waterman, B.R.; Belmont Jr, P.J.; Schoenfeld, A.J. Low back pain in the United States: Incidence and risk factors for presentation in the emergency setting. *Spine J.* **2012**, *12*, 63–70. [CrossRef]
59. Kohns, D.J.; Haig, A.J.; Uren, B.; Thompson, J.; Muraglia, K.A.; Loar, S.; Share, D.; Shedden, K.; Spires, M.C. Clinical predictors of the medical interventions provided to patients with low back pain in the emergency department. *J. Back Musculoskelet. Rehabil.* **2018**, *31*, 197–204. [CrossRef]
60. Mills, A.M.; Shofer, F.S.; Boulis, A.K.; Holena, D.N.; Abbuhl, S.B. Racial disparity in analgesic treatment for ED patients with abdominal or back pain. *Am. J. Emerg. Med.* **2011**, *29*, 752–756. [CrossRef]
61. Burgess, D.J.; Crowley-Matoka, M.; Phelan, S.; Dovidio, J.F.; Kerns, R.; Roth, C.; Saha, S.; van Ryn, M. Patient race and physicians' decisions to prescribe opioids for chronic low back pain. *Soc. Sci. Med.* **2008**, *67*, 1852–1860. [CrossRef]
62. Singhal, A.; Tien, Y.-Y.; Hsia, R.Y. Racial-ethnic disparities in opioid prescriptions at emergency department visits for conditions commonly associated with prescription drug abuse. *PLoS ONE* **2016**, *11*, e0159224. [CrossRef] [PubMed]
63. Trost, Z.; Sturgeon, J.; Guck, A.; Ziadni, M.; Nowlin, L.; Goodin, B.; Scott, W. Examining injustice appraisals in a racially diverse sample of individuals with chronic low back pain. *J. Pain* **2019**, *20*, 83–96. [CrossRef]
64. Lee, P.; Le Saux, M.; Siegel, R.; Goyal, M.; Chen, C.; Ma, Y.; Meltzer, A.C. Racial and ethnic disparities in the management of acute pain in US emergency departments: Meta-analysis and systematic review. *Am. J. Emerg. Med.* **2019**, *37*, 1770–1777. [CrossRef]
65. Adinkrah, E.; Bazargan, M.; Wisseh, C.; Assari, S. Medication complexity among disadvantaged African American seniors in Los Angeles. *Pharmacy* **2020**, *8*, 86. [CrossRef]
66. Bazargan, M.; Smith, J.L.; King, E.O. Potentially inappropriate medication use among hypertensive older African-American adults. *BMC Geriatr.* **2018**, *18*, 238. [CrossRef]
67. Kuo, Y.-F.; Raji, M.A.; Markides, K.S.; Ray, L.A.; Espino, D.V.; Goodwin, J.S. Inconsistent use of diabetes medications, diabetes complications, and mortality in older Mexican Americans over a 7-year period: Data from the Hispanic established population for the epidemiologic study of the elderly. *Diabetes Care* **2003**, *26*, 3054–3060. [CrossRef] [PubMed]
68. Raji, M.A.; Ostir, G.V.; Markides, K.S.; Espino, D.V.; Goodwin, J.S. Potentially inappropriate medication use by elderly Mexican Americans. *Ann. Pharmacother.* **2003**, *37*, 1197–1202. [CrossRef]
69. Bazargan, M.; Smith, J.; Saqib, M.; Helmi, H.; Assari, S. Associations between polypharmacy, self-rated health, and depression in African American older adults; mediators and moderators. *Int. J. Environ. Res. Public Health* **2019**, *16*, 1574. [CrossRef] [PubMed]
70. Murray, M.D.; Hendrie, H.C.; Lane, K.A.; Zheng, M.; Ambuehl, R.; Li, S.; Unverzagt, F.W.; Callahan, C.M.; Gao, S. Antihypertensive medication and dementia risk in older adult african americans with hypertension: A prospective cohort study. *J. Gen. Intern. Med.* **2018**, *33*, 455–462. [CrossRef]
71. Hu, D.; Juarez, D.T.; Yeboah, M.; Castillo, T.P. Interventions to increase medication adherence in African-American and Latino populations: A literature review. *Hawai'i J. Med. Public Health* **2014**, *73*, 11.
72. Parker, W. Health Literacy Among Elderly Hispanics and Medication Usage. Ph.D. Thesis, Walden University, Minneapolis, MN, USA, 2016.
73. Assari, S.; Bazargan, M. Polypharmacy and Psychological Distress May Be Associated in African American Adults. *Pharmacy* **2019**, *7*, 14. [CrossRef]
74. Assari, S.; Wisseh, C.; Saqib, M.; Helmi, H.; Bazargan, M. Polypharmacy and Depressive Symptoms in US-Born Mexican American Older Adults. *Psych* **2019**, *1*, 38. [CrossRef]
75. Forrester, S.N.; Taylor, J.L.; Whitfield, K.E.; Thorpe, R.J. Advances in Understanding the Causes and Consequences of Health Disparities in Aging Minorities. *Curr. Epidemiol. Rep.* **2020**, *7*, 59–67. [CrossRef]
76. Green, C.R.; Anderson, K.O.; Baker, T.A.; Campbell, L.C.; Decker, S.; Fillingim, R.B.; Kaloukalani, D.A.; Lasch, K.E.; Myers, C.; Tait, R.C. The unequal burden of pain: Confronting racial and ethnic disparities in pain. *Pain Med.* **2003**, *4*, 277–294. [CrossRef]

77. Craig, K.D.; Holmes, C.; Hudspith, M.; Moor, G.; Moosa-Mitha, M.; Varcoe, C.; Wallace, B. Pain in persons who are marginalized by social conditions. *Pain* **2020**, *161*, 261. [CrossRef]
78. Wallace, B.; Varcoe, C.; Holmes, C.; Moosa-Mitha, M.; Moor, G.; Hudspith, M.; Craig, K.D. Towards health equity for people experiencing chronic pain and social marginalization. *Int. J. Equity Health* **2021**, *20*, 53. [CrossRef] [PubMed]
79. Bergen, G.; Stevens, M.R.; Burns, E.R. Falls and Fall Injuries Among Adults Aged >/=65 Years—United States, 2014. *MMWR. Morb. Mortal. Wkly. Rep.* **2016**, *65*, 993–998. [CrossRef] [PubMed]
80. Kwon, S.C.; Han, B.H.; Kranick, J.A.; Wyatt, L.C.; Blaum, C.S.; Stella, S.Y.; Trinh-Shevrin, C. Racial and ethnic difference in falls among older adults: Results from the California Health Interview Survey. *J. Racial Ethn. Health Disparities* **2017**, *5*, 271–278. [CrossRef] [PubMed]
81. Marshall, L.M.; Litwack-Harrison, S.; Cawthon, P.M.; Kado, D.M.; Deyo, R.A.; Makris, U.E.; Carlson, H.L.; Nevitt, M.C.; Group, S.o.O.F.R. A prospective study of back pain and risk of falls among older community-dwelling women. *J. Gerontol. Ser. A Biomed. Sci. Med. Sci.* **2016**, *71*, 1177–1183. [CrossRef]
82. Knox, P.J.; Coyle, P.C.; Pugliese, J.M.; Pohlig, R.T.; Sions, J.M.; Hicks, G.E. Hip osteoarthritis signs and symptoms are associated with increased fall risk among community-dwelling older adults with chronic low back pain: A prospective study. *Arthritis Res. Ther.* **2021**, *23*, 71. [CrossRef]
83. Marshall, L.M.; Litwack-Harrison, S.; Makris, U.E.; Kado, D.M.; Cawthon, P.M.; Deyo, R.A.; Carlson, N.L.; Nevitt, M.C.; Group, O.F.i.M.S.R. A prospective study of back pain and risk of falls among older community-dwelling men. *J. Gerontol. Ser. A Biomed. Sci. Med Sci.* **2017**, *72*, 1264–1269. [CrossRef]
84. Munch, T.; Harrison, S.L.; Barrett-Connor, E.; Lane, N.E.; Nevitt, M.C.; Schousboe, J.T.; Stefanick, M.; Cawthon, P.M. Pain and falls and fractures in community-dwelling older men. *Age Ageing* **2015**, *44*, 973–979. [CrossRef]
85. de Rekeneire, N.; Visser, M.; Peila, R.; Nevitt, M.C.; Cauley, J.A.; Tylavsky, F.A.; Simonsick, E.M.; Harris, T.B. Is a fall just a fall: Correlates of falling in healthy older persons. The Health, Aging and Body Composition Study. *J. Am. Geriatr. Soc.* **2003**, *51*, 841–846. [CrossRef] [PubMed]
86. Hanlon, J.T.; Landerman, L.R.; Fillenbaum, G.G.; Studenski, S. Falls in African American and white community-dwelling elderly residents. *J. Gerontol. A Biol. Sci. Med. Sci.* **2002**, *57*, M473–M478. [CrossRef] [PubMed]
87. Karter, A.J.; Laiteerapong, N.; Chin, M.H.; Moffet, H.H.; Parker, M.M.; Sudore, R.; Adams, A.S.; Schillinger, D.; Adler, N.S.; Whitmer, R.A.; et al. Ethnic Differences in Geriatric Conditions and Diabetes Complications Among Older, Insured Adults With Diabetes: The Diabetes and Aging Study. *J. Aging Health* **2015**, *27*, 894–918. [CrossRef]
88. Faulkner, K.A.; Cauley, J.A.; Zmuda, J.M.; Landsittel, D.P.; Nevitt, M.C.; Newman, A.B.; Studenski, S.A.; Redfern, M.S. Ethnic Differences in the Frequency and Circumstances of Falling in Older Community-Dwelling Women. *J. Am. Geriatr. Soc.* **2005**, *53*, 1774–1779. [CrossRef] [PubMed]
89. Means, K.M.; O'Sullivan, P.S.; Rodell, D.E. Balance, mobility, and falls among elderly African American women. *Am. J. Phys. Med. Rehabil.* **2000**, *79*, 30–39. [CrossRef]
90. Quandt, S.A.; Stafford, J.M.; Bell, R.A.; Smith, S.L.; Snively, B.M.; Arcury, T.A. Predictors of falls in a multiethnic population of older rural adults with diabetes. *J. Gerontol. A Biol. Sci. Med. Sci.* **2006**, *61*, 394–398. [CrossRef]
91. Nicklett, E.J.; Taylor, R.J.; Rostant, O.; Johnson, K.E.; Evans, L. Biopsychosocial Predictors of Fall Events Among Older African Americans. *Res. Aging* **2017**, *39*, 501–525. [CrossRef]
92. Lunde, L.-H.; Pallesen, S.; Krangnes, L.; Nordhus, I.H. Characteristics of sleep in older persons with chronic pain: A study based on actigraphy and self-reporting. *Clin. J. Pain* **2010**, *26*, 132–137. [CrossRef]
93. George, K.M.; Peterson, R.L.; Gilsanz, P.; Mungas, D.M.; Glymour, M.M.; Mayeda, E.R.; Whitmer, R.A. Racial/ethnic differences in sleep quality among older adults: Kaiser Healthy Aging and Diverse Life Experiences (KHANDLE) Study. *Ethn. Dis.* **2020**, *30*, 469. [CrossRef] [PubMed]
94. Ancoli-Israel, S.; Klauber, M.R.; Stepnowsky, C.; Estline, E.; Chinn, A.; Fell, R. Sleep-disordered breathing in African-American elderly. *Am. J. Respir. Crit. Care Med.* **1995**, *152*, 1946–1949. [CrossRef] [PubMed]
95. Bazargan, M. Self-reported sleep disturbance among African-American elderly: The effects of depression, health status, exercise, and social support. *Int. J. Aging Hum. Dev.* **1996**, *42*, 143–160. [CrossRef]
96. Baker, T.A.; Whitfield, K.E. Sleep behaviors in older African American females reporting nonmalignant chronic pain: Understanding the psychosocial implications of general sleep disturbance. *J. Women Aging* **2014**, *26*, 113–126. [CrossRef]
97. Xiao, Q.; Blot, W.J.; Matthews, C.E. Weekday and weekend sleep duration and mortality among middle-to-older aged White and Black adults in a low-income southern US cohort. *Sleep Health* **2019**, *5*, 521–527. [CrossRef] [PubMed]
98. Arun, B. Effects of myofascial release therapy on pain related disability, quality of sleep and depression in older adults with chronic low back pain. *Int. J. Physiother Res.* **2014**, *2*, 318–323.
99. Koffel, E.; McCurry, S.M.; Smith, M.T.; Vitiello, M.V. Improving pain and sleep in middle-aged and older adults: The promise of behavioral sleep interventions. *Pain* **2019**, *160*, 529. [CrossRef] [PubMed]
100. Campbell, P.; Hope, K.; Dunn, K.M. The pain, depression, disability pathway in those with low back pain: A moderation analysis of health locus of control. *J. Pain Res.* **2017**, *10*, 2331. [CrossRef]
101. Sharpe, L.; McDonald, S.; Correia, H.; Raue, P.J.; Meade, T.; Nicholas, M.; Arean, P. Pain severity predicts depressive symptoms over and above individual illnesses and multimorbidity in older adults. *BMC Psychiatry* **2017**, *17*, 166. [CrossRef] [PubMed]

102. Zis, P.; Daskalaki, A.; Bountouni, I.; Sykioti, P.; Varrassi, G.; Paladini, A. Depression and chronic pain in the elderly: Links and management challenges. *Clin. Interv. Aging* **2017**, *12*, 709. [CrossRef] [PubMed]
103. Docking, R.E.; Fleming, J.; Brayne, C.; Zhao, J.; Macfarlane, G.J.; Jones, G.T. Epidemiology of back pain in older adults: Prevalence and risk factors for back pain onset. *Rheumatology* **2011**, *50*, 1645–1653. [CrossRef]
104. Reid, M.C.; Williams, C.S.; Concato, J.; Tinetti, M.E.; Gill, T.M. Depressive symptoms as a risk factor for disabling back pain in community-dwelling older persons. *J. Am. Geriatr. Soc.* **2003**, *51*, 1710–1717. [CrossRef]
105. Smith, P.D.; Becker, K.; Roberts, L.; Walker, J.; Szanton, S.L. Associations among pain, depression, and functional limitation in low-income, home-dwelling older adults: An analysis of baseline data from CAPABLE. *Geriatr. Nurs.* **2016**, *37*, 348–352. [CrossRef] [PubMed]
106. Baker, T.A.; Buchanan, N.T.; Corson, N. Factors influencing chronic pain intensity in older black women: Examining depression, locus of control, and physical health. *J. Women's Health* **2008**, *17*, 869–878. [CrossRef]
107. Bierman, A.; Lee, Y. Chronic pain and psychological distress among older adults: A national longitudinal study. *Res. Aging* **2018**, *40*, 432–455. [CrossRef] [PubMed]
108. Zafar, F.; Qasim, Y.F.; Farooq, M.U.; Shamael, I.; Khan, I.U.; Khan, D.H. The frequency of different risk factors for lower back pain in a tertiary care hospital. *Cureus* **2018**, *10*, e3183. [CrossRef]
109. Bruehl, S.; Olsen, R.B.; Tronstad, C.; Sevre, K.; Burns, J.W.; Schirmer, H.; Nielsen, C.S.; Stubhaug, A.; Rosseland, L.A. Chronic pain-related changes in cardiovascular regulation and impact on comorbid hypertension in a general population: The Tromsø study. *Pain* **2018**, *159*, 119–127. [CrossRef]
110. Dominick, K.L.; Ahern, F.M.; Gold, C.H.; Heller, D.A. Health-related quality of life among older adults with arthritis. *Health Qual. Life Outcomes* **2004**, *2*, 5. [CrossRef]
111. Clemens, J.Q.; Link, C.L.; Eggers, P.W.; Kusek, J.W.; Nyberg, L.M.; McKinlay, J.B.; Investigators, B.S. Prevalence of painful bladder symptoms and effect on quality of life in black, Hispanic and white men and women. *J. Urol.* **2007**, *177*, 1390–1394. [CrossRef]
112. Masel, M.C.; Graham, J.E.; Reistetter, T.A.; Markides, K.S.; Ottenbacher, K.J. Frailty and health related quality of life in older Mexican Americans. *Health Qual. Life Outcomes* **2009**, *7*, 70. [CrossRef] [PubMed]
113. Jakobsson, U.; Hallberg, I.R.; Westergren, A. Exploring determinants for quality of life among older people in pain and in need of help for daily living. *J. Clin. Nurs.* **2007**, *16*, 95–104. [CrossRef]
114. Jakobsson, U.; Hallberg, I.R. Pain and quality of life among older people with rheumatoid arthritis and/or osteoarthritis: A literature review. *J. Clin. Nurs.* **2002**, *11*, 430–443. [CrossRef] [PubMed]
115. Oster, G.; Harding, G.; Dukes, E.; Edelsberg, J.; Cleary, P.D. Pain, medication use, and health-related quality of life in older persons with postherpetic neuralgia: Results from a population-based survey. *J. Pain* **2005**, *6*, 356–363. [CrossRef] [PubMed]
116. Cedraschi, C.; Luthy, C.; Allaz, A.-F.; Herrmann, F.; Ludwig, C. Low back pain and health-related quality of life in community-dwelling older adults. *Eur. Spine J.* **2016**, *25*, 2822–2832. [CrossRef] [PubMed]
117. Machado, L.A.; Viana, J.U.; da Silva, S.L.; Couto, F.G.; Mendes, L.P.; Ferreira, P.H.; Ferreira, M.L.; Dias, J.; Dias, R.C. Correlates of a recent history of disabling low back pain in community-dwelling older persons. *Clin. J. Pain* **2018**, *34*, 515–524. [CrossRef] [PubMed]
118. Mutubuki, E.; Beljon, Y.; Maas, E.; Huygen, F.; Ostelo, R.; van Tulder, M.; van Dongen, J. The longitudinal relationships between pain severity and disability versus health-related quality of life and costs among chronic low back pain patients. *Qual. Life Res.* **2020**, *29*, 275–287. [CrossRef]
119. de Souza, I.M.B.; Sakaguchi, T.F.; Yuan, S.L.K.; Matsutani, L.A.; do Espírito-Santo, A.d.S.; Pereira, C.A.d.B.; Marques, A.P. Prevalence of low back pain in the elderly population: A systematic review. *Clinics* **2019**, *74*. [CrossRef] [PubMed]
120. Frilander, H.; Solovieva, S.; Mutanen, P.; Pihlajamäki, H.; Heliövaara, M.; Viikari-Juntura, E. Role of overweight and obesity in low back disorders among men: A longitudinal study with a life course approach. *BMJ Open* **2015**, *5*, e007805. [CrossRef] [PubMed]
121. Hussain, S.M.; Urquhart, D.M.; Wang, Y.; Shaw, J.E.; Magliano, D.J.; Wluka, A.E.; Cicuttini, F.M. Fat mass and fat distribution are associated with low back pain intensity and disability: Results from a cohort study. *Arthritis Res. Ther.* **2017**, *19*, 26. [CrossRef] [PubMed]
122. Ewald, S.C.; Hurwitz, E.L.; Kizhakkeveettil, A. The effect of obesity on treatment outcomes for low back pain. *Chiropr. Man. Ther.* **2016**, *24*, 48. [CrossRef]
123. Crespo, N.C.; Behar, A.I.; Juarez, J.; Elder, J.P. Multilevel Risk Factors Associated with Obesity in Older Latinos Ages 65 Years and Over. *Curr. Geriatr. Rep.* **2018**, *7*, 296–300. [CrossRef]
124. Yu, C.Y. Racial/Ethnic and Income Differences in Obesity Among Older Adults: The Role of Leisure-time Physical Activity and Neighborhood Social Cohesion. *J. Phys. Act Health* **2017**, *14*, 169–175. [CrossRef]
125. Flegal, K.M.; Kruszon-Moran, D.; Carroll, M.D.; Fryar, C.D.; Ogden, C.L. Trends in obesity among adults in the United States, 2005 to 2014. *JAMA* **2016**, *315*, 2284–2291. [CrossRef] [PubMed]
126. Andersen, R.E.; Crespo, C.J.; Bartlett, S.J.; Bathon, J.M.; Fontaine, K.R. Relationship between body weight gain and significant knee, hip, and back pain in older Americans. *Obes. Res.* **2003**, *11*, 1159–1162. [CrossRef] [PubMed]

International Journal of
Environmental Research and Public Health

MDPI

Article

Influence of Stabilization Techniques Used in the Treatment of Low Back Pain on the Level of Kinesiophobia

Przemysław Filipczyk [1,*], Karolina Filipczyk [2] and Edward Saulicz [3]

[1] Faculty of Health Sciences, Jan Długosz University in Czestochowa, 42-200 Czestochowa, Poland
[2] Department of Psychiatry, Faculty of Medical Sciences in Zabrze, Medical University of Silesia, 42-612 Katowice, Poland; k.filipczyk@o2.pl
[3] Departhment of Kinesiotheraphy and Special Methods in Physiotheraphy, The Physical Education Academy Jerzy Kukuczka in Katowice, 40-065 Katowice, Poland; e.saulicz@awf.katowice.pl
* Correspondence: filipmail@o2.pl; Tel.: +48-34-365-59-83

Abstract: The aim of this study was to try to compare the effectiveness of manual therapy techniques in combination with stabilization techniques: the so-called Australian method and the Neurac method in relation to pain sensations and the level of kinesiophobia. A total of 69 people were examined, divided into three groups of 23 people each. The Visual Analogue Scale was used to assess the antalgic effect, and the Kinesiophobia Causes Scale questionnaire was used to assess the level of kinesiophobia. Patients improved over four weeks, during which they were assessed three times. The evaluation of the desired parameters was also performed over a 24-week period to assess long-term performance. Stabilization techniques are an effective extension of manual therapy techniques in patients with low back pain. People in the groups additionally improved in terms of stabilization techniques, which are characterized by a lower level of kinesiophobia. Its lowest level was found in the group additionally improved with the Neurac method. In the long-term study, the level of kinesiophobia in this group was still maintained at a reduced level. The use of stabilization techniques involving patients in action may significantly affect the level of kinesiophobia, and thus have a much wider effect than just pain reduction.

Keywords: low back pain; manual therapy; kinesiophobia; fear of pain; Australian method; Neurac

Citation: Filipczyk, P.; Filipczyk, K.; Saulicz, E. Influence of Stabilization Techniques Used in the Treatment of Low Back Pain on the Level of Kinesiophobia. *IJERPH* **2021**, *18*, 6393. https://doi.org/10.3390/ijerph18126393

Academic Editors: Paul B. Tchounwou and Byoung-Hee Lee

Received: 22 April 2021
Accepted: 11 June 2021
Published: 13 June 2021

1. Introduction

The main psychological factor influencing the reappearance and persistence of pain is fear [1–3]. As early as in the 1980s, Philips [4] wrote that the fear and avoidance of pain sensations resulted in a behaviour pattern that would lead to an exaggerated perception of pain itself. In patients with acute lumbar pain, it has been found that the fear of symptoms is also associated with reduced participation in daily tasks, especially those with different lifting tasks [5]. These people also perceive their disability as a much more serious phenomenon [6,7]. The feeling of fear intensifies the more and more frequent use of sick leave, which in the long run also creates a risk of job loss [8].

In patients with acute pain, which at the beginning had an increased level of fear of pain sensations, there are more intense experiences related to the self-assessment of their disability [9]. In these people, the avoidance of activities that may intensify the pain sensation is even more pronounced. This is explained, among others, by the Fear Avoidance Model (FAM) [10]. The main assumption of this model is the fear of pain. Fear and avoidance of pain are an important feature of the development of chronic pain in many patients with musculoskeletal disorders. This model suggests that high psychological anxiety will be associated with poor clinical outcomes, potentially resulting in depressive symptoms, increased pain intensity, greater physical disturbance and the development of the further disability of the patient [11]. On the other hand, the chronic condition will have a negative impact on the quality of life [12]. More fear being related to pain may increase the susceptibility to the formation of new pain episodes. Even in currently healthy people, the risk of further ailments increases, which leads to the chronicity of the phenomenon [13]. It has been proven that people suffering from chronic lower back pain avoid risky situations, and if they decide to engage in any activity, it is performed with a very high level of caution, which directly affects the quality of a given motor task [14]. Such conduct will affect the dysfunction of the lumbar muscles, which may potentially lead to even greater limitations in physical activity and an increase in pain sensations [15].

These psychological aspects related to fear are a condition known as kinesiophobia. It is defined as an excessive, irrational and weakening fear of movement and physical activity, resulting in increased sensitivity to repeated, painful damage [16]. The patient can sometimes more or less consciously capture the phenomenon of kinesiophobia in themselves by increasing attention directed at themselves. It is worth mentioning that patients are very often accompanied by "catastrophic thinking", the greater the pain stimulus caused [17]. It is worth mentioning here that the greater the catastrophic thinking is, the greater the pain stimulus that triggered this thinking will be. Further increase in this phenomenon will also be related to the repetition of the pain stimulus, e.g., during the next painful procedure [17]. The solution to these spirals of abnormalities may be to slowly activate the patients to break the described closed circles that drive the negative phenomenon. The gradual progression of exercise programs is nothing else than the gradual exposure to a fear-inducing stimulus described by Davey [18]. According to the author, it is this approach that will be the most effective component of the treatment of people suffering from excessive anxiety and phobias. Moderate, and above all, gradual exposure to an exercise stimulus, which is also an anxiety factor, may allow the patient to be sure that the therapy is effective, so that the fear of movement is unjustified.

The direction which allows for the slow activation of the patient in the rehabilitation process can be an interesting solution, especially if it is combined with methods commonly recognized as very effective in antalgic action, such as elements of manual therapy, the Neurac method or the so-called Australian method [19,20].

The aim of the study was to evaluate the effectiveness of the antalgic effect and to assess the impact of stabilization techniques on the phenomenon of kinesiophobia.

2. Materials and Methods

2.1. Participants

The participants of the experiment (P) were patients with non-specific lower back pain. Sixty-nine people (37 women and 32 men) participated in the study. The two research groups and the control group each consisted of 23 people at the beginning of the experiment. The total number of people who completed the full cycle of the research experiment program was 46 people (24 women and 22 men), i.e., 66.66% of the respondents who started the experiment. Figure 1 shows the CONSORT Flow Diagram with the transparent reporting of trials at each stage of the experiment. People who did not complete the research process most often explained their deviation from the program by the lack of time caused by the multitude of professional and home duties. At the time of the commencement of the improvement process, the research groups were characterized by a similar level of pain intensity. The analysis of variance (Wilkas test) for the pain scale (VAS) in individual research groups showed that the mean values of the pain scales were not differentiated between the groups (p = 0.135). The participants were divided into 3 groups: manual therapy group (TM, n = 23), Neurac group (N, n = 23), and Australian group (A, n = 23).

Patients were selected for research groups on the basis of systematic random selection. This means that the group improved only with manual therapy techniques consisted of participants who were the first to apply for the study (No. 1 of the respondent), and then every third person participating in the research. Subsequently, it was the persons who reported for the tests numbered 4, 7, 10, etc. A similar random interval concerned the second group-treated with manual therapy techniques and by the Neurac method. In this case, the people who signed up for the group were assigned to group tests of numbers 2, 5,

8., etc. For the third group—treated with techniques of manual therapy and the so-called Australian method—people from numbers 3, 6, 9 were targeted., etc. When applying for the study, 4 people did not meet the criteria for inclusion in the study (3 people had comorbid mental disorders, and one person dropped out). This interval continued until we obtained 23 individuals in each personal research group, giving a total of 69 in the experimental group of patients complaining of non-specific pain ailments.

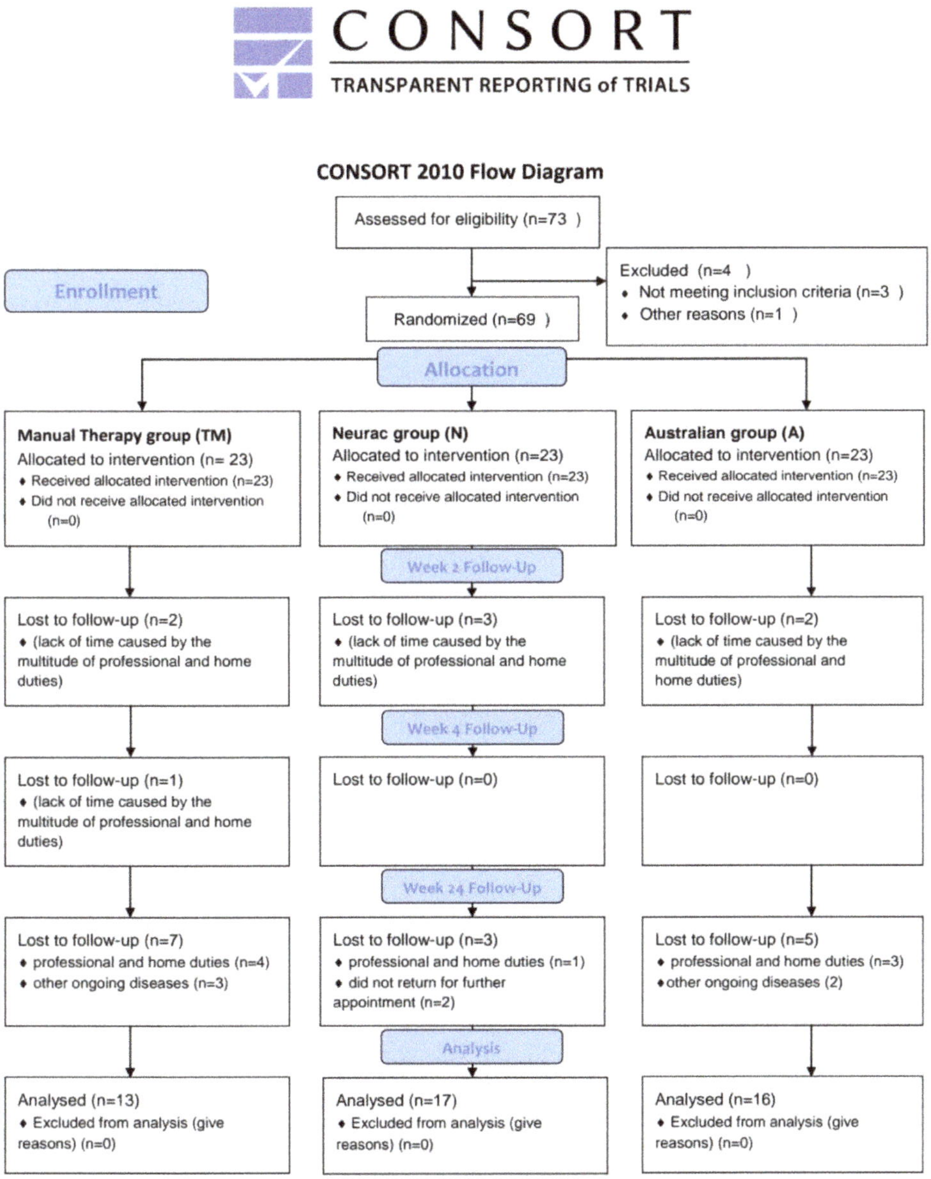

Figure 1. Transparent reporting of trials—CONSORT Flow Diagram.

2.2. Ethical Statement

All research procedures were carried out in accordance with the medical experiment design, which received a positive opinion from Resolution No. 6/2013 of 25 April 2013., by University Bioethics Committee for Scientific Research at the Academy of Physical Education "Jerzy Kukuczka" in Katowice.

2.3. Study Design

Study design was set as a randomized controlled trial.

2.4. Procedure

A diagram of the flow of progress through the study phases in all groups such as enrollment, assignment, intervention, observation, and data analysis is shown in Figure 1.

2.5. Intervention

The improvement of the experiment participants was carried out for a period of 4 weeks. During this period, 10 therapeutic meetings were held. They were held on average every 3 days. At that time, there were 4 measurement meetings: before (week 0), during (week 2) and after the rehabilitation period (week 4), and one further study after 6 months (week 24) from the end of therapy.

2.5.1. Manual Therapy Techniques (TM)

The techniques were used in all 3 groups. Therapeutic techniques were used: passive mobilization of intervertebral joints, active mobilization with the participation of the patient, high velocity low amplitude (HVLA) manipulations, traction of the entire lumbar and individual segments, and transverse massage of the paraspinal muscles. The duration of a single therapeutic visit was approximately 15 min. The work with the patient was as follows. After the examination and study of the problem, a trial therapy consisting of mobilization was performed. After the end of the first series, a test was carried out to check what effect the applied technique had on the pain symptoms. If it was possible to reduce them, this technique was used in five series: 10 mobilizations plus a break. The series were repeated five times, and then another test was carried out to see how much pain was reduced in the subject. At the end of the visit, the patient was shown basic exercises that they could safely and independently perform at home [21,22]. These were, among others, hip raises while lying on the back, hip raises combined with taking one leg off the ground, or a relieving position while lying on the back with legs placed on the ball.

In the remaining groups (Neurac method and the so-called Australian method), manual therapy was combined with stabilization training. The single visit time in these groups was increased to approximately 30 min. Manual therapy techniques took 15 min (as in the control group). The remaining time was devoted to stabilization techniques. The combination of manual therapy methods with stabilization methods took place starting with the first visit of the patient. The results regarding the General Index of Kinesiophobia and the VAS scale are presented in the Figures 2 and 3.

2.5.2. Neurac Method (N)

In the group treated by the Neurac (N) method, four key elements of this method were used, namely exercises using the patient's own weight, working in suspension; the use of vibrations on selected parts of the body; gradual increase in the load by increasing the workload; and avoiding provocation or increasing pain during exercise. At the beginning of the improvement, relieving elements were used in the form of, for example, elastic lines, which enabled the exercise to be performed. At this stage, the greatest attention was paid to learning the cooperation of various antagonistic muscle groups, shaping the sense of position and joint kinaesthesia, stimulation of neuromuscular activity and exercises in functional movement patterns. The therapy took into account isometric muscle tension, which was maintained as long as the exercise was correctly performed and the patient did not report any pain sensations. Four basic exercises for the core muscles were used, which were carried out using standard principles of exercise progression, allowing for the adjustment of difficulties and loads in each of them. These included: change in the arm length of the acting force; changes in the position of the patient's body in relation to the suspension point; changing the height of the slings or lines; changing the level of instability; performing additional movements; and the application of additional weights. While working with the patient, the following operating strategy was used: the exercises were performed from 3 to 6 repetitions in 2–4 series. The intervals between the series were approximately 30–60 s. New exercises were started at such a level of difficulty that the exercising person could correctly perform the motor task and without experiencing pain. The level of difficulty was passed to the next level when the patient was able to perform the assumed number of repetitions in the last series at a lower level, while maintaining the appropriate quality of movement [23]. The aforementioned 4 basic elements were carried out through unloading exercises performed in closed kinematic chains, taking into account the patient's body weight and unstable ground through the use of slings. In addition, in the exercises, mechanical disturbances consisting of shaking the lines and slings were used to further enhance the instability effect. The greatest attention was devoted to learning the cooperation of various antagonistic muscle groups, shaping the sense of position and joint kinaesthesia, the stimulation of neuromuscular activity and exercises in functional movement patterns [24]. The therapy took into account isometric muscle tension, which was maintained as long as the exercise was correctly performed and the patient did not report any pain sensations. Increasing the exercise time in which the subject was able to hold the position increased with the repetitions, and at the same time, the time in which the position was held increased. This progression was a way of moving to further stages of work according to the Neurac method, thus taking into account the so-called the progression ladder [25]. The results regarding the General Index of Kinesiophobia and the VAS scale are presented in the Figures 4 and 5.

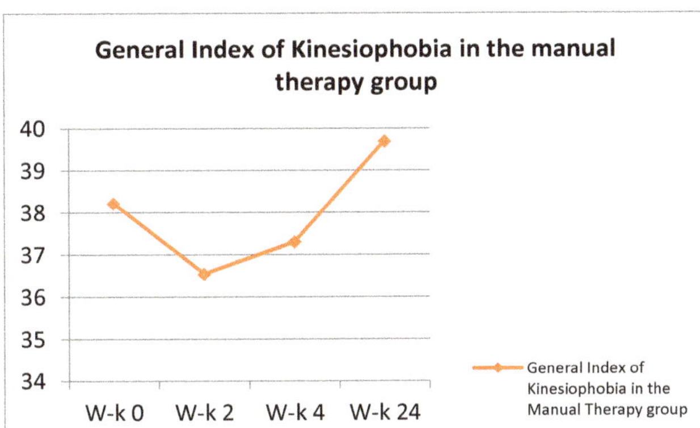

Figure 2. General Index of Kinesiophobia in the manual therapy group.

2.5.3. The So-Called Australian Method (A)

It is mainly based on the research of C. Richardson, G. Jull, P. Hodges and J. Hides conducted at the University of Queensland in Brisbane, proving that the stabilization mechanism of the lumbar spine and pelvis is disturbed in people with pain and overload in the lower spine [26]. Therapy with the use of the so-called Australian method was conducted based on the methodology developed by Peter B O'Sullivan [27]. Rehabilitation in group A began with learning to tighten the deep abdominal muscles in the supine position with bent lower limbs, in the forward lying position and in a supported kneeling position. The way to repeat the exercise and build the ability to self-check whether the tension is correct was by palpation. The patient was instructed to gently and slowly tense the abdominal walls so that their lower part was pulled in and a slight tension appeared under the fingers. If a person showed a problem with this task, additional variants of learning to activate the transverse abdominal muscle were used, e.g., sono-feedback. The described method of therapy was its first stage, lasting about 2–3 meetings, i.e., until the patients mastered the technique of the exercises. The stabilization training in group A consisted of 3 stages. In the first (described above), local segmental control was launched by activating and training the local muscle system.

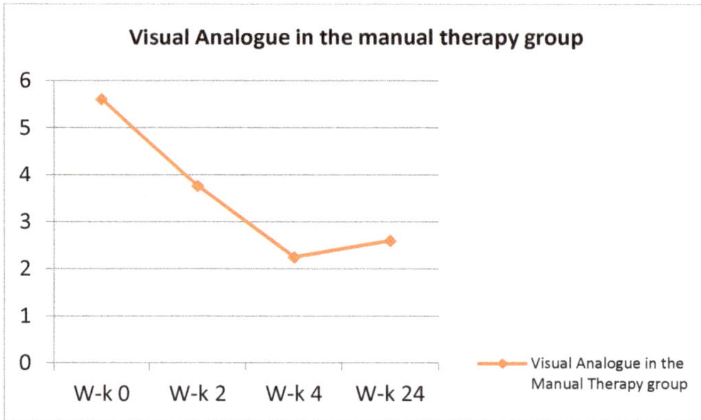

Figure 3. Visual Analogue in the manual therapy group.

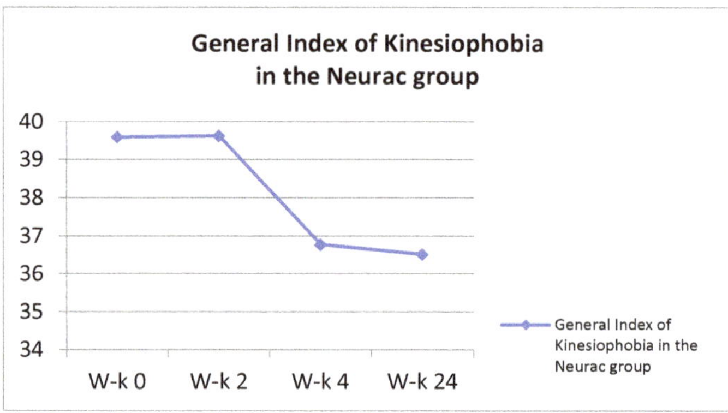

Figure 4. General Index of Kinesiophobia in the Neurac group.

In the second stage, participants learned segmental control in closed kinematic chains. The aim of this stage was to further develop the segmental control of individual joints by activating and training local muscles in conjunction with the muscles of the anti-gravity system. Biofeedback was used in all exercises thanks to the possibilities offered by a stabilizer device equipped with a blood pressure meter [28]. An exemplary exercise at this stage, performed in the supination position, was to maintain a constant pressure in the stabilizer while retracting the abdomen for approximately 10–15 s. The pressure in the inflated device was 40 mmHg. The device was placed under the lumbar spine (approximately at the height from S2 to L1 [29]. Increasing pressure means the excessive activity of the rectus abdominis and external oblique muscles; similarly, reducing the pressure in the "stabilizer" will mean a weakening of the activity of the deep muscles. There were also exercises in the pronation position. In this case, the exercise cushion was placed so that its lower line was aligned with the line formed by the upper front hip spikes. In the supine position, pulling the abdomen should reduce the pressure in the "stabilizer" (at the initial value of 70 mmHg) by 6–10 mmHg and stay at this level for about 10–15 s. At this stage, the subjects were taught to control the deep muscles while performing closed-chain movements. The suggested activity consisted, for example, in bending the knee in a supination position [30].

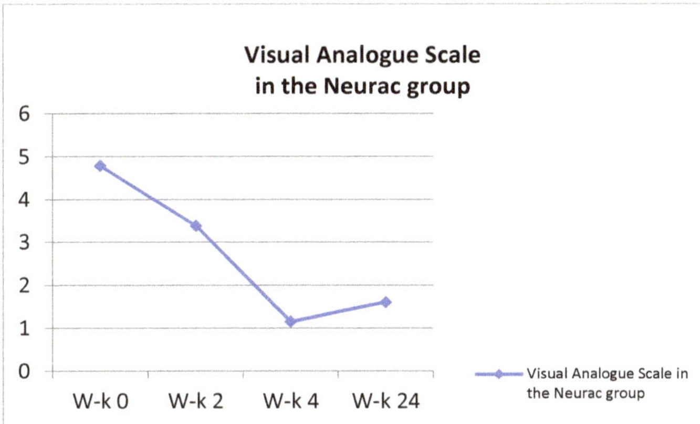

Figure 5. Visual Analogue in the Neurac group.

In the third stage of improvement, the subjects learned through exercises how to correctly segment control in open kinematic chains. At this stage, the described exercises were performed by the patients under the additional load of lifting the straightened lower limb. The re-education of the tension of the deep muscles of the lower torso in the standing and sitting position was a necessary development of stabilization exercises. At each stage of work with the patient, the therapist observed the work of the subject in terms of correctness and conveyed comments when performing the task incorrectly. During the exercises, a similar work pattern was maintained at all stages. The patient performed approximately 8–10 repetitions of each exercise. The break between them was about 10 s, and the break between other exercises was selected individually and was usually about 2 min. The results regarding the General Index of Kinesiophobia and the VAS scale are presented in the Figures 6 and 7.

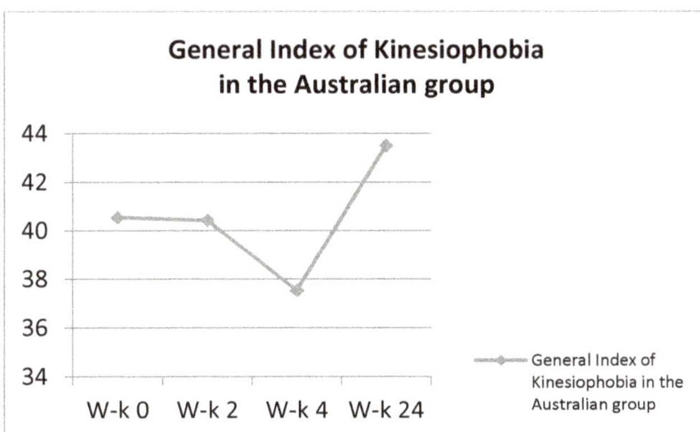

Figure 6. General Index of Kinesiophobia in the Australian group.

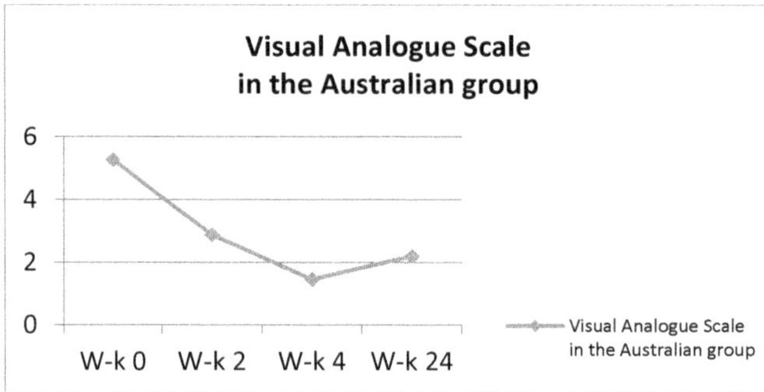

Figure 7. Visual Analogue in the Australian group.

2.6. Outcome Measures

To determine the level of kinesiophobia, we used a questionnaire that allowed to assess the degree of kinesiophobia in patients participating in the experiment: the Kinesiophobia Causes Scale (KCS) questionnaire [31]. This questionnaire is used to determine the barriers to physical activity in two domains: biological and psychological. Thanks to the answers given (on a scale of 0–100), presented as a percentage, it is possible to determine the level of the intensity of barriers to physical activity. The biological domain (DB) is the average of the factors-parameters: morphological, stimulation demand, energy resources and the strength of biological drives. The psychological domain (DP) is the average of points from factors such as: self-acceptance, motor skills, well-being and social influence. The overall kinesiophobia index (OWK) is half the sum of the biological and psychological domain points. The total score of the questionnaire ranges from 0 to 100 and can be expressed as a percentage of kinesiophobic behaviour. The higher the KSC index in the subjects, the higher the level of kinesiophobia [32].

To determine the level of pain intensity. The effectiveness of the antalgic effect was assessed using one of the most accurate pain scales—the Visual Analogue Scale (VAS). The scale is a 10 cm line. The patient marks a point on it which indicates the intensity of pain from 0—no pain at all; to 10—the strongest pain imaginable. [33].

2.7. Statistical Analysis

All the collected numerical data were subjected to statistical analysis in STATISTICA version 12.0 by StatSoft. The basic descriptive statistics procedures were performed, in which mean values (x), standard deviation (sd) as well as minimum (min) and maximum (max) values were calculated. ANOVA was used to compare the clinical status between the groups analysed on the basis of the questionnaires and the pain scale. Levene's test was also performed as a control. If it showed significant differences in the homogeneity of variance; then, the Kruskall–Wallis test was performed instead of ANOVA. In both cases, a post hoc test was performed. The Bonferroni post hoc test was also used to compare the intergroup effects, inter-study effects and interactions between groups.

3. Results

Characteristics of the respondents from particular groups in terms of basic biometric data are presented in Table 1.

The analysis of variance for repeated measurements of the VAS showed that the mean values of the obtained results were not differentiated between the groups ($p = 0.12$). The interaction effect is not statistically significant ($p = 0.78$), so the effectiveness of therapy in individual research groups was similar.

Table 1. Characteristics of the respondents from particular groups in terms of basic biometric data.

Feature	TM	A	N	p-Value
Gender	13 W (18.84%) 10 M (14.49%)	13 W (18.84%) 10 M (14.49%)	12 W (17.39%) 10 M (15.94%)	0.943 [1]
Age	50.4 ± 10.3 32–64	45.4 ± 9.11 27–64	37.2 ± 10.4 20–60	<0.001 * [2]
Body height	169.7 ± 10.0 152–187	171.4 ± 7.1 159–187	173.9 ± 9.7 154–190	0.288 [2]
Body weight	77.3 ± 14.9 50–106	73 ± 9.3 52–89	70.9 ± 13.7 50–95	0.243 [2]
BMI	26.68 ± 3.8 16.9–32.9	24.9 ± 3.2 19.3–34.8	23.3 ± 3.1 18.8–29.7	0.005 * [2]

[1]—Pearson Chi-square test; [2]—ANOVA test; TM—manual therapy group; N—Neurac group; A—Australian group; W—woman; M—Man; * statistically significant result.

The therapeutic effect of changing the pain sensation at particular stages of the study is statistically significant ($p < 0.00$). Post hoc analysis showed statistically significant differences in the baseline (week 0) and final study (week 4), and the baseline (week 0) and late study (week 24) in each study group ($p < 0.001$ for each group). In the initial and intermediate study, there were differences between the TM and A groups ($p = 0.015$). However, these were not present in the N group ($p = 0.055$). The results are presented in Figure 8.

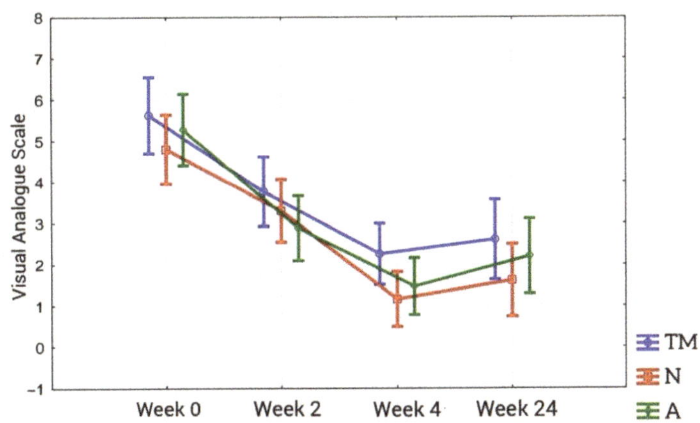

Figure 8. The course of variability of pain intensity according to the VAS scale in individual stages of therapy in relation to research groups.

There are no statistically significant differences between the final examination and the follow-up examination in the TM and A groups. However, in the N group, these differences are statistically significant.

For the Kinesiophobia Causes Scale (KCS): The analysis of variance with the ANOVA test for all components of the questionnaire showed a statistically significant difference between the individual measurements (biological dominant) and in the group-study interaction (psychological dominant). Further analysis with the Bonferroni post hoc test showed no statistically significant differences from the biological dominant and the overall kinesiophobia index for both intergroup effects, inter-study effects and group-study interaction. On the other hand, for the psychological dominant, the post hoc test revealed a statistically significant intergroup difference regarding distant measurements ($p = 0.044$) between group N and group A. The results are presented in Figure 9. The results concerning the kinesiophobia index are presented in Table 2.

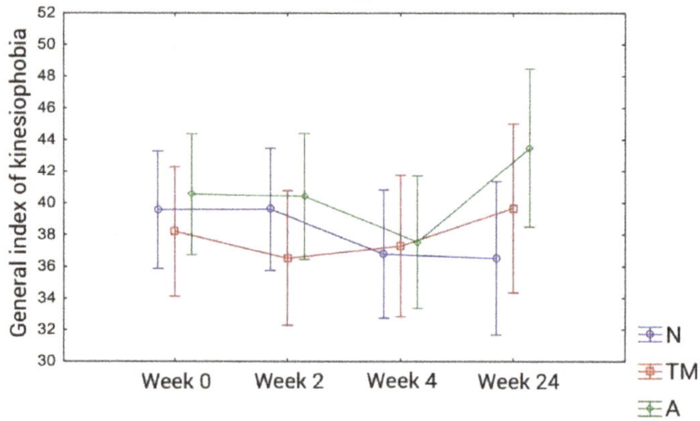

Figure 9. Mean values of the General Index of Kinesiophobia in individual groups.

Table 2. Therapeutic effect in relation to the KCS questionnaire.

Dependent Variables	*TM*				*N*				*A*				ANOVA Test *p*-Value		
	Wk 0	**Wk 2**	**Wk 4**	**Wk 24**	**Wk 0**	**Wk 2**	**Wk 4**	**Wk 24**	**Wk 0**	**Wk 2**	**Wk 4**	**Wk 24**	**Group**	**Research**	**Interaction**
DB	32.43 ± 2.22	31.36 ± 2.76	29.19 ± 2.46	32.78 ± 3.16	34.82 ± 2.02	32.83 ± 2.51	29.82 ± 2.23	31.49 ± 2.87	32.25 ± 2.08	32.40 ± 2.59	27.90 ± 2.30	32.84 ± 2.96	0.936	**0.035** *	0.926
DP	44.00 ± 2.52	41.70 ± 2.51	45.39 ± 2.64	46.59 ± 3.18	44.35 ± 2.28	46.42 ± 2.28	43.74 ± 2.40	41.55 ± 2.89	48.86 ± 2.35	48.47 ± 2.35	47.18 ± 2.47	54.13 ± 2.97	0.134	**0.366**	**0.006** *
OWK	**38.21 ± 2.03**	**36.53 ± 2.11**	**37.29 ± 2.21**	**39.68 ± 2.65**	**39.59 ± 1.84**	**39.62 ± 1.92**	**36.78 ± 2.01**	**36.52 ± 2.40**	**40.55 ± 1.90**	**40.43 ± 1.97**	**37.54 ± 2.07**	**43.49 ± 2.47**	**0.510**	**0.120**	**0.153**

DB—biological dominant; DP—psychological dominant; OWK—General Index of Kinesiophobia; Wk—Week 0–24; * statistically significant result.

4. Discussion

The results of the presented study confirm the hypothesis that the therapy that combines various stabilization techniques with manual therapy techniques is an effective form of work with the patient.

Physical activity affects our body in many ways. Many studies show that being active can alleviate symptoms of depression, reduce psychotropic medications, and provide additional therapeutic benefits beyond psychotherapy.

There are many studies showing that adapted physical activity, mainly through specific exercise, is very beneficial for the patient in a biomedical sense. It improves strength, supports the healing process, and even the fusion of bones, ligaments or muscles [34].

In the context of this work, however, exercise is primarily associated with psychological benefits, in contrast to bed rest and other forms of immobility, which are simply disastrous for structures such as discs, muscles, joints, bones and ligaments [35]. An interesting study similar to the presented study compares the therapeutic benefits of stabilization exercises in patients with various levels of anxiety measured before being active. They showed that in the case of people in which higher levels of kinesiophobia have been detected, a regular exercise program was more effective than standard medical care [36]. In the experiment, the subjects used various therapeutic methods. However, only in the N method were they led in a seemingly greater load, such as stabilization exercises in suspension. Perhaps it was this fact which contributed to the better results presented by the subjects from the N group in the distant study. The better effect in the N group may be due to the fact that in this method of rehabilitation, the patient is gradually "exposed" to greater effort, and the gradual progression of the difficulty level of the exercises helps to get rid of the fear of this activity. At the same time, this approach seems to perfectly fit into the assumptions of cognitive behavioural techniques, which use situations that are "dangerous" or "threatening" in the opinion of the patient, while explaining what exactly the fear avoidance model is. Then, adapted exercise tasks are introduced based on the hierarchy of fear-inducing situations [37]. In the Neurac method, however, we refer to this hierarchy as the "progression ladder".

Using the patient's individual symptoms, beliefs and methods of avoidance, it is possible to clearly explain the operation of the vicious circle that maintains the problem of pain for a long time. In the other methods used in the experiment, the subjects were not so exposed to effort. They did not risk confrontation with a greater load, which is that of performing stabilization exercises while suspended, but performed tasks in stable positions, such as lying or sitting. One could consider whether the so-called Australian method supplemented with manual therapy techniques or the lack of any additional method and using only manual therapy measures would yield similar results. Considering only the fear of being active, this may be the case.

Nevertheless, it should be remembered that working with a patient suffering from LBP is a multi-factor task in which pain plays the leading role. Pain, as the most pronounced sensation, is often the only one that forces the patient to seek help. Additionally, for the pain factor itself, each of the methods used in the experiment works effectively, reducing it to a significant extent. Moreover, in each of the research groups, the effectiveness of the therapy in terms of pain relief was similar, even in a long-term study. Kinesiophobia can also develop the so-called "disuse syndrome". In other words, the fear of pain will be responsible for the deterioration of physical condition and the formation of abnormal movement patterns, often seen in patients with chronic pain. Of course, this condition will not only be influenced by the fear of being active, but also by avoiding activity, and thus weakening physical strength [38].

The fear of being active further increases the patient's excessive vigilance, the vigilance focused on avoiding the source of the threat, but also the sources that may cause (according to the patient) a similar effect. Excessive vigilance occurs when a person is constantly engaged in searching for potential sources of danger in the surrounding environment. Of course, this condition will be associated with a decrease in the willingness to expose oneself during any activity. Additionally, the person will only show selective participation in activities which, according to them, may be associated with threat. Both the mere avoidance of participation in certain activities and the described excessive vigilance reduce anxiety, but only in the short term. In the long run, these may be counterproductive [39].

Determining the level of kinesiophobia in patients before starting the rehabilitation process may be of fundamental importance for a multi-disciplinary view of the problem of improving non-specific lower back pain. People with higher levels of kinesiophobia respond well to coping strategies for everyday situations. A simple and clear commentary on the way of performing certain activities, but also explaining the phenomenon of kinesiophobia itself, discussing the strategy of not avoiding activity, may turn out to be a milestone in improving LBP patients.

Limitations

The topic related to the choice of the type of combination therapy, taking into account the goal of not only the reduction in pain, but also the reduction in kinesiophobia in patients with LBP, is still on the verge. Additionally, the issue itself requires a lot of analysis. It should also be remembered that the group of people who have completed the experiment process is relatively small, which does not enable us to confidently draw far-reaching conclusions about kinesiophobia. Nevertheless, this seems to be an interesting topic and worth developing in the context of reducing kinesiophobia in LBP patients.

The baseline characteristics showed significant differences in age and BMI in group N. Although the statistical analysis did not show significant differences in the baseline studies between the groups in the context of the assessment of the VAS scale and the KCS questionnaire, it should be taken into account when developing these studies. The lower BMI index and the age of the respondents could have contributed to the better acceptance of the load during exercise by this group of respondents. Therefore, increasing the number of subjects, and reducing the number of subjects, could prove to be valuable.

5. Conclusions

1. The therapy that combines various stabilization techniques with manual therapy techniques is an effective form of work with the patient. This method of therapy gives a long-lasting analgesic effect.

2. The effect of reducing the level of kinesiophobia is visible in each of the research groups (N and A), as well as in the control group (TM). However, it was only in the N group, where the patients were challenged by more demanding exercises compared to other research groups, that the effect lasted the longest and was visible in the distant study.

Author Contributions: Conceptualization, P.F. and E.S.; methodology, P.F.; software, P.F. and E.S.; validation, P.F., K.F. and E.S.; formal analysis, P.F.; investigation, P.F.; resources, K.F.; data curation, P.F.; writing—original draft preparation, P.F.; writing—review and editing, P.F. and E.S.; visualization, P.F.; supervision, P.F.; project administration, P.F.; funding acquisition, P.F., E.S. All authors have read and agreed to the published version of the manuscript.

Funding: This research received no external funding.

Institutional Review Board Statement: Not applicable.

Informed Consent Statement: Not applicable.

Data Availability Statement: The data that support the findings of this study are available from the corresponding author, upon reasonable request.

Conflicts of Interest: The authors declare no conflict of interest.

References

1. Ledoux, E.; Dubois, J.D.; Descarreaux, M. Physical and psychosocial predictors of functional trunk capacity in older adults with and without low back pain. *J. Manip. Physiol. Ther.* **2012**, *35*, 338–345. [CrossRef]
2. Rottermund, J.; Knapik, A.; Saulicz, E.; Myśliwiec, A.; Saulicz, M.; Rygiel, K.A.; Linek, P. Back and neck pain among school teachers in Poland and its correlations with physical activity. *Occup. Med.* **2015**, *66*, 771–778. [CrossRef]
3. Leeuw, M.; Goossens, M.E.; Linton, S.J.; Crombez, G.; Boersma, K.; Vlaeyen, J.W. The fear-avoidance model of musculoskeletal pain: Current state of scientific evidence. *J. Behav. Med.* **2007**, *30*, 77–94. [CrossRef] [PubMed]
4. Philips, H.C. Avoidance behaviour and its role in sustaining chronic pain. *Behav. Res. Ther.* **1987**, *25*, 273–279. [CrossRef]
5. Kemani, M.; Hägg, O.; Jakobsson, M.; Lundberg, M. Fear of Movement Is Related to Low Back Disability During a Two-Year Period in Patients Who Have Undergone Elective Lumbar Spine. *Surg. World Neurosurg.* **2020**, *137*, 416–424. [CrossRef]
6. Markfelder, T.; Pauli, P. Fear of pain and pain intensity: Meta-analysis and systematic review. *Psychol. Bull.* **2020**, *146*, 411–450. [CrossRef] [PubMed]
7. Swinkels-Meewisse, I.E.; Roelofs, J.; Oostendorp, R.A.; Verbeek, A.L.; Vlaeyen, J.W.S. Acute low back pain: Pain-related fear and pain catastrophizing influence physical performance and perceived disability. *Pain* **2006**, *120*, 36–43. [CrossRef] [PubMed]
8. Steenstra, I.; Munhall, C.; Irvin, E.; Oranye, N.; Passmore, S.; Van Eerd, D.; Mahood, Q.; Hogg-Johnson, S. Systematic Review of Prognostic Factors for Return to Work in Workers with Sub Acute and Chronic Low Back Pain. *J. Occup. Rehabil.* **2017**, *27*, 369–381. [CrossRef] [PubMed]
9. Larsson, C.; Hansson, E.; Sundquist, K.; Jakobsson, U. Kinesiophobia and its relation to pain characteristics and cognitive affective variables in older adults with chronic pain. *BMC Geriatr.* **2016**, *16*, 1–128. [CrossRef]
10. Seekatz, B.; Meng, K.; Faller, H. Depressivity as mediator in the fear-avoidance model: A path analysis investigation of patients with chronic back pain. *Schmerz* **2013**, *27*, 612–618. [CrossRef]
11. Crombez, G.; Eccleston, C.; Van Damme, S.; Vlaeyen, J.; Karoly, P. Fear-Avoidance Model of Chronic Pain The Next Generation. *Clin. J. Pain* **2012**, *28*, 475–483. [CrossRef] [PubMed]
12. Sawatzky, R.; Liu-Ambrose, T.; Miller, W.; Marra, C. Physical activity as a mediator of the impact of chronic conditions on quality of life in older adults. *Health Qual. Life Outcomes* **2007**, *5*, 1–11. [CrossRef]

13. Calatayud, J.; Guzmán-González, B.L.; Andersen, L.; Cruz-Montecinos, C.; Morell, M.T.; Roldán, R.; Ezzatvar, Y.; Casaña, J. Effectiveness of a Group-Based Progressive Strength Training in Primary Care to Improve the Recurrence of Low Back Pain Exacerbations and Function: A Randomised Trial. *Int. J. Environ. Res. Public Health* **2020**, *17*, 8326. [CrossRef] [PubMed]
14. Esteve, R.; Marcos, E.; Reyes-Pérez, A.; López-Martínez, A.E.; Ramírez-Maestre, C. Pain Acceptance Creates an Emotional Context That Protects against the Misuse of Prescription Opioids: A Study in a Sample of Patients with Chronic Noncancer. *Int. J. Environ. Res. Public Health* **2021**, *18*, 3054. [CrossRef]
15. Suri, P.; Fry, A.L.; Gellhorn, A.C. Do muscle characteristics on lumbar spine magnetic resonance imaging or computed tomography predict future low back pain, physical function, or performance? A systematic review. *Am. Acad. Phys. Med. Rehabil.* **2015**, *12*, 1269–1281. [CrossRef] [PubMed]
16. Xu, Y.; Song, Y.; Sun, D.; Fekete, G.; Gu, Y. Effect of Multi-Modal Therapies for Kinesiophobia Caused by Musculoskeletal Disorders: A Systematic Review and Meta-Analysis. *Int. J. Environ. Res. Public Health* **2020**, *17*, 9439. [CrossRef]
17. Sullivan, M.J.; Thibault, P.; Andrikonyte, J.; Butler, H.; Catchlove, R.; Lariviere, C. Psychological influences on repetition- induced summation of activity-related pain in patients with chronic low back pain. *Pain* **2009**, *141*, 70–78. [CrossRef]
18. Davey, G.C.L. *Phobias. A Handbook of Theory, Research and Treatment*; Wiley: Chichester, UK, 1997.
19. Buchmann, J.; Baumann, N.; Meng, K.; Semrau, J.; Kuhl, J.; Pfeifer, K.; Kazén, M.; Vogel, H.; Faller, H. Endurance and avoidance response patterns in pain patients: Application of action control theory in pain research. *PLoS ONE* **2021**, *25*, 16.
20. Vibe Fersum, P.; O'Sullivan, P.; Skouen, J.S. Efficacy of classification-based cognitive functional therapy in patients with non-specific chronic low back pain: A randomized controlled trial. *Eur. J. Pain* **2013**, *17*, 916–928. [CrossRef]
21. Hidalgo, B.; Hall, T.; Bossert, J.; Dugeny, A.; Cagnie, B.; Pitance, L. The efficacy of manual therapy and exercise for treating non-specific neck pain: A systematic review. *J. Back Musculoskelet Rehabil.* **2017**, *30*, 1149–1169. [CrossRef]
22. Hidalgo, B.; Detrembleur, C.; Hall, T.; Mahaudens, P.; Nielens, H. The efficacy of manual therapy and exercise for different stages of non-specific low back pain: An update of systematic reviews. *J. Man. Manip. Ther.* **2014**, *22*, 59–74. [CrossRef] [PubMed]
23. Unsgaard-Tøndel, M.; Fladmark, A.; Salvesen, Ø. Motor Control Exercises, Sling Exercises, and General Exercises for Patients With Chronic Low Back Pain: A Randomized Controlled Trial with 1-Year Follow-up. *Phys. Ther.* **2010**, *90*, 1426–1440. [CrossRef]
24. Mueller, J.; Niederer, D. Dose-response-relationship of stabilisation exercises in patients with chronic non-specific low back pain: A systematic review with meta-regression. *Sci. Rep.* **2020**, *10*, 16921. [CrossRef] [PubMed]
25. Kirkesola, G. Neurac—A new treatment method for long-term musculoskeletal pain. *J. Fysioter.* **2009**, *76*, 16–25.
26. Richardson, C.A.; Jull, G.A. Muscle control—Pain control. What exercises would you prescribe? *Man. Ther.* **1995**, *1*, 2–104. [CrossRef] [PubMed]
27. O'Sullivan, P.B. Lumbar segmental "instability": Clinical presentation and specific stabilizing exercise management. *Man. Ther.* **2000**, *5*, 2. [CrossRef]
28. Glasoe, W.M.; Marquez, R.R.; Miller, W.J. Comparison of Standard Blood Pressure Cuff and a Commercially Available Pressure Biofeedback Unit While Performing Prone Lumbar Stabilization Exercise. *Orthop. Pract.* **2009**, *17*, 26–28.
29. Dafkou, K.; Kellis, E.; Ellinoudis, A.; Sahinis, C. The Effect of Additional External Resistance on Inter-Set Changes in Abdominal Muscle Thickness during Bridging Exercise. *J. Sports Sci. Med.* **2020**, *19*, 102–111.
30. Lima, P.O.; de Oliveira, R.R.; de Moura Filho, A.G.; Raposo, M.C.; Costa, L.O.; Laurentino, G.E.; Bodyw, J. Reproducibility of the pressure biofeedback unit in measuring transversus abdominis muscle activity in patients with chronic nonspecific low back pain. *Mov. Ther.* **2012**, *16*, 251–257. [CrossRef] [PubMed]
31. Knapik, A.; Saulicz, E.; Gnat, R. Kinesiophobia—Introducing a New Diagnostic Tool. *J. Hum. Kinet.* **2011**, *28*, 25–31. [CrossRef]
32. Knapik, A.; Saulicz, E.; Kuszewski, M. Płeć a poziom kinezjofobii w populacji osób dorosłych z południowej Polski. *Przegląd Med. Uniw. Rzesz. Nar. Inst. Leków Warszawie* **2012**, *3*, 277–287.
33. Chiarotto, A.; Maxwell, L.J.; Ostelo, R.W.; Boers, M.; Tugwell, P.; Terwee, C.B. Measurement Properties of Visual Analogue Scale, Numeric Rating Scale, and Pain Severity Subscale of the Brief Pain Inventory in Patients with Low Back Pain: A Systematic Review. *J. Pain* **2019**, *20*, 245–263. [CrossRef] [PubMed]
34. Kazeminia, M.; Salari, N.; Vaisi-Raygani, A.; Jalali, R.; Abdi, A.; Mohammadi, M.; Daneshkhah, A.; Hosseinian-Far, M.; Shohaimi, S. The effect of exercise on anxiety in the elderly worldwide: A systematic review and meta-analysis. *Health Qual. Life Outcomes* **2020**, *11*, 18–363. [CrossRef]
35. Comachio, J.; Magalhães, M.O.; Campos Carvalho, E.; Silva, A.P.M.; Marques, A.P. A cross-sectional study of associations between kinesiophobia, pain, disability, and quality of life in patients with chronic low back pain. *Adv. Rheumatol.* **2018**, 58. [CrossRef]
36. Klaber Moffett, J.A.; Carr, J.; Howarth, E.H. High fear-avoiders of physical activity benefit from an exercise program for patients with back pain. *Spine* **2004**, *29*, 1167–1179. [CrossRef]
37. Hasenbring, M.I.; Verbunt, J.A. Fear-avoidance and endurance-related responses to pain: New models of behavior and their consequences for clinical practice. *Clin. J. Pain* **2010**, *26*, 747–753. [CrossRef] [PubMed]
38. Schega, L.; Kaps, B.; Broscheid, K.C.; Bielitzki, R.; Behrens, M.; Meiler, K.; Drange, S.; Franke, J. Effects of a multimodal exercise intervention on physical and cognitive functions in patients with chronic low back pain (MultiMove): Study protocol for a randomized controlled trial. *BMC Geriatr.* **2021**, *21*, 151. [CrossRef] [PubMed]
39. Pinel, L.; Perez-Nieto, M.A.; Redondo, M.; Rodríguez-Rodríguez, L.; León, L. Anxiety, Reinforcement Sensitivity and Social Context in Accepting the Experience of Pain Among Rheumatoid Arthritis Patients. *Front. Psychiatry* **2020**, *11*. [CrossRef]

International Journal of
Environmental Research and Public Health

MDPI

Article

Seasonality of Back Pain in Italy: An Infodemiology Study

Jacopo Ciaffi [1,*], Riccardo Meliconi [1,2], Maria Paola Landini [3], Luana Mancarella [1], Veronica Brusi [1], Cesare Faldini [2,4] and Francesco Ursini [1,2]

[1] Medicine and Rheumatology Unit, IRCCS Istituto Ortopedico Rizzoli (IOR), 40136 Bologna, Italy; riccardo.meliconi@ior.it (R.M.); luana.mancarella@ior.it (L.M.); veronica.brusi@ior.it (V.B.); francesco.ursini2@unibo.it (F.U.)

[2] Department of Biomedical and Neuromotor Science (DIBINEM), University of Bologna, 40125 Bologna, Italy; cesare.faldini@ior.it

[3] Scientific Direction, IRCCS Istituto Ortopedico Rizzoli (IOR), 40136 Bologna, Italy; mariapaola.landini@ior.it

[4] 1st Orthopaedic and Traumatologic Clinic, IRCCS Istituto Ortopedico Rizzoli (IOR), 40136 Bologna, Italy

* Correspondence: jacopo.ciaffi@ior.it

Abstract: Background: E-health tools have been used to assess the temporal variations of different health problems. The aim of our infodemiology study was to investigate the seasonal pattern of search volumes for back pain in Italy. Methods: In Italian, back pain is indicated by the medical word "lombalgia". Using Google Trends, we selected the three search terms related to "lombalgia" with higher relative search volumes (RSV), (namely, "mal di schiena", "dolore alla schiena" and "dolore lombare"), representing the semantic preferences of users when performing web queries for back pain in Italy. Wikipedia page view statistics were used to identify the number of visits to the page "lombalgia". Strength and direction of secular trends were assessed using the Mann–Kendall test. Cosinor analysis was used to evaluate the potential seasonality of back pain-related RSV. Results: We found a significant upward secular trend from 2005 to 2020 for search terms "mal di schiena" ($\tau = 0.734$, $p < 0.0001$), "dolore alla schiena" ($\tau = 0.713$, $p < 0.0001$) and "dolore lombare" ($\tau = 0.628$, $p < 0.0001$). Cosinor analysis on Google Trends RSV showed a significant seasonality for the terms "mal di schiena" ($p_{cos} < 0.001$), "dolore alla schiena" ($p_{cos} < 0.0001$), "dolore lombare" ($p_{cos} < 0.0001$) and "lombalgia" ($p_{cos} = 0.017$). Cosinor analysis performed on views for the page "lombalgia" in Wikipedia confirmed a significant seasonality ($p_{cos} < 0.0001$). Both analyses demonstrated a peak of interest in winter months and decrease in spring/summer. Conclusions: Our infodemiology approach revealed significant seasonal fluctuations in search queries for back pain in Italy, with peaking volumes during the coldest months of the year.

Keywords: back pain; Google Trends; infodemiology; seasonality; Wikipedia

Citation: Ciaffi, J.; Meliconi, R.; Landini, M.P.; Mancarella, L.; Brusi, V.; Faldini, C.; Ursini, F. Seasonality of Back Pain in Italy: An Infodemiology Study. *IJERPH* **2021**, *18*, 1325. https://doi.org/10.3390/ijerph18031325

Academic Editors: Vincenzo Denaro, Sergio Iavicoli, Fabrizio Russo and Gianluca Vadalà

Received: 9 December 2020
Accepted: 28 January 2021
Published: 1 February 2021

Publisher's Note: MDPI stays neutral with regard to jurisdictional claims in published maps and institutional affiliations.

1. Introduction

In the dynamic environment of the web, the term *e-health* has been broadly used since late 1990s to encompass all aspects of the Internet and medicine [1]. Over the years, information and communication technology (ICT) for health-related matters has become an important source of implementation for traditional health resources [2]. E-health can facilitate access to services, reduce healthcare costs and improve the quality of patient care [3,4]. Technological advancements offered to policy makers, healthcare professionals and researchers represent an opportunity to stimulate positive health behaviours, promote prevention and simplify retrieval of information about diseases or treatments [5–7]. As of July 2020, 59% of the global population, almost 4.57 billion people, were actively using the Internet [8]. As shown in surveys conducted in the United States and Europe, searching the Internet for health information is becoming even easier with the rapidly spreading use of health apps on mobile devices [9]. At the same time, the huge amount of data generated can be collected, stored and communicated over the web to assist decision makers in tailoring precision medicine [10]. The concept of analysing Internet data to predict patterns of disease distribution is known as infodemiology, a term coined in the early 2000s, when the pioneering work of Professor Gunther Eysenbach showed the correlation between Google search volumes for key terms related to influenza and the number of influenza cases occurring one week later [11,12]. This was among the earliest demonstrations of how patterns of search queries and their changes over time reflected public interest towards a specific health-related topic. In the following years, data from Google searches obtained through the Google Trends interface have been used to investigate seasonal variations of different health problems such as restless leg syndrome [13,14], multiple sclerosis [15], depression [16] and bruxism [17]. More specifically, in the field of musculoskeletal diseases, Google Trends has been used to evaluate global interest and seasonality of searches for rheumatoid arthritis [18,19], systemic lupus erythematosus [20,21], ankylosing spondylitis [22], osteoarthritis [23], gout [24] and fibromyalgia [25]. Moreover, seasonality of common joint symptoms such as foot and ankle pain [26] or knee pain [27] has been assessed. Interestingly, recent studies outlined how, amongst searches for painful conditions, back pain was one of the top queries, raising the hypothesis of seasonal variations in search trends [28,29]. Back pain is indeed an extremely common complaint [30], to the point that in 2015 the estimated prevalence of activity-limiting low back pain was 7.3% globally, corresponding to 540 million people suffering from the condition [31]. These numbers imply that back pain, and in particular low back pain, now represents the first cause of disability in the world [31]. Back pain is experienced by individuals of all ages and has a considerable impact on global public health [32]. Therefore, as expected, the volume of web searches for back pain have increased worldwide in the last years [28] and, since technological advancements have allowed people to seek on the Internet healthcare information and advice about medicine and treatments, using Internet data to understand the interest in back pain could be of relevance for healthcare professionals. As of October 2020, Google led the search engine market in Italy with 96.3% of share [33]. Therefore, the analysis of search volumes performed with Google has a solid basis supporting its reliability. However, besides Google Trends, data also accumulated during Internet search activities on Wikipedia pages have been used for research purposes and to build epidemiological models [34,35]. An increasing number of investigations have used Google Trends and Wikipedia page view data to assess public interest in different health conditions and their seasonal variations, providing novel and complementary methods to integrate information from traditional sources [36–39]. Therefore, we decided to conduct an ecological infodemiology study using Internet data to track search patterns and identify variability in public interest for back pain in Italy, specifically aiming at determining periods with highest and lowest search popularity. Potential clinical implications of our study are primarily related to the application of retrieved data to guide preventative interventions for low back pain and to develop specific health programmes to effectively manage and reduce the burden of the condition.

2. Materials and Methods

2.1. Google Trends Data Availability

Google Trends (Google Inc., Menlo Park, CA, USA) [40] is a free instrument which allows researchers to explore data on the frequency of query strings utilized by users when performing a search in Google. Data are available from January 2004. To improve comparability between search terms, Google Trends data are adjusted to the overall volume of searches for a given time frame and geographical location and provided as a relative search volume (RSV) ranging from 0 to 100 [41]. To reduce selection bias, Google Trends automatically excludes duplicate queries performed by the same user (i.e., identical IP address) within a short period of time [41]. In Google Trends, words or phrases can be searched either as search terms (ST, search queries matching for all terms, similar to the application of the "AND" Boolean operator) or "topics" (TP), a group of related terms that share the same concept in any language [41]. Furthermore, each query in Google Trends generates a table of related searches, ranked for either percentage increase in RSV or relative frequency (0–100) [41]. The data obtained can be exported as comma-separated values (.CSV) files for further manipulation and analysis.

2.2. Wikipedia Page View Data Availability

Wikipedia page view statistics [42] is a free instrument available for all Wikipedia pages, which allows researchers to ascertain how many people have visited an article during a given time frame [43]. Data on visit counts are available from July 2015 and can be generated on a daily or monthly basis and according to language-specific Wikipedia projects or user platforms (desktop, mobile application, mobile web) [43]. Different from Google Trends, each time a user visits or reloads a Wikipedia page, a new count is generated [43]. The data retrieved can be exported as comma-separated values (.CSV) files for further manipulation and analysis.

2.3. Search Process and Data Retrieval

On 26 April 2020, we performed a Google Trends search to identify queries in the "health" category related to back pain in Italy on a month-by-month basis, from January 2004 to April 2020. To select the search terms or topic that best fitted the user preference in performing back pain-related searches, we followed a systematic approach. First of all, we searched the ST "lombalgia" (medical term used in Italy to describe low back pain) and "dolore" (corresponding to generic "pain" in the Italian language) and explored related searches after ranking for their RSV. Subsequently, we selected the three search terms that resulted in the higher RSVs ("mal di schiena", "dolore alla schiena", "dolore lombare") while at the same time reflecting a precise intention of the user to retrieve information on back pain. Furthermore, we performed a Wikipedia page view search to retrieve the number of visits to the page "Lombalgia" [44] from 1 July 2015 to 30 April 2020. Data were retrieved for searches from all platforms on the https://it.wikipedia.org/wiki/Pagina_principale.

2.4. Statistical Analysis

Reporting of results was compliant with a previously published Checklist for the Documentation of Google Trends Use [38]. To graphically evaluate the temporal trend for individual search terms, we plotted time series data using the Locally Weighted Scatter Plot Smoothing (LOESS) function [45]. LOESS is a nonparametric method where least squares regression is performed in local neighbourhood subsets of data, smoothing numerical vectors. LOESS curves are mainly used to reveal trends and cycles in data that might be difficult to model with a parametric curve. To assess the significance of temporal trends in back pain-related Google searches, the Mann–Kendall test [46,47] was used. The Mann–Kendall is a nonparametric test to statistically assess if there is a monotonic upward or downward trend of the variable of interest over time, and it is based on the difference in signs between earlier and later data points of a time series. The strength and direction of the trend is represented by the τ (tau) value (positive = upward, negative = downward). Cosinor analysis was used to evaluate the potential seasonality of back pain-related RSV. Cosinor-based techniques have been extensively adopted to analyse time series in chronobiology [48]. In short, cosinor analysis fits a sinusoid to the experimental data according to the following equation:

$$Y(t) = M + A\cos(2\pi t/\tau + \phi) + e(t) \quad (1)$$

where M is the MESOR (Midline Statistic of Rhythm, a rhythm-adjusted mean), A is the amplitude (a measure of the amplitude of the variation of the peak from the baseline), ϕ is the acrophase (a measure of the time of the peak), τ is the period (duration of one cycle) and e(t) is the error term. Since the included variables were continuous, a Poisson model was fitted. The cosine model consists of sine and cosine terms that together identify the sinusoid. Therefore, it contains two p-values, one of which was shown as recommended. To control the false discovery rate caused by multiple testing, the criterion of significance was set at $p < 0.025$. Data processing and analyses were conducted using the

R software, V.3.4.4 (R Core Team, Auckland, New Zealand) using the "season", "kendall" and "ggplot2" packages.

2.5. Ethical Considerations

Ethical approval was deemed unnecessary because data were provided in aggregate form with no person-identifiable information; further explanations on personal data protection are available on the Google privacy policy website [49].

3. Results

3.1. Secular Trend of Google Searches for Back Pain

Visual inspection of the time series plot with LOESS smoothing suggested an increasing secular trend of RSV for the search terms "mal di schiena" (Figure 1A), "dolore alla schiena" (Figure 1B) and "dolore lombare" (Figure 1C). No clearly identifiable trend was evident for the search term "lombalgia" (Figure 1D). Mann–Kendall analysis confirmed a significant upward secular trend for the search terms "mal di schiena" ($\tau = 0.734$, $p < 0.0001$), "dolore alla schiena" ($\tau = 0.713$, $p < 0.0001$) and "dolore lombare" ($\tau = 0.628$, $p < 0.0001$) and a significant, although less pronounced downward trend for the search term "lombalgia" ($\tau = -0.128$, $p = 0.009$).

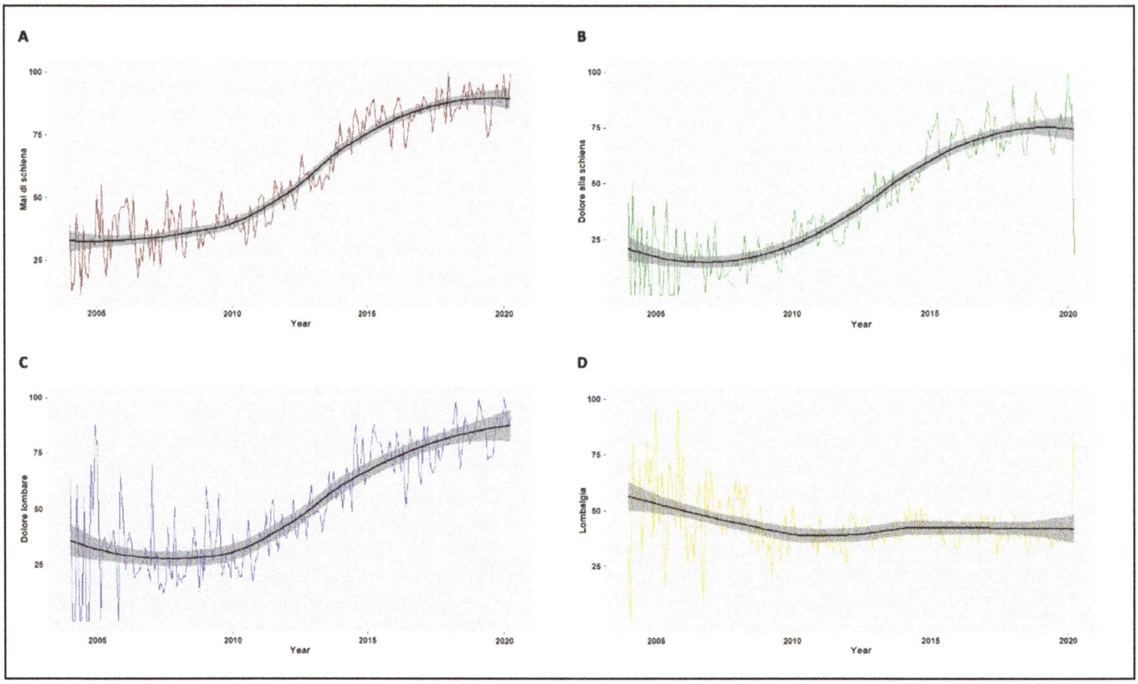

Figure 1. Time series plot Locally Weighted Scatter Plot Smoothing (LOESS) of relative search volumes (RSV) for the search terms "mal di schiena" (**A**), "dolore alla schiena" (**B**), "dolore lombare" (**C**), "lombalgia" (**D**) from January 2004 to April 2020.

3.2. Analysis of Seasonality

The cosinor analysis on RSV obtained from Google Trends searches based on 196 monthly observations showed a significant seasonality for the search terms "mal di schiena" ($A = 3.63$, $\phi = 11.8$, low point month = 5.8, $p_{cos} < 0.001$) (Figure 2A), "dolore alla schiena" ($A = 5.12$, $\phi = 1.3$, low point = 7.3, $p_{cos} < 0.0001$) (Figure 2B), "dolore lombare" ($A = 3.73$, $\phi = 12.3$, low point = 6.3, $p_{cos} < 0.0001$) (Figure 2C) and "lombalgia" (A = 1.69, ϕ = 12.4, low point = 6.4, $p_{cos} = 0.017$) (Figure 2D), demonstrating a peak in RSV for back pain-related search terms in the central winter months and a decrease at the beginning of summer.

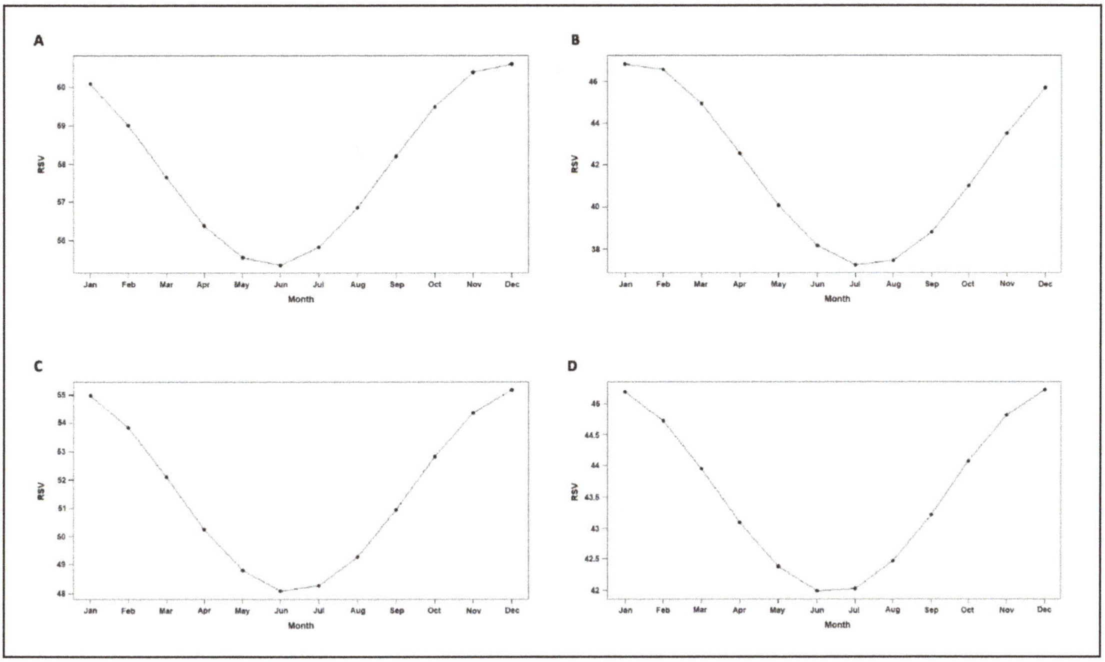

Figure 2. Seasonal fluctuations for the terms "mal di schiena" (**A**), "dolore alla schiena" (**B**), "dolore lombare" (**C**) and "lombalgia" (**D**) based on Cosinor analysis on Google Trends relative search volumes (RSV) from 196 monthly observations from January 2004 to April 2020.

Furthermore, we analysed the monthly views for the page "lombalgia" in Wikipedia. Cosinor analysis, performed on the available time frame (July 2015–April 2020, 58 observations) confirmed a significant seasonality (A = 117.33, ϕ = 10.2, low point = 4.2, $p_{cos} < 0.0001$) with a peak in early winter and a decrease in spring.

4. Discussion

E-health tools are increasingly being used in the healthcare sector by both patients and providers, with broad potential applications [50]. Low back pain is a multifactorial condition caused by the interaction of genetic, physical, psychological and occupational factors, with a complex pathophysiology [51]. The aim of our study was to investigate the seasonal variation in back pain-related searches in Italy through an infodemiology approach. Results obtained from the analysis of Google Trends search volumes and Wikipedia page view statistics demonstrated a significantly peaking interest towards back pain during the winter months.

Several mechanisms may account for the observed seasonality of back pain. First of all, it is already known and commonly accepted that meteorological variables such as temperature and humidity can modulate chronic pain [52–57]; in particular, in individuals with chronic low back pain, specific weather conditions (e.g., ambient temperature and vapor pressure) are assumed to affect pain perception [58]. Although these effects are small in magnitude, considering meteorological parameters as potential modulators of pain perception entails possible implications for clinical management of patients with chronic low back pain. To this purpose, environmental variables can be manipulated in the home or work setting in order to achieve optimal conditions for minimizing the impact on musculoskeletal pain.

On the other side, literature findings regarding the association between weather and back pain are still inconsistent, with studies showing no effects of meteorological parameters on the onset of back pain or on the intensity of symptoms during an exacerbation [59,60]. At polar latitudes, less chronic musculoskeletal pain has been described in winter compared to summer [61], but the variation, although statistically significant, was minimal, with questionable clinical relevancy. Moreover, hospital admissions for back pain exacerbations were found to be unrelated to average temperature or rainfall [62].

Physical activity is a well-recognized barrier to the development of musculoskeletal pain. A recent meta-analysis of cross-sectional studies demonstrated that people engaged in medium level physical activity have a 10% lower risk of low back pain when compared to those engaged in low level physical activity [63]. A clear seasonality has been demonstrated in average levels of physical activity, with leisure-time energy expenditure approximately 15–20% higher during spring and summer [64]. On this basis, another potential explanation of the observed seasonal variation in back pain may be attributable to a more sedentary lifestyle during the winter months. Complementarily, cold season is associated with a change in eating patterns (resulting in a higher caloric intake), contributing to an increase in body weight peaking in winter [65]. Obesity, in turn, is a well-recognized risk factor for the development of back pain [66].

Another possible explanation for our findings is the correlation between back pain and vitamin D status. It is well established that vitamin D plasma levels are influenced by seasonal variations, with highest values reported in summer [67]. A relationship between vitamin D status and low back pain has been extensively demonstrated in literature, with stronger associations observed in younger women and those with markedly reduced levels of vitamin D [68]. Severity of vitamin D deficiency has also been postulated to be causally associated with chronic low back pain, with increasing intensity of symptoms in patients with lower vitamin D levels [69,70].

Furthermore, it is possible to speculate that seasonal fluctuations observed in the RSV for back pain in Italy may be influenced also by other factors, for instance the differences in footwear used by the population between winter and summer, with potential implications on the development of postural back pain symptoms [71,72].

However, it could be argued that the retrieved seasonal pattern might be explained by the highest use of the web during winter months when people spend more time at home. Anyway, results obtained through the Google Trends platform represent relative volumes normalized on overall search engine traffic and therefore less likely to be influenced by reduced web usage; furthermore, it should be pointed out that several earlier studies focusing on different health topics showed seasonal variations with Internet searches peaking in summer [13,24,26,73,74].

Additionally, a limitation of our research, intrinsically related to the nature of infodemiology studies, is that the analysis of Google Trends is only able to capture the search behaviour in a given period of time and geographic area, representing the interest people have in a topic, which in the case of back pain does not necessarily reflect the actual number of individuals experiencing the symptom. As shown by previous evidence, web queries may be influenced by other variables not included in our study. News regarding celebrities may have a significant impact on the public opinion, with a consequent increase in web search volumes [19,75,76]. Moreover, global initiatives to raise awareness on medical conditions may influence Internet search query activity [77] and, in the specific case of low back pain, we can't exclude that World Spine Day taking place every year on 16 October might have contributed to the increase in interest observed in early winter. The major strengths of our study are the large amount of data, the long observation period and the analysis of two complementary sources, namely, Google Trends and Wikipedia. Nevertheless, some limitations need to be acknowledged. We analysed the seasonality of searches for back pain only in Italy; therefore, our results cannot be generalized to other countries. However, the value of our methodology consisted in the use of common language "everyday words", requiring a deep knowledge of local semantics for back pain symptoms in different geographical and linguistic areas. Moreover, web data do not allow researchers to gather demographic information about the users, thus precluding the opportunity to stratify our findings in specific subpopulations and analyse differences between men and women or in different age groups. Hence, our results can only be interpreted at a general Italian population level.

5. Conclusions

In conclusion, our infodemiology approach revealed significant seasonal fluctuations in search queries for back pain in Italy, with peaking volumes during the coldest months of the year. According to previous literature, weather conditions, physical activity or vitamin D status might explain, at least in part, the observed seasonal pattern. Despite introducing an intriguing hypothesis, our findings should be considered preliminary and require clinical validation; furthermore, possible mechanisms behind the seasonality of back pain need to be further elucidated. If confirmed in future studies, the winter-peaking seasonality of back pain could be exploited for implementing prophylactic strategies, e.g., encouraging seasonally tailored physical activity programs and calorie restriction or promoting recognition and management of vitamin D deficiency when appropriate.

Author Contributions: Conceptualization, J.C. and F.U.; methodology, R.M. and F.U.; validation, L.M. and M.P.L.; formal analysis, F.U. and C.F.; data curation, L.M.; writing—original draft preparation, J.C. and V.B.; writing—review and editing, R.M., C.F. and M.P.L.; supervision, R.M. and C.F. All authors have read and agreed to the published version of the manuscript.

Funding: This research received no external funding.

Institutional Review Board Statement: Not applicable.

Informed Consent Statement: Not applicable.

Data Availability Statement: No new data were created in this study. Data sharing is not applicable to this article because the data presented are openly available on the web.

Conflicts of Interest: The authors declare no conflict of interest.

References

1. Eysenbach, G. What is e-health? *J. Med. Internet Res.* **2001**, *3*, E20. [CrossRef] [PubMed]
2. Pagliari, C.; Sloan, D.; Gregor, P.; Sullivan, F.; Detmer, D.; Kahan, J.P.; Oortwijn, W.; MacGillivray, S. What is eHealth (4): A scoping exercise to map the field. *J. Med. Internet Res.* **2005**, *7*, e9. [CrossRef] [PubMed]
3. Jung, M.L.; Loria, K. Acceptance of Swedish e-health services. *J. Multidiscip. Healthc.* **2010**, *3*, 55–63. [CrossRef] [PubMed]
4. González, M.E.; Quesada, G.; Urrutia, I.; Gavidia, J.V. Conceptual design of an e-health strategy for the Spanish health care system. *Int. J. Health Care Qual. Assur.* **2006**, *19*, 146–157. [CrossRef]
5. Srivastava, S.; Pant, M.; Abraham, A.; Agrawal, N. The Technological Growth in eHealth Services. *Comput. Math. Methods Med.* **2015**, *2015*, 894171. [CrossRef] [PubMed]
6. Kampmeijer, R.; Pavlova, M.; Tambor, M.; Golinowska, S.; Groot, W. The use of e-health and m-health tools in health promotion and primary prevention among older adults: A systematic literature review. *BMC Health Serv. Res.* **2016**, *16* (Suppl. S5), 290. [CrossRef]
7. Strzelecki, A. Google Medical Update: Why Is the Search Engine Decreasing Visibility of Health and Medical Information Websites? *Int. J. Environ. Res. Public Health* **2020**, *17*, 1160. [CrossRef]
8. Global Digital Population as of October 2020. Available online: https://www.statista.com/statistics/617136/digital-population-worldwide/ (accessed on 11 November 2020).
9. Bach, R.L.; Wenz, A. Studying health-related internet and mobile device use using web logs and smartphone records. *PLoS ONE* **2020**, *15*, e0234663. [CrossRef]
10. Pastorino, R.; De Vito, C.; Migliara, G.; Glocker, K.; Binenbaum, I.; Ricciardi, W.; Boccia, S. Benefits and challenges of Big Data in healthcare: An overview of the European initiatives. *Eur. J. Public Health* **2019**, *29*, 23–27. [CrossRef]
11. Eysenbach, G. Infodemiology: Tracking flu-related searches on the web for syndromic surveillance. *AMIA Annu. Symp. Proc.* **2006**, *2006*, 244–248.
12. Eysenbach, G. Infodemiology and infoveillance: Framework for an emerging set of public health informatics methods to analyze search, communication and publication behavior on the Internet. *J. Med. Internet Res.* **2009**, *11*, e11. [CrossRef] [PubMed]

13. O'Keeffe, S.T. Summertime blues? A re-examination of the seasonality of web searches for restless legs and leg cramps. *Sleep Med.* **2017**, *37*, 119–123. [CrossRef] [PubMed]
14. Ingram, D.G.; Plante, D.T. Seasonal trends in restless legs symptomatology: Evidence from Internet search query data. *Sleep Med.* **2013**, *14*, 1364–1368. [CrossRef] [PubMed]
15. Moccia, M.; Palladino, R.; Falco, A.; Saccà, F.; Lanzillo, R.; Brescia Morra, V. Google Trends: New evidence for seasonality of multiple sclerosis. *J. Neurol. Neurosurg Psychiatry* **2016**, *87*, 1028–1029. [CrossRef]
16. Yang, A.C.; Huang, N.E.; Peng, C.K.; Tsai, S.J. Do seasons have an influence on the incidence of depression? The use of an internet search engine query data as a proxy of human affect. *PLoS ONE* **2010**, *5*, e13728. [CrossRef]
17. Kardeş, S.; Kardeş, E. Seasonality of bruxism: Evidence from Google Trends. *Sleep Breath* **2019**, *23*, 695–701. [CrossRef]
18. Wu, G.C.; Tao, S.S.; Zhao, C.N.; Mao, Y.M.; Wu, Q.; Dan, Y.L.; Pan, H.F. Leveraging Google Trends to investigate the global public interest in rheumatoid arthritis. *Rheumatol. Int.* **2019**, *39*, 1439–1444. [CrossRef]
19. Mahroum, N.; Bragazzi, N.L.; Sharif, K.; Gianfredi, V.; Nucci, D.; Rosselli, R.; Brigo, F.; Adawi, M.; Amital, H.; Watad, A. Leveraging Google Trends, Twitter, and Wikipedia to Investigate the Impact of a Celebrity's Death From Rheumatoid Arthritis. *J. Clin. Rheumatol.* **2018**, *24*, 188–192. [CrossRef]
20. Wu, G.C.; Cao, F.; Shen, H.H.; Hu, L.Q.; Hu, Y.; Sam, N.B. Global public interest in systemic lupus erythematosus: An investigation based on internet search data. *Lupus* **2019**, *28*, 1435–1440. [CrossRef]
21. Pan, H.F.; Wang, P.; Wu, G.C.; Zou, Y.F.; Xu, Z.; Ye, D.Q.; Hu, W. Seasonal variation in systemic lupus erythematosus and rheumatoid arthritis: An ecological study based on internet searches. *Autoimmun. Rev.* **2019**, *18*, 825–827. [CrossRef]
22. Mei, Y.J.; Mao, Y.M.; Cao, F.; Wang, T.; Li, Z.J. Using internet search data to explore the global public concerns in ankylosing spondylitis. *Postgrad. Med. J.* **2020**. [CrossRef] [PubMed]
23. Jellison, S.S.; Bibens, M.; Checketts, J.; Vassar, M. Using Google Trends to assess global public interest in osteoarthritis. *Rheumatol. Int.* **2018**, *38*, 2133–2136. [CrossRef] [PubMed]
24. Kardeş, S. Seasonal variation in the internet searches for gout: An ecological study. *Clin. Rheumatol.* **2019**, *38*, 769–775. [CrossRef] [PubMed]
25. Bragazzi, N.L.; Amital, H.; Adawi, M.; Brigo, F.; Watad, S.; Aljadeff, G.; Amital, D.; Watad, A. What do people search online concerning the "elusive" fibromyalgia? Insights from a qualitative and quantitative analysis of Google Trends. *Clin. Rheumatol.* **2017**, *36*, 1873–1878. [CrossRef]
26. Telfer, S.; Woodburn, J. Let me Google that for you: A time series analysis of seasonality in internet search trends for terms related to foot and ankle pain. *J. Foot Ankle Res.* **2015**, *8*, 27. [CrossRef]
27. Dewan, V.; Sur, H. Using google trends to assess for seasonal variation in knee injuries. *J. Arthrosc. Jt. Surg.* **2018**, *5*, 175–178. [CrossRef]
28. Kamiński, M.; Łoniewski, I.; Marlicz, W. "Dr. Google, I am in Pain"-Global Internet Searches Associated with Pain: A Retrospective Analysis of Google Trends Data. *Int. J. Environ. Res. Public Health* **2020**, *17*, 954. [CrossRef]
29. Yamaguchi, Y.; Lee, D.; Nagai, T.; Funamoto, T.; Tajima, T.; Chosa, E. Googling Musculoskeletal-Related Pain and Ranking of Medical Associations' Patient Information Pages: Google Ads Keyword Planner Analysis. *J. Med. Internet Res.* **2020**, *22*, e18684. [CrossRef]
30. Hoy, D.; Bain, C.; Williams, G.; March, L.; Brooks, P.; Blyth, F.; Woolf, A.; Vos, T.; Buchbinder, R. A systematic review of the global prevalence of low back pain. *Arthritis Rheum* **2012**, *64*, 2028–2037. [CrossRef]
31. Global, regional, and national incidence, prevalence, and years lived with disability for 310 diseases and injuries, 1990-2015: A systematic analysis for the Global Burden of Disease Study 2015. *Lancet* **2016**, *388*, 1545–1602. [CrossRef]
32. Hartvigsen, J.; Hancock, M.J.; Kongsted, A.; Louw, Q.; Ferreira, M.L.; Genevay, S.; Hoy, D.; Karppinen, J.; Pransky, G.; Sieper, J.; et al. What low back pain is and why we need to pay attention. *Lancet* **2018**, *391*, 2356–2367. [CrossRef]
33. Market Share Held by the Leading Search Engines in Italy between September 2019 and September 2020. Available online: https://www.statista.com/statistics/623043/search-engines-ranked-by-market-share-in-italy/ (accessed on 11 November 2020).
34. Hickmann, K.S.; Fairchild, G.; Priedhorsky, R.; Generous, N.; Hyman, J.M.; Deshpande, A.; Del Valle, S.Y. Forecasting the 2013–2014 influenza season using Wikipedia. *PLoS Comput. Biol.* **2015**, *11*, e1004239. [CrossRef] [PubMed]
35. Sciascia, S.; Radin, M. What can Google and Wikipedia can tell us about a disease? Big Data trends analysis in Systemic Lupus Erythematosus. *Int. J. Med. Inform.* **2017**, *107*, 65–69. [CrossRef] [PubMed]
36. Cervellin, G.; Comelli, I.; Lippi, G. Is Google Trends a reliable tool for digital epidemiology? Insights from different clinical settings. *J. Epidemiol. Glob. Health* **2017**, *7*, 185–189. [CrossRef]
37. Brownstein, J.S.; Freifeld, C.C.; Madoff, L.C. Digital disease detection—Harnessing the Web for public health surveillance. *N. Engl. J. Med.* **2009**, *360*, 2153–2155, 2157. [CrossRef] [PubMed]
38. Nuti, S.V.; Wayda, B.; Ranasinghe, I.; Wang, S.; Dreyer, R.P.; Chen, S.I.; Murugiah, K. The use of google trends in health care research: A systematic review. *PLoS ONE* **2014**, *9*, e109583. [CrossRef]
39. Mavragani, A.; Ochoa, G.; Tsagarakis, K.P. Assessing the Methods, Tools, and Statistical Approaches in Google Trends Research: Systematic Review. *J. Med. Internet Res.* **2018**, *20*, e270. [CrossRef]
40. Google Trends. Available online: https://trends.google.com/ (accessed on 26 April 2020).
41. Google Trends Help Center. Available online: https://support.google.com/trends/ (accessed on 26 April 2020).
42. Wikipedia: Page View Statistics. Available online: https://tools.wmflabs.org/pageviews/ (accessed on 26 April 2020).

43. Pageviews Analysis—Documentation. Available online: https://meta.wikimedia.org/wiki/Pageviews_Analysis (accessed on 26 April 2020).
44. Wikipedia: Lombalgia. Available online: https://it.wikipedia.org/wiki/Lombalgia (accessed on 26 April 2020).
45. Cleveland, W.S.; Devlin, S.J. Locally Weighted Regression: An Approach to Regression Analysis by Local Fitting. *J. Am. Stat. Assoc.* **1988**, *83*, 15. [CrossRef]
46. Mann, H.B. Nonparametric tests against trend. *Econometrica* **1945**, *13*, 15. [CrossRef]
47. Kendall, M.G. *Rank Correlation Methods*, 4th ed.; Griffin: London, UK, 1975.
48. Cornelissen, G. Cosinor-based rhythmometry. *Theor. Biol. Med Model.* **2014**, *11*, 16. [CrossRef]
49. Google Privacy & Terms. Available online: https://policies.google.com/ (accessed on 26 April 2020).
50. Penedo, F.J.; Oswald, L.B.; Kronenfeld, J.P.; Garcia, S.F.; Cella, D.; Yanez, B. The increasing value of eHealth in the delivery of patient-centred cancer care. *Lancet Oncol.* **2020**, *21*, e240–e251. [CrossRef]
51. Vlaeyen, J.W.S.; Maher, C.G.; Wiech, K.; Van Zundert, J.; Meloto, C.B.; Diatchenko, L.; Battié, M.C.; Goossens, M.; Koes, B.; Linton, S.J. Low back pain. *Nat. Rev. Dis. Primers* **2018**, *4*, 52. [CrossRef] [PubMed]
52. Shutty, M.S., Jr.; Cundiff, G.; DeGood, D.E. Pain complaint and the weather: Weather sensitivity and symptom complaints in chronic pain patients. *Pain* **1992**, *49*, 199–204. [CrossRef]
53. Jamison, R.N.; Anderson, K.O.; Slater, M.A. Weather changes and pain: Perceived influence of local climate on pain complaint in chronic pain patients. *Pain* **1995**, *61*, 309–315. [CrossRef]
54. Roth-Isigkeit, A.; Thyen, U.; Stöven, H.; Schwarzenberger, J.; Schmucker, P. Pain among children and adolescents: Restrictions in daily living and triggering factors. *Pediatrics* **2005**, *115*, e152–e162. [CrossRef]
55. Timmermans, E.J.; van der Pas, S.; Schaap, L.A.; Sánchez-Martínez, M.; Zambon, S.; Peter, R.; Pedersen, N.L.; Dennison, E.M.; Denkinger, M.; Castell, M.V.; et al. Self-perceived weather sensitivity and joint pain in older people with osteoarthritis in six European countries: Results from the European Project on OSteoArthritis (EPOSA). *BMC Musculoskelet. Disord.* **2014**, *15*, 66. [CrossRef]
56. Moldofsky, H. Chronobiological influences on fibromyalgia syndrome: Theoretical and therapeutic implications. *Baillieres Clin. Rheumatol.* **1994**, *8*, 801–810. [CrossRef]
57. Hawley, D.J.; Wolfe, F.; Lue, F.A.; Moldofsky, H. Seasonal symptom severity in patients with rheumatic diseases: A study of 1,424 patients. *J. Rheumatol.* **2001**, *28*, 1900–1909.
58. McGorry, R.W.; Hsiang, S.M.; Snook, S.H.; Clancy, E.A.; Young, S.L. Meteorological conditions and self-report of low back pain. *Spine* **1998**, *23*, 2096–2102. [CrossRef]
59. Steffens, D.; Maher, C.G.; Li, Q.; Ferreira, M.L.; Pereira, L.S.; Koes, B.W.; Latimer, J. Effect of weather on back pain: Results from a case-crossover study. *Arthritis Care Res.* **2014**, *66*, 1867–1872. [CrossRef]
60. Beilken, K.; Hancock, M.J.; Maher, C.G.; Li, Q.; Steffens, D. Acute Low Back Pain? Do Not Blame the Weather—A Case-Crossover Study. *Pain Med.* **2017**, *18*, 1139–1144. [CrossRef] [PubMed]
61. Abeler, K.; Sand, T.; Friborg, O.; Bergvik, S. Seasonality in pain, sleep and mental distress in patients with chronic musculoskeletal pain at latitude 69° N. *Chronobiol. Int.* **2020**, 1–12. [CrossRef]
62. Okwerekwu, G.; Brooks, F.; Spolton-Dean, C.; Khurana, A.; Manoj-Thomas, A.; Cordell-Smith, J. Is there a seasonal variation of acute admissions for back pain. *Spine J.* **2015**, *15*, S76. [CrossRef]
63. Alzahrani, H.; Mackey, M.; Stamatakis, E.; Zadro, J.R.; Shirley, D. The association between physical activity and low back pain: A systematic review and meta-analysis of observational studies. *Sci. Rep.* **2019**, *9*, 8244. [CrossRef]
64. Pivarnik, J.M.; Reeves, M.J.; Rafferty, A.P. Seasonal variation in adult leisure-time physical activity. *Med. Sci. Sports Exerc.* **2003**, *35*, 1004–1008. [CrossRef]
65. Ma, Y.; Olendzki, B.C.; Li, W.; Hafner, A.R.; Chiriboga, D.; Hebert, J.R.; Campbell, M.; Sarnie, M.; Ockene, I.S. Seasonal variation in food intake, physical activity, and body weight in a predominantly overweight population. *Eur. J. Clin. Nutr.* **2006**, *60*, 519–528. [CrossRef]
66. Zhang, T.T.; Liu, Z.; Liu, Y.L.; Zhao, J.J.; Liu, D.W.; Tian, Q.B. Obesity as a Risk Factor for Low Back Pain: A Meta-Analysis. *Clin. Spine Surg.* **2018**, *31*, 22–27. [CrossRef]
67. Holick, M.F. Environmental factors that influence the cutaneous production of vitamin D. *Am. J. Clin. Nutr.* **1995**, *61*, 638s–645s. [CrossRef]
68. Zadro, J.; Shirley, D.; Ferreira, M.; Carvalho-Silva, A.P.; Lamb, S.E.; Cooper, C.; Ferreira, P.H. Mapping the Association between Vitamin D and Low Back Pain: A Systematic Review and Meta-Analysis of Observational Studies. *Pain Physician* **2017**, *20*, 611–640.
69. Gokcek, E.; Kaydu, A. Assessment of Relationship between Vitamin D Deficiency and Pain Severity in Patients with Low Back Pain: A Retrospective, Observational Study. *Anesth Essays Res.* **2018**, *12*, 680–684. [CrossRef]
70. Panwar, A.; Valupadas, C.; Veeramalla, M.; Vishwas, H.N. Prevalence of vitamin D deficiency in chronic and subacute low back pain patients in India: A triple-arm controlled study. *Clin. Rheumatol.* **2018**, *37*, 1367–1374. [CrossRef] [PubMed]
71. Vieira, E.R.; Brunt, D. Does wearing unstable shoes reduce low back pain and disability in nurses? A randomized controlled pilot study. *Clin. Rehabil.* **2016**, *30*, 167–173. [CrossRef] [PubMed]
72. Bai, D.Y.; Yuan, Z.G.; Shao, J.J.; Zhu, T.; Zhang, H.J. Unstable shoes for the treatment of lower back pain: A meta-analysis of randomized controlled trials. *Clin. Rehabil.* **2019**, *33*, 1713–1721. [CrossRef] [PubMed]

73. Liu, F.; Allan, G.M.; Korownyk, C.; Kolber, M.; Flook, N.; Sternberg, H.; Garrison, S. Seasonality of Ankle Swelling: Population Symptom Reporting Using Google Trends. *Ann. Fam. Med.* **2016**, *14*, 356–358. [CrossRef]
74. Zhang, X.; Dang, S.; Ji, F.; Shi, J.; Li, Y.; Li, M.; Jia, X.; Wan, Y.; Bao, X.; Wang, W. Seasonality of cellulitis: Evidence from Google Trends. *Infect. Drug Resist.* **2018**, *11*, 689–693. [CrossRef]
75. Tijerina, J.D.; Morrison, S.D.; Nolan, I.T.; Parham, M.J.; Richardson, M.T.; Nazerali, R. Celebrity Influence Affecting Public Interest in Plastic Surgery Procedures: Google Trends Analysis. *Aesthetic Plast. Surg.* **2019**, *43*, 1669–1680. [CrossRef]
76. Kaleem, T.; Malouff, T.D.; Stross, W.C.; Waddle, M.R.; Miller, D.H.; Seymour, A.L.; Zaorsky, N.G.; Miller, R.C.; Trifiletti, D.M.; Vallow, L. Google Search Trends in Oncology and the Impact of Celebrity Cancer Awareness. *Cureus* **2019**, *11*, e5360. [CrossRef]
77. Havelka, E.M.; Mallen, C.D.; Shepherd, T.A. Using Google Trends to assess the impact of global public health days on online health information seeking behaviour in Central and South America. *J. Glob. Health* **2020**, *10*, 010403. [CrossRef]

International Journal of
Environmental Research and Public Health

MDPI

Article

What Psychosocial and Physical Characteristics Differentiate Office Workers Who Develop Standing-Induced Low Back Pain? A Cross-Sectional Study

Beatriz Rodríguez-Romero [1,*], Michelle D Smith [2], Alejandro Quintela-del-Rio [3] and Venerina Johnston [2,4]

1 Psychosocial Intervention and Functional Rehabilitation Research Group, Department of Physiotherapy, Medicine and Biomedical Sciences, Faculty of Physiotherapy, Universidade da Coruña, Campus de A Coruña, 15071 A Coruna, Spain
2 School of Health and Rehabilitation Sciences, The University of Queensland, Brisbane 4072, Australia; m.smith5@uq.edu.au (M.D.S.); v.johnston@uq.edu.au (V.J.)
3 Psychosocial Intervention and Functional Rehabilitation Research Group, Department of Mathematics, Faculty of Physiotherapy, Universidade de A Coruna, 15071 A Coruna, Spain; alejandro.quintela@udc.es
4 Recover Injury Research Centre, The University of Queensland, Brisbane 4006, Australia
* Correspondence: beatriz.romero@udc.es; Tel.: +34-981-167000 (ext. 5841)

Received: 30 July 2020; Accepted: 25 September 2020; Published: 28 September 2020

Abstract: This study examines demographic, physical and psychosocial factors associated with an increase in low back pain (LBP) during a one-hour standing task. A cross-sectional survey with 40 office workers was conducted. The primary outcome was pain severity during a one-hour standing task recorded every 15 min using a 100 mm Visual Analogue Scale (VAS). Participants were defined as pain developers (PD), if they reported a change in pain of ≥10 mm from baseline, or non-pain developers (NPD). Physical outcomes included participant-rated and examiner-rated trunk and hip motor control and endurance. Self-report history of LBP, physical activity, psychosocial job characteristics, general health and pain catastrophising were collected. Fourteen participants were PD. Hip abduction, abdominal and spinal muscle endurance was lower for PD ($p \leq 0.05$). PD had greater self-reported difficulty performing active hip abduction and active straight leg raise tests ($p \leq 0.04$). Those reporting a lifetime, 12 month or 7-day history of LBP ($p < 0.05$) and lower self-reported physical function ($p = 0.01$) were more likely to develop LBP during the standing task. In conclusion, a history of LBP, reduced trunk and hip muscle endurance and deficits in lumbopelvic/hip motor control may be important to consider in office workers experiencing standing-induced LBP.

Keywords: low back pain (LBP); standing position; musculoskeletal pain; sedentary behaviour

1. Introduction

Low back pain (LBP) is recognised as a serious health concern and is the leading cause of activity limitation and work absenteeism internationally [1]. It is a multi-factorial and heterogeneous condition including complex interactions between physical, psychological, social and comorbid health factors that is best explained by a biopsychosocial framework [2].

Regarding workplace factors and the occurrence of LBP, prolonged sitting is commonly cited in the literature, but the available evidence has not confirmed a consistent association between occupational sitting and LBP [3,4]. Nevertheless, office workers spend over two-thirds of their workday seated and this prolonged and uninterrupted sedentary behaviour [5] is a recognised public health concern [6]. The office workplace has been identified as a key setting for targeting a reduction of this ubiquitous behaviour [5]. Explicit recommendations have been provided, such as breaking up of seated-based work with standing-based work, the use of sit-stand desks or taking of short active standing breaks, to promote the avoidance of prolonged sedentary periods [7].

Prolonged standing has been associated with the development of musculoskeletal disorders including LBP for some adults [8,9]. However, causality between occupational standing and LBP could not be shown and not all people who are exposed to prolonged standing will develop LBP [10].

Previous studies [11] have used an induced pain standing task [12] to understand the characteristics that predispose a person to developing LBP during prolonged standing, but these factors are not well understood. Researchers have suggested an association between (i) reduced hip abductor muscle endurance [13,14]; (ii) reduced endurance or strength of the trunk muscles [13,15–18]; and (iii) movement control dysfunction [15,17,19] and the development of LBP during standing. However, the evidence is inconsistent and studies have used university students and community volunteers, rather than office workers [13,15,19–22].

The importance of psychosocial factors in the management of LBP is well documented in the general population [23] and in office workers [24]. However, to our knowledge, psychosocial factors have not been considered in relation to the development of standing-induced LBP in office workers. This may be because most studies investigating a prolonged standing task are conducted in young volunteers with mean age 24 ± 3 years (rather than office workers) without a history of musculoskeletal problems [13,15–18]. These inclusion criteria omit a large part of the population who have a history of LBP and do not target the population of office workers. Thus, the population of people who are often given and/or may benefit from standing workstations have been neglected from much of the previous research in this area.

The aim of this study was to identify individual, physical and psychosocial factors associated with the development (or increased severity) of LBP during a one-hour standing task. Given the findings of previous studies, we hypothesised that specific individual (history of LBP), physical (deficits in motor control and muscle endurance) and psychosocial factors (e.g., pain catastrophising and high job demands) would be associated with the development (or increased severity) of LBP during a one-hour standing task. Outcomes from this study will provide clinicians with useful information about the relationship between biopsychosocial variables and the potential development of LBP in office workers using height-adjustable workstations.

2. Materials and Methods

2.1. Participants

A convenience sample of forty office workers was recruited from the local university and surrounding community via flyers, electronic noticeboards and social media to participate in this cross-sectional laboratory study. Inclusion criteria were office-based workers who worked ≥ 30 h/week mostly sitting at a computer and were aged ≥ 18 years. Participants were excluded if they were pregnant or less than six months postpartum, had any major trauma or surgery to the spine or lower limb in the last 12 months or had a diagnosis of neurological or systemic pathology. An online survey was used to assess eligibility. The study was carried out in Brisbane (Australia), at the University of Queensland. The sample size was determined based on a power analysis, setting a probability of type I error of 0.05 (alpha), a power of 0.8 (1—probability of type II error), in order to detect a moderate effect size (Cohen's d = 0.5) with the G*Power (v.3.1.) software [25], and the minimum sample size is $n = 34$. Based on previous work [15,21], a total of 40 participants was considered sufficient to detect differences of medium effect size, considering a significance level of 0.05 and a statistical power of 0.8. The study was conducted in accordance with the Declaration of Helsinki. The University of Queensland Human Research Ethics Committee B approved this study. This project complies with the provisions contained in the National Statement on Ethical Conduct in Human Research and complies with the regulations governing experimentation on humans (Registration: EC00457; Approval Number: #2017000666). This study was registered in the Protocol Registration and Results System (PRS) (NCT03678623).

2.2. Study Procedure

Participants completed a series of self-report questionnaires and attended laboratory testing conducted by a trained physiotherapist. Testing was undertaken according to a standardised protocol in the following order: (i) height and weight (ii) tests of movement control, (iii) trunk and hip muscle endurance tests, and (iv) one-hour standing task.

2.3. Physical Outcome Measures

One-hour standing task: the standing task consisted of participants standing for one-hour while performing their usual computer-based work [21,26]. Participants stood within a rectangular floor space (122 cm × 61 cm) with their body fist-width away from a height-adjustable workstation. The workstation was standardised to each participant so that the desk height was 5–6 cm below the lateral epicondyle, the computer monitor was at arm's length from the body, and the top of the computer monitor was at eye level. Participants were allowed to shift their weight as often as desired but were asked to keep both feet on the ground the majority of the time. The participant was not allowed to lean on the workstation with their arms, legs or trunk. During the standing task, participants recorded the location of any pain they experienced on a body map and rated pain intensity on a 100 mm Visual Analogue Scale (VAS) at baseline, at 15 min intervals and at the end of the test. The 100 mm was anchored by "no pain" (0 mm) and "worst pain imaginable" (100 mm) [27]. The investigator verbally asked the participants to rate their pain, without access to their previous scores. The test was terminated after one hour or if the participant requested the test to be stopped or experienced pain ≥ 70 mm on the VAS.

Tests of motor control (Figure S1): the active hip abduction (AHAbd) [28], active straight leg raise (ASLR) [29] and prone knee flexion (PKF) [30] tests, which have previously established reliability, were used to assess control of movement. Each test was repeated three times on each side in random order.

In the AHAbd test, participants adopted a side-lying position and were asked to raise their top leg towards the ceiling as far as possible, keeping the knee extended, the lower limb aligned with the trunk and the pelvis aligned in the frontal plane. The difficulty in performing the test was self-rated by the participants on a 0–5 scale (ranging from 0 "no difficulty at all" to 5 "unable to lift the limb") [28]. Quality of movement and ability to maintain the pelvis in the frontal plane were rated by the examiner on a 4-point scale (ranging from 0 "able to maintain pelvis position" to 3 "severe loss of frontal plane pelvis alignment") [28]. The average participant-rated and examiner-rated score for each leg was used for analysis.

To perform the ASLR test, participants were positioned supine and asked to lift their leg 20 cm above the table with the knee straight and hold for 10 s. Participants rated the difficulty of performing the ASLR on a 0–5 scale (ranging from 0 "no difficulty at all" to 5 "unable to lift the limb") [29]. The examiner scored the ASLR quality of movement by awarding one point for each of the following deviations: pelvic rotation toward the raised leg, tremor of the raised leg, slow speed of performance and verbal or nonverbal expression of difficulty by the subject. Possible scores ranged from 0 (no deviations present) to 5 (all deviations present or unable to raise the leg off the table). Scores for the left and right sides were combined to form a total score out of 10 for participant-rated and examiner-rated outcomes [29].

For the PKF, participants were positioned prone and asked to bend their knee as far as possible without moving their back [30]. The test was rated by the examiner as "correct" (a score of 0) when the knee was flexed to at least 90° without movement of the low back and pelvis, or "not correct" (a score of 1) when the low back moved into extension or rotation during knee flexion [30].

Tests of muscle endurance (Figure S2): trunk and hip muscle endurance was assessed using the following tests with published protocols and established test–retest reliability: abdominal (flexor) muscle endurance [31], side-bridge (right and left) [31], supine bridge [32], isometric hip abduction (right and left) [33] and Biering–Sorensen [31]. The tests were performed in random order, except the Biering–Sorensen, which was performed last. The objective of each test was to hold a static position for as long as possible. Each test was performed once and the time for which the position was held (seconds) was used for analysis. The test was stopped when the participant could not maintain the position (e.g., fell below horizontal for the isometric hip abduction or Biering–Sorensen tests) or elected to stop due to pain. The Biering–Sorensen test was also stopped if the maximal test time (240 s) was reached.

For the abdominal endurance test [31], participants were positioned sitting with their trunk supported in 60° of flexion, knees and hips flexed to 90° and arms crossed over the chest and feet secured. Trunk support was removed, and the participant was required to hold this position. To perform the side bridge tests [31], participants assumed a side-lying position with knees extended and the top foot in front of the bottom foot. The participant lifted their body off the ground to support their weight on their lower forearm and feet. For the supine bridge test [32], participants lay supine with knees flexed to 90°, soles of the feet on a 20 cm narrow base and hands by the ears. Participants raised their pelvis off the table and held a position with the shoulders, hips and knees in a straight line. If the participant held the position for 2 min, the dominant knee was extended and the test continued in single-leg support [32]. The isometric hip abduction test was performed with the participant in side-lying position with the pelvis secured to the plinth with straps, the bottom leg flexed at the hip and knee and a weight of 7.5% of body mass on the top ankle [33]. The participant lifted the top leg to horizontal and held this position. An inclinometer secured to the leg 10 cm superior to the lateral femoral condyle monitored any deviation from horizontal. The Biering–Sorensen test was performed in prone position. The participant's pelvis, hips and knees were secured to a table. The trunk and upper limbs were positioned off of the table and supported on a chair [31]. To start the test, the chair was removed, and the participant was required to maintain a horizontal trunk position. Deviation from horizontal was measured using an inclinometer positioned on the mid-axillary line of the trunk at the level of the scapular.

Prior to data collection, intra-rater reproducibility for the physical tests was evaluated. The strength of agreement (measured with intraclass correlation coefficients (ICC)) was moderate (ICC: 0.4–0.6) for the AHAbd test examiner rating, ASLR examiner and participant rating and PKF test; substantial (ICC: 0.5–0.8) for the AHAbd test participant rating; and almost perfect (ICC: 0.8–1.0) for the abdominal endurance, right and left side-bridge, Biering–Sorensen test, supine-bridge and right and left isometric hip abduction test. Bland–Altman charts confirmed the results provided by the corresponding ICCs.

2.4. Self-Report Outcome Measures

Self-reported measures included demographics; history and severity of LBP history; total and occupational physical activity; psychosocial job characteristics; general health; and pain catastrophising.

The Nordic Musculoskeletal Questionnaire (NMQ) [34] was used to record lifetime, 12-month and 7-day prevalence of LBP. Participants were shown a body map with the low back region shaded and asked to indicate if they had experienced pain in that region ("Yes"/"No" response options) in any of the timeframes. Severity of LBP in the last 7 days was assessed with a 100 mm VAS [27].

The short form International Physical Activity Questionnaire (IPAQ) is a reliable and valid method of measuring self-report physical activity [35]. Total physical activity in MET—minutes/week—was calculated based on the sum of walking, moderate-intensity and vigorous-intensity physical activity [35]. The Occupational Sitting and Physical Activity Questionnaire (OSPAQ) was utilised to estimate time spent sitting, standing and walking at work [36]. It has acceptable reliability and validity with objective measures. Participants describe the proportion of a typical workday spent sitting and doing other physical activity. Results are expressed in minutes per day.

The abbreviated Job Content Questionnaire (JCQ) [37] was used to assess psychosocial job characteristics. It includes four domains: job control, psychological job demands, social support and physical demands. Each item was rated using a 4-point Likert. Higher scores reflect lower job control, lower psychological job demands, lower social support and lower physical demands.

The Short-Form 12 (SF-12) evaluated general health status in eight dimensions (physical function (PF), role-physical (RP), bodily pain (BP), general health (GH), vitality (VT), social function (SF), role-emotional (RE) and mental health (MH) [38], which are also reduced to Physical (PCS) and Mental (MCS) Component Summary scores. Values above or below 50 (the normative score from the general population) are interpreted as better or worse than the reference population, respectively.

The Pain Catastrophising Scale was used to assess propensity for pain catastrophising [39]. The scale contains a 5-point Likert scale (0 = "not at all" to 4 = "all the time") on which participants rate their thoughts or feelings when experiencing pain. Scores ranged from 0 to 52 and greater scores indicate a greater degree of catastrophising.

2.5. Data Analysis

Participants were defined as either a pain developer (PD) or non-pain developer (NPD) based on their response to the standing paradigm protocol. PD reported a change of ≥10 mm in LBP from baseline throughout the standing paradigm. NPD were participants who did not report any change in symptoms or reported a change of <10 mm on the VAS scale throughout the standing task.

Studies investigating criteria for minimally clinically important difference (MCID) scores for pain VAS have been conducted across a range of diagnoses and populations [40]. While MCID scores are typically used to detect improvements in symptoms in response to treatment, minimal detectable change (MDC) may also be used to investigate changes in pain. Based on previous studies [11,19,20,26], the decision was made to use a change of ≥10 mm on the pain VAS at any time between start and end of the test as the cut-off point to categorise participants as PD or NPD.

2.6. Statistical Analysis

Statistical analyses were performed using RStudio software (R version 3.5.2, R Foundation for Statistical Computing, Vienna, Austria), with the significance level set at $p < 0.05$. Significant differences in quantitative variables between PD and NPD groups were assessed using Student's *t*-test or Mann–Whitney's U test (depending on the normality of the corresponding variables). The normality assumption was verified through Shapiro–Wilk's test and graphically observed by means of kernel density plots. In the case of categorical variables, differences between sub-groups were assessed using the chi-square or Fisher's exact test. Missing data were not considered in the computations.

Because of the small sample in some subgroups, the existence of significant and non-spurious differences was analysed using a Bayesian hypothesis test [41]. A Bayesian testing procedure proceeds in a similar way to a classical hypothesis test but, in this case, the *p*-value for adopting a decision is replaced by the named Bayes factor. This number is the ratio of probabilities of the null hypothesis and the alternative, considering the statistical information provided by the sample data. The obtained numbers were interpreted using the original Kass and Raftery's categories [41]: Bf < 1 implies that the data are more likely under the alternative hypothesis (H1) than under the null. The strength of the likelihood can be interpreted as follows: between 1 and 0.33, anecdotal evidence; from 0.33 to 0.1, moderate evidence; from 0.1 to 0.033, strong; between 0.033 and 0.01, very strong; and <0.01, extreme.

3. Results

Forty office workers (22 females; mean age: 37.4 ± 6.6 years) were included in the study (Figure 1). Fourteen participants (35%) were classified as PD (due to an increase in LBP intensity of ≥10 mm during the standing paradigm), and 26 participants (65%) were NPD.

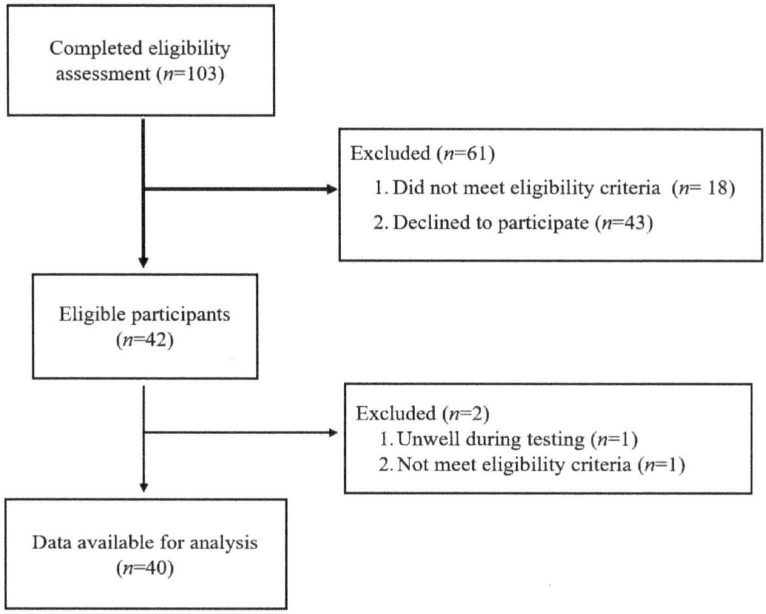

Figure 1. Flow diagram.

Table 1 provides the demographics, anthropometrics and self-report measures for PD and NPD groups. There were no statistically significant differences in age, sex, anthropometrics, total or occupational physical activity, psychosocial job characteristics (JCQ) or pain catastrophising between groups (all $p > 0.12$). A larger proportion of participants who developed LBP during the standing task (PD) reported experiencing LBP previously in their lifetime (64%), in the last year (93%) and in the last 7 days (79%) than those who did not experience LBP during the standing task (NPD) (all $p < 0.05$). LBP severity at the start of testing ($p = 0.68$) was similar for PD and NPD groups. The mental and physical component summary scores of the SF-12, and all dimensions except physical function (55.8 ± 2.9 vs. 49.8 ± 10, $p = 0.01$), were similar between PD and NPD participants (all $p > 0.06$).

Table 1. Differences in demographics and self-report measures for non-pain developers (NPD) and pain developers (PD).

	NPD (n = 26)	PD (n = 14)		
			p	Bf
Age (years), mean ± SD	36.3 ± 8.6	39.4 ±11.5	0.51	2.2
Sex (female), n (%)	13 (50.0)	9 (64.3)	0.39	
BMI (kg/m^2), mean ± SD	26.2 ± 4.3	26.5 ± 7.8	0.49	3.1
IPAQ, MET—min/week (median)	2138.5 ± 1897.5	2357.0 ± 978	0.80	2.7
History of LBP				
LBP Lifetime			0.04	0.35
No (n, %)	18 (69.2)	5 (35.7)		
Yes (n, %)	8 (30.8)	9 (64.3)		
LBP last year			0.03	0.31
No (n, %)	10 (38.5)	1 (7.1)		
Yes (n, %)	16 (61.5)	13 (92.9)		
LBP last 7 days			0.02	0.14
No (n, %)	16 (61.5)	3 (21.4)		
Yes (n, %)	10 (38.5)	11 (78.6)		
LBP min 0			0.08	-
No (n, %)	24 (92.3)	10 (71.4)		
Yes (n, %)	2 (7.7)	4 (28.6)		
LBP severity, last 7 days (0–100), mean ± SD	30.4 (31.2)	51.2 (21.8)	0.08	0.86
LBP severity, start of testing (0–100), mean ± SD	12.7 (3.7)	10.0 (7.5)	0.68	-
OSPAQ, mean ± SD				
Sitting per week (min)	2126.1 ± 517.6	1929.9 ± 175.8	0.09	1.50
Sitting per workday (min)	430.6 ± 102.8	391.9 ± 85	0.16	1.80
Standing per week (min)	281 ± 250.2	221.2 ± 190.8	0.64	2.50
Standing per workday (min)	57.1 ± 50.3	40.5 ± 30	0.54	1.9
Walking per week (min)	179.3 ± 144.2	210.4 ± 281.2	0.68	2.9
Walking per workday (min)	36.3 ± 28.6	39.7 ± 43.1	0.67	3.0
PCS, mean ± SD				
Rumination	3.5 ±3.4	5.3 ±4.2	0.18	1.32
Magnification	1.7 ±1.3	2.9 ±2.9	0.29	0.84
Helplessness	2.7 ±3.7	4.6 ±4.8	0.16	1.51
PCS total (0–52)	7.8 ±7.2	12.8 ±10.2	0.12	0.93
JCQ, mean ± SD				
Job Control (24–96)	46.7 ± 13.2	42.4 ± 10.5	0.31	2.0
Psychological Job Demands (3–12)	8 ± 1.7	7.6 ± 1.4	0.34	2.5
Social Support (4–16)	7.5 ± 1.7	7.1 ± 1.8	0.62	2.8
Physical Demands (2–8)	7.5 ± 0.8	7.6 ± 0.9	0.40	2.9
SF-12, Normal-Based Scores (50 ± 10)				
Vitality	52.1 ± 9.1	54.0 ± 9.3	0.29	2.68
Social Function	51.1 ± 7.9	49.3 ± 9.1	0.44	2.64
Role Emotional	43.9 ± 12.5	47.0 ± 11.7	0.34	2.48
Mental Health	48.1 ± 9.6	53.1 ± 6.5	0.09	0.96
Mental Component Summary	45.6 ± 9.7	51.4 ± 8.7	0.06	0.83
Physical Function	55.8 ± 2.9	49.8 ± 10	0.01	0.13
Role Physical	49.8 ± 9.3	49.3 ± 12.9	0.66	3.10
Bodily Pain	52.5 ± 5.8	47.4 ± 10.5	0.12	0.69
General Health	52.2 ± 8.8	49.9 ± 12.1	0.72	2.59
Physical Component Summary	55.4 ± 5.0	48.8 ± 10.6	0.09	0.21

p-values indicate statistically significant differences in having had a prior history of LBP (lifetime, last year, last 7 days) and physical function between groups. Bf, Bayesian factor; BMI, body mass index; IPAQ, International Physical Activity Questionnaire; MET, Metabolic Equivalent of Task (computed as the sum of walking, moderate-intensity and vigorous-intensity physical activity); LBP, low back pain; OSPAQ, Occupational Sitting and Physical Activity Questionnaire; min, minutes; PCS, Pain Catastrophising Scale; JCQ; Job Content Questionnaire.

The median holding times for the trunk and hip muscle endurance tests were lower for PD than NPD for the abdominal endurance test, Biering–Sorensen test, isometric hip abduction test (bilaterally) and the left side bridge test (all $p \leq 0.05$; Table 2, Figure 2). Participants who were PD self-reported having greater difficulty in performing the AHAbd and ASLR tests than NPD (all $p < 0.05$; Table 2). The examiner scores for AHAbd, ASLR and PKB were not different between groups (all $p > 0.06$).

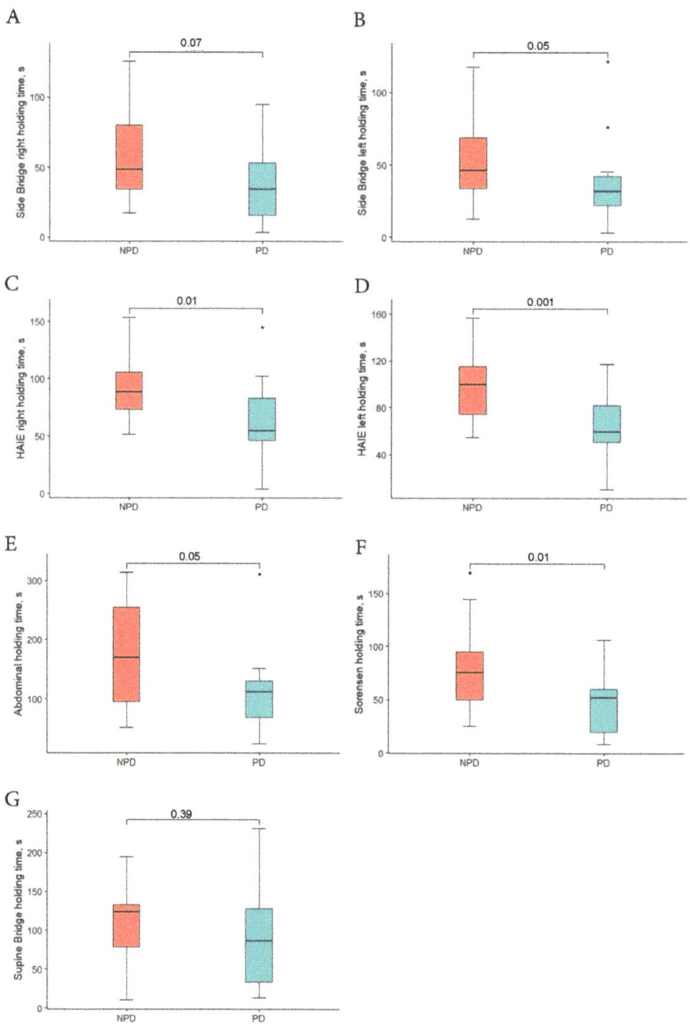

Figure 2. Endurance test. (**A**) Side bridge right side, (**B**) side bridge left side, (**C**) HAIE right side, (**D**) HAIE left side, (**E**) abdominal, (**F**) Sorensen, (**G**) supine bridge holding times for individual NPD (n = 26) and PD (n = 14). In the box plot, red boxes represented NPD and blue boxes represent PD; horizontal bold lines show the medians, and the box limits indicate the 25th and 75th percentiles. Whiskers extend one and a half times the length of the box (interquartile range), and dots represent the outliers. p obtained from the Mann–Whitney U test or t-Student test for independent samples. NPD, non-pain developers; PD, pain developers; s, seconds; HAIE, Hip Abductor Isometric Endurance test.

Table 2. Differences in physical outcome measures for non-pain developers (NPD) and pain developers (PD).

	NPDs (n = 26)	PDs (n = 14)		
			p	Bf
Tests of trunk and hip muscle endurance				
Side bridge right side (s)	57.3 ± 28.9	39.4 ± 29.9	0.07	0.84
Side bridge left side (s)	52.6 ± 25.6	38.1 ± 30.1	0.05	1.14
Isometric hip abduction (right leg) (s)	90.7 ± 25.6	62.4 ± 36.5	0.01	0.15
Isometric hip abduction (left leg) (s)	97.8 ± 27.9	62.2 ± 32.2	0.001	0.02
Supine bridge (s)	101.1 ± 51	89.2 ± 61.3	0.39	2.64
Abdominal (s)	172.1 ± 86.4	112.0 ± 69.6	0.05	0.47
Sorensen (s)	78.6 ± 36.3	48.7 ± 31.9	0.01	0.25
Tests of movement control				
AHAbd, right side, examiner score (0–3)	1.4 ± 0.7	1.6 ± 1	0.69	2.74
AHAbd, left side, examiner score (0–3)	1.6 ± 0.6	1.8 ± 0.9	0.33	1.99
AHAbd, right side, participant score (0–5)	1.5 ± 0.9	2.2 ± 0.9	0.04	0.56
AHAbd, left side, participant score (0–5)	1.5 ± 0.9	2.4 ± 1.1	0.03	0.21
ASLR, total examiner score (0–10)	2.5 ± 1	2.8 ± 1.1	0.42	2.38
ASLR, total participant score (0–10)	2.2 ± 1.4	3.4 ± 1.7	0.03	0.39
PKF, right side, examiner score (0–1)	0.7 ± 0.4	0.6 ± 0.4	0.39	2.30
PKF, left side, examiner score (0–1)	0.4 ± 0.4	0.6 ± 0.4	0.06	0.81

Data are presented as mean ± SD. p-values indicate statistically significant differences in trunk and hip muscle endurance (side bridge left side, isometric hip abduction, abdominal and Sorensen test) and in movement control (AHAbd and ASLR participant score). Bf, Bayesian factor; s, seconds; AHAbd, active hip abduction; ASLR, active straight leg raise; PKF, prone knee flexion (0 = correct; 1 = not correct).

A post-hoc analysis was done to explore the effect of history of LBP in the past year on trunk and hip muscle endurance and control of movement measures (Table 3). Analyses indicate that individuals with a history of LBP have lower examiner ratings for control of movement during the ASLR (p = 0.01) and PKF (left side only; p = 0.006) tests.

Table 3. Differences in physical outcome measures for participants with and without history of lower back pain (LBP) in the past year.

	LBP—Last Year (n = 29)	No LBP—Last Year (n = 11)	p
Test of trunk and hip muscle endurance			
Side bridge right side (s)	48.9 ± 29.9	56.7 ± 31.6	0.52
Side bridge left side (s)	44.0 ± 26.0	56.9 ± 31.2	0.31
Isometric hip abduction (right leg) (s)	78.5 ± 36.2	86.8 ± 19.6	0.67
Isometric hip abduction (left leg) (s)	81.1 ± 35.3	96.5 ± 27.7	0.06
Supine bridge (s)	88.1 ± 57.4	120.3 ± 38.3	0.09
Abdominal (s)	143.2 ± 78.9	171.8 ± 100.6	0.44
Sorensen (s)	66.4 ± 36.7	72.4 ± 40.4	0.52
Test of movement control			
AHAbd, right side, examiner score (0–3)	1.5 ± 0.8	1.2 ± 0.7	0.14
AHAbd, left side, examiner score (0–3)	1.7 ± 0.7	1.5 ± 0.7	0.32
AHAbd, right side, participant score (0–5)	1.9 ± 0.9	1.3 ± 0.8	0.10
AHAbd, left side, participant score (0–5)	1.9 ± 1.1	1.4 ± 0.8	0.15
ASLR, total examiner score (0–10)	2.9 ± 0.9	1.9 ± 1.0	0.01
ASLR, total participant score (0–10)	2.6 ± 1.6	2.3 ± 1.5	0.99
PKF, right side, examiner score (0–1)	0.7 ± 0.4	0.5 ± 0.5	0.12
PKF, left side, examiner score (0–1)	0.6 ± 0.4	0.2 ± 0.3	0.006

Data are presented as mean ± SD. s, seconds; AHAbd, active hip abduction; ASLR, active straight leg raise; PKF, prone knee flexion.

4. Discussion

The purpose of this study was to identify demographic, physical and psychosocial factors associated with the development of LBP during a one-hour standing task in office workers. In our study, 35% of office workers developed LBP or experienced an increase in LBP intensity when standing for one hour. Our hypotheses that PD would have deficits in muscle endurance and control of lumbopelvic movement than NPD were partly supported. PD exhibited significantly reduced endurance of the trunk and hip muscles during the abdominal, side-bridge, hip abduction and trunk extensor tests, and deficits in lumbopelvic movement control, when performing an AHAbd and ASLR test compared to NPD. We correctly hypothesised that participants with a history of LBP would be more likely to develop standing-induced LBP that those without an LBP history. Our data suggest that there is a relationship between lifetime, yearly or weekly history of LBP and the development or worsening of standing-induced LBP.

Our findings of deficits in the side bridge and isometric hip abduction endurance tests, and participant-rated difficulty performing AHAbd test, in office workers who were PD suggest that decreased hip abductor endurance and/or strength may be associated with standing-induced LBP. This finding is consistent with previous studies that report decreased hip abduction [14] and side bridge [13] endurance in university students who are PD compared to those who are NPD, and differences in the AHAbd test between people who do and do not develop pain during prolonged standing [15,17,19]. Further, similar to our data, Nelson-Wong et al. [19] found that participant self-rated difficulty performing the AHAbd test was better able to predict those who developed pain during prolonged standing than the examiner-rated score. Thus, together with previous work in other populations, our data in office workers suggest that tests of hip abductor muscle function may be clinically useful for individuals who stand for prolonged periods, such office workers transitioning to a sit–stand workstation.

Our study identified that PD had poorer trunk (abdominal and lumbar) muscle strength/endurance and greater self-rated difficulty performing an ASLR than NPD. In contrast to our findings, the few studies that have assessed trunk muscle function in relation to standing-induced LBP did not find that deficits of endurance/strength [19,22] or ASLR performance [15,19] discriminated between PD and NPD. One key difference between our study and this previous work is the population studied. Our sample included individuals with and without a history of LBP, whereas other studies have included asymptomatic individuals without previous LBP. Deficits in trunk muscle function and ASLR performance have been reported in individuals with LBP [30,42]. Nelson-Wong et al. [15,43] found differences in co-activation of the trunk flexor and extensor muscles groups during a prolonged standing task in PD and NPD and suggested that trunk muscle activation may be related to standing-induced LBP. The relationship between trunk muscle strength/endurance, muscle activation patterns and standing-induced LBP requires further investigation.

Our results demonstrate that having had a prior history of LBP differentiated between PD during one-hour standing and NPD. Experiencing an episode of LBP in the last 12 months, which is a predictor of recurrence of LBP in the general population [44] and in office workers [24,45], was significantly associated with the development of pain during standing. This highlights the importance of determining LBP history in office workers who may want to transition to a sit–stand workstation. Deficits in hip [46] and trunk [42] muscle strength/endurance and control [30] have been shown to be different in individuals with and without LBP. Analyses of differences in these outcomes between study participants with and without a history of LBP in the past year identified differences in examiner-rated lumbopelvic control during ASLR and PKB tests, but no differences in muscle strength/endurance or perceived task difficulty. Thus, it is unlikely that deficits in trunk and hip muscle function in PD are purely due to a greater proportion of PD having a history of LBP.

There were no differences in most self-report variables measured between study groups, except for better physical function for NPD compared to PD. Similar to our findings, pain catastrophising was not found to be related to standing-induced LBP in individuals without a history of LBP [15,16,18]. Previous studies have found psychological characteristics predictive of LBP in an occupational setting [45]. However, the low severity of LBP in our participants, positive job control and social support and low psychological and physical demands may explain the lack of differences in psychological characteristics. It is also possible that psychological factors are more relevant in a clinical population rather than a "healthy" working population. The small sample size or ceiling effect of measures could explain our findings.

There are some limitations to this study to consider. Firstly, clinical assessments were performed by a single physical therapist. While inter-rater reliability for each measure was substantial to almost perfect, it is unknown whether the rating of movement control would be similar between different examiners. Secondly, the length of the standing task may have been insufficient to trigger differences in some outcomes. A recent systematic review suggested that clinically relevant levels of LBP were reached after 71 min of standing [8]; however, previous work by our research group and others suggests that as little as 15 min is sufficient to induce LBP [47]. Thirdly, while the sample size was based on a power analysis and on previous studies, it was small for some subgroups of variables and only medium effect sizes could be determined. Bayesian analysis was performed to address this concern.

Findings from this study suggest that a history of LBP and deficits in hip abductor and trunk muscle strength/endurance and participant-rated difficulty performing tasks that challenge lumbopelvic control are associated with the development of standing-induced LBP in office workers. Clinicians could consider assessing hip and trunk muscle function in office workers who are planning to transition to a sit–stand workstation. It is important to note that office workers in this study stood for one hour without sitting, resulting in 35% of participants developing standing-induced LBP. It is recommended that office workers seeking advice regarding the use of height adjustable workstations be advised to gradually upgrade time spent standing to enable the hip and trunk muscles to adjust to increased demands during standing [48] and change position regularly [11,49].

5. Conclusions

Office workers who reported LBP during a one-hour standing task in this cross-sectional study presented with decreased hip and trunk muscle endurance, self-reported difficulty with tasks that challenge lumbopelvic control and an increased likelihood of having a history of LBP than NPD. Findings suggest that these outcomes may be important to evaluate in office workers who are transitioning to standing workstations. Prospective research is needed to determine if these measures are able to identify office workers who develop standing LBP when transitioning to a sit–stand workstation. While this was not a treatment study, and data were collected at one point in time, the researchers propose that addressing hip and trunk muscle deficits may help to reduce the risk of developing LBP when using a standing workstation. Future research should look at the effect of addressing hip and trunk muscle deficits on standing induced LPB in office workers.

Supplementary Materials: The following are available online at http://www.mdpi.com/1660-4601/17/19/7104/s1, Figure S1: Motor Control Test. (A) AHAbd test, (B) ASLR test, (C) PKF test. Figure S2: Test of muscle endurance. (A) Abdominal, (B) Sidebridge, (C) Supine Bridge, (D) Isometric hip abduction, (E) Biering-Sorensen test.

Author Contributions: Conceptualization, V.J., M.D.S., and B.R.-R.; formal analysis, A.Q.-d.-R. and B.R.-R.; investigation, B.R.-R., V.J., and M.D.S.; methodology, V.J., M.D.S., B.R.-R., and A.Q.-d.-R.; project administration, V.J.; resources, V.J., B.R.-R.; supervision, V.J.; visualisation, B.R.-R., M.D.S., A.Q.-d.-R., and V.J.; writing—original draft: B.R.-R. and V.J.; writing—review and editing: V.J., M.D.S., B.R.-R., and A.Q.-d.-R. All authors have read and agreed to the published version of the manuscript.

Funding: This research received no external funding.

Acknowledgments: The authors thank all the participants who took part in the study.

Conflicts of Interest: The authors declare no conflict of interest.

References

1. Hoy, D.; March, L.; Brooks, P.; Blyth, F.; Woolf, A.; Bain, C.; Williams, G.; Smith, E.; Vos, T.; Barendregt, J.; et al. The global burden of low back pain: Estimates from the Global Burden of Disease 2010 study. *Ann. Rheum. Dis.* **2014**, *73*, 968–974. [CrossRef]
2. Hartvigsen, J.; Hancock, M.J.; Kongsted, A.; Louw, Q.; Ferreira, M.L.; Genevay, S.; Hoy, D.; Karppinen, J.; Pransky, G.; Sieper, J.; et al. What low back pain is and why we need to pay attention. *Lancet* **2018**, *391*, 2356–2367. [CrossRef]
3. Roffey, D.M.; Wai, E.K.; Bishop, P.; Kwon, B.K.; Dagenais, S. Causal assessment of occupational sitting and low back pain: Results of a systematic review. *Spine J.* **2010**, *10*, 252–261. [CrossRef] [PubMed]
4. De Carvalho, D.E.; de Luca, K.; Funabashi, M.; Breen, A.; Wong, A.Y.L.; Johansson, M.S.; Ferreira, M.L.; Swab, M.; Neil Kawchuk, G.; Adams, J.; et al. Association of Exposures to Seated Postures With Immediate Increases in Back Pain: A Systematic Review of Studies With Objectively Measured Sitting Time. *J. Manip. Physiol. Ther.* **2020**, *43*, 1–12. [CrossRef]
5. Ryan, C.G.; Grant, P.M.; Dall, P.M.; Granat, M.H. Sitting patterns at work: Objective measurement of adherence to current recommendations. *Ergonomics* **2011**, *54*, 531–538. [CrossRef]
6. Patterson, R.; McNamara, E.; Tainio, M.; de Sá, T.H.; Smith, A.D.; Sharp, S.J.; Edwards, P.; Woodcock, J.; Brage, S.; Wijndaele, K. Sedentary behaviour and risk of all-cause, cardiovascular and cancer mortality, and incident type 2 diabetes: A systematic review and dose response meta-analysis. *Eur. J. Epidemiol.* **2018**, *33*, 811–829. [CrossRef]
7. Buckley, J.P.; Hedge, A.; Yates, T.; Copeland, R.J.; Loosemore, M.; Hamer, M.; Bradley, G.; Dunstan, D.W. The sedentary office: An expert statement on the growing case for change towards better health and productivity. *Br. J. Sport. Med.* **2015**, *49*, 1357–1362. [CrossRef]
8. Coenen, P.; Willenberg, L.; Parry, S.; Shi, J.W.; Romero, L.; Blackwood, D.M.; Maher, C.G.; Healy, G.N.; Dunstan, D.W.; Straker, L.M. Associations of occupational standing with musculoskeletal symptoms: A systematic review with meta-Analysis. *Br. J. Sports Med.* **2018**, *52*, 174–181. [CrossRef]
9. Waters, T.R.; Dick, R.B. Evidence of health risks associated with prolonged standing at work and intervention effectiveness. *Rehabil. Nurs.* **2015**, *40*, 148–165. [CrossRef]
10. Roffey, D.M.; Wai, E.K.; Bishop, P.; Kwon, B.K.; Dagenais, S. Causal assessment of occupational standing or walking and low back pain: Results of a systematic review. *Spine J.* **2010**, *10*, 262–272. [CrossRef]
11. Coenen, P.; Parry, S.; Willenberg, L.; Shi, J.W.; Romero, L.; Blackwood, D.M.; Healy, G.N.; Dunstan, D.W.; Straker, L.M. Associations of prolonged standing with musculoskeletal symptoms—A systematic review of laboratory studies. *Gait Posture* **2017**, *58*, 310–318. [CrossRef] [PubMed]
12. Villafañe, J.H.; Gobbo, M.; Peranzoni, M.; Naik, G.; Imperio, G.; Cleland, J.A.; Negrini, S. Validity and everyday clinical applicability of lumbar muscle fatigue assessment methods in patients with chronic non-specific low back pain: A systematic review. *Disabil. Rehabil.* **2016**, *38*, 1859–1871. [CrossRef] [PubMed]
13. Marshall, P.W.M.; Patel, H.; Callaghan, J.P. Gluteus medius strength, endurance, and co-activation in the development of low back pain during prolonged standing. *Hum. Mov. Sci.* **2011**, *30*, 63–73. [CrossRef] [PubMed]
14. Viggiani, D.; Callaghan, J.P. Hip Abductor Fatigability and Recovery Are Related to the Development of Low Back Pain During Prolonged Standing. *J. Appl. Biomech.* **2018**, *34*, 39–46. [CrossRef]
15. Nelson-Wong, E.; Callaghan, J.P. Is muscle co-activation a predisposing factor for low back pain development during standing? A multifactorial approach for early identification of at-risk individuals. *J. Electromyogr. Kinesiol.* **2010**, *20*, 256–263. [CrossRef]
16. Sorensen, C.J.; George, S.Z.; Callaghan, J.P.; Van Dillen, L.R. Psychological Factors Are Related to Pain Intensity in Back-Healthy People Who Develop Clinically Relevant Pain During Prolonged Standing: A Preliminary Study. *PM R* **2016**, *8*, 1031–1038. [CrossRef]
17. Sorensen, C.J.; Johnson, M.B.; Norton, B.J.; Callaghan, J.P.; Van Dillen, L.R. Asymmetry of lumbopelvic movement patterns during active hip abduction is a risk factor for low back pain development during standing. *Hum. Mov. Sci.* **2016**, *50*, 38–46. [CrossRef]
18. Hwang, C.T.; Van Dillen, L.R.; Haroutounian, S. Do Changes in Sensory Processing Precede Low Back Pain Development in Healthy Individuals? *Clin. J. Pain* **2018**, *34*, 525–531. [CrossRef]

19. Nelson-Wong, E.; Timothy, F.; Callaghan, J.P. Development of active hip abduction screening test for identifying occupational low back pain. *J. Orthop. Sports Phys. Ther.* **2009**, *39*, 649–657. [CrossRef]
20. Bussey, M.D.; Kennedy, J.E.; Kennedy, G. Gluteus medius coactivation response in field hockey players with and without low back pain. *Phys. Ther. Sport* **2016**, *17*, 24–29. [CrossRef]
21. Nelson-Wong, E.; Callaghan, J.P. Repeatability of Clinical, Biomechanical, and Motor Control Profiles in People with and without Standing-Induced Low Back Pain. *Rehabil. Res. Pract.* **2010**, *2010*, 289278. [CrossRef] [PubMed]
22. Ringheim, I.; Austein, H.; Indahl, A.; Roeleveld, K. Postural strategy and trunk muscle activation during prolonged standing in chronic low back pain patients. *Gait Posture* **2015**, *42*, 584–589. [CrossRef] [PubMed]
23. Manchikanti, L.; Singh, V.; Falco, F.J.E.; Benyamin, R.M.; Hirsch, J.A. Epidemiology of Low Back Pain in Adults. *Neuromodulation Technol. Neural Interface* **2014**, *17*, 3–10. [CrossRef] [PubMed]
24. Janwantanakul, P.; Sitthipornvorakul, E.; Paksaichol, A. Risk Factors for the Onset of Nonspecific Low Back Pain in Office Workers: A Systematic Review of Prospective Cohort Studies. *J. Manip. Physiol. Ther.* **2012**, *35*, 568–577. [CrossRef]
25. Erdfelder, E.; FAul, F.; Buchner, A.; Lang, A.G. Statistical power analyses using G*Power 3.1: Tests for correlation and regression analyses. *Behav. Res. Methods* **2009**, *41*, 1149–1160. [CrossRef]
26. Sorensen, C.J.; Johnson, M.B.; Callaghan, J.P.; George, S.Z.; Van Dillen, L.R. Validity of a Paradigm for Low Back Pain Symptom Development During Prolonged Standing. *Clin. J. Pain* **2015**, *31*, 652–659. [CrossRef]
27. Von Korff, M.; Jensen, M.P.; Karoly, P. Assessing global pain severity by self-report in clinical and health services research. *Spine* **2000**, *25*, 3140–3151. [CrossRef]
28. Davis, A.M.; Bridge, P.; Miller, J.; Nelson-Wong, E. Interrater and intrarater reliability of the active hip abduction test. *J. Orthop. Sports Phys. Ther.* **2011**, *41*, 953–960. [CrossRef]
29. Mens, J.M.; Vleeming, A.; Snijders, C.J.; Koes, B.W.; Stam, H.J. Reliability and validity of the active straight leg raise test in posterior pelvic pain since pregnancy. *Spine* **2001**, *26*, 1167–1171. [CrossRef]
30. Luomajoki, H.; Kool, J.; de Bruin, E.D.; Airaksinen, O. Movement control tests of the low back; evaluation of the difference between patients with low back pain and healthy controls. *BMC Musculoskelet Disord* **2008**, *9*, 170. [CrossRef]
31. McGill, S.M.; Childs, A.; Liebenson, C. Endurance times for low back stabilization exercises: Clinical targets for testing and training from a normal database. *Arch. Phys. Med. Rehabil.* **1999**, *80*, 941–944. [CrossRef]
32. Schellenberg, K.L.; Lang, J.M.; Chan, K.M.; Burnham, R.S. A clinical tool for office assessment of lumbar spine stabilization endurance: Prone and supine bridge maneuvers. *Am. J. Phys. Med. Rehabil.* **2007**, *86*, 380–386. [CrossRef] [PubMed]
33. Van Cant, J.; Dumont, G.; Pitance, L.; Demoulin, C.; Feipel, V. Test-retest reliability of two clinical tests for the assessment of hip abductor endurance in healthy females. *Int. J. Sports Phys. Ther.* **2016**, *11*, 24–33. [PubMed]
34. Kuorinka, I.; Jonsson, B.; Kilbom, A.; Vinterberg, H.; Biering-Sørensen, F.; Andersson, G.; Jørgensen, K. Standardised Nordic questionnaires for the analysis of musculoskeletal symptoms. *Appl. Erg.* **1987**, *18*, 233–237. [CrossRef]
35. Silsbury, Z.; Goldsmith, R.; Rushton, A. Systematic review of the measurement properties of self-report physical activity questionnaires in healthy adult populations. *BMJ Open* **2015**, *5*, e008430. [CrossRef]
36. Jancey, J.; Tye, M.; McGann, S.; Blackford, K.; Lee, A.H. Application of the Occupational Sitting and Physical Activity Questionnaire (OSPAQ) to office based workers. *BMC Public Health* **2014**, *14*, 762. [CrossRef]
37. Karasek, R.; Brisson, C.; Kawakami, N.; Houtman, I.; Bongers, P.; Amick, B. The Job Content Questionnaire (JCQ): An instrument for internationally comparative assessments of psychosocial job characteristics. *J. Occup. Health Psychol.* **1998**, *3*, 322–355. [CrossRef]
38. Gandek, B.; Ware, J.E.; Aaronson, N.K.; Apolone, G.; Bjorner, J.B.; Brazier, J.E.; Bullinger, M.; Kaasa, S.; Leplege, A.; Prieto, L.; et al. Cross-validation of item selection and scoring for the SF-12 Health Survey in nine countries: Results from the IQOLA Project. International Quality of Life Assessment. *J. Clin. Epidemiol.* **1998**, *51*, 1171–1178. [CrossRef]
39. Sullivan, M.J.; Bishop, S.R.; Pivik, J. The pain catastrophizing scale: Development and validation. *Psychol. Assess.* **1995**, *74*, 524–532. [CrossRef]
40. Ostelo, R.W.; Deyo, R.A.; Stratford, P.; Waddell, G.; Croft, P.; Von Korff, M.; Bouter, L.M.; de Vet, H.C. Interpreting change scores for pain and functional status in low back pain: Towards international consensus regarding minimal important change. *Spine* **2008**, *33*, 90–94. [CrossRef]

41. Kass, R.E.; Raftery, A.E. Bayes factors. *J. Am. Stat. Assoc.* **1995**, *90*, 773–795. [CrossRef]
42. Demoulin, C.; Vanderthommen, M.; Duysens, C.; Crielaard, J.M. Spinal muscle evaluation using the Sorensen test: A critical appraisal of the literature. *Jt. Bone Spine* **2006**, *73*, 43–50. [CrossRef] [PubMed]
43. Nelson-Wong, E.; Gregory, D.E.; Winter, D.A.; Callaghan, J.P. Gluteus medius muscle activation patterns as a predictor of low back pain during standing. *Clin. Biomech.* **2008**, *23*, 545–553. [CrossRef] [PubMed]
44. da Silva, T.; Mills, K.; Brown, B.T.; Herbert, R.D.; Maher, C.G.; Hancock, M.J. Risk of Recurrence of Low Back Pain: A Systematic Review. *J. Orthop. Sport. Phys. Ther.* **2017**, *47*, 305–313. [CrossRef]
45. Taylor, J.B.; Goode, A.P.; George, S.Z.; Cook, C.E. Incidence and risk factors for first-time incident low back pain: A systematic review and meta-analysis. *Spine J.* **2014**, *14*, 2299–2319. [CrossRef]
46. de Sousa, C.S.; de Jesus, F.L.A.; Machado, M.B.; Ferreira, G.; Ayres, I.G.T.; de Aquino, L.M.; Fukuda, T.Y.; Gomes-Neto, M. Lower limb muscle strength in patients with low back pain: A systematic review and meta-analysis. *J. Musculoskelet. Neuronal Interact.* **2019**, *19*, 69–78.
47. Gallagher, K.M.; Campbell, T.; Callaghan, J.P. The influence of a seated break on prolonged standing induced low back pain development. *Ergonomics* **2014**, *57*, 555–562. [CrossRef]
48. Johnston, V.; Gane, E.M.; Brown, W.; Vicenzino, B.; Healy, G.N.; Gilson, N.; Smith, M.D. Feasibility and impact of sit-stand workstations with and without exercise in office workers at risk of low back pain: A pilot comparative effectiveness trial. *Appl. Ergon.* **2019**, *76*, 82–89. [CrossRef]
49. Agarwal, S.; Steinmaus, C.; Harris-Adamson, C. Sit-stand workstations and impact on low back discomfort: A systematic review and meta-analysis. *Ergonomics* **2018**, *61*, 538–552. [CrossRef]

International Journal of
Environmental Research and Public Health

MDPI

Systematic Review

Effect of Home Exercise Training in Patients with Nonspecific Low-Back Pain: A Systematic Review and Meta-Analysis

Chloé Quentin [1], Reza Bagheri [2], Ukadike C. Ugbolue [3], Emmanuel Coudeyre [4], Carole Pélissier [5,6], Alexis Descatha [7,8], Thibault Menini [9], Jean-Baptiste Bouillon-Minois [10,*] and Frédéric Dutheil [11]

1 CHU Clermont-Ferrand, Department of Family Medicine, Faculty of Medicine, Université Clermont Auvergne, F-63000 Clermont-Ferrand, France; chloe.quentin@etu.uca.fr
2 Department of Exercise Physiology, University of Isfahan, Isfahan 81746-73441, Iran; will.fivb@yahoo.com
3 School of Health and Life Sciences, Institute for Clinical Exercise & Health Science, University of the West of Scotland, Glasgow G720LH, UK; u.ugbolue@uws.ac.uk
4 Physical Medicine and Rehabilitation, INRAE, UNH, CHU Clermont-Ferrand, Université Clermont Auvergne, F-63000 Clermont-Ferrand, France; ecoudeyre@chu-clermontferrand.fr
5 UMRESTTE, IFSTTAR, Université Lyon, 42100 Saint Etienne, France; carole.pelissier@chu-st-etienne.fr
6 UMRESTTE, IFSTTAR, Université St Etienne, 42100 Saint Etienne, France
7 Inserm, EHESP, Irset (Institut de Recherche en santé, Environnement et Travail), CAPTV-CDC, Université Angers, CHU Angers, F-49000 Angers, France; alexis.descatha@inserm.fr
8 Inserm, EHESP, Irset (Institut de Recherche en santé, Environnement et Travail), CAPTV-CDC, Université Rennes, F-49000 Angers, France
9 Department of Family Medicine, Faculty of Medicine, CHU Clermont-Ferrand, ACCePPT, Université Clermont Auvergne, F-63000 Clermont Ferrand, France; thibault.menini@uca.fr
10 Department of Emergency, CNRS, LaPSCo, Physiological and Psychosocial Stress, CHU Clermont-Ferrand, Université Clermont Auvergne, F-63000 Clermont-Ferrand, France
11 Occupational and Environmental Medicine, CNRS, LaPSCo, Physiological and Psychosocial Stress, CHU Clermont-Ferrand, WittyFit, Université Clermont Auvergne, F-63000 Clermont-Ferrand, France; fdutheil@chu-clermontferrand.fr
* Correspondence: jbb.bouillon@gmail.com; Tel.: +33-6-7436-0423; Fax: +33-4-7327-4649

Citation: Quentin, C.; Bagheri, R.; Ugbolue, U.C.; Coudeyre, E.; Pélissier, C.; Descatha, A.; Menini, T.; Bouillon-Minois, J.-B.; Dutheil, F. Effect of Home Exercise Training in Patients with Nonspecific Low-Back Pain: A Systematic Review and Meta-Analysis. *IJERPH* **2021**, *18*, 8430. https://doi.org/10.3390/ijerph18168430

Academic Editors: Sergio Iavicoli, Vincenzo Denaro, Gianluca Vadalà and Fabrizio Russo

Received: 28 June 2021
Accepted: 5 August 2021
Published: 10 August 2021

Abstract: Background: Exercise therapy is recommended to treat non-specific low back pain (LBP). Home-based exercises are promising way to mitigate the lack of availability of exercise centers. In this paper, we conducted a systemic review and meta-analysis on the effects of home-based exercise on pain and functional limitation in LBP. Method: PubMed, Cochrane, Embase and ScienceDirect were searched until April 20th, 2021. In order to be selected, studies needed to report the pain and functional limitation of patients before and after home-based exercise or after exercise both in a center and at-home. Random-effect meta-analyses and meta-regressions were conducted. Results: We included 33 studies and 9588 patients. We found that pain intensity decreased in the exclusive home exercise group (Effect size = −0.89. 95% CI −0.99 to −0.80) and in the group which conducted exercise both at-home and at another setting (−0.73. −0.86 to −0.59). Similarly, functional limitation also decreased in both groups (−0.75. −0.91 to −0.60, and −0.70, −0.92 to −0.48, respectively). Relaxation and postural exercise seemed to be ineffective in decreasing pain intensity, whereas trunk, pelvic or leg stretching decreased pain intensity. Yoga improved functional limitation. Supervised training was the most effective method to improve pain intensity. Insufficient data precluded robust conclusions around the duration and frequency of the sessions and program. Conclusion: Home-based exercise training improved pain intensity and functional limitation parameters in LBP.

Keywords: musculoskeletal disorders; lumbalgia; physical activity; prevention; public health

1. Introduction

Low back pain (LBP) is a major public health issue [1,2], commonly described as pain and discomfort, localized below the costal margin and above the inferior gluteal folds, with or without leg pain [3]. Non-specific LBP is defined as LBP not attributed to a recognizable known specific pathology (e.g., inflammatory, tumoral or infectious process) [3]. Pain intensity and functional limitation are major factors in the prognosis of LBP [4]. Exercise therapy is recommended as first-line treatment [2,5,6]. However, the availability of centers for exercise therapy is lacking in the public health system [2,5]. Considering that the home is the most accessible setting [7], home-based exercise may be of particular interest in the management of LBP [7]. To facilitate the comparison of results between studies and to enable the pooling of data in this systematic review, an international multidisciplinary panel recommended, inter alia, pain intensity and functional limitation as core outcomes [3,8–10]. To our knowledge, to date no meta-analysis has assessed the effects of home-based exercise on pain intensity and functional limitation in LBP. At exercise centers some meta-analyses have suggested that a reduction in the risk of LBP could be achieved via various aerobic and resistance exercise training sessions, pilates and stabilization/motor control [11]. However, a European recommendation highlighted the absence of a clear consensus on the best exercise therapy [2]. For home-based exercise in LBP, data are scarce. Although data are lacking around the effectiveness of home-based exercise, as well as data regarding the optimal intensity, frequency and duration of exercise, supervised exercise seemed to produce the best outcomes in exercise centers [2,5]. Individual characteristics, such as age, sex, or education [3], may also influence responses to the home-based exercise program.

In light of this, we conducted a systematic review and a meta-analysis in order to assess the effect of home-based exercise on the pain intensity and functional limitations in LBP. The secondary objectives of this study were to assess the influence of the types and modalities of home-based exercise, and to investigate the putative influence of sociodemographic and characteristics of patients in the treatment of LBP.

2. Materials and Methods

2.1. Literature Search

We reviewed all studies reporting on the effect of home-based exercise training on nonspecific LBP (i.e., LBP not consecutive to a specific pathology such as inflammatory, tumoral or infectious process) [3]. Animal studies were excluded. The PubMed, Cochrane Library, Embase and ScienceDirect databases were searched until 20 April 2021, with the following keywords: low back pain AND (exercise OR physical) AND home (details of the specific search strategy used within each database are available in Appendix A). The search was not limited to specific years and no language restrictions were applied. To be included, articles were required to simultaneously meet the five following inclusion criteria: (1) randomized controlled trials (RCTs); (2) population $\geq$ 16 years old; (3) with non-specific LBP (chronic or not); (4) evaluation of at least one of our main clinically relevant outcome (i.e., pain intensity or functional limitation); and (5) studies including home-based exercise therapy. Home exercise programs are defined as a series of exercises that patients complete at home for therapeutic gains or to improve physical capacity. Home exercises are designed to be practical, accessible and feasible so that patients can maximize efforts. We excluded those studies which assessed patients with specific LBP (i.e., caused by a specific cause such as pregnancy or pathological entities). Conference papers, congress, and seminars were excluded. In addition, the reference lists of all publications meeting the inclusion criteria were manually searched in order to identify any further studies not found through the electronic search. Ancestry searches were also completed on previous reviews to locate other potentially eligible primary studies. Two authors (Chloé Quentin and Reza Bagheri) conducted the literature searches, reviewed the abstracts and, based on the selection criteria, determined the suitability of the articles for inclusion, and extracted the data. When necessary, disagreements were resolved with the inclusion of a third author (Frédéric Dutheil).

2.2. Data Extraction

The data collected included: (1) characteristics of the study, including the first author's name, publication year, country and continent, study design, outcomes of included articles, and number of participants; (2) characteristics of individuals, such as the mean age, sex (percentage of males), weight, height, and body mass index, percentage of smokers and regularly physical active individuals, education and marital status, and duration of complaints; (3) characteristics of the intervention, such as whether the intervention was supervised (totally supervised/partially supervised/not supervised), standardized or individualized (partially or fully), the type of intervention (education, aerobic exercise, stretching, strengthening, relaxation, postural exercise, yoga, other exercises), the frequency and duration of sessions, the duration of the program, and the location of training (home or other setting); and (4) characteristics of our main outcomes, such as the type of assessment of pain intensity and functional limitation, and measures (mean and standard deviation) before and after the training.

2.3. Quality of Assessment

We used the Scottish Intercollegiate Guidelines Network (SIGN) criteria designed for randomized clinical trials to check the quality of included articles. The checklist consists of 10 items. We gave a general quality score for each included study based on main causes of bias. We used 4 possibilities for scoring each item (yes, no, can't say or not applicable) [12].

2.4. Statistical Considerations

We conducted meta–analyses on the effect of LBP exercise on pain intensity and functional limitation. P-values less than 0.05 were considered statistically significant. For the statistical analysis, we used Stata software (version 16, StataCorp, College Station, TX, USA) [13–15]. The main characteristics were synthetized for each study population and reported as a mean ± standard deviation (SD) for continuous variables and number (%) of the categorical variables. First, we conducted random-effect meta-analyses (using the DerSimonian and Laird approach [16,17]) on the effect of home-based exercise for LBP, by comparing levels of pain intensity or functional limitation after the training program versus baseline levels (i.e., before exercise). The results were expressed as effect sizes (ES, standardized mean differences—SMD). ES is a unitless measure centered at zero if pain intensity or functional limitation did not differ between after and before the training program. A negative ES denoted an improvement in the pain intensity or functional limitation of the patient (i.e., decreased levels of pain intensity or functional limitation after exercise compared to before). An ES of −0.8 reflects a large effect, −0.5 a moderate effect, and −0.2 a small effect. Following this, we conducted meta-analyses stratified on: (1) the location of the training program (exclusively home, or home plus another setting); (2) characteristics of intervention, whether it was supervised (totally supervised/partially supervised) or not, and standardized or individualized (partially or fully). We computed the aforementioned meta-analysis using all the measurement time. To verify the strength of our results, we computed sensitivity analyses using only the median time of follow-up and then using only the last time of follow-up. We evaluated heterogeneity in the study results by examining forest plots, confidence intervals (CI) and I-squared (I^2). I^2 values are a common metric used to measure heterogeneity between studies and are easily interpretable. I^2 values range from 0 to 100%, and are considered low at <25%, modest at 25–50%, and high at >50% [18]. For example, a significant heterogeneity could be linked to the characteristics of the studies, such as sociodemographic, or the characteristics of the intervention. We searched for potential publication bias using funnel plots of all the aforementioned meta-analyses, in order to conduct further sensitivity analyses by excluding studies that were not evenly distributed around the base of the funnel. When possible (where there was a sufficient sample size), meta–regressions were proposed in order to study the associations between changes in pain intensity or functional limitation, and clinically relevant parameters such as sociodemographic (age, sex, body mass index, etc.), and the characteristics of the intervention (e.g., type of exercise, supervised or not, standardized or individualized, frequency and duration of sessions, and duration of programs). The results were expressed as regression coefficients and 95% CI.

3. Results

An initial search produced 24,699 possible articles. The removal of duplicates and use of the selection criteria reduced the number of articles reporting the effect of home-based exercise on LBP patients to 33 articles [5,19–49] (Figure 1). All included articles were written in English. The main characteristics of the studies are described in Table 1.

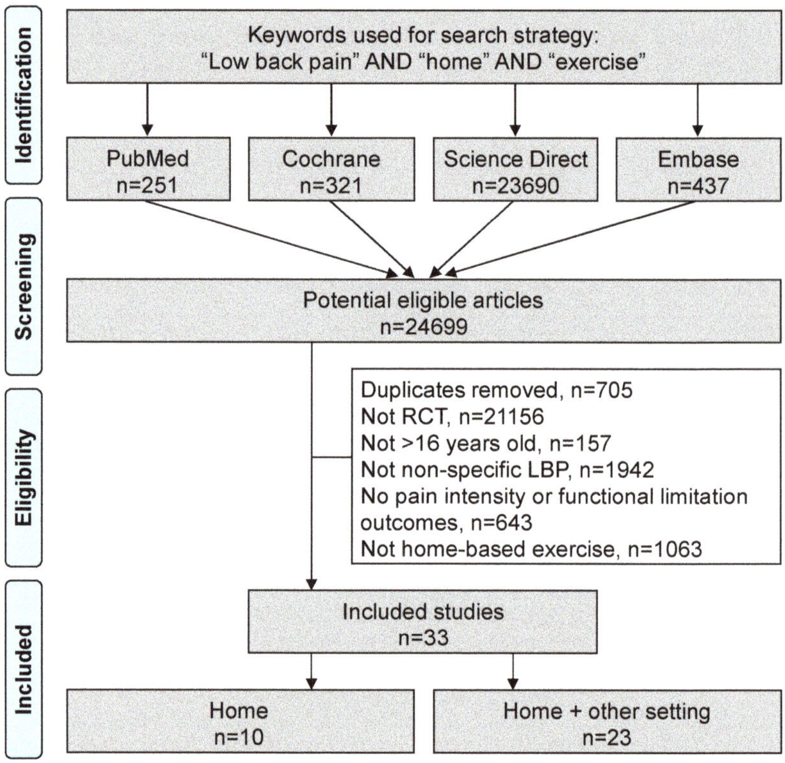

Figure 1. Flow chart.

Table 1. Characteristics of included studies.

Study			*Characteristics of Indivuals*			*Exercise Setting*		*Type of Intervention*								*Volume of*	*Exercise*		*Outco*	*Mes*
Study	**Design**	**Coun-try**	**n gp**	**Age (Mean ± SD)**	**Sex (% Men)**	**Home**	**Other**	**Stren-gth**	**Stret-ching**	**Rela-xation**	**Aero-bic**	**Edu-cation**	**Postu-ral**	**Yoga**	**Other**	**Number of weeks**	**n Session/Week**	**Duration of Session**	**Pain**	**Functionnal Disability**
Alp 2014	RCT	Turkey	24	48 ± 27.5	0	X		X	X							6 weeks	3	45–60 min	VAS	RMDQ
			24	51 ± 39.8	0		X	X	X								7	?		
Ben Salah Frih 2009	RCT	Tuni-sia	54	34.7 ± 1.14	24.1	X	X	X	X							4 weeks	7	30 min	VAS	
			53	36.9 ± 1.29	26.4		X	X	X						Proprioception exercise	4 weeks	3	90 min		
Bernadelli 2020	RCT	Italy	51	50.5 ± 9.7	25.5		X	X	X			X				7 weeks	1	30 min	—	RMDQ
			50	51.9 ± 8.1	16	X			X			X								
Bertozzi 2015	RCT	Italy	20	42.7 ± 8.7	-	X	X		X	X			X			5 weeks	2	60 min	VAS	RMDQ
			20	47.5 ± 7.5	-	X			X	X			X				?	?		
Bronfort 2014	RCT	USA	96	57.1 ± 12	41	X	X	X				X	X			12 weeks	7	?	NPRS	RMDQ
			96	57.7 ± 11.9	32	X		X				X	X							
Bronfort 2011	RCT	USA	100	44.5 ± 11.8	43			X	X		X					12 weeks	2	60 min	NPRS	RMDQ
			100	45.2 ± 10.8	44										Spinal manipulative therapy	12 weeks	2	60 min		
			101	45.6 ± 10.3	41.6	X		X	X			X				12 weeks	7	60 min		
Buttagat 2019	RCT	Thai-land	11	39.7 ± 17.4	27.2	X			X						Thai self-massage	4 weeks	3	?	VAS	ODI
			11	41.3 ± 15.8	9	X						X				?	?	?		
Chaléat-Valayer 2016	RCT	France	171	47.1 ± 8.5	23	X	X	X	X			X	X		rhythmic exercise	2 years	7	10 min	VAS	QBPDS
Descarreaux 2002	RCT	Ca-nada	10	33.1	70	X		X								3 weeks	14	?	VAS	ODI
			20	35.0	35	X		X	X							6 weeks	14	?		
Ewert 2009	RCT	Ger-many	100	37.9 ± 11.6	8.0	X	X	X	X	X	X	X	X		cognitive-behavioral approach/stress control + warm up	13 weeks	2	60 min	WHYMPI	sf36-pcs
			102	41.1 ± 10.8	5.9	X	X	X	X	X	X				Warm up	13 weeks	?	?		

Table 1. *Cont.*

Study			*Characteristics of Indivuals*			*Exercise Setting*		*Type of Intervention*								*Volume of*	*Exercise*		*Outco*	*Mes*
Study	**Design**	**Coun-try**	**n gp**	**Age (Mean ± SD)**	**Sex (% Men)**	**Home**	**Other**	**Stren-gth**	**Stret-ching**	**Rela-xation**	**Aero-bic**	**Edu-cation**	**Postu-ral**	**Yoga**	**Other**	**Number of weeks**	**n Session/ Week**	**Duration of Session**	**Pain**	**Functionnal Disability**
Frost 1998	RCT	En-gland	36	34.2 ± 9.4	-	X	X		X	X	X	X	X			4 weeks	2	?		MODI
			35	38.5 ± 9.3	-	X				X		X	X				?	?		
Garcia 2013	RCT	Brasil	74	54.2 ± 1.57	31.0	X	X		X			X	X			24 weeks	7	240 min	VAS	RMDQ
			74	53.7 ± 1.53	21.6	X	X	X				X	X							
Goode 2018	RCT	En-gland	20	69.6 ± 3.5	95	X		X	X	X	X	X								
			20	69.5 ± 4,0	90	X		X	X	X	X	X			Activity pacing + Cognitive restructuring	12 weeks	?	?		RMDQ
			20	71.9 ± 6.5	95	X			X											
Groessl 2017	RCT	USA	76	53.3 ± 12.7	73	X	X							X	Breathing	12 weeks	7	60 min	BPI	RMDQ
			76	53.6 ± 13.9	75											none	-	-		
Haufe 2017	RCT	Ger-many	112	43.5 ± 9.7	-	X	X	X								20 weeks	3	20 min		
			114	41.9 ± 10.6	-											20 weeks	3	20 min	VAS	ODI
			10	49.9 ± 8.8	80	X	X	X	X		X	X				6 weeks	2	20 min		
Ibrahim 2018	RCT	Nige-ria	10	48.5 ± 14.9	70	X	X	X	X		X	X				6 weeks	2	20 min	NPRS	ODI
			10	50.3 ± 9.1	90	X	X		X		X	X					1	?		
Iversen 2018	RCT	Nor-way	37	43 ± 13	46	X	X	X	X		X	X				12 weeks	4	?		
			37	47 ± 11	41	X	X	X	X	X		X			ball games + body awareness + circle training	12 weeks	4.5	?	NPRS	ODI
			34	45 ± 15	55											none	-	-		
Kanas 2018	RCT	Brasil	17	-	-	X		X	X		X					8 weeks	3	?	NPRS	RMDQ
			13	-	-	X	X	X	X		X									
Kendall 2015	RCT	Ca-nada	40	33	55	X		X			X	X			RTUS	6 weeks	?	?	VAS	ODI
		Aus-tralia	40	41	40	X		X			X	X			RTUS					

Table 1. *Cont.*

Study			*Characteristics of Indivuals*			*Exercise Setting*		*Type of Intervention*								*Volume of*	*Exercise*		*Outco*	*Mes*
Study	**Design**	**Country**	**n gp**	**Age (Mean ± SD)**	**Sex (% Men)**	**Home**	**Other**	**Strength**	**Stretching**	**Relaxation**	**Aerobic**	**Education**	**Postural**	**Yoga**	**Other**	**Number of weeks**	**n Session/ Week**	**Duration of Session**	**Pain**	**Functionnal Disability**
	RCT	Turkey	20	37.1 ± 6.5	21.1	X	X	X	X		X	X					10	?		
Koldaş 2008			20	41.5 ± 8.3	22.2	X	X	X	X			X				6 weeks	7	?	VAS	RMDQ
			20	42.1 ± 9.5	22.2	X		X	X			X					7	?		
	RCT	Finland	29	39.9 ± 8.9	-		X	X							warm-up + Balance + coordination	12 weeks	?	?		
Kuukkanen 2000			29	39.9 ± 7.9	-	X		X								12 weeks	?	?		ODI
			28	39.9 ± 7.9	-											none	-	-		
Kuukkanen 2007	RCT	Finland	28	40 ± 7.9	46.4	X		X	X							12 weeks	7	?	Borg CR-10 scale	ODI
	RCT	England	98	44.1 ± 16.2	-	X										5 weeks	?	?		
Miller 2007			137	43.7 ± 14.8	-	X										5 weeks	?	?		RMDQ
			150	44.9 ± 15.4	-											-	-	-		
Michalsen 2016	RCT	Germany	32	55.5 ± 10.6	-	X	X								Jyoti meditaion	8 weeks	7	25 min	VAS	RMDQ
		Germany	36	54.8 ± 10.6	-	X		X	X									20 min		
Neyaz 2019	RCT	India	35	33	-	X	X	X	X	X		X				6 weeks	7	30–35 min	DVPRS	RMDQ
			35	38	-	X	X		X	X										
Shirado 2010	RCT	Japan	103	42.0 ± 11.6	47.5	X		X	X			X				8 weeks	14	?	VAS	
			98	42.5 ± 12.3	40.8	X									Massage		?			
	RCT	USA	81	72.5 ± 5.6	43.2	X	X	X	X		X	X			Balance exercise + massage		2	45–60 min		
Schulz 2019			80	73.6 ± 5.3	52.5	X	X	X	X		X	X			Balance exercise + massage	12 weeks	1	60 min	NPRS	RMDQ
			80	74.7 ± 5.6	50.0	X		X	X		X	X					1	15 min		
	RCT	USA	127	46.4 ± 10.4	-	X	X			X				X	breathing		?	75 min		
Saper 2017			129	46.4 ± 11,0	-	X	X				X	X				12 weeks	7	60 min	NPRS	RMDQ
			64	44.2 ± 10.8	-							X					?	?		

Table 1. *Cont.*

Study	*Study*		*Characteristics of Indivuals*			*Exercise Setting*		*Type of Intervention*								*Volume of*	*Exercise*		*Outco*	*Mes*
Study	**Design**	**Country**	**n gp**	**Age (Mean ± SD)**	**Sex (% Men)**	**Home**	**Other**	**Strength**	**Stretching**	**Relaxation**	**Aerobic**	**Education**	**Postural**	**Yoga**	**Other**	**Number of weeks**	**n Session/ Week**	**Duration of Session**	**Pain**	**Functionnal Disability**
Sakuma 2012	RCT	Japan	67	32.6 ± 11.5	-	X				X				X		2 weeks	7	9 min	VAS	
			31	35.8 ± 13												none	-	-		
Tottoli 2019	RCT	Brazil	72	-			X	X	X				X			6 weeks	2	50 min	VAS	BQBPSDQ
			72			X		X	X				X		Warm up					
Wajswelner 2012	RCT	Australia	44	49.3 ± 14.1		X	X						X		Breathing	24 weeks		60 min	NPRS	
			43	48.9 ± 16.4		X	X	X	X						Swiss ball					
Winter 2015	RCT	Australia	12	45.9 ± 13.3		X			X								5			
			13	48.9 ± 7.2		X			X							6 weeks	5		NPRS	MODI
			13	38.3 ± 12.8		X		X	X								3			
Zadro 2019	RCT	Australia	30	68.8 ± 5.5	40.0	X		X	X		X					8 weeks	3	60 min	NPRS	RMDQ
			30	67.8 ± 6	56.6											none				

3.1. Quality of Assessment

Overall, the methodological quality of the included studies was good, with an average score of 75% for items meeting the criteria of the SIGN checklist, ranging from 40% [5,44] to from 90% [32,42,50]. All studies failed to include a blind assessment. All studies reported achieving ethical approval (Figure 2).

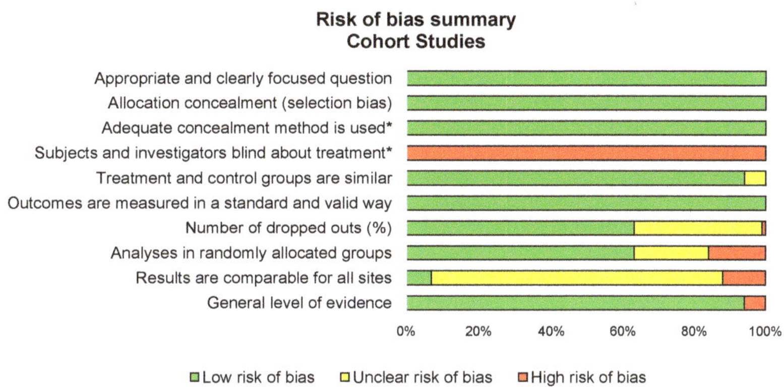

Figure 2. Summary of methodological qualities of included studies using the SIGN checklist.

3.2. Study Designs of Included Articles

The included studies were published between 1998 [28] and 2020 [51] and conducted in various geographic locations, with all continents represented (Europe [19,25,27,28,30,32,34,36–39,48,51], North America [22,26,31,35,42,50], South America [5,29,44], Asia [23,24,40,41,43], Africa [20], and Oceania [35,45,46,49]. All the studies were RCT. Thirteen studies were monocentric [5,19–24,29–31,33,42] and eleven were multicentric [25,27,32,35,37,38,40,41,43,50,51]. Studies were single-blind [19,23–25,28–30,32,33,35] or not blind randomized trials [5,22,27,47,51].

3.3. Inclusion and Exclusion Criteria of Included Studies

All the included studies included adults, except one study that also included participants over 16 years old [39]. Globally, participants were recruited using electronic, newspaper or local advertisements [23,25,26,33,35,42,44–47,49,50] or from consultations with specialists or general practitioners [5,19,20,25,40,41,49]. Some studies included specific populations, such as healthcare workers [25,27,51], sedentary older adults [30,42], poultry industry slaughterers [21], only women [19], or individuals from a rural community [33].

3.4. Population

The sample sizes of the studies ranged from 22 [23] to 385 participants [39]. We included a total of 9,588 LBP patients. The ages of the participants were reported in all except two of the studies [5,44]. The mean age of LBP patients undertaking home-based exercise training was 49.3 years (95% CI 45.5 to 52.9), with ages ranging from 32.6 ± 11.5 [43] to 74.7 ± 6.0 [42] years old. Gender was not reported in 14 of the selected studies [5,21,22,24,34,38,40,43,44,46,47,50]. The mean proportion of men was 18% (95% CI 0.15 to 0.20), with the proportion of men in the studies ranging from 0 [19] to 53% [42]. The BMI of participants was reported in 16 of the selected studies. The mean BMI was 29.5 kg/m^2 (95% CI 28.3 to 26.9), with BMIs ranging from 21.5 ± 2.7 [33] to 32.7 ± 7.4 [50]. Other parameters were seldomly reported. The education status of participants was reported in eight studies [27,29,31,33,34,42,49], but degrees were not expressed in the same way across most of the studies. Smoking status was only reported in six studies [25,28,29,42,43,49], with a mean proportion of smokers of 4% (95% CI 2 to 5%), ranging from 0% [47] to 11% [29]. Only three studies mentioned leisure physical activity [20,29,43], with a mean proportion of regularly active patients of 10% (95% CI 6 to 15%), and proportions varying from 3 [43] to 18% [29].

3.5. Intervention: Characteristics of Exercise

3.5.1. Type of Exercise

Nearly all of the selected studies (26 studies i.e., 79%) used strength-based exercises, mostly combined with other exercise [5,19,20,22–25,27,29,30,33–38,40–42,44–46,48,51]. Only two of the studies did not [26,32]. Education was included in sixteen of the studies [22,23,25,27–30,32–36,40–42,47,50,51], stretching in twenty-three studies [5,19–21,23–37,40–42,44–49,51], aerobic exercise in thirteen studies [5,22,24,27,28,30,32–36,42,49,50], postural exercise in eight studies [21,25,27–29,44,45,47], relaxation in eight studies [21,27,28,30,34,40,43,50], and yoga in four studies [26,29,30,52]. Other exercises were only occasionally reported, including Thai self-massage [23], spinal manipulative therapy [22], stress-controlling techniques and a behavioral approach [27], Jyoti meditation [48], breathing [31,50], and ball games associated with body awareness and circle training [34].

3.5.2. Duration of Intervention

The duration of the invention was reported in all of the selected studies, with the programs lasting an average of 11.4 weeks, and varying from 2 weeks [43] to 2 years [25].

3.5.3. Frequency and Duration of Sessions

On average, the studies reported 4.8 sessions per week—ranging from 1 [42,51] to 14 [26] sessions per week—with each session lasting 63 min—ranging from 9 [43] to 240 min [29]. The frequency of sessions was not reported in four of selected studies [30,35,39,45], and the duration of sessions was not reported in in thirteen studies [5,23,28,30,34–39,41,46,47].

3.5.4. Standardization

Standardization was not reported in seven of the selected studies [23,30,35,39,41,47,51]. The exercises were standardized in twenty of the studies [5,19,21,25,28,29,31–37,40,43,45,46,48–50], partially standardized in six of the studies [27,29,38,42,44,47], and individualized in five of the studies [20,24,26,27,45].

3.5.5. Supervision

Supervision was reported in all except two of the studies [23,34]. Exercises were fully supervised in eight studies [19,20,33,39,44,45,47,51], partially supervised in twenty-one studies [5,20,21,25,27–34,36,37,40–42,44,47,48,50], and not supervised in ten studies [19,28,35,36,42,43,46,48,49,51].

3.5.6. Location

The location was reported in all of the studies. Exercises took place exclusively at home in twenty-one studies [5,19,21–23,26,28,30,35–38,41–44,46–49,51] and both at home and another setting (workplace, health center or training center) in eighteen studies [5,20,21,25,27–29,31–34,36,40,42,45,47,48,50].

3.6. Outcomes-Pain Intensity and Functional Limitations

Before and after the physical exercise program, the pain intensity was assessed in 27 studies [5,19–27,29,31–37,40–50] and functional limitations in 28 studies [5,19,21–35,37–40,42,44,46–51]. Participants' pain intensity was evaluated in most studies using a visual analogue scale [19–21,23,25,26,29,32,35,36,41,43,44,48], or a numeric pain rating scale [5,22,24,33,34,42,45–47,49,50]. Other pain assessments were made using the West Haven-Yale Multidimensional Pain Inventory [27], the Brief Pain Inventory [31], the Borg CR-10 scale [37], and the Defense and Veterans Pain Rating Scale [40]. Functional disability was evaluated in most of the studies using the Roland-Morris Disability [5,19,21,22,29–31,36,39,40,42,47–51], the Oswestry Disability Index [23,32–35,37,38] or its modified version [24,28,46], except in three studies that used the French version of the Quebec back pain disability scale [25], the Short Form Health Survey physical component scale [27], and the Quebec Back Pain Disability Scale Questionnaire [44]. Assessments of the outcomes were made at a median time of three months after the beginning of the exercise program, ranging from half a month [43] to five years [37]. Some studies reported outcomes at different times during the exercise protocol [5,21,22,25,27–29,31,36,37,40,42,43,45,47,48,51].

3.7. Meta-Analysis on the Effect of Home-Based Exercise

Overall, home-based exercise training decreased pain intensity (effect size = −0.89, 95% CI −0.99 to −0.80) and decreased functional limitation (−0.73%, −0.86 to −0.59) for participants, regardless of an exclusive at-home location or not. Pain intensity decreased in a similar proportion between an exclusive at-home setting (−0.97, −1.14 to −0.79) and a combination of exercises at home and in another setting (−0.89, −0.96 to −0.74). Similarly, functional limitation also improved in both settings (−0.69, −0.93 to −0.46; and −0.93, −1.34 to −0.52, respectively) (Figures 3 and 4, and Appendix B).

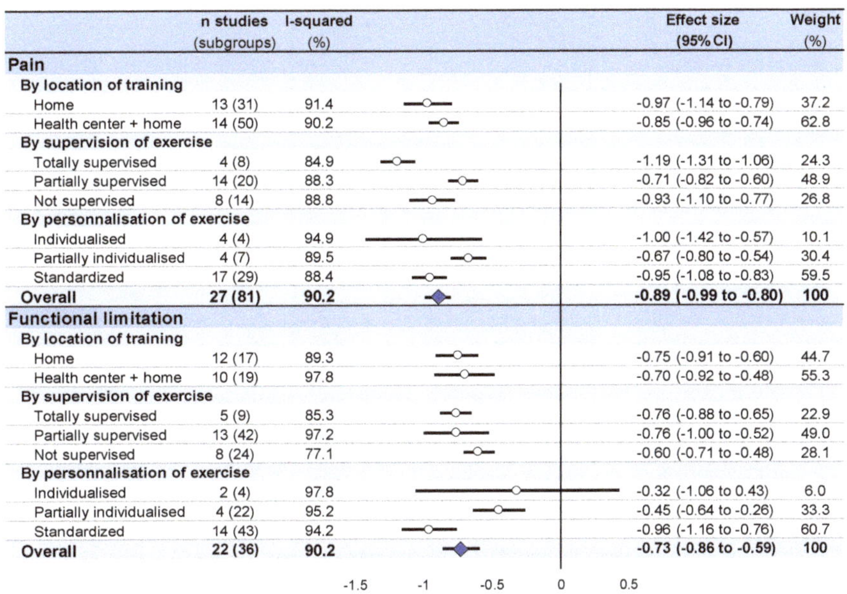

Figure 3. Summary of meta-analysis on the effect of home-based exercise on pain intensity and functional limitation, stratified by setting (exclusive home-based training versus home-based and other setting), supervision, and standardization of training.

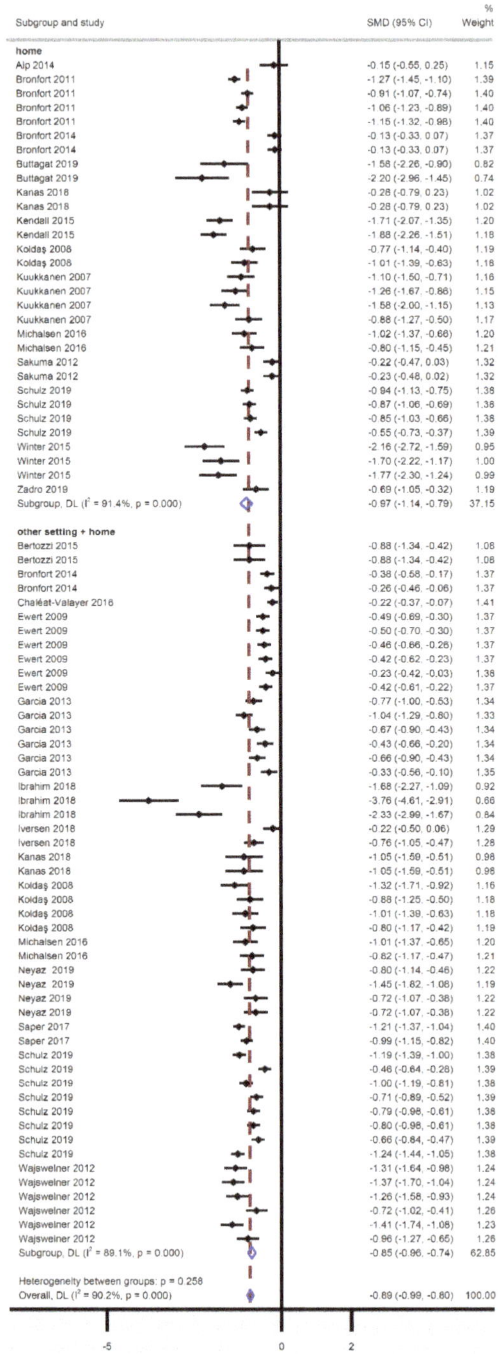

Figure 4. Meta-analysis on the effect of home-based exercise on pain intensity, stratified by setting (exclusive home-based training versus home-based and other setting).

3.8. Stratification by Characteristics of Training

Stratification by the supervision of training demonstrated a decrease in pain intensity and an improvement of functional limitation regardless of the characteristics of the training. For pain intensity, a totally supervised training seemed to be the most effective in terms of decreasing the pain intensity (effect size = −1.19, 95% CI −1.31 to −1.06; versus −0.71, −0.82 to −1.06 for partially supervised training and −0.93; −1.1 to −0.77 for unsupervised training). For functional limitation, improvement seemed similar regardless of the level of supervision of training (−0.76, −0.88 to −0.65 for totally supervised training; −0.76; −1.0 to −0.52 for partially supervised training, and −0.60, −0.71 to −0.48 for unsupervised training). Concerning the standardization of training, both for pain intensity and functional limitation, a standardized protocol seemed to produce the greatest benefits (−0.95, −1.31 to −1.08 for pain intensity, and −0.96, −1.16 to −0.76 for functional limitation), whereas a partially individualized program had the lowest benefits (−0.67, −0.8 to −0.54 for pain intensity and −0.45, −0.64 to −0.26 for functional limitation). An individualized program demonstrated very wide confidence intervals for pain intensity (−1.00, −1.42 to −0.57) and was not significant for functional limitation (−0.32; −1.06 to 0.43) (Figure 4).

3.9. Metaregressions

There was no difference in the improvement of pain intensity and functional limitation depending on the setting (exclusive home-based exercise vs. home-based exercise combined with exercises in a center, $p = 0.66$). Totally supervised exercise produced better benefits regarding pain intensity when compared with partially individualized (coefficient −0.50, 95% CI −0.74 to −0.25) or not supervised (−0.28, −0.58 to −0.01) programs, with no influence on functional limitation. Standardized protocols had better benefits when compared to partially individualized training in relation to both pain intensity (−0.29, −0.55 to −0.03) and functional limitation (−0.51, −0.84 to −0.18). Despite most studies mixing different types of exercise, metaregressions demonstrated an improvement in pain intensity for pelvic (−0.63, −0.92 to −0.30), leg (−0.27, −0.52 to −0.02) and trunk (−0.36, −0.65 to −0.07) stretching. In contrast, pain intensity seemed to be exacerbated by relaxation (0.38, 0.12 to 0.65). Postural exercises seemed to have a deleterious effect on both pain intensity (0.35, 0.11 to 0.59) and functional limitation (0.24, 0.00 to 0.68). Yoga improved functional limitation (−0.94, −1.67 to −0.2). The volume of training (the frequency and duration of sessions, and duration of programs) was not associated with an improvement in pain intensity or functional limitation. A longer follow-up was associated with a decrease in pain intensity (effect size = −0.03. 95% CI −0.05 to −0.01) and functional limitation (−0.06; −0.10 to −0.02). Men were more likely to improve pain intensity (−0.16, −0.23 to −0.08, per 10% men) and functional limitation (−0.23, −0.30 to −0.16, per 10%men) to a greater extent than women. The training program was less effective for decreasing pain intensity in people with a higher body mass (1.01, 0.19 to 1.85, per 10 $kg{\cdot}m^2$). Other parameters, such as age, education status, smoking, or duration of symptoms, were not associated with decreased pain intensity and functional limitation (Figure 5).

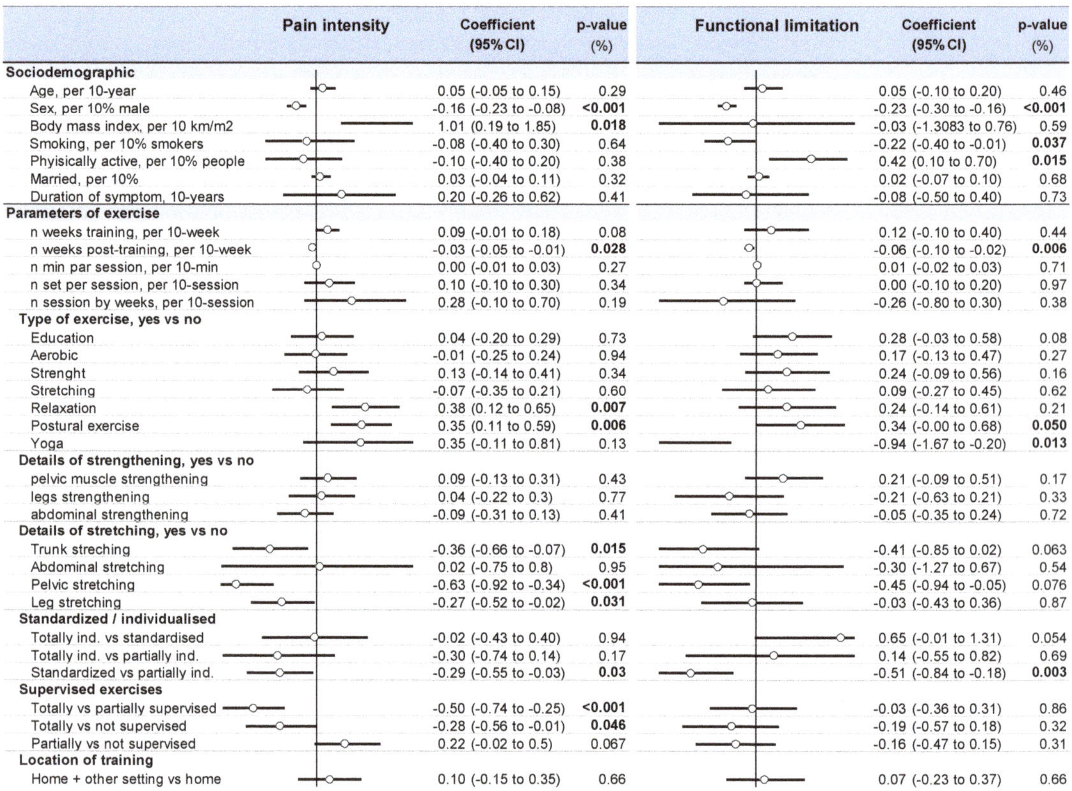

Figure 5. Metaregressions i.e., putative influencing variables on pain intensity and functional limitation following home-based exercise in LBP.

3.10. Sensitivity Analyses

We computed sensitivity analyses using only the median time of follow-up (i.e., three months (Appendix C for meta-analysis and Appendix D for metaregressions)), and the last time of follow-up (Appendices E and F). All the results were similar. Because of the wide heterogeneity of the selected studies (all I-squared are >80%), we failed to reperform all the aforementioned meta–analyses after the exclusion of studies that were not evenly distributed around the base of the funnel (Appendix G).

4. Discussion

The main findings of this research were that home-based exercise training improved pain intensity and functional limitation in LBP patients, regardless the modality of exercises. Supervised training and standardized training improved pain intensity to the greatest extent, independently of the influence of the duration and frequency of the training. Training was less beneficial for women and for patients with a high body mass index.

4.1. The Benefits of Home Exercise Training on LBP Patients

This study is the first systematic review and meta-analysis of studies investigating the effectiveness of home exercise programs on pain and functional limitation in patients with LBP. Structured center-based programs have the advantage that the amount and quality of the training can be controlled and supervised, but these programs are expensive, limiting their implementation possibilities [7]. Moreover, many adults experience barriers to attending such programs, including a lack of affinity with the culture of fitness centers [7].

Home-based exercises are especially valuable because they require fewer resources [52] and less time from health institutions and health practitioners [7]. Our meta-analysis showed strong evidence that physical exercise training can take place at home to improve LBP, even though we found no studies comparing the same training program between home and another setting. Studies comparing home-based exercise to a control group without exercise [25,31,38,43] showed an improvement in pain intensity and functional limitation. Thus, home-based exercise training could be a cost-effective intervention in the treatment of LBP. If multiple short bouts of moderate-intensity physical exercise produce significant training effects [53], learning to integrate physical activity into daily life can become a main goal in the treatment of LBP. Moreover, home-based exercise also improves other comorbidities, such as knee osteoarthritis [54], obesity [55], depression [56], gait speed in people with Parkinson's disease [57], chronic obstructive pulmonary disease [58] and reduces the risk of cardiovascular mortality [59].

4.2. Which Type of Exercise Training?

There is overwhelming evidence that regular physical activity is associated with reduced LBP [2–4,6,52,60]. The most appropriate exercise intervention is still unknown. Opinions differ over the optimal exercise modalities used to treat LBP. The "active ingredient" of exercise programs is largely unknown, although various exercise options are available [2,60,61]. Considerably more research is required in order to allow for the development and promotion of a wider variety of low cost, but effective exercise programs [2,3]. Our metaregression demonstrated the benefits of pelvic, leg and trunk stretching, in line with the literature [53]. However, drawing any firm conclusions on the best type of exercise is impossible because most studies integrated strength and aerobic training, precluding further analysis (e.g., there was a lack of reference groups that omitted either strength or aerobic training). The predominance of strength in the selected studies is due to the high level of proof of its efficacy in the treatment of LBP [11,53], whereas the benefits aerobic exercises are more under debate [3]. Easily-performed exercises produced noticeable benefits and supported adherence to home-based exercise programs [62]. While aerobic training was easily achievable at home [59], strength training may require more supervision, at least at the beginning [63]. However, strength training is still achievable at home in a wide range of pathologies [52,57]. Despite conflicting results in the literature [2,60] relaxation and postural exercise seemed ineffective in reducing LBP, as well as education alone [60,64]. Similarly to center-based exercise [60], we found that yoga improved functional limitation, as previous studies also showed. Yoga usually combines a wide variety of exercises channeled towards improving strength and flexibility [31], which may explain its positive effects on reducing LBP [3]. Finally, considering the high impact of long-term adherence to exercise [62], the appropriateness of exercise programs may be best determined by the preferences of both the patient [6] and therapist [2].

4.3. Supervision, Standardization, Frequency, and Duration of Exercise Training

Several authors claimed that supervised exercise therapy had proven to be effective in reducing pain and improving functional performance in the treatment of patients with non-specific LBP [2,5,6,29,60] whereas others showed that supervision did not significantly influence final outcomes [5,22,42] but did enhance participants' satisfaction with care [42]. We showed that the best improvements in LBP were achieved through supervised exercise training. It is important to note that all the home-based exercise programs were prescribed by a physiotherapist or health professional with a degree-level qualification in exercise prescription. The majority of the home-based exercise programs in our review incorporated partial supervision [5,20,21,25,27–34,36,37,40–42,44,47,48,50], (i.e., the use of a variety of methods, including home visits by the therapist, occasional group-based sessions at a center or telephones calls). Some publications suggested that external sources of reinforcement, like monitoring, may serve to influence physical behavior [65]. In addition, applying supervised home-based exercise is possible to achieve in many ways, helping to optimize effectiveness of the training [66]. Importantly, our results were in favor of standardized exercise compared to individualized exercise, which may be discordant with the literature based on training in centers [2,26]. This may be explained by the fact that easily-performed standardized exercises can promote a better adherence [62], and could be more in line with home exercise, whereas individualized exercise may be more in line with practice in a center. Lastly, we failed to demonstrate an influence of the volume of exercise on reducing LBP, despite a strong dose-response relationship between physical activity and its overall benefits [2,3]. The absence of such a significant influence on our study may be due to the wide variety of exercise interventions available, and the inconsistency of the intensity and duration of exercise [3].

4.4. Predictors of Pain Intensity Improvements

Although no gender differences were found in relation to pain improvement after exercise in most publication [53,60], our study found strong evidence that males with LBP benefited the most from exercise training. Even if the included studies did not report on observance of exercise, women may lack the time to engage in a daily routine of training [67,68]. Fractionalization of an exercise bout into multiple bouts spread across the day may produce greater benefits and allow for greater adherence [69]. Interestingly, some studies also reported a higher prevalence of LBP in women [1,70]. In our meta regressions, age was not associated with pain improvement. This suggests that home-based exercise, even late in life, can be effective [71]. We also demonstrated that the benefits of exercise were less effective in individuals with a higher body mass index, in line with the literature [53]. Furthermore, individuals using medication, those with no heavy physical demands at work and individual recovery expectations are important parameters influencing the prognostics of LBP [1,4,53], although this was seldomly mentioned in the studies included in our review.

4.5. Limitations

Our study has some limitations. All studies were randomized and patients were not blinded to the interventions. Several biases could have been introduced via the literature search and selection procedure. We conducted the meta-analyses on only published articles, therefore they were theoretically exposed to publication bias. Meta-analyses also inherit the limitations of the individual studies of which they are composed. The availability of some individual characteristics limits the ability to assess all potential treatment effect. Only 11 studies [22–24,26,30,35,39,43,46,49,51] had groups exercising only at home. Hence, comparisons between the efficacy of home-based training versus training in a center cannot reflect a high level of proof—even if we only included randomized trials. Similarly, the lack of studies using a control group without exercise precluded further comparisons. Another major limitation of our meta-analysis is the lack of data on physical activity levels, as well as on medications used. Additionally, the heterogeneity between the study protocols and evaluation may have impacted the results. Some short time-frames (two weeks [43]) may also have been too short for a therapeutic effect. Some studies included targeted population [21,25,27,30,31,33,43,51], however the large sample size of over 10,000 individuals of all ages and categories promotes the generalizability of our results. Even if the weight of studies requires careful thought, because some studies had several measurement interval times and training duration, sensitivity analyses based on median or last time of follow-up time demonstrated similar results. Moreover, our method had the advantage of avoiding selection bias [72].

5. Conclusions

From the literature, it is concluded that home training can be successful if the training is done at home with friends from a community group taking part. Home-based exercise training improved pain intensity and functional limitation parameters in participants experiencing LBP. Supervised training and a standardized program seemed beneficial, although insufficient data precluded drawing any robust conclusions around the duration and frequency of sessions. Further dedicated randomized controlled trials in which information about the type and characteristics of home-based exercise are included are warranted.

Author Contributions: Conceptualization, C.Q. and F.D.; methodology, J.-B.B.-M. and F.D.; software and analysis, F.D.; validation, all authors; writing—original draft preparation, C.Q. and F.D.; writing—review and editing, all authors; supervision, F.D.; funding acquisition, N/A. All authors have read and agreed to the published version of the manuscript.

Funding: This research received no external funding.

Institutional Review Board Statement: Not applicable.

Informed Consent Statement: Not applicable.

Data Availability Statement: All available data are included in this article.

Conflicts of Interest: The authors declare no conflict of interest.

Appendix A

Details for the search strategy used within each database

Pubmed

("low back pain"[MH] OR "low back pain*"[TW] OR "lumbago"[TW] OR "low back ache"[TW] OR "lower back pain*"[TW] OR "pain, low back"[TW] OR "low backache"[TW] OR "back pain, low"[TW] OR "back pain, lower"[TW] OR "Low back syndrome"[TW] OR "Lumbalgia"[TW] OR "Lumbar pain"[TW] OR "LBP"[TW])

AND

("exercise"[MH] OR "exercise therapy"[MH] OR "exercise movement techniques"[MH] OR exercise[TW] OR exercises[TW] OR "activity physical"[TW] OR "physical activities"[TW] OR "physical activity"[TW] OR "sports"[TW] OR "sport" [TW])

AND

(Homes[TW] OR "housing"[TW]OR Home[TW] OR domiciliary[TW])

Embase

('low back pain'/exp OR 'low back pain *': ti,ab,kw OR lumbago:ti,ab,kw OR 'low back ache': ti,ab,kw OR 'lower back pain *': ti,ab,kw OR 'pain, low back': ti,ab,kw OR 'low backache': ti,ab,kw OR 'back pain, low': ti,ab,kw OR 'back pain, lower': ti,ab,kw OR 'low back syndrome': ti,ab,kw OR lumbalgia:ti,ab,kw OR 'lumbar pain': ti,ab,kw OR lbp:ti,ab,kw

AND

('exercise'/exp OR 'kinesiotherapy'/exp OR 'physical activity'/exp OR 'sport'/exp OR exercise *: ti,ab,kw OR 'exercise therapy': ti,ab,kw OR 'exercise movement techniques': ti,ab,kw OR 'activity physical': ti,ab,kw OR 'physical activities': ti,ab,kw OR 'physical activity': ti,ab,kw OR sports:ti,ab,kw OR sport:ti,ab,kw)

AND

(homes:ti,ab,kw OR home:ti,ab,kw OR housing:ti,ab,kw OR domiciliary:ti,ab,kw)

Cochrane

("low back pain" OR "low back pain*" OR "lumbago" OR "low back ache" OR "lower back pain*" OR "pain, low back" OR "low backache" OR "back pain, low" OR "back pain, lower" OR "Low back syndrome" OR "Lumbalgia" OR "Lumbar pain" OR "LBP")

AND ("exercise" OR "exercise therapy" OR "exercise movement techniques" OR exercise OR exercises OR "activity physical" OR "physical activities" OR "physical activity" OR "sports" OR "sport")

AND (Homes OR "housing" OR Home OR domiciliary)

Science-Direct

"low back pain" AND "exercise" AND "home".

Appendix B

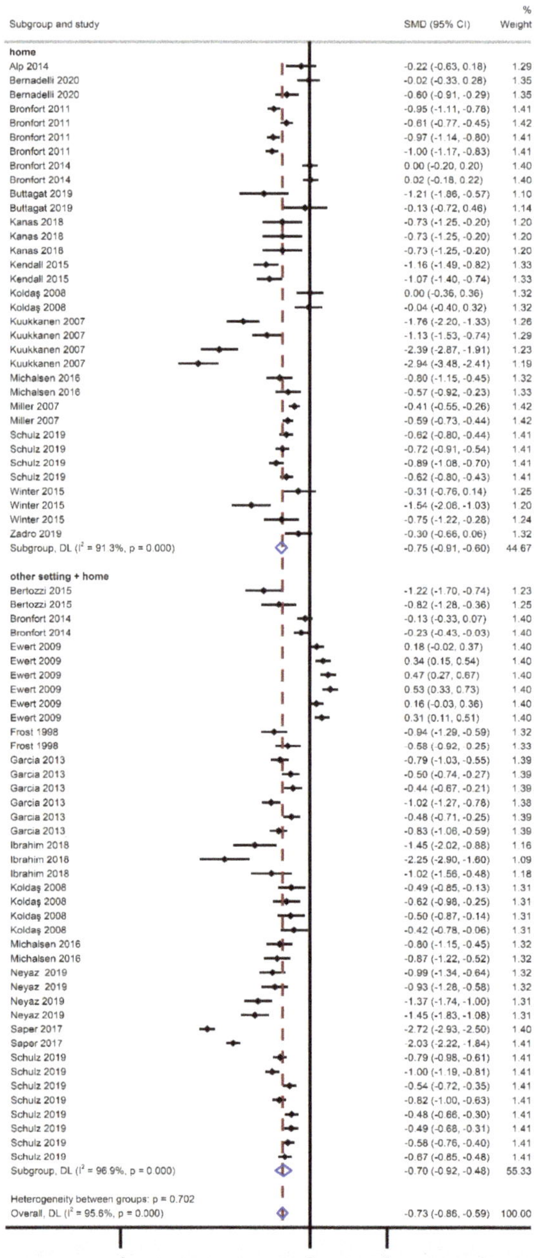

Figure A1. Meta-analysis on the effect of home-based exercise on functional limitation, stratified by setting (exclusive home-based training versus home-based and other setting).

Appendix C

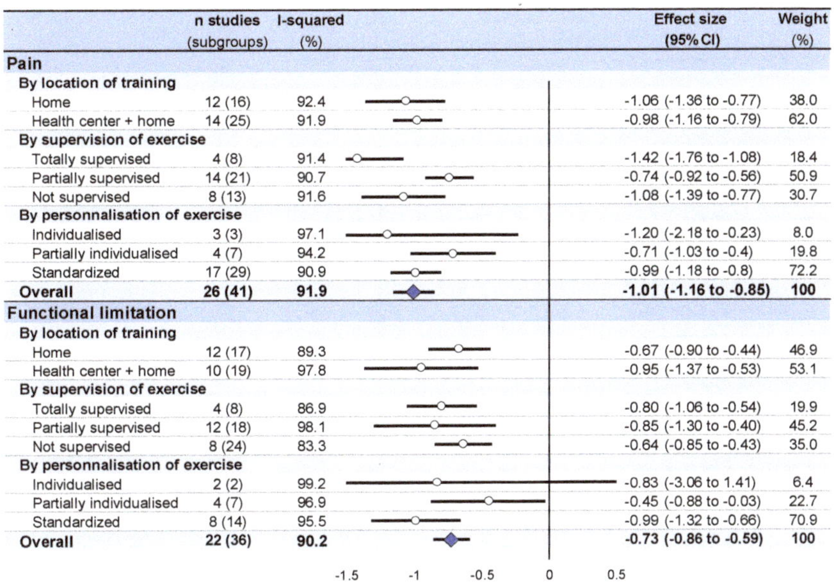

Figure A2. Summary of meta-analysis on the effect of home-based exercise on pain intensity and functional limitation, stratified by setting (exclusive home-based training versus home-based and other settings), supervision, and standardization of training), using only the median time of follow-up.

Appendix D

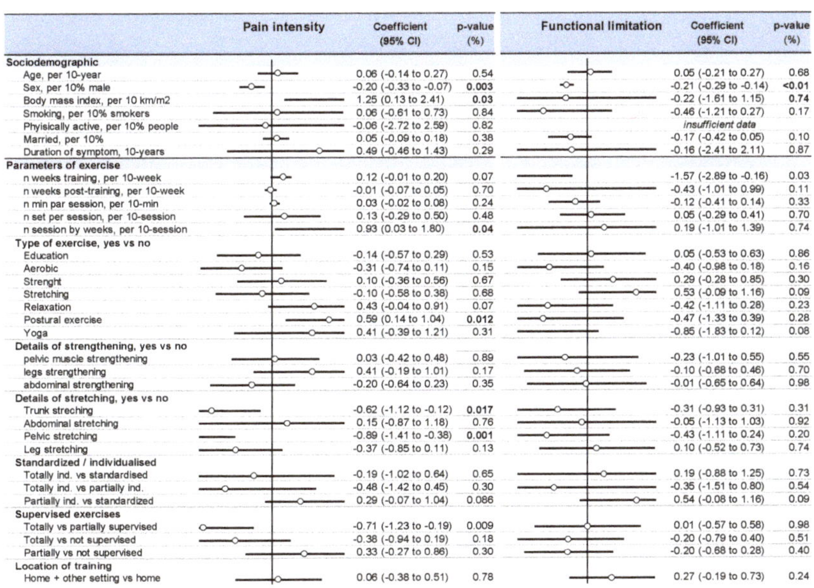

Figure A3. Metaregressions (i.e., putative influencing variables on pain intensity and functional limitation following home-based exercise in LBP), using only the median time of follow-up.

Appendix E

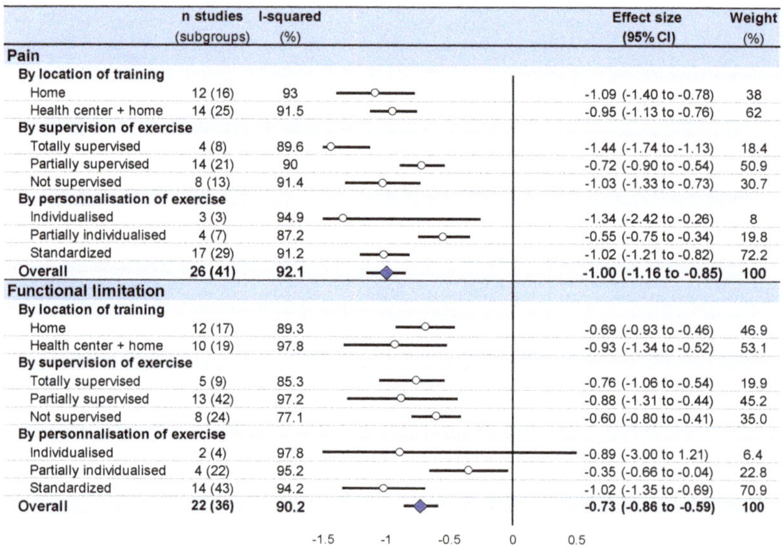

Figure A4. Summary of meta-analysis on the effect of home-based exercise on pain intensity and functional limitation, stratified by setting (exclusive home-based training versus home-based and other settings, supervision, and standardization of training), using only the last time of follow-up.

Appendix F

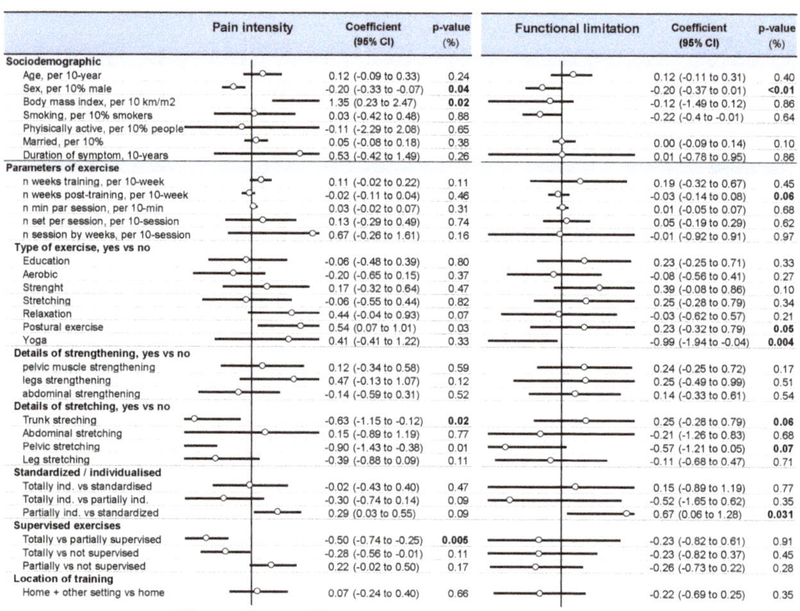

Figure A5. Metaregressions (i.e., putative influencing variables on pain intensity and functional limitation following home-based exercise in LBP), using only the last time of follow-up.

Appendix G

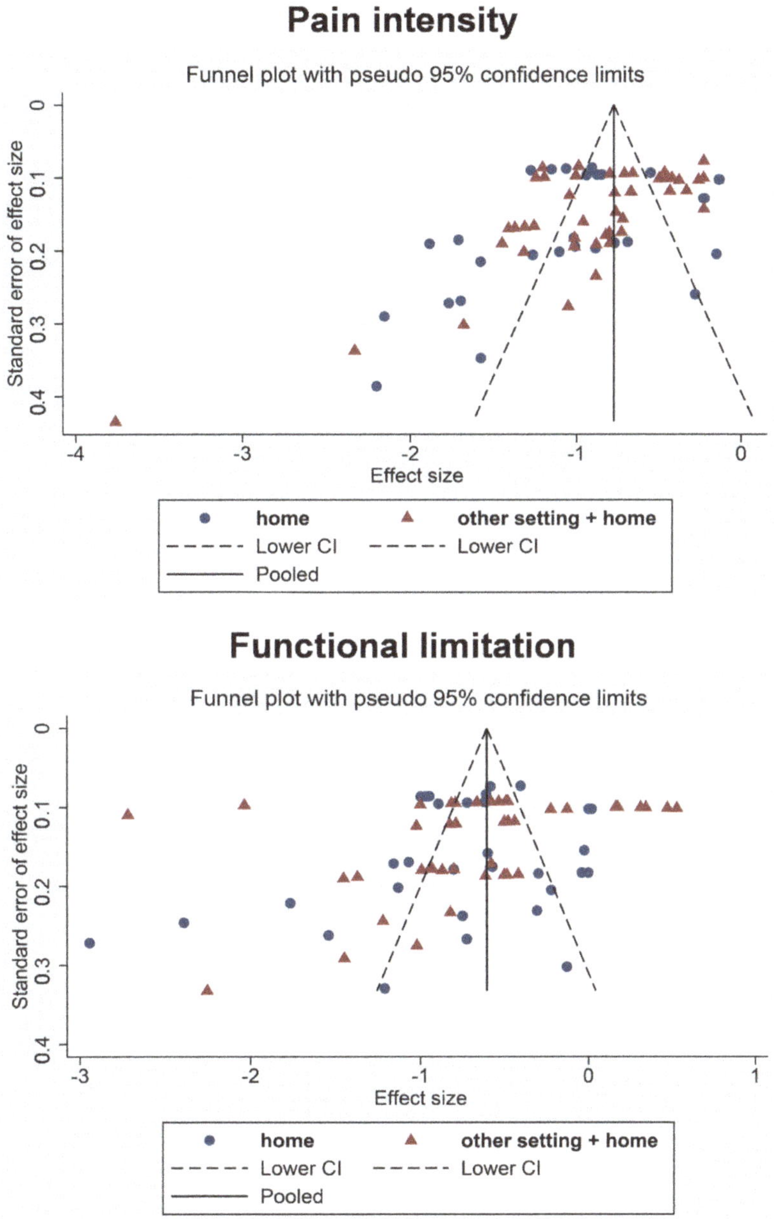

Figure A6. Funnel plot (meta-funnel) for effect size of home-based exercise on pain intensity and functional limitation, in low back pain. Each dot represents a single study, with its corresponding effect size (x-axis) and its associated standard error of the effect estimate (y-axis). Large high-powered studies are placed towards the top, and smaller low-powered studies towards the bottom. The plot should ideally resemble a pyramid or inverted funnel, with scatter due to sampling variation. Studies outside the funnel plot are likely to present bias.

References

1. Hoy, D.; Brooks, P.; Blyth, F.; Buchbinder, R. The Epidemiology of low back pain. *Best Pract. Res. Clin. Rheumatol.* **2010**, *24*, 769–781. [CrossRef]
2. Airaksinen, O.; Brox, J.I.; Cedraschi, C.; Hildebrandt, J.; Klaber-Moffett, J.; Kovacs, F.; Mannion, A.F.; Reis, S.; Staal, B.; Ursin, H.; et al. Chapter 4 European guidelines for the management of chronic nonspecific low back pain. *Eur. Spine J.* **2006**, *15*, s192–s300. [CrossRef]
3. Searle, A.; Spink, M.; Ho, A.; Chuter, V. Exercise interventions for the treatment of chronic low back pain: A systematic review and meta-analysis of randomised controlled trials. *Clin. Rehabil.* **2015**, *29*, 1155–1167. [CrossRef]
4. Hayden, J.A.; Wilson, M.N.; Riley, R.D.; Iles, R.; Pincus, T.; Ogilvie, R. Individual recovery expectations and prognosis of outcomes in non-specific low back pain: Prognostic factor review. *Cochrane Database Syst. Rev.* **2019**. [CrossRef]
5. Kanas, M.; Faria, R.S.; Salles, L.G.; Sorpreso, I.C.E.; Martins, D.; Da Cunha, R.A.; Wajchenberg, M. Home-based exercise therapy for treating non-specific chronic low back pain. *Rev. Assoc. Médica Bras.* **2018**, *64*, 824–831. [CrossRef]
6. Balagué, F.; Mannion, A.F.; Pellise, F.; Cedraschi, C. Non-specific low back pain. *Lancet* **2012**, *379*, 482–491. [CrossRef]
7. Opdenacker, J.; Delecluse, C.; Boen, F. A 2-Year Follow-Up of a Lifestyle Physical Activity Versus a Structured Exercise Intervention in Older Adults. *J. Am. Geriatr. Soc.* **2011**, *59*, 1602–1611. [CrossRef] [PubMed]
8. Chiarotto, A.; Boers, M.; Deyo, R.A.; Buchbinder, R.; Corbin, T.P.; Costa, L.O.; Foster, N.E.; Grotle, M.; Koes, B.W.; Kovacs, F.M.; et al. Core outcome measurement instruments for clinical trials in nonspecific low back pain. *Pain* **2017**, *159*, 481–495. [CrossRef] [PubMed]
9. Bombardier, C. Outcome Assessments in the Evaluation of Treatment of Spinal Disorders. *Spine* **2000**, *25*, 3100–3103. [CrossRef]
10. Ostelo, R.W.J.G.; Deyo, R.A.; Stratford, P.; Waddell, G.; Croft, P.; Von Korff, M.; Bouter, L.M.; Henrica, C. Interpreting change scores for pain and functional status in low back pain: Towards international consensus regarding minimal important change. *Spine* **2008**, *33*, 90–94. [CrossRef]
11. Owen, P.J.; Miller, C.T.; Mundell, N.L.; Verswijveren, S.J.J.M.; Tagliaferri, S.D.; Brisby, H.; Bowe, S.J.; Belavy, D.L. Which specific modes of exercise training are most effective for treating low back pain? Network meta-analysis. *Br. J. Sports Med.* **2019**, *54*, 1279–1287. [CrossRef]
12. Harbour, R.; Miller, J. A new system for grading recommendations in evidence based guidelines. *BMJ* **2001**, *323*, 334–336. [CrossRef]
13. D'Azy, C.B.; Pereira, B.; Chiambaretta, F.; Dutheil, F. Oxidative and Anti-Oxidative Stress Markers in Chronic Glaucoma: A Systematic Review and Meta-Analysis. *PLoS ONE* **2016**, *11*, e0166915. [CrossRef]
14. Lanhers, C.; Pereira, B.; Naughton, G.; Trousselard, M.; Lesage, F.-X.; Dutheil, F. Creatine Supplementation and Upper Limb Strength Performance: A Systematic Review and Meta-Analysis. *Sports Med.* **2016**, *47*, 163–173. [CrossRef] [PubMed]
15. Courtin, R.; Pereira, B.; Naughton, G.; Chamoux, A.; Chiambaretta, F.; Lanhers, C.; Dutheil, F. Prevalence of dry eye disease in visual display terminal workers: A systematic review and meta-analysis. *BMJ Open* **2016**, *6*, e009675. [CrossRef] [PubMed]
16. D'Azy, C.B.; Pereira, B.; Naughton, G.; Chiambaretta, F.; Dutheil, F. Antibioprophylaxis in Prevention of Endophthalmitis in Intravitreal Injection: A Systematic Review and Meta-Analysis. *PLoS ONE* **2016**, *11*, e0156431. [CrossRef]
17. Lanhers, C.; Pereira, B.; Naughton, G.; Trousselard, M.; Lesage, F.-X.; Dutheil, F. Creatine Supplementation and Lower Limb Strength Performance: A Systematic Review and Meta-Analyses. *Sports Med.* **2015**, *45*, 1285–1294. [CrossRef] [PubMed]
18. Tricco, A.C.; Lillie, E.; Zarin, W.; O'Brien, K.; Colquhoun, H.; Kastner, M.; Levac, D.; Ng, C.; Sharpe, J.P.; Wilson, K.; et al. A scoping review on the conduct and reporting of scoping reviews. *BMC Med. Res. Methodol.* **2016**, *16*, 1–10. [CrossRef]
19. Alp, A.; Mengi, G.; Avsaroglu, A.H.; Mert, M.; Sigirli, D. Efficacy of Core-Stabilization Exercise and Its Comparison with Home-Based Conventional Exercise in Low Back Pain Patients. *Bursa Uludag Univ. Res. Inf. Syst.* **2014**. [CrossRef]
20. Frih, Z.B.S.; Fendri, Y.; Jellad, A.; Boudoukhane, S.; Rejeb, N. Efficacy and treatment compliance of a home-based rehabilitation programme for chronic low back pain: A randomized, controlled study. *Ann. Phys. Rehabil. Med.* **2009**, *52*, 485–496. [CrossRef] [PubMed]
21. Bertozzi, L.; Villafañe, J.H.; Capra, F.; Reci, M.; Pillastrini, P. Effect of an Exercise Programme for the Prevention of Back and Neck Pain in Poultry Slaughterhouse Workers. *Occup. Ther. Int.* **2014**, *22*, 36–42. [CrossRef] [PubMed]
22. Bronfort, G.; Maiers, M.J.; Evans, R.L.; Schulz, C.; Bracha, Y.; Svendsen, K.H.; Grimm, R.H.; Owens, E.F.; Garvey, T.A.; Transfeldt, E.E. Supervised exercise, spinal manipulation, and home exercise for chronic low back pain: A randomized clinical trial. *Spine J.* **2011**, *11*, 585–598. [CrossRef] [PubMed]
23. Buttagat, V.; Techakhot, P.; Wiriya, W.; Mueller, M.; Areeudomwong, P. Effectiveness of traditional Thai self-massage combined with stretching exercises for the treatment of patients with chronic non-specific low back pain: A single-blinded randomized controlled trial. *J. Bodyw. Mov. Ther.* **2020**, *24*, 19–24. [CrossRef]
24. Chhabra, H.S.; Sharma, S.; Verma, S. Smartphone app in self-management of chronic low back pain: A randomized controlled trial. *Eur. Spine J.* **2018**, *27*, 2862–2874. [CrossRef]
25. Chaléat-Valayer, E.; Denis, A.; Abelin-Genevois, K.; Zelmar, A.; Siani-Trebern, F.; Touzet, S.; Bergeret, A.; Colin, C.; Fassier, J.B. Long-term effectiveness of an educational and physical intervention for preventing low-back pain recurrence: A randomized controlled trial. *Scand. J. Work Environ. Health* **2016**, *42*, 510–519. [CrossRef]
26. Descarreaux, M.; Normand, M.C.; Laurencelle, L.; Dugas, C. Evaluation of a specific home exercise program for low back pain. *J. Manip. Physiol. Ther.* **2002**, *25*, 497–503. [CrossRef]

27. Ewert, T.; Limm, H.; Wessels, T.; Rackwitz, B.; Von Garnier, K.; Freumuth, R.; Stucki, G. The Comparative Effectiveness of a Multimodal Program Versus Exercise Alone for the Secondary Prevention of Chronic Low Back Pain and Disability. *PM&R* **2009**, *1*, 798–808. [CrossRef]
28. Frost, H.; Lamb, S.; Moffett, K.A.J.; Fairbank, T.J.; Moser, S.J. A fitness programme for patients with chronic low back pain: 2-year follow-up of a randomised controlled trial. *Pain* **1998**, *75*, 273–279. [CrossRef]
29. Garcia, A.N.; Costa, L.D.C.M.; da Silva, T.M.; Gondo, F.L.B.; Cyrillo, F.N.; Costa, R.A.; Costa, L. Effectiveness of Back School Versus McKenzie Exercises in Patients with Chronic Nonspecific Low Back Pain: A Randomized Controlled Trial. *Phys. Ther.* **2013**, *93*, 729–747. [CrossRef]
30. Goode, A.P.; Taylor, S.S.; Hastings, S.N.; Stanwyck, C.; Coffman, C.J.; Allen, K.D. Effects of a Home-Based Telephone-Supported Physical Activity Program for Older Adult Veterans with Chronic Low Back Pain. *Phys. Ther.* **2018**, *98*, 369–380. [CrossRef]
31. Groessl, E.J.; Liu, L.; Chang, D.G.; Wetherell, J.L.; Bormann, J.E.; Atkinson, J.H.; Baxi, S.; Schmalzl, L. Yoga for Military Veterans with Chronic Low Back Pain: A Randomized Clinical Trial. *Am. J. Prev. Med.* **2017**, *53*, 599–608. [CrossRef] [PubMed]
32. Haufe, S.; Wiechmann, K.; Stein, L.; Kück, M.; Smith, A.; Meineke, S.; Zirkelbach, Y.; Duarte, S.R.; Drupp, M.; Tegtbur, U. Low-dose, non-supervised, health insurance initiated exercise for the treatment and prevention of chronic low back pain in employees. Results from a randomized controlled trial. *PLoS ONE* **2017**, *12*, e0178585. [CrossRef]
33. Ibrahim, A.A.; Akindele, M.O.; Ganiyu, S.O. Motor control exercise and patient education program for low resource rural community dwelling adults with chronic low back pain: A pilot randomized clinical trial. *J. Exerc. Rehabil.* **2018**, *14*, 851–863. [CrossRef]
34. Iversen, V.M.; Vasseljen, O.; Mork, P.J.; Berthelsen, I.R.; Børke, J.-B.B.; Berheussen, G.F.; Tveter, A.T.; Salvesen, Ø.; Fimland, M.S. Resistance training in addition to multidisciplinary rehabilitation for patients with chronic pain in the low back: Study protocol. Contemp. *Clin. Trials Commun.* **2017**, *6*, 115–121. [CrossRef]
35. Kendall, K.D.; Emery, C.A.; Wiley, J.P.; Ferber, R. The effect of the addition of hip strengthening exercises to a lumbopelvic exercise programme for the treatment of non-specific low back pain: A randomized controlled trial. *J. Sci. Med. Sport* **2015**, *18*, 626–631. [CrossRef]
36. Doğan, K.; Tur, B.S.; Kurtaiş, Y.; Atay, M.B. Comparison of three different approaches in the treatment of chronic low back pain. *Clin. Rheumatol.* **2008**, *27*, 873–881. [CrossRef] [PubMed]
37. Kuukkanen, T.; Mälkiä, E.; Kautiainen, H.; Pohjolainen, T. Effectiveness of a home exercise programme in low back pain: A randomized five-year follow-up study. *Physiother. Res. Int. J. Res. Clin. Phys. Ther.* **2007**, *12*, 213–224. [CrossRef] [PubMed]
38. Kuukkanen, T.M.; Mälkiä, E.A. An experimental controlled study on postural sway and therapeutic exercise in subjects with low back pain. *Clin. Rehabil.* **2000**, *14*, 192–202. [CrossRef]
39. Miller, J.; Stanley, I.; Moore, K. Videotaped exercise instruction: A randomised controlled trial in musculoskeletal physiotherapy. *Physiother. Theory Pract.* **2004**, *20*, 145–154. [CrossRef]
40. Neyaz, O.; Sumila, L.; Nanda, S.; Wadhwa, S. Effectiveness of Hatha Yoga Versus Conventional Therapeutic Exercises for Chronic Nonspecific Low-Back Pain. *J. Altern. Complement. Med.* **2019**, *25*, 938–945. [CrossRef]
41. Shirado, O.; Doi, T.; Akai, M.; Hoshino, Y.; Fujino, K.; Hayashi, K.; Marui, E.; Iwaya, T. Multicenter Randomized Controlled Trial to Evaluate the Effect of Home-Based Exercise on Patients with Chronic Low Back Pain. *Spine* **2010**, *35*, E811–E819. [CrossRef] [PubMed]
42. Schulz, C.; Evans, R.; Maiers, M.; Schulz, K.; Leininger, B.; Bronfort, G. Spinal manipulative therapy and exercise for older adults with chronic low back pain: A randomized clinical trial. *Chiropr. Man Ther.* **2019**, *27*, 1–14. [CrossRef]
43. Sakuma, Y.; Sasaki-Otomaru, A.; Ishida, S.; Kanoya, Y.; Arakawa, C.; Mochizuki, Y.; Seiishi, Y.; Sato, C. Effect of a Home-Based Simple Yoga Program in Child-Care Workers: A Randomized Controlled Trial. *J. Altern. Complement. Med.* **2012**, *18*, 769–776. [CrossRef]
44. Tottoli, C.R.; van Tulder, M.; da Silva, E.N.; Marques, Y.A.; Martins, W.R.; Carregaro, R.L. Effectiveness and cost-effectiveness of Pilates versus home-based exercises in individuals with chronic non-specific low back pain: Randomised controlled trial protocol. *Eur. J. Physiother.* **2019**, *23*, 95–101. [CrossRef]
45. Wajswelner, H.; Metcalf, B.; Bennell, K. Clinical Pilates versus General Exercise for Chronic Low Back Pain. *Med. Sci. Sports Exerc.* **2012**, *44*, 1197–1205. [CrossRef]
46. Winter, S. Effectiveness of targeted home-based hip exercises in individuals with non-specific chronic or recurrent low back pain with reduced hip mobility: A randomised trial. *J. Back Musculoskelet. Rehabil.* **2015**, *28*, 811–825. [CrossRef]
47. Bronfort, G.; Hondras, M.A.; Schulz, C.; Evans, R.L.; Long, C.R.; Grimm, R. Spinal Manipulation and Home Exercise with Advice for Subacute and Chronic Back-Related Leg Pain. *Ann. Intern. Med.* **2014**, *161*, 381–391. [CrossRef]
48. Michalsen, A.; Kunz, N.; Jeitler, M.; Brunnhuber, S.; Meier, L.; Lüdtke, R.; Büssing, A.; Kessler, C. Effectiveness of focused meditation for patients with chronic low back pain—A randomized controlled clinical trial. Complement. *Ther. Med.* **2016**, *26*, 79–84. [CrossRef]
49. Zadro, J.; Shirley, D.; Simic, M.; Mousavi, S.J.; Ceprnja, D.; Maka, K.; Sung, J.; Ferreira, P. Video-Game–Based Exercises for Older People with Chronic Low Back Pain: A Randomized Controlledtable Trial (GAMEBACK). *Phys. Ther.* **2018**, *99*, 14–27. [CrossRef] [PubMed]
50. Saper, R.B.; LeMaster, C.; Delitto, A.; Sherman, K.J.; Herman, P.M.; Sadikova, E.; Stevans, J.; Keosaian, J.E.; Cerrada, C.J.; Femia, A.L.; et al. Yoga, Physical Therapy, or Education for Chronic Low Back Pain. *Ann. Intern. Med.* **2017**, *167*, 85–94. [CrossRef]

51. Bernardelli, G.; Vigna, L.; Nava, C.; Colonna, V.D.G.; Andersen, L.L.; Consonni, D.; Riboldi, L. Physical Activity in Healthcare Workers with Low Back Pain. *J. Occup. Environ. Med.* **2020**, *62*, e245–e249. [CrossRef]
52. Anwer, S.; Alghadir, A.; Brismée, J.-M. Effect of Home Exercise Program in Patients with Knee Osteoarthritis. *J. Geriatr. Phys. Ther.* **2016**, *39*, 38–48. [CrossRef]
53. Hayden, J.A.; van Tulder, M.; Tomlinson, G. Systematic Review: Strategies for Using Exercise Therapy to Improve Outcomes in Chronic Low Back Pain. *Ann. Intern. Med.* **2005**, *142*, 776–785. [CrossRef] [PubMed]
54. DeBusk, R.F.; Stenestrand, U.; Sheehan, M.; Haskell, W.L. Training Effects of Long Versus Short Bouts of Exercise in Healthy Subjects. *Am. J. Cardiol.* **1990**, *65*, 1010–1013. [CrossRef]
55. Kim, H.; Reece, J.; Kang, M. Effects of Accumulated Short Bouts of Exercise on Weight and Obesity Indices in Adults: A Meta-Analysis. *Am. J. Health Promot.* **2020**, *34*, 96–104. [CrossRef]
56. Sukhato, K.; Lotrakul, M.; Dellow, A.; Ittasakul, P.; Thakkinstian, A.; Anothaisintawee, T. Efficacy of home-based non-pharmacological interventions for treating depression: A systematic review and network meta-analysis of randomised controlled trials. *BMJ Open* **2017**, *7*, e014499. [CrossRef]
57. Flynn, A. Home-based prescribed exercise improves balance-related activities in people with Parkinson's disease and has ben-efits similar to centre-based exercise: A systematic review. *J. Physiother.* **2019**, *65*, 189–199. [CrossRef]
58. Wuytack, F.; Devane, D.; Stovold, E.; McDonnell, M.; Casey, M.; McDonnell, T.J.; Gillespie, P.; Raymakers, A.; Lacasse, Y.; McCarthy, B. Comparison of outpatient and home-based exercise training programmes for COPD: A systematic review and meta-analysis. *Respirology* **2017**, *23*, 272–283. [CrossRef]
59. Anderson, L.; Thompson, D.; Oldridge, N.; Zwisler, A.-D.; Rees, K.; Martin, N.; Taylor, R.S. Exercise-based cardiac rehabilitation for coronary heart disease. *Cochrane Database Syst. Rev.* **2016**, *1*, CD001800. [CrossRef]
60. Van Middelkoop, M.; Rubinstein, S.M.; Verhagen, A.P.; Ostelo, R.; Koes, B.; van Tulder, M.W. Exercise therapy for chronic nonspecific low-back pain. *Best Pract. Res. Clin. Rheumatol.* **2010**, *24*, 193–204. [CrossRef] [PubMed]
61. Karlsson, M.; Bergenheim, A.; Larsson, M.E.H.; Nordeman, L.; Van Tulder, M.; Bernhardsson, S. Effects of exercise therapy in patients with acute low back pain: A systematic review of systematic reviews. *Syst. Rev.* **2020**, *9*, 1–25. [CrossRef]
62. Saner, J.; Bergman, E.M.; de Bie, R.A.; Sieben, J.M. Low back pain patients' perspectives on long-term adherence to home-based exercise programmes in physiotherapy. *Musculoskelet. Sci. Pract.* **2018**, *38*, 77–82. [CrossRef] [PubMed]
63. Lacroix, A.; Hortobágyi, T.; Beurskens, R.; Granacher, U. Effects of Supervised vs. Unsupervised Training Programs on Balance and Muscle Strength in Older Adults: A Systematic Review and Meta-Analysis. *Sports Med.* **2017**, *47*, 2341–2361. [CrossRef]
64. Ferreira, M.L.; Smeets, R.J.; Kamper, S.J.; Ferreira, P.H.; Machado, L.A. Can we explain heterogeneity among randomized clinical trials of exercise for chronic back pain? A meta-regression analysis of randomized controlled trials. *Phys. Ther.* **2010**, *90*, 1383–1403. [CrossRef]
65. Alschuler, K.N.; Hoodin, F.; Murphy, S.L.; Rice, J.; Geisser, M.E. Factors contributing to physical activity in a chronic low back pain clinical sample: A comprehensive analysis using continuous ambulatory monitoring. *Pain* **2011**, *152*, 2521–2527. [CrossRef] [PubMed]
66. Mura, G. Active exergames to improve cognitive functioning in neurological disabilities: A systematic review and meta-analysis. *Eur. J. Phys. Rehabil. Med.* **2018**, *54*, 13. [CrossRef]
67. Brownson, R.C.; Baker, E.A.; Housemann, R.A.; Brennan, L.K.; Bacak, S.J. Environmental and Policy Determinants of Physical Activity in the United States. *Am. J. Public Health* **2001**, *91*, 1995–2003. [CrossRef] [PubMed]
68. Medina-Mirapeix, F.; Escolar-Reina, P.; Gascón-Cánovas, J.J.; Montilla-Herrador, J.; Jimeno-Serrano, F.J.; Collins, S.M. Predictive factors of adherence to frequency and duration components in home exercise programs for neck and low back pain: An observational study. *BMC Musculoskelet. Disord.* **2009**, *10*, 155. [CrossRef]
69. Murphy, M.H.; Lahart, I.; Carlin, A.; Murtagh, E. The Effects of Continuous Compared to Accumulated Exercise on Health: A Meta-Analytic Review. *Sports Med.* **2019**, *49*, 1585–1607. [CrossRef] [PubMed]
70. Bener, A.; El-Rufaie, O.F.; Siyam, A.; Abuzeid, M.S.O.; Toth, F.; Lovasz, G. Epidemiology of low back pain in the United Arab Emirates. *APLAR J. Rheumatol.* **2004**, *7*, 189–195. [CrossRef]
71. Byberg, L. Total mortality after changes in leisure time physical activity in 50 year old men: 35 year follow-up of population based cohort. *BMJ* **2009**. [CrossRef] [PubMed]
72. Dutheil, F.; Comptour, A.; Morlon, R.; Mermillod, M.; Pereira, B.; Baker, J.S.; Charkhabi, M.; Clinchamps, M.; Bourdel, N. Autism spectrum disorder and air pollution: A systematic review and meta-analysis. *Environ. Pollut.* **2021**, *278*, 116856. [CrossRef] [PubMed]

International Journal of
Environmental Research and Public Health

MDPI

Study Protocol

The Relationship between Reactive Balance Control and Back and Hamstring Strength in Physiotherapists with Non-Specific Back Pain: Protocol for a Cross-Sectional Study

Erika Zemková [1,2,*], Eva Ďurinová [3], Andrej Džubera [4], Henrieta Horníková [1], Juraj Chochol [4], Jana Koišová [3], Michaela Šimonová [3] and Ludmila Zapletalová [3]

1 Department of Biological and Medical Sciences, Faculty of Physical Education and Sports, Comenius University in Bratislava, 814 69 Bratislava, Slovakia; henrieta.hornikova@uniba.sk

2 Sports Technology Institute, Faculty of Electrical Engineering and Information Technology, Slovak University of Technology, 812 19 Bratislava, Slovakia

3 Faculty of Health Sciences, University of Ss. Cyril and Methodius in Trnava, 917 01 Trnava, Slovakia; eva.durinova@ucm.sk (E.Ď.); jana.koisova@ucm.sk (J.K.); michaela.simonova@ucm.sk (M.Š.); zapletalovalidi@gmail.com (L.Z.)

4 Department of Neurosurgery, Slovak Medical University and University Hospital—St. Michael's Hospital, 811 08 Bratislava, Slovakia; andrej.dzubera@nsmas.sk (A.D.); juraj.chochol@nsmas.sk (J.C.)

* Correspondence: erika.zemkova@uniba.sk; Tel.: +421-2-901-79-905

Citation: Zemková, E.; Ďurinová, E.; Džubera, A.; Horníková, H.; Chochol, J.; Koišová, J.; Šimonová, M.; Zapletalová, L. The Relationship between Reactive Balance Control and Back and Hamstring Strength in Physiotherapists with Non-Specific Back Pain: Protocol for a Cross-Sectional Study. *IJERPH* **2021**, *18*, 5578. https://doi.org/10.3390/ijerph18115578

Academic Editors: Vincenzo Denaro, Sergio Iavicoli, Fabrizio Russo and Gianluca Vadalà

Received: 15 March 2021
Accepted: 21 May 2021
Published: 23 May 2021

Abstract: Back pain is one of the most costly disorders among the worldwide working population. Within that population, healthcare workers are at a high risk of back pain. Though they often demonstrate awkward postures and impaired balance in comparison with healthy workers, there is no clear relationship between compensatory postural responses to unpredictable stimuli and the strength of related muscle groups, in particular in individuals with mild to moderate back pain. This paper presents a study protocol that aims to evaluate the relationship between peak anterior to peak posterior displacements of the center of pressure (CoP) and corresponding time from peak anterior to peak posterior displacements of the CoP after sudden external perturbations and peak force during a maximum voluntary isometric contraction of the back and hamstring muscles in physiotherapists with non-specific back pain in its early stages. Participants will complete the Oswestry Disability Questionnaire. Those that rate their back pain on the 0–10 Low Back Pain Scale in the ranges 1–3 (mild pain) and 4–6 (moderate pain) will be considered. They will undergo a perturbation-based balance test and a test of the maximal isometric strength of back muscles and hip extensors. We assume that by adding tests of reactive balance and strength of related muscle groups in the functional testing of physiotherapists, we would be able to identify back problems earlier and more efficiently and therefore address them well before chronic back disorders occur.

Keywords: back problems; Low Back Pain Scale; maximum voluntary isometric contraction; postural stability; unexpected external postural perturbations

1. Introduction

In 2019, the WHO reported that among musculoskeletal conditions, low back pain (LBP) is the leading cause of disability worldwide. In particular, work-related back pain is a major cause of reduced productivity and increased disability of workers, which places a significant financial burden on healthcare systems. Of all the professions, healthcare workers are at the highest risk of back problems [1–10]. This is due to overexertion of the back while handling patients, which may lead to awkward spinal postures and impaired static and dynamic balance.

Poor balance in itself may contribute to more severe back pain. Indeed, most LBP patients demonstrate impaired postural stability. For instance, differences exist in postural control strategy, the CoP displacement, and muscle activation patterns between people with and without non-specific LBP [11]. A greater postural instability, denoted by higher CoP velocity and excursions, has been found in people with non-specific LBP compared to healthy controls [12]. Their impaired postural stability seems to be associated with the presence of pain, whereas it is not related to the exact location and pain duration. There is no relationship between the magnitude of CoP excursions and the pain intensity.

Therefore, there is a need to identify more appropriate variables associated with back pain and to specify test conditions tailored to the requirements for assessment of healthcare workers with non-specific back pain in its early stages. Both the CoP and center of mass (CoM) variables should be measured in dynamic conditions with higher task demands. Standing on a spring-supported platform or a foam surface during testing is more efficient in identifying within- and between-group differences when compared to tests of static balance [13]. A good discriminatory accuracy in differentiating between sedentary and physically active young adults was also reported during a perturbation-based balance tests [14]. On the other hand, there were no significant differences in the magnitudes of CoP and CoM displacements in the mediolateral direction during unexpected surface translation in individuals with chronic LBP and healthy controls [15]. However, those with chronic LBP showed an earlier peak CoM displacement and later onset of initial CoP displacement [15]. Nonetheless, readjustment of postural stability in the anteroposterior direction after perturbations elicited by weight unloading has yet to be investigated.

The lumbar extensors and hip musculature may be an important factor in determining the motor control dysfunctions, such as limited balance, that arise in chronic LBP [16]. Lumbar extension strength has been shown to correlated significantly with Star Excursion Balance Test scores and explained ~19.3% to ~37.8% of its variance in the chronic LBP group and ~9.5% to ~16.9% in the asymptomatic group [16].

Traditionally, good isometric endurance of back muscles was sought to prevent first-time LBP occurrence [17]. The Sørensen test has been frequently used [18]. Such a measurement has been recommended as a standard for lifting tasks [19]. This was based on evidence that lower isometric strength is associated with LBP. However, the risk of back problems increases threefold when the requirements for lifting tasks are beyond or equal to their strength capacity. Static strength measurements underestimate the loads on the spine under dynamic conditions. The predicted spinal loads are 33–60% less under static than dynamic conditions depending on the lifting technique [20]. The recruitment patterns of trunk muscles and thus the internal loading of the spine are also different under these two conditions. Therefore, tests performed under dynamic conditions seem to be more appropriate for healthy as well as LBP populations. An exercise in the form of a deadlift to high pull that involves the major muscle groups in the lower and upper body may be able to simulate lifting tasks [21]. In such a case, a predictor of lifting performance with light loads might be a peak rate of force development (RFD) produced during a maximum voluntary isometric contraction (MVC) of the back muscles [22]. Thus, besides MVC peak force generated by back muscles, the subjects' ability to produce a maximum force in a short period of time should by determined to obtain further insight into the loaded lifting performance in those prone to LBP.

In comparison with these frequently used strength and endurance tests of back muscles, less attention has been paid to the assessment of hamstring strength in LBP individuals and the associations with variables of postural and core stability measured in more challenging conditions. Though there is the belief that an association exists between a stronger body core and shorter reaction time, or stronger lower limbs and faster recovery, there is also an opposite view that suggests that stronger lower limbs contribute to slower recovery [23]. Therefore, a question remains as to whether reactive balance control after unpredictable stimuli is associated with the strength of relevant muscle groups in the high-risk population of healthcare workers.

This paper presents a protocol for a cross-sectional study that will investigate the relationship between peak anterior to peak posterior displacements of the CoP and corresponding time from peak anterior to peak posterior displacements of the CoP after sudden external perturbations and peak force during MVCs of the back and hamstring muscles in physiotherapists with non-specific back pain, in particular those with mild to moderate back pain. We will also investigate how the ability to produce maximum force in a short period of time during MVCs of the back and hamstring muscles relates to postural responses to externally induced perturbations in a control group without back pain.

2. Materials and Methods

2.1. Study Design

This cross-sectional study is designed to evaluate associations among measures of a perturbation-based balance test and back and hamstring strength tests in physiotherapists with non-specific back pain. The study will be implemented and reported in line with the Standard Protocol Items: Recommendations for Interventional Trials (SPIRIT) statement. The study design is illustrated in Figure 1.

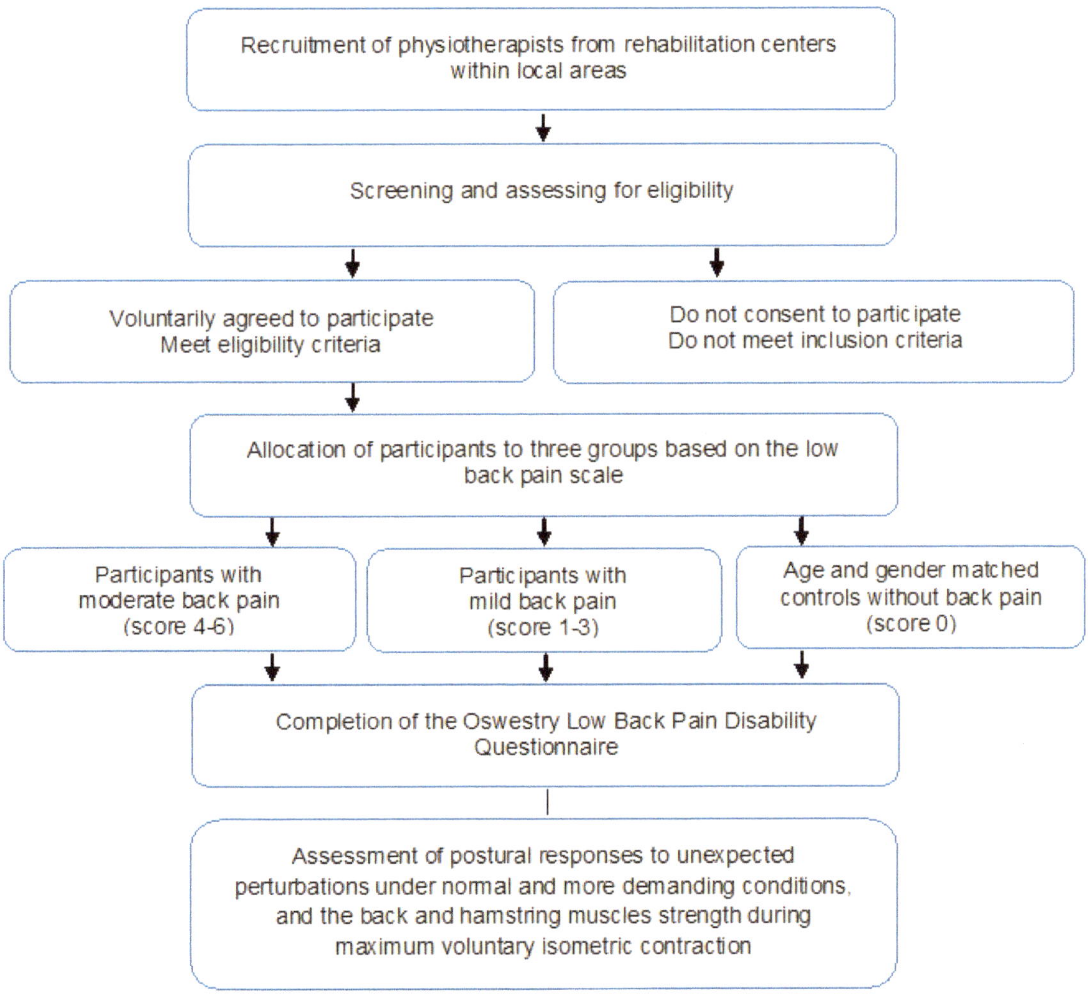

Figure 1. Flowchart of the study design.

2.2. Participants and Setting

A group of 82 female and male physiotherapists with non-specific back pain will be recruited from rehabilitation centers within local areas. The control group will include 41 participants. These effect sizes are sufficient to identify significant interaction effects. The contact with these centers and the necessary support for recruitment of subjects and data collection has been will be provided. We expect approximately 90% of eligible physiotherapists to consent to participate in the study. The timetable will be specified when the COVID-19 pandemic is over.

2.3. Inclusion and Exclusion Criteria

Subjects will be screened for eligibility by the research team members. Those who report non-specific back pain [24,25] for more than 12 weeks [26–29] will be eligible. Inclusion criteria for LBP individuals and healthy controls will require no history of orthopedic or neurological conditions that could influence balance. Participants scoring <7 out of 10 on a numeric pain rating scale will be eligible for inclusion. Those who have previously undergone medically invasive procedures for back pain or have a diagnosis that may explain their low back symptoms (e.g., former injury or illness within the last 12 months) will be excluded. Exclusion criteria will also include pain not primarily generated from the musculoskeletal system, infections, diabetes, and pregnancy. As the strength of hamstring muscles will also be evaluated, participants who will report any other current muscular, joint, or neurological conditions affecting lower limb function will be excluded.

2.4. Allocation

Participants will be divided into two groups according to the Numeric Rating Scale (mean of three assessments: current LBP, the usual/mean LBP within the last two weeks, and the worst LBP within the last two weeks), which is widely used in medical settings to obtain information about the level of a patient's pain. The scale ranges from 0 (no pain at all) to 10 (unbearable pain). Participants experiencing mild pain (pain score 1–3), which is easy to manage physically as well as psychologically and does not interfere with most daily activities, and moderate pain (pain score 4–6), which interferes with daily activities and requires changes to manage pain symptoms during the last three months, will be considered. A control group of matched age, gender, and sample size will include those reporting no back pain.

2.5. Sample Size Estimation

The statistical power was calculated using the software program G*power 3.1 for Mac OS X. The calculation of the sample size was conducted with $\alpha = 0.05$ (5% change of type I error) and $1 - \beta = 0.80$ (power 80%) and using the results from our previous measurements. This indicated that a sample size of 34 individuals per group is needed to identify significant interaction effects. To reach this target sample size and achieve sufficient participant enrolment, 20% will be added to allow for dropouts.

2.6. Procedures

Participants will be asked to avoid exercises of higher intensity prior to the investigation. Before assessment, they will be given information on each testing protocol and the instructions during measurements. To eliminate learning effects, they will be required to practice a whole procedure beforehand. Tests will be conducted by the same examiners at the same time of the day for all participants. In comparison with frequently used laboratory tests, the use of portable and user-friendly diagnostic systems suited for testing in field conditions will be preferred in the present study. Using easy to administer tests that can reveal in a relatively short time period the impairment of postural and core stability and reduction in strength of particular muscle groups in LBP individuals could increase their applications in practice.

2.7. Data Collection and Management

The study will comprise the collection of personal demographic and outcome data. The questionnaire data will be transferred to a database and checked for correctness by research team members to ensure their quality. All information and outcome data will be stored in password-protected computers, which will be accessible by authorized research team members. Data management during the project and after the project's completion will provide information on how data will be collected, documented, stored, and archived.

2.8. Descriptive Measures

Upon arrival at the laboratory, the participants' characteristics will be summarized. These will include age, height, body mass, body composition, gender, and additional information related to back problems such as the amount of daily practice with clients, the type and duration of sporting activities, previous injuries, diseases, and so forth.

2.9. Primary Outcomes

Oswestry Disability Index

Participants will complete the Oswestry Low Back Pain Disability Questionnaire, which gives a subjective score of the level of function and/or disability in daily activities [30]. It is considered the 'gold standard' of low back functional outcome tools [30]. The Oswestry Disability Questionnaire data are reliable and have a scale of adequate width to reliably reveal worsening or improvement in most subjects [31]. The questionnaire consists of two parts—Oswestry Disability Index (ODI) version 2 and Visual Analogue Scale (VAS). The ODI questionnaire is designed to give information about how back or leg pain affects a patient's ability to manage in everyday life. The ODI consists of 10 questions with a single choice of 6 answers. The VAS is a measurement instrument to measure the strength of pain experienced by a patient. Using a ruler, the score is determined by measuring the distance (mm) on a 10 cm line between 'no pain at all' and 'my pain is as bad as it could be'. The patient marks a score from 0–10. A higher score indicates greater pain intensity. The pain VAS is a unidimensional measure of pain intensity, widely used in diverse adult populations. The combination of ODI and VAS is accepted worldwide and used in clinical practice because it is easy to use among patients and physicians.

2.10. Secondary Outcomes

2.10.1. Postural Responses to Unexpected Perturbations

Participants will perform a perturbation-based balance test. They will be instructed to perform the test under two conditions: (1) bipedal and (2) tandem stance. Participants' eyes will be focused on a spot on the wall at their eye level. They will be asked to stand barefoot on a force plate, a shoulder width apart, while their arms will be held in front horizontally. They will be asked to hold a bar with a 2 kg fixed load in their hands. After the test initiation, a signal from the computer will trigger a random load release over a 5 s period, therefore the participant will receive no cues as to when the perturbation will occur. The load release will produce a sudden change in the external forces acting on the participant that will lead to a slight anterior and a greater posterior CoP displacement. This postural perturbation will cause only a sway response, so the participant will not need to perform a step to maintain balance. The perturbation will be determined by peak anterior and peak posterior displacements within 1 s following the load drop. The recording will end 2–3 s following the load drop.

Three trials of each test condition will be randomly performed. The primary outcome measures will include peak anterior to peak posterior displacement of the CoP and the time from peak anterior to peak posterior displacement of the CoP. Secondary measures will include peak anterior displacement of the CoP, peak posterior displacement of the CoP, the time to peak anterior displacement of the CoP, and the time to peak posterior displacement of the CoP. These variables will be monitored by means of the FiTRO Sway Check (FiTRONiC, SVK). This system registers the actual force in the corners of the force plate and calculates an instant CoP position (sampling rate of 100 Hz, 12 bit AD signal conversion, resolution of the CoP position less than 0.1 mm, measuring range of 0–1000 N/s, non-linearity of +/−0.02% FS, combined error of 0.03%, sensitivity of 2 mV/V +/− 0.25%, overload capacity of 150%/sensor). Additionally, a ratio of the CoP anterior to the CoP posterior values will be estimated. The reliability of parameters of the perturbation-based balance test is good to excellent, with low SEM (7.1–10.7%) and high ICC values (0.78–0.92) [16]. This test is also sensitive in discriminating between sedentary and physically active young and early to late middle-aged adults. Good discriminatory accuracy of these variables is indicated by an area under the ROC curve >0.80 [14].

2.10.2. Maximal Strength of the Back and Hamstring Muscles

Before the test begins, the participants will be warmed up by performing two submaximal isometric contractions so as to become accustomed to the procedure. Afterwards, they will be placed into the proper position with knee and hip angles of 141° and 124°, respectively, set up by a handheld goniometer. This position corresponds to the portion of the clean lift during which the highest values of power are achieved [32]. A handlebar of this device will be attached to a floor-mounted load cell. Its height above the floor will be determined for each participant during a familiarization trial. Once the participants are in position, they will initiate the contraction after a countdown of "3, 2, 1, pull". They will perform three maximal MVCs as forcefully as possible for at least 5 s. Participants will be provided with verbal encouragement at each trial. A minimum of a 2 min rest period will be given between MVC efforts. The visual feedback on the instantaneous force will be provided in real time on a monitor positioned in front of the research team member. Force will be measured using the FiTRO Back Dynamometer (FiTRONiC, SVK). Analog signals will be AD converted and sampled at a rate of 1000 Hz. Peak force will be analyzed.

The same system will be used for measurement of the hamstring strength. Participants, after a warm-up, will perform three 5 s isometric contractions with maximal effort on each leg at 90° of knee flexion while lying on the rehabilitation bed in the prone position. This knee flexion angle is based on findings that showed high reliability of isometric posterior lower limb muscle force [33]. In a randomized design, the left and right leg peak force will be registered. The best result from the three attempts will be taken for the analysis.

The control group without back pain will undergo the same procedures, however, these participants will also perform three MVCs as quickly and as forcefully as possible for at least 3 s. In this case, the peak RFD in addition to peak force during MVCs of the back and hamstring muscles will be analyzed.

Assessment of postural control after sudden external perturbations and maximal isometric strength of the back and hamstring muscles in participants will be completed by the measurement of their low back and hamstring flexibility.

2.11. Patient and Public Involvement

The public and patients will be not directly involved in the present study. Local medical centers will provide support for recruitment of physiotherapists with non-specific back pain. Test results will be provided to participants on request and the overall outcomes will be available to them on completion of the study.

2.12. Ethics and Dissemination

The procedures described are in accordance with the ethical standards as laid down by the 1964 Helsinki Declaration and its later amendments. Participants will be verbally informed of the main study objective, procedures, risks and benefits, confidentiality, and the voluntary nature of their participation and provided an opportunity to ask questions. Prior to inclusion, written informed consent will be obtained. Projects were approved by the ethics committee of the Faculty of Physical Education and Sports, Comenius University in Bratislava (Nos. 4/2017 and 1/2020).

The findings obtained will be publicly available in particular journals. Research results will also be presented at scientific conferences and disseminated outside and/or within related healthcare centers and/or universities.

2.13. Statistical Analysis

Statistical analysis of the data obtained will be conducted using the SPSS program for Windows, version 24.0 (SPSS, Inc., Chicago, IL, USA). The normality hypothesis will be tested using the Kolmogorov–Smirnov test. A parametric analysis will be conducted if the data are normally distributed. Associations between variables of the perturbation-based balance test and maximal isometric back and hamstring muscle strength in participants with mild to moderate back pain will be assessed using Pearson's product moment correlation coefficient (small: $r = 0.10$–<0.30, medium: $r = 0.30$–<0.50, large: $r = \geq 0.50$) [34]. Similarly, the correlation between RFD produced during MVCs of the back and hamstring muscles and postural responses to externally induced perturbations in the control group will be determined. A standard multiple regression analysis will be performed to investigate which of these variables of the muscle strength could be significant predictors of compensatory postural responses to unpredictable stimuli in individuals without back pain. The amount of variance explained will be determined by the coefficient of determination (r^2). A multivariable logistic regression model will be used to determine associations of back pain (intensity and duration) with the covariates (age, gender, body mass, and eventually psychological factors and/or medical history of participants). The significance level will be set at $\alpha = 5\%$. Data will be presented as the mean (standard deviation).

3. Discussion

Therapists in health care, including physiotherapists, physical therapists, and athletic therapists, are at high risk of developing LBP [1,3,35–37]. The occurrence of back problems will likely rise as patients become heavier with increasing obesity. Therefore, more efforts should be directed toward their prediction by regular assessment of physical factors associated with back pain in its early stages. Among them are those associated with awkward spinal postures, impaired static and dynamic balance, and reduced core and lower limb strength. However, conflicting evidence exists on associations among these performance determinants and the occurrence of back problems.

For instance, a significant relationship was demonstrated between the percentage of time spent in awkward postures in the sagittal plane (trunk flexion $\geq 45°$) and in the frontal plane (lateral bend $\geq 20°$) and LBP in hospital nurses [38]. However, a systematic review by Roffey et al. [39] revealed that there is no association between awkward postures and LBP and that there is no temporal relationship. Weak associations and no evidence for other aspects of causality in certain specific subcategories were demonstrated in few studies [39]. Therefore, it is unlikely that awkward occupational postures are independently causative of LBP in the working population [39].

Further, individuals with LBP often exhibit altered responses to sudden perturbations. In particular, delayed trunk muscle responses [40–42], decreased amplitudes of muscle activation [40,43], and increased co-contraction [40] are associated with the occurrence of LBP. However, most studies investigating postural responses to perturbations applied to the body's trunk have been limited to electromyographic (EMG) recordings while less attention has been paid to the assessment of muscle strength and power.

The trunk and hip muscle strength contributes to lumbar spine stability, particularly during functional tasks. Additionally, a low level of the trunk muscle co-contraction is important for the body's core stability [44]. Such a level of stiffness provides necessary stability against minor perturbations. In particular, direction-specific muscle reflex responses play an essential role in the stability of the body's core when encountering unexpected postural perturbations [44].

The evidence has demonstrated that shortening, weakness, and/or muscle stiffness of the pelvic and lumbar regions contribute to LBP. There is an association between decreased stability and an increased risk of low back or knee injury [44]. These individuals demonstrate impaired postural stability, abnormal recruitment patterns of the trunk muscles, and delayed muscle reflex responses to unexpected trunk unloading [44]. Additionally, lower strength of knee extensors and hip abductor/extensors has been reported in LBP patients than healthy controls [45–49]. Provided that the muscles of lower limbs, particularly the hip muscles, play an important role in the stability of the lumbar spine [50], it is likely that lumbar instability and inefficient lumbopelvic motor control, together with weakness in hip muscles, may contribute to LBP development. However, a relationship between hip strength and back pain has not been found [51,52].

Thus far, the association of hip weakness with the presence of non-specific LBP has not been sufficiently investigated. Although several studies have demonstrated the efficiency of trunk and lower limb strengthening exercises for improvement of core stability [53–55], less attention has been paid to core stability in relation to the strength of particular muscle groups. This is due to a paucity of easy to administer tests using portable and user-friendly diagnostic systems suited for testing in field conditions.

Our study will address this gap in the available research. It will provide insight into the association between the ability of subjects to maintain postural stability after an unexpected perturbation and maximal isometric strength of the back and hamstring muscles. Our attention will be paid to healthcare workers, namely physiotherapists with mild to moderate non-specific back pain in whom it is hard to reveal slight impairments of the ability to regulate CoM motion using less sophisticated non-laboratory techniques. Identifying the relationship between these factors will help us to design a test battery tailored specifically for this high-risk population using tests that can be performed in field conditions. The limitation will be the sample consisting of mainly female participants due to larger number of women working in the healthcare sector. However, the LBP prevalence rate in this population is high, with the majority of cases occurring after starting work [3,56].

In addressing LBP prevention, the ability to produce maximum force in a short period of time during MVCs of the back and hamstring muscles will be assessed in the control group without back pain. This method could provide feedback about whether those with higher peak RFD during MVC of relevant muscle groups are capable of responding effectively to unexpected postural perturbations. It can be implemented in healthy subjects who may benefit from such testing by predicting their LBP risk. Revealing impaired postural and/or core stability and reduced strength of relevant muscle groups could support self-help strategies in the prevention of back problems by the application of hip–trunk stabilization and strengthening exercises in their daily activities. This may contribute to a reduction in chronic back problems and consequently lowering healthcare system costs.

Author Contributions: Conceptualization, E.Z., E.Ď., A.D., H.H., J.C., J.K., M.Š., and L.Z.; methodology, E.Z., E.Ď., A.D., H.H., J.C., J.K., M.Š., and L.Z.; writing—original draft preparation, E.Z.; writing—review and editing, E.Z., E.Ď., A.D., H.H., J.C., J.K., M.Š., and L.Z. All authors have read and agreed to the published version of the manuscript.

Funding: This research received no external funding.

Institutional Review Board Statement: The procedures described are in accordance with the ethical standards on human experimentation stated in the 1964 Helsinki Declaration and its later amendments. Projects were approved by the ethics committee of the Faculty of Physical Education and Sports, Comenius University in Bratislava (Nos. 4/2017 and 1/2020).

Informed Consent Statement: Not applicable.

Acknowledgments: This work was supported by the Slovak Research and Development Agency under the contract No. APVV-15-0704; the Scientific Grant Agency of the Ministry of Education, Science, Research and Sport of the Slovak Republic and the Slovak Academy of Sciences (No. 1/0089/20); and the Cross-border Co-operation Programme INTERREG V-A SK-CZ/2018/06 (No. 304011P714) co-financed by the European Regional Development Fund.

Conflicts of Interest: The authors declare no conflict of interest.

References

1. Hogan, D.A.M.; O'Sullivan, L.W.; Nolan, S.; Greiner, B.A. Are Irish therapists at heightened risk for low back pain? *Occup. Med. (Lond).* **2016**, *66*, 351–357. [CrossRef]
2. Mehrdad, R.; Narges Shams-Hosseini, N.S.; Aghdaei, S.; Yousefian, M. Prevalence of low back pain in health care workers and comparison with other occupational categories in Iran: A systematic review. *Iran J. Med. Sci.* **2016**, *41*, 467–478.
3. Milhem, M.; Kalichman, L.; Ezra, D.; Alperovitch-Najenson, D. Work-related musculoskeletal disorders among physical therapists: A comprehensive narrative review. *Int. J. Occup. Med. Environ. Health.* **2016**, *29*, 735–747. [CrossRef] [PubMed]
4. Awosan, K.J.; Yikawe, S.S.; Oche, O.M.; Oboirien, M. Prevalence, perception and correlates of low back pain among healthcare workers in tertiary health institutions in Sokoto, Nigeria. *Ghana Med. J.* **2017**, *51*, 164–174.
5. Çınar-Medeni, Ö.; Elbasan, B.; Duzgun, I. Low back pain prevalence in healthcare professionals and identification of factors affecting low back pain. *J. Back Musculoskelet. Rehabil.* **2017**, *30*, 451–459. [CrossRef]
6. Riccò, M.; Pezzetti, F.; Signorelli, C. Back and neck pain disability and upper limb symptoms of home healthcare workers: A case-control study from Northern Italy. *Int. J. Occup. Med. Environ. Health.* **2017**, *30*, 291–304. [CrossRef]
7. Şimşek, S.; Yağcı, N.; Şenol, H. Prevalence of and risk factors for low back pain among healthcare workers in Denizli. *Agri* **2017**, *29*, 71–78. [CrossRef] [PubMed]
8. Alnaami, I.; Awadalla, N.J.; Alkhairy, M.; Alburidy, S.; Alqarni, A.; Algarni, A.; Alshehri, R.; Amrah, B.; Alasmari, M.; Mahfouz, A.A. Prevalence and factors associated with low back pain among health care workers in southwestern Saudi Arabia. *BMC Musculoskelet. Disord.* **2019**, *20*, 56. [CrossRef]
9. Mekonnen, T.H. Work-related factors associated with low back pain among nurse professionals in East and West Wollega Zones, Western Ethiopia, 2017: A cross-sectional study. *Pain Ther.* **2019**, *8*, 239–247. [CrossRef] [PubMed]
10. Mroczek, B.; Łubkowska, W.; Jarno, W.; Jaraczewska, E.; Mierzecki, A. Occurrence and impact of back pain on the quality of life of healthcare workers. *Ann. Agric. Environ. Med.* **2020**, *27*, 36–42. [CrossRef]
11. Koch, C.; Hänsel, F. Non-specific low back pain and postural control during quiet standing: A systematic review. *Front. Psychol.* **2019**, *10*, 586. [CrossRef]
12. Ruhe, A.; Fejer, R.; Walker, B. Center of pressure excursion as a measure of balance performance in patients with non-specific low back pain compared to healthy controls: A systematic review of the literature. *Eur. Spine J.* **2011**, *20*, 358–368. [CrossRef] [PubMed]
13. Zemková, E.; Kováčiková, Z.; Jeleň, M.; Neumannová, K.; Janura, M. Postural and trunk responses to unexpected perturbations depend on the velocity and direction of platform motion. *Physiol. Res.* **2016**, *65*, 769–776. [CrossRef]
14. Zemková, E.; Štefániková, G.; Muyor, J.M. Load release balance test under unstable conditions effectively discriminates between physically active and sedentary young adults. *Hum. Mov. Sci.* **2016**, *48*, 142–152. [CrossRef]
15. Henry, S.M.; Hitt, J.R.; Jones, S.L.; Bunn, J.Y. Decreased limits of stability in response to postural perturbations in subjects with low back pain. *Clin. Biomech. (Bristol, Avon)* **2006**, *21*, 881–892. [CrossRef]
16. Behennah, J.; Conway, R.; Fisher, J.; Osborne, N.; Steele, J. The relationship between balance performance, lumbar extension strength, trunk extension endurance, and pain in participants with chronic low back pain, and those without. *Clin. Biomech.* **2018**, *53*, 22–30. [CrossRef]
17. Biering-Sørensen, F. Physical measurements as risk indicators for low-back trouble over a one-year period. *Spine (Phila Pa 1976)* **1984**, *9*, 106–119. [CrossRef]
18. Demoulin, C.; Vanderthommen, M.; Duysens, C.; Crielaard, J.M. Spinal muscle evaluation using the Sorensen test: A critical appraisal of the literature. *Joint Bone Spine.* **2006**, *73*, 43–50. [CrossRef] [PubMed]
19. Karwowski, W.; Marras, W.S. (Eds.) *The Occupational Ergonomics Handbook*; CRC Press: Boca Raton, FL, USA, 1999.
20. Rosecrance, J.C.; Cook, T.M.; Golden, N.S. A comparison of isometric strength and dynamic lifting capacity in men with work-related low back injuries. *J. Occup. Rehabil.* **1991**, *1*, 197–205. [CrossRef]
21. Zemková, E.; Cepková, A.; Uvaček, M.; Hamar, D. A new method to assess the power performance during a lifting task in young adults. *Measurement* **2016**, *91*, 460–467. [CrossRef]
22. Zemková, E.; Poór, O.; Pecho, J. Peak rate of force development and isometric maximum strength of back muscles are associated with power performance during load-lifting tasks. *Am. J. Men's Health* **2019**, *13*, 1–8. [CrossRef]
23. Jones-Hershinow, C. An evaluation of the relationship between core endurance and lower extremity strength, and the ability to recover after perturbation. *UNC's Undergrad. Res. J.* **2016**, *5*, 31–43.
24. Balagué, F.; Mannion, A.F.; Pellisé, F.; Cedraschi, C. Non-specific low back pain. *Lancet* **2012**, *379*, 482–491. [CrossRef]
25. Maher, C.; Underwood, M.; Buchbinder, R. Non-specific low back pain. *Lancet* **2017**, *389*, 736–747. [CrossRef]
26. Burton, A.K.; Tillotson, K.M.; Main, C.J.; Hollis, S. Psychosocial predictors of outcome in acute and subchronic low back trouble. *Spine (Phila Pa 1976)* **1995**, *20*, 722–728. [CrossRef]
27. Dunn, K.M.; Jordan, K.; Croft, P.R. Characterizing the course of low back pain: A latent class analysis. *Am. J. Epidemiol.* **2006**, *163*, 754–761. [CrossRef]
28. Tamcan, O.; Mannion, A.F.; Eisenring, C.; Horisberger, B.; Elfering, A.; Müller, U. The course of chronic and recurrent low back pain in the general population. *Pain* **2010**, *150*, 451–457. [CrossRef] [PubMed]

29. Axén, I.; Bodin, L.; Bergström, G.; Halasz, L.; Lange, F.; Lövgren, P.W.; Rosenbaum, A.; Leboeuf-Yde, C.; Jensen, I. Clustering patients on the basis of their individual course of low back pain over a six month period. *BMC Musculoskelet. Disord.* **2011**, *12*, 99. [CrossRef]
30. Fairbank, J.C.; Pynsent, P.B. The Oswestry Disability Index. *Spine.* **2000**, *25*, 2940–2952. [CrossRef] [PubMed]
31. Davidson, M.; Keating, J.L. A comparison of five low back disability questionnaires: Reliability and responsiveness. *Phys. Ther.* **2002**, *82*, 8–24. [CrossRef] [PubMed]
32. Garhammer, J. A review of power output studies of Olympic and powerlifting: Methodology, performance prediction, and evaluation tests. *J. Strength Cond. Res.* **1993**, *7*, 76–89. [CrossRef]
33. McCall, A.; Nedelec, M.; Carling, C.; Le Gall, F.; Berthoin, S.; Dupont, G. Reliability and sensitivity of a simple isometric posterior lower limb muscle test in professional football players. *J. Sports Sci.* **2015**, *33*, 1298–1304. [CrossRef] [PubMed]
34. Cohen, J. A power primer. *Psychol. Bull.* **1992**, *112*, 155–159. [CrossRef] [PubMed]
35. Mierzejewski, M.; Kumar, S. Prevalence of low back pain among physical therapists in Edmonton, Canada. *Disabil. Rehabil.* **1997**, *19*, 309–317. [CrossRef]
36. Holder, N.L.; Clark, H.A.; DiBlasio, J.M.; Hughes, C.L.; Scherpf, J.W.; Harding, L.; Shepard, K.F. Cause, prevalence, and response to occupational musculoskeletal injuries reported by physical therapists and physical therapist assistants. *Phys. Ther.* **1999**, *79*, 642–652. [CrossRef] [PubMed]
37. Chung, S.H.; Her, J.G.; Ko, T.; Ko, J.; Kim, H.; Lee, J.S.; Woo, J. Work-related musculoskeletal disorders among Korean physical therapists. *J. Phys. Ther. Sci.* **2013**, *25*, 55–59. [CrossRef]
38. Nourollahi, M.; Afshari, D.; Dianat, I. Awkward trunk postures and their relationship with low back pain in hospital nurses. *Work* **2018**, *59*, 317–323. [CrossRef] [PubMed]
39. Roffey, D.M.; Wai, E.K.; Bishop, P.; Kwon, B.K.; Dagenais, S. Causal assessment of awkward occupational postures and low back pain: Results of a systematic review. *Spine J.* **2010**, *10*, 89–99. [CrossRef]
40. Radebold, A.; Cholewicki, J.; Panjabi, M.M.; Patel, T.C. Muscle response pattern to sudden trunk loading in healthy individuals and in patients with chronic low back pain. *Spine (Phila Pa 1976)* **2000**, *25*, 947–954. [CrossRef]
41. Cholewicki, J.; Silfies, S.P.; Shah, R.A.; Greene, H.S.; Reeves, N.P.; Alvi, K.; Goldberg, B. Delayed trunk muscle reflex responses increase the risk of low back injuries. *Spine (Phila Pa 1976)* **2005**, *30*, 2614–2620. [CrossRef]
42. Reeves, N.P.; Cholewicki, J.; Milner, T.E. Muscle reflex classification of low-back pain. *J. Electromyogr. Kinesiol.* **2005**, *15*, 53–60. [CrossRef]
43. MacDonald, D.; Moseley, G.L.; Hodges, P.W. People with recurrent low back pain respond differently to trunk loading despite remission from symptoms. *Spine (Phila Pa 1976)* **2010**, *35*, 818–824. [CrossRef] [PubMed]
44. Borghuis, J.; Hof, A.L.; Lemmink, K.A. The importance of sensory-motor control in providing core stability: Implications for measurement and training. *Sports Med.* **2008**, *38*, 893–916. [CrossRef]
45. de Sousa, C.S.; de Jesus, F.L.A.; Machado, M.B.; Ferreira, G.; Ayres, I.G.T.; de Aquino, L.M.; Fukuda, T.Y.; Gomes-Neto, M. Lower limb muscle strength in patients with low back pain: A systematic review and meta-analysis. *J. Musculoskelet. Neuronal. Interact.* **2019**, *19*, 69–78.
46. Nourbakhsh, M.R.; Arab, A.M. Relationship between mechanical factors and incidence of low back pain. *J. Orthop. Sports Phys. Ther.* **2002**, *32*, 447–460. [CrossRef]
47. Tsai, Y.-S.; Sell, T.C.; Smoliga, J.M.; Myers, J.B.; Learman, K.E.; Lephart, S.M. A comparison of physical characteristics and swing mechanics between golfers with and without a history of low back pain. *J. Orthop. Sports Phys. Ther.* **2010**, *40*, 430–438. [CrossRef] [PubMed]
48. Cai, C.; Kong, P.W. Low back and lower-limb muscle performance in male and female recreational runners with chronic low back pain. *J. Orthop. Sports Phys. Ther.* **2015**, *45*, 436–443. [CrossRef] [PubMed]
49. Cooper, N.A.; Scavo, K.M.; Strickland, K.J.; Tipayamongkol, N.; Nicholson, J.D.; Bewyer, D.C.; Sluka, K.A. Prevalence of gluteus medius weakness in people with chronic low back pain compared to healthy controls. *Eur. Spine J.* **2016**, *25*, 1258–1265. [CrossRef]
50. Marshall, P.W.; Patel, H.; Callaghan, J.P. Gluteus medius strength, endurance, and co-activation in the development of low back pain during prolonged standing. *Hum. Mov. Sci.* **2011**, *30*, 63–73. [CrossRef] [PubMed]
51. Nadler, S.F.; Malanga, G.A.; Bartoli, L.A.; Feinberg, J.H.; Prybicien, M.; Deprince, M. Hip muscle imbalance and low back pain in athletes: Influence of core strengthening. *Med. Sci. Sports Exerc.* **2002**, *34*, 9–16. [CrossRef]
52. Nadler, S.F.; Malanga, G.A.; DePrince, M.; Stitik, T.P.; Feinberg, J.H. The relationship between lower extremity injury, low back pain, and hip muscle strength in male and female collegiate athletes. *Clin. J. Sport Med.* **2000**, *10*, 89–97. [CrossRef]
53. Coulombe, B.J.; Games, K.E.; Neil, E.R.; Eberman, L.E. Core stability exercise versus general exercise for chronic low back pain. *J. Athl. Train.* **2017**, *52*, 71–72. [CrossRef] [PubMed]
54. Bernet, B.A.; Peskura, E.T.; Meyer, S.T.; Bauch, P.C.; Donaldson, M.B. The effects of hip-targeted physical therapy interventions on low back pain: A systematic review and meta-analysis. *Musculoskelet. Sci. Pract.* **2019**, *39*, 91–100. [CrossRef]
55. Kim, B.; Yim, J. Core stability and hip exercises improve physical function and activity in patients with non-specific low back pain: A randomized controlled trial. *Tohoku J. Exp. Med.* **2020**, *251*, 193–206. [CrossRef] [PubMed]
56. Karahan, A.; Kav, S.; Abbasoglu, A.; Dogan, N. Low back pain: Prevalence and associated risk factors among hospital staff. *J. Adv. Nurs.* **2009**, *65*, 516–524. [CrossRef] [PubMed]

MDPI
St. Alban-Anlage 66
4052 Basel
Switzerland
Tel. +41 61 683 77 34
Fax +41 61 302 89 18
www.mdpi.com

International Journal of Environmental Research and Public Health Editorial Office
E-mail: ijerph@mdpi.com
www.mdpi.com/journal/ijerph

www.ingramcontent.com/pod-product-compliance
Ingram Content Group UK Ltd.
Pitfield, Milton Keynes, MK11 3LW, UK
UKHW060123010826
14089UKWH00006B/110

9 783036 544557